DIFFERENTIAL GEOMETRY

By Heinrich W. Guggenheimer
POLYTECHNIC INSTITUTE OF NEW YORK

DOVER PUBLICATIONS, INC., NEW YORK

Published in Canada by General Publishing Company, Ltd., 30 Lesmill Road, Don Mills, Toronto, Ontario.
Published in the United Kingdom by Constable and Company, Ltd., 10 Orange Street, London WC2H 7EG.

This Dover edition, first published in 1977, is an unabridged and corrected republication of the work originally published by the McGraw-Hill Book Company, Inc., New York, in 1963.

International Standard Book Number: 0-486-63433-7
Library of Congress Catalog Card Number: 76-27496

Manufactured in the United States of America
Dover Publications, Inc.
180 Varick Street
New York, N. Y. 10014

PREFACE

L'esprit de la Géometrie moderne est d'élever toûjours les veritez soit anciennes, soit nouvelles à la plus grande universalité qu'il se puisse.

B. FONTENELLE, *Histoire de l'Académie Royale des Sciences*, 1704, p. 60.

This is a text of local differential geometry as an application of advanced calculus and linear algebra. The book presupposes only a fair knowledge of matrix algebra and of advanced calculus of functions of several real variables.

Because of the weakness of geometry in the usual undergraduate curriculum, the author, from his own experience, has found it necessary to give a general introduction to the study of geometry. Therefore, the text starts with a detailed treatment of plane euclidean geometry. No geometric knowledge is presupposed beyond the elements of euclidean and analytic geometry.

Geometry, as understood in this book, is based on the theory of continuous groups. The author has tried to make the group-theoretic structure, the "Erlanger Program," as explicit as possible. An introduction to the local theory of Lie groups and transformation groups is given in Chaps. 5 to 7. On the other hand, all topological questions have been strictly excluded, although it is always indicated where and how topological methods are needed to solve problems raised by the local theory. As far as possible, results in global differential geometry have been included if they may be proved by simple arguments of advanced calculus.

As Professor Nomizu has written, " . . . one thing seems certain. That is that the work of Elie Cartan on connections, holonomy groups, and homogeneous spaces, is the source of all that is interesting in contemporary differential geometry." The treatment of differential geometry in this book is an introduction to and commentary on the work of Elie Cartan. As far as possible, both tensor calculus and exterior differentiation have been replaced by matrix algebraic operations. Especially in the theory of frames of different orders, exterior differentiation has been replaced by a certain matrix operation; it has been retained only for the obtention of integrability conditions. These changes simplify many proofs, particularly of integration theorems, and give an intuitive meaning to all operations. They also allow a neat treatment of differentiability hypotheses at all stages.

Although it is impossible to give an introduction to tensor calculus without the appearance of some centipede tensors, the matrix approach shortens most derivations without detrimental effect on the acquisition of computational skills. Tensor algebra appears only toward the middle of the book; the changes in notation that become necessary at that stage are rather natural and should not result in any confusion.

The author has striven not only to write a rigorous exposition of differential geometry but to give the student a feeling for the beauty and interest of rigor. The text is concerned mostly with questions of method. Many interesting results, both classical and modern, will also be found in the more than 800 exercises. As a matter of principle, the book contains more detail than the instructor will be able to cover in a one-year course, thus furnishing material for study assignments necessary in a beginning graduate course.

Each chapter is followed by a list of references. They are intended to cover the material presented both in the text and in the exercises. The author has tried to give those references which can be used most easily for further studies; very often they do not represent the first sources. A number of eighteenth and nineteenth century references have been included. The rewriting in modern terminology of these papers has been a very stimulating experience for the author.

The logical order of the chapters is rather rigid, except that Secs. 3-2, 3-3, 3-5, and Chap. 4 may be omitted without impairing an understanding of the remainder of the text. Likewise, affine differential geometry (Secs. 7-3 and 8-3 and Chap. 12) and line geometry (Secs. 8-2 and 10-6) form entities in themselves, although probably a study of at least one of these two geometries is necessary for a real understanding of the group-theoretic approach to differential geometry. Projective differential geometry, deemed unsuitable for an introductory course, has not been included. For the same reason, complex variables have been avoided.

The topological basis of modern differential geometry is given in Sec. 9-4. The definition of tangent spaces follows rather closely Professor Ehresmann's theory of jets. This definition is simpler than the usual one in terms of infinitesimal transformations, and it works as well for C^r manifolds as for C^∞ ones.

Riemann geometry and generalized spaces present a special problem to the writer of a text on modern differential geometry. While a good grasp of the fundamental ideas and an adequate computational technique are essential, the interesting and profound questions belong to global geometry. Therefore I have tried to present the fundamentals and to show how the techniques of the preceding chapters are applied in Riemann and more general geometries. The guiding criterion was to help the student to understand both local (computational) and global (group-theoretic) work on generalized spaces. If a problem is amenable to both the tensor calculus and the group-theoretic methods, the latter has been preferred. For the final, most general definition of a connection I did not follow Ehresmann's definition in terms of vector-valued forms in principal fiber spaces. The definition used, essentially due to Knebelman, is easily written without use of the language of algebraic topology; it is at least as general as the fiber-space approach. Because of the previous broad treatment of all necessary techniques, the exposition in the last chapters can be condensed.

A textbook never can be an encyclopedia. My guideline in the selection of material was to present classical problems with modern methods.

Heinrich W. Guggenheimer

CONTENTS

1 ELEMENTARY DIFFERENTIAL GEOMETRY

1-1. Curves

In the first chapters of this book we study plane differential geometry. We start with an investigation of the various definitions of a curve. Our intuitive notion of a "curve" contains so many different features that it is necessary to introduce a number of concepts in order to arrive at an exact definition that is neither too broad nor too narrow for our purposes. We will see that different branches of differential geometry deal with different notions of a curve. A detailed discussion of the various definitions of a curve will also lead to a better understanding of the theory of surfaces and higher-dimensional spaces in later parts of the book.

The idea of a curve which we are trying to formalize is that of a piece of wire that has been twisted and stretched into some odd shape but has not been torn apart. Mathematically, the wire becomes an interval of the real-number line, and the operation performed on it is a continuous map.

Definition 1-1. *A plane Peano curve is a continuous map of the unit interval* [0,1] *into the plane.*

We will use the standard notations I for the unit interval and R^2 for the plane. A Peano curve may be given in an easily understood shorthand as

$$\mathbf{f}: I \to R^2$$

A point in I, that is, a parameter value $0 \leq t \leq 1$, has as its image $\mathbf{f}(t)$ a point in R^2. In some cartesian system of coordinates, the Peano curve is given by an ordered pair of real-valued, continuous functions

$$\mathbf{f}(t) = (x_1(t), x_2(t))$$

In at least one important aspect our analytic definition does not agree with the intuitive geometric picture that we want to formalize. A Peano curve defines the set of points covered by it; it is not itself a point set.

***Example* 1-1.** The four curves

$$\mathbf{f}_1(t) = (\cos 2\pi t, \sin 2\pi t) \qquad \mathbf{f}_2(t) = (\cos 2\pi t^2, \sin 2\pi t^2)$$
$$\mathbf{f}_3(t) = (\sin 2\pi t, \cos 2\pi t) \qquad \mathbf{f}_4(t) = (\cos 4\pi t, \sin 4\pi t)$$

are distinct Peano curves. The point set defined in the plane by any one of the four curves is the unit circle S^1: $x_1^2 + x_2^2 = 1$.

From a geometric point of view, the maps $\mathbf{f}_1$ and $\mathbf{f}_2$ should define the same curve. Under $\mathbf{f}_3$, the unit circle is endowed with a negative orientation, and under $\mathbf{f}_4$ there are two values t for each point on the circle. This example shows that we need a suitable notion of equivalence of Peano curves, with the understanding that equivalent curves should represent the same "geometric object." The *geometric* properties of the object will then be those properties of the mapping function $\mathbf{f}$ that are common to all equivalent maps. A similar process is used in euclidean geometry where we consider as abstractly identical all congruent figures, although they may differ by their position in the plane. We shall introduce several types of equivalence; we shall identify some curves defining the same point set with different parametrizations; and sometimes we shall identify point sets that may be brought into one another by some motion or transformation of the plane in itself. It will be one of our major results that geometry is the study of invariants of certain "equivalences" or "identifications."

Definition 1-2. *A continuous curve defined by a Peano curve* $\mathbf{f}$ *is the set of all maps* $\mathbf{g}(t) = \mathbf{f}(H(t))$, *where* H *is a one-to-one continuous map of* I *onto itself, and* $H(0) = 0$.

A one-to-one map H such that both H and its inverse $\overset{-1}{H}$ are continuous is called a *homeomorphism.* If H is defined on a compact set (such as I), the continuity of H implies that of $\overset{-1}{H}$. Since $\mathbf{f}(t) = \mathbf{g}(\overset{-1}{H}(t))$, the geometric object "continuous curve $\mathbf{f}$" is uniquely defined by any one of the functions $\mathbf{g}$ in the set. Any such *admissible* function $\mathbf{g} = \mathbf{f}H$ will be called a *representative* of the continuous curve. In example 1-1, the maps $\mathbf{f}_1$ and $\mathbf{f}_2$ represent the same continuous curve, since $t \rightarrow t^2$ is a homeomorphism of I onto itself that leaves the origin fixed.

***Example* 1-2.** The continuous curve "segment of $y = 2x$ bounded by the origin (0,0) and the point (1,2), outward orientation," may be represented, for example, by $\mathbf{g}_1(t) = (t,2t)$ or by $\mathbf{g}_2(t) = (t^2,2t^2)$ or by

$$\mathbf{g}_3(t) = (e^{1-1/t},2e^{1-1/t}) \qquad \text{for } t > 0,\ \mathbf{g}_3(0) = (0,0)$$

A continuous curve may also be given by a map $\mathbf{f}(u)$, $a \leq u \leq b$, of any closed interval $[a,b]$ into the plane, since $u = a + (b - a)t$ is a one-to-one continuous map of I onto $[a,b]$. The same remark will hold true for all other classes of curves to be introduced.

A Peano curve is *closed* if $\mathbf{f}(0) = \mathbf{f}(1)$. Our definition of a continuous curve implies $\mathbf{g}(0) = \mathbf{g}(1) = \mathbf{f}(0) = \mathbf{f}(1)$ for all representatives of a closed curve. This means that we give a special significance to the starting point $\mathbf{f}(0)$. For example, the function

$$\mathbf{f}_5(t) = (\cos 2\pi(t + \tfrac{1}{2}), \sin 2\pi(t + \tfrac{1}{2}))$$

will not represent the same circle as does $\mathbf{f}_1(t)$ in example 1-1, although both describe the unit circle with counterclockwise orientation. To overcome this difficulty, we refer closed curves not to the unit interval I but to the unit circle S^1 given by $\mathbf{f}_1$ of example 1-1. The parameter t measures the 2π-th part of the arc. We may look at S^1 as the image of I in which we have identified the points 0 and 1. A homeomorphism H of S^1 onto itself is *orientation-preserving* if on I it is given by a monotone increasing function.

Definition 1-3. *A closed Peano curve is a continuous map of the unit circle S^1 into the plane R^2. A closed continuous curve defined by a closed Peano curve* $\mathbf{f}$ *is the set of all functions* $\mathbf{g}(t) = \mathbf{f}(H(t))$, *$H$ being an orientation-preserving homeomorphism of S^1 onto itself.*

According to this definition, both $\mathbf{f}_5$ and $\mathbf{f}_1$ represent the same circle as a closed continuous curve, the connecting homeomorphism being a rotation of S^1 by π. The same map may now represent two distinct geometric beings, viz., a continuous and a closed continuous curve. The two geometric objects may well have different geometric properties.

Actually, we shall use the definitions only for very special continuous curves, since it turns out that any plane continuum (i.e., a compact connected set) can be parametrized so as to become a Peano curve. Peano's famous first example of a pathological "curve" is the parametrization of the entire unit square, obtained in the following way: Divide the square into four parts as shown in Fig. 1-1*a*. Each smaller square is subdivided into four parts and numbered as shown in Fig. 1-1*b*. This process is repeated indefinitely. If the real number $t \in I$ is given in its expansion to the base 4, $t = a_1/4 + a_2/4^2 + \cdots + a_n/4^n + \cdots$, $a_n = 0, 1, 2, 3$, the image of t is the unique limit point $\mathbf{p}(t)$ contained in the nested sequence of squares with labels a_1, a_1a_2, $a_1a_2a_3$, ... The mapping $t \to \mathbf{p}(t)$ is continuous. For any $\epsilon > 0$, take k such that $4^k\epsilon > \sqrt{2}$. If $|t^* - t| < \delta(\epsilon) = 1/4^k$, the first $k - 1$ digits in the expansions of t and t^* coincide; hence $\mathbf{p}(t^*)$ and $\mathbf{p}(t)$ are in the same square of the $(k - 1)$-th subdivision, and the distance between them is less than the diameter $\sqrt{2}/4^k < \epsilon$ of such a square.

In one respect, the definitions we have given are too narrow. Many interesting curves are not compact, e.g., a straight line or a logarithmic spiral. Since we are interested in the properties of curves only in the neighborhood of some point, we replace any unbounded curve by a

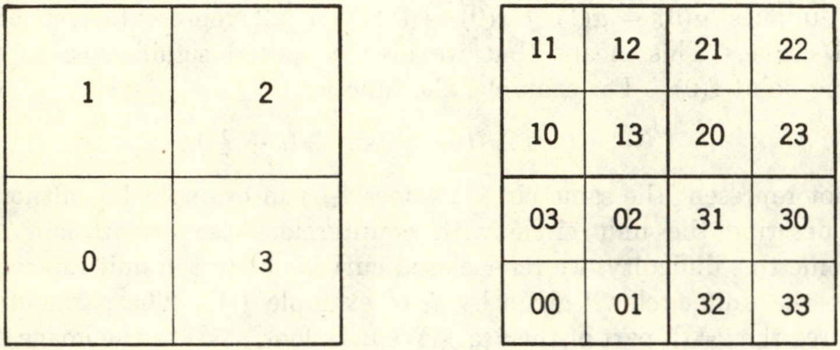

1	2
0	3

(*a*)

11	12	21	22
10	13	20	23
03	02	31	30
00	01	32	33

(*b*)

Fig. 1-1

sufficiently large closed segment on it. For example, a line $\mathbf{a} + \mathbf{b}t$ may be treated through its segments $I_\lambda = \{\mathbf{a} + \mathbf{b}\lambda t\}$, $0 \leq t \leq 1$, where λ is a real number.

We will use boldface for vectors and for curves as vector functions. Points in the plane that are not identified with vectors will be denoted by capitals.

A *Jordan* curve is an equivalence class of homeomorphisms of I into R^2 (or of S^1 into R^2 in the case of closed curves). Though widely used in topology, this notion again is unsuitable for our purposes, as it is both too narrow and too wide—too narrow because it excludes curves with double points; too wide because Jordan curves may not have a tangent at any point. An example of the latter is $\mathbf{x}(t) = (t,f(t))$, where $f(t)$ is a continuous, nondifferentiable function defined on I (see exercise 1-1, Prob. 11). In our final definition, we shall have to include differentiability, and we shall have to find our way between admitting area-filling curves and excluding all singularities.

Definition 1-4. *A map* $\mathbf{f}(t) = (x_1(t),x_2(t))\colon I \to R^2$ *is* C^n *if for any* $t_0 \in I$ *there exists a neighborhood of* t_0 *in* I *such that the restriction of* $\mathbf{f}$ *to that neighborhood is a homeomorphism given by n-times continuously differentiable functions* $x_1(t)$, $x_2(t)$. *A* C^n *curve defined by a* C^n *map* $\mathbf{f}$ *is* *the set of all functions* $\mathbf{g}(t) = \mathbf{f}(H(t))$, *H being a homeomorphism of I onto itself which is an n-times differentiable function,* $H(0) = 0$.

Continuous curves will be called C^0 curves. Closed C^n curves are defined similarly from maps $S^1 \to R^2$. By the Heine-Borel theorem† there is a finite number of neighborhoods covering I so that $\mathbf{f}$ is a homeomorphism in each of them. This means that a C^n curve is a finite union of differentiable Jordan curves. Nevertheless, a C^n curve may have an infinity of double points. An example is given at the end of this section. A C^n curve may also have cusps, as is shown by the *cycloid*

$$\mathbf{f}(t) = (4\pi t - \sin 4\pi t, 1 - \cos 4\pi t)$$

The first and second derivatives of the mapping functions have simple geometric interpretations in euclidean geometry. As a consequence, we shall deal mostly with C^2 curves. For convex arcs, however, some considerations of curvature involving second derivatives may be circumvented.

Definition 1-5. *A set is convex if with any two points it contains the segment defined by the two points* (Fig. 1-2*a*). *A curve with endpoints*

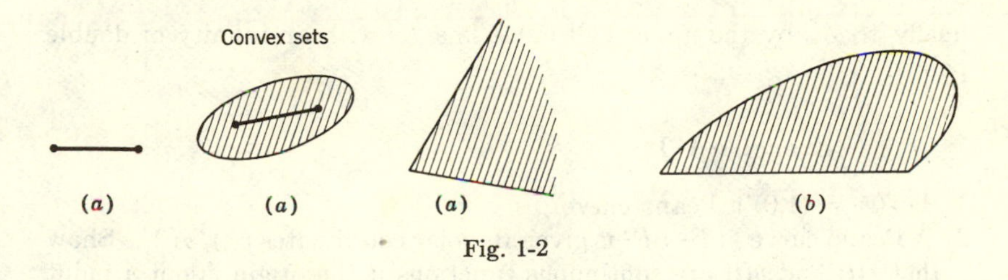

Fig. 1-2

P_0P_1 *is convex if its point set, together with the segment* P_0P_1, *bounds a convex set in* R^2 (Fig. 1-2*b*). *A* K^n *curve is a* C^n *curve such that for any of its maps* $\mathbf{f}$ *and any* $t \in I$ *there exists an* $\epsilon > 0$ *for which* $\mathbf{f}(t)$ *restricted to* $[t - \epsilon, t + \epsilon]$ *defines a convex curve.*

A K^n curve is the union of a finite number of convex curves, joined together without introducing inflections or cusps. (Inflections and cusps are points at which unions of convex curves cease to be K^n. In a cusp, the "tangent vector" vanishes. The treatment of singularities is avoided in this text.) Such a curve may contain an infinity of double points.

***Example* 1-3.** On a unit circle, denote by A_k the endpoint of the arc of length $(1 - 2^{-k})\pi$ measured from (1,0). These points are joined by parabolical arcs having as tangents at A_k the tangents to the circle.

† See R. C. Buck, "Advanced Calculus," p. 82, McGraw-Hill Book Company, Inc., New York, 1956.

The curve obtained in starting from A_0, going to $A_\infty = (-1,0)$ through the parabolic arcs, returning to A_0 by the lower half circle, and then

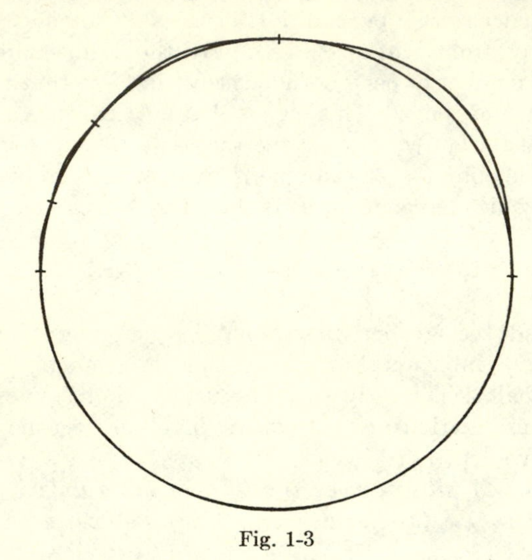

Fig. 1-3

finally to A_∞ by the upper half circle is a K^1 with an infinity of double points.

Exercise 1-1

1. Is $\mathbf{f}(t) = (1,0)$ a Peano curve?
2. A Peano curve $\mathbf{f}: I \to R^2$ is given in polar coordinates $r(t)$, $\theta(t)$. Show that $r(t)$ and $\theta(t)$ are continuous functions if the origin is not a point on the curve.
3. Prove that a real function $H(t)$ defined on I is a homeomorphism of I onto itself, if it is strictly monotone and continuous and if either $H(0) = 0$, $H(1) = 1$ or $H(0) = 1$, $H(1) = 0$.
4. How many distinct closed continuous curves are given by the mappings of example 1-1?
5. Represent the graph of a continuous function $x_2 = f(x_1)$, $a \leq x_1 \leq b$, as a Peano curve.
6. Show that any homeomorphism of the circle S^1 onto itself may be obtained as $R\overset{-1}{\mathbf{f}_1}H\overset{-1}{\mathbf{f}_1}$, where $\mathbf{f}_1$ is given in example 1-1, H is a homeomorphism of I onto itself, and R is a rotation of the plane about the origin.
7. Any number $m/4^n (m,n$ integers) has two expansions to the base 4, a finite one $a_1/4 + \cdots + a_n/4^n$ and an infinite one $a_1/4 + \cdots +$

$a_{n-1}/4^{n-1} + (a_n - 1)/4^n + 3/4^{n+1} + 3/4^{n+2} + \cdots$. Show that both expansions define the same point on Peano's curve. Draw the approximating squares for $\mathbf{p}(\frac{1}{4})$.

8. Prove that on Peano's curve $\mathbf{p}(\frac{1}{6}) = \mathbf{p}(\frac{1}{2})$.

9. Peano's map of I onto the unit square is not one-to-one. Show that any point on a side of a square in Peano's construction will be a double point of the curve, and any point which is a vertex of a square will be a quadruple point.

10. A *logarithmic spiral* is the graph of $r = ce^{-m\theta}$ in polar coordinates. The graph is not a compact set. Given any two points $(r_0,\theta_0;r_1,\theta_1)$ of the spiral, represent by a Peano curve the segment defined on the spiral by the two points.

11. We define a function $y(t)(t \in I)$ by a limit process.

(*a*) $y_0(t) = t$.

(*b*) Assume $y_n(t)$ to be defined. We define y_{n+1} first on the points subdividing I into 3^{n+1} equal parts. Let

$$t_0 = \frac{a_1}{3} + \cdots + \frac{a_n}{3^n}$$

$$t_1 = \frac{a_1}{3} + \cdots + \frac{a_n}{3^n} + \frac{1}{3^{n+1}}$$

$$t_2 = \frac{a_1}{3} + \cdots + \frac{a_n}{3^n} + \frac{2}{3^{n+1}}$$

$$t_3 = \frac{a_1}{3} + \cdots + \frac{a_n + 1}{3^n}$$

Then define

$$y_{n+1}(t_0) = y_n(t_0) \qquad y_{n+1}(t_3) = y_n(t_3)$$
$$y_{n+1}(t_1) = y_n(t_0) + \tfrac{2}{3}[y_n(t_3) - y_n(t_0)]$$
$$y_{n+1}(t_2) = y_n(t_0) + \tfrac{1}{3}[y_n(t_3) - y_n(t_0)]$$

In the intervals between these division points the function is defined as the linear function joining the relative values in the division points.

(i) Draw graphs of $y_0(t)$, $y_1(t)$, $y_2(t)$.

(ii) Show that the sequence of functions $y_n(t)$ converges to a continuous nowhere-differentiable function $y(t)$.

12. Give a description of the curve of example 1-3 in terms of mapping functions $x_1(t)$, $x_2(t)$.

13. Give the complete formal definition of a closed C^n (K^n) curve.

14. Find an example of a K^n curve with an infinity of self-intersections.

15. Given two curves $\mathbf{f}(t)$, $\mathbf{g}(t)$, $\mathbf{f}(1) = \mathbf{g}(0)$. The curve $\mathbf{h}(t)$,

$$\mathbf{h}(t) = \begin{cases} \mathbf{f}(2t) & t \le \frac{1}{2} \\ \mathbf{g}(2t - 1) & t \ge \frac{1}{2} \end{cases}$$

is the *sum* of **f** and **g**. Prove that the point set of **h** is the union of the point sets of **f** and **g** and that **h** is C^0 if both **f** and **g** are C^0.

16. Find a C^0 map $I \to R^2$ which is not C^1 and whose image set

$$\{P|\mathbf{f}(t) = P\}$$

is a unit circle.

17. A differentiable homeomorphism on an open interval has a differentiable inverse on a closed subinterval. Prove this statement and discuss its application to Def. 1-4.

1-2. Vector and Matrix Functions

It is assumed that the reader is familiar with the fundamentals of vector and matrix algebra. For typographical convenience, a column vector will usually be written as a row in braces:

$$\begin{pmatrix} x_1 \\ x_2 \\ \cdots \\ x_n \end{pmatrix} = \{x_1,x_2, \ . \ . \ . \ ,x_n\}$$

Row vectors are indicated by parentheses:

$$\{x_1,x_2, \ . \ . \ . \ ,x_n\}^t = (x_1,x_2, \ . \ . \ . \ ,x_n)$$

The symbol ^{t} represents the transpose of a matrix.

A system of basis vectors $\mathbf{e}_1, \ . \ . \ . \ , \mathbf{e}_n$ of an n-dimensional vector space is called a *frame* in that space and is noted as a column vector of row vectors $\{\mathbf{e}_1, \ . \ . \ . \ ,\mathbf{e}_n\}$. Coordinate vectors are row vectors. A vector, i.e., an element of a vector space, is the matrix product of a coordinate vector and a frame:

$$\mathbf{x} = \sum_{i=1}^{n} x_i\mathbf{e}_i = (x_1, \ . \ . \ . \ ,x_n)\{\mathbf{e}_1, \ . \ . \ . \ ,\mathbf{e}_n\}$$

If the elements of a matrix $A = (a_{ij}(u))$ are differentiable functions of a variable u, the derivative of A is the matrix of the derivatives:

$$\frac{dA}{du} = \left(\frac{da_{ij}}{du}\right)$$

Derivatives with respect to a "general" parameter u will also be written with the "dot" symbol, $dA/du = A^{\cdot}$. The "prime" symbol will be reserved for differentiation with respect to certain "invariant" parameters which we shall introduce later. Such an invariant parameter will often be denoted by s: $dA/ds = A'$.

It follows immediately from the definition that differentiation and transposition commute:

$$(A^t)^{\cdot} = (A^{\cdot})^t \tag{1-1}$$

If a C^n map is represented by its coordinate vector in some *fixed* frame

$$\mathbf{f}(t) = (x_1(t), x_2(t))$$

the derived vector $\mathbf{f}^{\cdot}(t) = (x_1^{\cdot}, x_2^{\cdot})$ is parallel to the tangent of the curve at the point $\mathbf{f}(t)$; the tangent itself is the set of points with coordinates $\mathbf{f}(t) + \lambda\mathbf{f}^{\cdot}(t)$.

The matrix product is bilinear; hence the Leibniz rule holds:

$$[A(u)B(u)]^{\cdot} = A^{\cdot}(u)B(u) + A(u)B^{\cdot}(u) \tag{1-2}$$

The *scalar product* (dot product) of two row vectors $\mathbf{a}$ and $\mathbf{b}$ is defined as

$$\mathbf{a} \cdot \mathbf{b} = \mathbf{a}\mathbf{b}^t \tag{1-3}$$

Here we identify a 1×1 matrix with the unique element it contains. A 1×1 matrix is always identical to its transpose; hence

$$\mathbf{a} \cdot \mathbf{b} = \mathbf{b} \cdot \mathbf{a}$$

The scalar product of column vectors is similarly defined. The *length* $|\mathbf{a}|$ of a vector $\mathbf{a}$ is the square root of its dot square:

$$|\mathbf{a}| = (\mathbf{a} \cdot \mathbf{a})^{\frac{1}{2}}$$

The *angle* of two vectors $\mathbf{a}$, $\mathbf{b}$ is defined by

$$\cos \angle(\mathbf{a},\mathbf{b}) = \frac{\mathbf{a} \cdot \mathbf{b}}{|\mathbf{a}|\,|\mathbf{b}|}$$

The absolute value of the cosine is never greater than 1 by virtue of the *Cauchy inequality*

$$(\mathbf{a} \cdot \mathbf{b})^2 \leq |\mathbf{a}|^2|\mathbf{b}|^2$$

or, in components,

$$(\Sigma a_i b_i)^2 \leq (\Sigma a_i^2)(\Sigma b_i^2)$$

Two nonzero vectors are *orthogonal* if their dot product is zero. The following result is fundamental:

Lemma 1-6. *The derived vector* $\mathbf{a}^{\cdot}(u)$ *of a differentiable vector* $\mathbf{a}(u)$ *of constant length is orthogonal to* $\mathbf{a}(u)$.

By hypothesis, $\mathbf{a}(u) \cdot \mathbf{a}(u) = \text{const}$; hence

$$\mathbf{a}^{\cdot}(u) \cdot \mathbf{a}(u) + \mathbf{a}(u) \cdot \mathbf{a}^{\cdot}(u) = 2\mathbf{a}^{\cdot}(u) \cdot \mathbf{a}(u) = 0$$

A square matrix is *orthogonal* if its transpose is its inverse:

$$AA^t = A^tA = U \tag{1-4}$$

Here U stands for the unit matrix $U = (\delta_{ij})$, $\delta_{ii} = 1$, $\delta_{ij} = 0$ if $i \neq j$. Transformation by an *orthogonal* matrix A leaves the dot product invariant, $\mathbf{a}A \cdot \mathbf{b}A = \mathbf{a}AA^t\mathbf{b}^t = \mathbf{a}\mathbf{b}^t = \mathbf{a} \cdot \mathbf{b}$. Especially, $|\mathbf{a}| = |\mathbf{a}A|$, and $\mathbf{a} \cdot \mathbf{b} = 0$ implies $\mathbf{a}A \cdot \mathbf{b}A = 0$ for orthogonal A. The determinant of an orthogonal matrix is ± 1. If it is $+1$, the matrix is a *rotation*. Every 2×2 rotation may be written as

$$\begin{pmatrix} \cos\theta & \sin\theta \\ -\sin\theta & \cos\theta \end{pmatrix}$$

Of special importance is the rotation by $\pi/2$, the matrix $J = \begin{pmatrix} 0 & 1 \\ -1 & 0 \end{pmatrix}$. The *cross product* (vector product) of two *plane* vectors is

$$\mathbf{a} \times \mathbf{b} = \mathbf{a} \cdot \mathbf{b}J^t$$

If A is a rotation, $J = A^tJA$.

Definition 1-7. *The Cartan matrix of a differentiable nonsingular square matrix* $A(u)$ *is* $C(A) = \dot{A}A^{-1}$.

From the definition we have at once

$$C(AB) = C(A) + AC(B)A^{-1} \tag{1-5}$$

This formula contains much of differential geometry.

Lemma 1-8. *The Cartan matrix of an orthogonal matrix function is skew-symmetric,* $C(A)^t = -C(A)$.

PROOF: Differentiate (1-4),

$$(AA^t)^{\cdot} = \dot{A}A^t + A\dot{A}^{t} = \dot{A}A^{-1} + (\dot{A}A^{-1})^t = 0$$

since $A^t = A^{-1}$ by hypothesis. The last equation proves the lemma. We shall see later on that it admits a converse.

The integral of a matrix function is the matrix of the integrals of its elements:

$$\int_{t_0}^{t} (a_{ij}(u))\,du = \left(\int_{t_0}^{t} a_{ij}(u)\,du\right)$$

Exercise 1-2

1. Show that $\mathbf{a} \times \mathbf{b} = -\mathbf{b} \times \mathbf{a}$.
2. If A is a *rotation*, show that $\mathbf{a} \times \mathbf{b} = \mathbf{a}A \times \mathbf{b}A$.
3. Let k be a constant number. Show that $C(kA) = C(A)$.

◆ 4. $A = \begin{pmatrix} r\cos\theta & r\sin\theta \\ -r\sin\theta & r\cos\theta \end{pmatrix}$, r const, θ independent variable. Compute $C(A)$.

◆ This symbol precedes exercises for which answers are given at the end of the book.

5. For differentiable nonsingular $A(u)$, prove that

$$dA^{-1}/du = -A^{-1}(dA/du)A^{-1}$$

(HINT: Differentiate $AA^{-1} = U$.)

6. Prove that $C(A^{-1}) = -C(A^t)^t$. (Use Prob. 5.)

7. Use Prob. 6 to give an alternative proof for lemma 1-8.

8. Prove that $C(AM) = C(A)$ if M is a constant, nonsingular matrix.

9. Show that the process of taking the Cartan matrix is associative: $C(A \cdot BC) = C(AB \cdot C) = C(ABC)$.

10. Prove that $\mathbf{a} \cdot \mathbf{b} = |\mathbf{a}|\,|\mathbf{b}| \cos \angle(\mathbf{a},\mathbf{b})$, $\mathbf{a} \times \mathbf{b} = |\mathbf{a}|\,|\mathbf{b}| \sin \angle(\mathbf{a},\mathbf{b})$.

1-3. Some Formulas

Let $\mathbf{f}(t)$ be a C^1 curve. We represent $\mathbf{f}(t)$ either by its cartesian coordinates $x_1(t)$, $x_2(t)$ with respect to some fixed axes, or by polar coordinates $r^2(t) = x_1^2 + x_2^2$, $\phi(t) = \arctan x_2/x_1$.

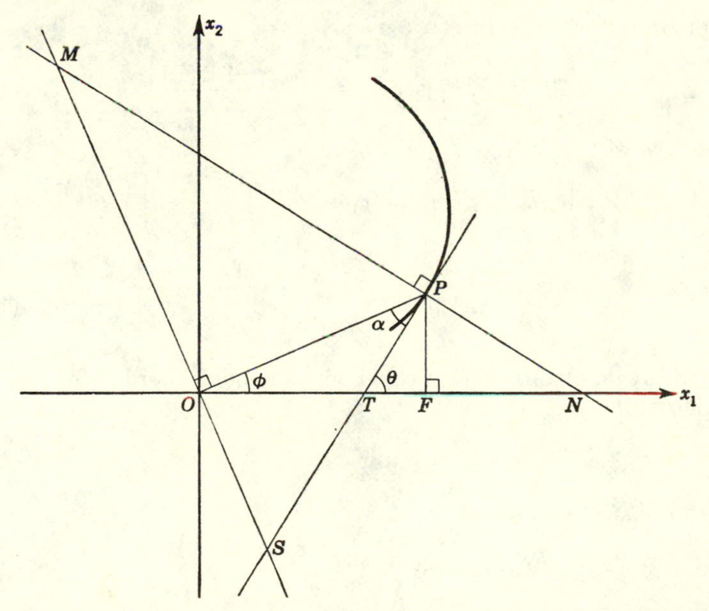

Fig. 1-4

Through a point $P = \mathbf{f}(t_0)$ we draw the tangent to $\mathbf{f}$ as the line with slope $\dot{x}_2(t_0)/\dot{x}_1(t_0)$. The normal to $\mathbf{f}$ at P is the line through P with slope $-\dot{x}_1(t_0)/\dot{x}_2(t_0)$. Let T be the point of intersection of the tangent with the x_1 axis, N the point of intersection of the normal with the same axis, and F the projection of P on x_1, that is, the point $F(x_1(t_0),0)$ (Fig. 1-4).

O is the origin of the system of coordinates; the radius vector of P is the directed segment OP. We draw also the normal to OP at O; it intersects the tangent at S and the normal at M. Let $\alpha = \angle OPT$ be the angle between the radius vector and the tangent, and θ the angle between the $+x_1$ axis and the tangent. By definition, $\tan\theta = x_2^{\cdot}/x_1^{\cdot}$, $\tan\phi = x_2/x_1$; hence

$$\tan\alpha = \tan(\theta - \phi) = \frac{x_1 x_2^{\cdot} - x_1^{\cdot} x_2}{x_1 x_1^{\cdot} + x_2 x_2^{\cdot}} \tag{1-6}$$

In polar coordinates, $x_1 = r\cos\phi$, $x_2 = r\sin\phi$, and

$$\tan\alpha = \frac{r\phi^{\cdot}}{r^{\cdot}} \tag{1-7}$$

The segment TF is the *subtangent*. Its length is

$$|TF| = |FP|\,|\cot\theta| = \left|\frac{x_2 x_1^{\cdot}}{x_2^{\cdot}}\right| \tag{1-8}$$

The length of the *tangent* is

$$|TP| = |TF|\,|\sec\theta| = \left|\frac{x_2(x_1^{\cdot 2} + x_2^{\cdot 2})^{\frac{1}{2}}}{x_2^{\cdot}}\right| \tag{1-9}$$

that of the *normal*

$$|PN| = |TP|\,|\tan\theta| = \left|\frac{x_2(x_1^{\cdot 2} + x_2^{\cdot 2})^{\frac{1}{2}}}{x_1^{\cdot}}\right| \tag{1-10}$$

and of the *subnormal*

$$|FN| = \left|\frac{x_2 x_2^{\cdot}}{x_1}\right| \tag{1-11}$$

Turning now to quantities connected with the polar coordinates, we have the *polar subtangent*

$$|OS| = r\,|\tan\alpha| = \left|\frac{r^2\phi^{\cdot}}{r^{\cdot}}\right| \tag{1-12}$$

the *polar tangent*

$$|PS| = r\,|\sec\alpha| = \left|\frac{r(r^{\cdot 2} + r^2\phi^{\cdot 2})^{\frac{1}{2}}}{r^{\cdot}}\right| \tag{1-13}$$

the *polar subnormal*

$$|OM| = r\,|\cot\alpha| = \left|\frac{r^{\cdot}}{\phi^{\cdot}}\right| \tag{1-14}$$

and the *polar normal*

$$|PM| = r\,|\csc\alpha| = \frac{(r^{\cdot 2} + r^2\phi^{\cdot 2})^{\frac{1}{2}}}{|\phi^{\cdot}|} \tag{1-15}$$

Exercise 1-3

1. Simplify formulas (1-6) to (1-15) for graphs of functions $x_2 = f(x_1)$, respectively, $r = r(\phi)$.

◆ **2.** Compute the tangent, subtangent, normal, and subnormal for the ellipse and the parabola.

3. The graph of $r = c\phi$ (c = const) is known as the *spiral of Archimedes.* Show that its polar subnormal is $|c|$ and that its polar subtangent is $r^2/|c|$. Derive from this a construction of the tangent to the spiral with straightedge and compass, given O and P.

◆ **4.** Find all curves with a constant polar subnormal.

5. A *hyperbolic spiral* is the graph of $r\phi = c$ ($c > 0$, const). Show that $\alpha = -\phi$, that its polar subtangent is c, and that its polar subnormal is r^2/c. Indicate a construction of the tangent to the hyperbolic spiral by straightedge and ruler, given O and P. Show that $x_2 = c$ is an asymptote for $\phi \to 0$.

6. Describe exactly to which category of curves introduced in Sec. 1-1 belong the spirals defined in Probs. 3 and 5.

7. As P of a curve $\mathbf{f}(t)$ varies, the perpendicular from O meets PS at the *podal curve of* **f** *with respect to the pole O.* Show that the podal curve of a parabola $r = p/(1 + \cos\phi)$ relative to the pole $O(2p,O)$ is the graph of $r = p/(2\cos\phi/3)$ (*Maclaurin's trisectrix*).

* **8.** Show that the podal curve of an **f** with respect to a pole O is the image of the polar reciprocal of **f** for any circle of center O under an inversion relative to the same circle (theorem of Teixeira). (HINT: Let P' be any point on the polar reciprocal. Its polar is a tangent of f; on this tangent the point inverse to P' is S.)

9. An important family of curves is described by $r = a\sin^{1/n} n\phi$ (*spirals of Maclaurin*). For $n < 0$, the curves are not compact. Discuss the following special cases: $n = 2$ (lemniscate), $n = 1$ (circle), $n = \frac{1}{2}$ (cardioid), $n = -\frac{1}{2}$ (parabola), $n = -2$ (equilateral hyperbola). n is the *order* of the spiral.

10. Show that the podal curve of a spiral of Maclaurin of order n with respect to the origin of the coordinates is a spiral of Maclaurin of order $n/(n+1)$ (see Probs. 7 and 9).

11. The *conchoid* of $r = r(\phi)$ is $r^* = r(\phi) + a$. The conchoid of a straight line is the *conchoid of Nicomedes,* so-called for its form ($\kappa\acute{o}\gamma\chi\eta$ = shell). Show that the normal to the conchoid always passes through the endpoint M of the polar subnormal of the corresponding point on the original curve.

◆**12.** Find all curves whose subtangent is constant.

◆**13.** Find all curves whose tangent is of constant length.

* Exercises with an asterisk presuppose some knowledge of other parts of geometry.

◆**14.** Find the curves for which the ratio of the polar subtangent to the polar subnormal is constant.

References

Bouligand, G.: "Introduction à la géométrie infinitésimale directe," Librairie Vuibert, Paris, 1932.

Locher-Ernst, L.: "Einführung in die freie Geometrie ebener Kurven," Birkhäuser Verlag, Basel, 1952.

Menger, K.: "Kurventheorie," B. G. Teubner Verlagsgesellschaft, mbH, Berlin, 1932.

Peano, Giuseppe: "Opere scelte," vol. 1, nos. 24, 138, 29, Edizioni Cremonese, Rome, 1957.

Peano, Giuseppe: "Applicazioni geometriche del calcolo infinitesimale," Bocca, Turin, 1887.

2

CURVATURE

2-1. Arc Length

In our definition, a C^n curve is a set of an infinity of maps. We will try to single out a unique representative of every curve in a geometrically significant way. In euclidean geometry this is done by referring a curve to its arc length as a parameter.

Definition 2-1. *A curve* $\mathbf{x}(s)$ *is said to be defined as a function of its arc length if the tangent vector* $\mathbf{x}'(s) = d\mathbf{x}/ds$ *is a unit vector,* $|\mathbf{x}'(s)| = 1$.

In terms of the arc length, the map $\mathbf{x}(s)$ is not defined on the unit interval I but on some other interval $[0,L]$, L being the *total length* of the curve. For any C^n curve, $n \geq 1$, the arc length s and hence also the function $\mathbf{x}(s)$ are unique. For let $\mathbf{f}(u)$ be any representative of the given curve. The tangent has length $|d\mathbf{f}/du| = (x_1^{\cdot 2} + x_2^{\cdot 2})^{\frac{1}{2}}$; hence s is obtained uniquely (up to an additive constant) from

$$\frac{ds}{du} = (x_1^{\cdot 2} + x_2^{\cdot 2})^{\frac{1}{2}}$$

s is completely determined if we put $s(0) = 0$. In the next three chapters, we shall always refer a curve to its arc length s. Differentiation with respect to this special parameter is indicated by a prime.

The arc length defined formally by a property of the tangent vector may be given a more geometric interpretation, as a limit of lengths of polygonal curves inscribed into an arc. Given two points of a curve, $P_0 = \mathbf{f}(u_0)$, $P_1 = \mathbf{f}(u_1)$, take a finite subdivision of the interval $[u_0,u_1]$, Σ: $u_0 = t_0 < t_1 < t_2 < \cdots < t_N = u_1$. The sum of the lengths of the straight segments $\mathbf{f}(t_i)\mathbf{f}(t_{i+1})$ is

$$L(\Sigma) = \sum_{i=0}^{N-1} |\mathbf{f}(t_{i+1}) - \mathbf{f}(t_i)|$$
$$= \sum_{i=0}^{N-1} [(x_1(t_{i+1}) - x_1(t_i))^2 + (x_2(t_{i+1}) - x_2(t_i))^2]^{\frac{1}{2}}$$

Definition 2-2. *The length of the arc P_0P_1 is $L(P_0,P_1) = \sup_\Sigma L(\Sigma)$. A curve is rectifiable if all its arcs are of finite length.*

The next theorem says that Defs. 2-1 and 2-2 are coherent.

Theorem 2-3. *A C^1 curve is rectifiable and on it the length of an arc is measured by the arc length, $L(P_0,P_1) = s(u_1) - s(u_0) = \int_{u_0}^{u_1} (x_1'^2 + x_2'^2)^{\frac{1}{2}}\,dt$.*

Let m_{ij} and M_{ij} be, respectively, the minimum and the maximum of $x_i'(u)$ in the interval $t_j \leq u \leq t_{j+1}$. By the mean-value theorem,

$$\sum_{j=0}^{N-1} (m_{1j}^2 + m_{2j}^2)^{\frac{1}{2}}(t_{j+1} - t_j) \leq L(\Sigma) \leq \sum_{j=0}^{N-1} (M_{1j}^2 + M_{2j}^2)^{\frac{1}{2}}(t_{j+1} - t_j)$$

Given a sequence of subdivisions Σ such that $\max\,(t_{j+1} - t_j) \to 0$, both the right-hand side and the left-hand side of this inequality tend to $\int_{u_0}^{u_1} (x_1'^2 + x_2'^2)^{\frac{1}{2}}\,dt$. The subdivision Σ^* is said to be *finer* than Σ if all division points of Σ are also division points of Σ^*. The triangle inequality implies $L(\Sigma) \leq L(\Sigma^*)$. The previous argument shows that for each Σ we may find a finer Σ^* such that $\left| \int_{u_0}^{u_1} (x_1'^2 + x_2'^2)^{\frac{1}{2}}\,dt - L(\Sigma^*) \right| < \epsilon$ and $\sup_\Sigma L(\Sigma) - L(\Sigma^*) < \epsilon$ for any given ϵ.† One has, as a corollary, that *the length of an arc is independent of the representation of the curve.*

There are rectifiable curves that are not C^1. A K^0 curve is a union of a finite number of convex arcs, each of which may be taken sufficiently small so as to permit the introduction of a system of cartesian axes that will make $x_2(t)$ a convex (or concave) function of $x_1(t)$. $[x_2(x_1(t_1)) - x_2(x_1(t_0))]/(x_1(t_1) - x_1(t_0))$ then is a nondecreasing (nonincreasing) function of $x_1(t_1)$, and its absolute value will have a finite maximum M at one of its endpoints. Hence $L(\Sigma) \leq (1 + M^2)^{\frac{1}{2}}(x_1(u_1) - x_1(u_0))$. The set of numbers $L(\Sigma)$ is bounded; it has a finite least upper bound $L(P_0,P_1)$. By a more detailed study of the derivatives and the integrals of convex functions one might prove the whole theorem 2-3 for K^0 curves. We shall see several times later that a convexity property may be substituted for one order of differentiability in our theorems.

There is a simple analytic characterization of rectifiable curves. For a

† This proof is based on the definition of the Riemann integral as the limit of sequences. If one adopts the definition in terms of the upper and lower Darboux integral, the theorem is an immediate consequence of the double inequality for $L(\ )$.

subdivision Σ of an interval $[u_0,u_1]$ and a function $F(u)$ defined on that interval, we define the *variation of F on* Σ as

$$v_\Sigma(F) = \sum_{i=0}^{N-1} |F(t_{i+1}) - F(t_i)|$$

and the *variation of F on the interval* $[u_0,u_1]$ as

$$v_{[u_0,u_1]}(F) = \sup_\Sigma v_\Sigma(F)$$

A function is of *bounded variation* if its variation on any closed interval is finite. Then

$$v_\Sigma(x_1) \le L(\Sigma) \qquad v_\Sigma(x_2) \le L(\Sigma)$$

and, by the "triangle inequality" $(a^2 + b^2)^{\frac{1}{2}} \le |a| + |b|$,

$$L(\Sigma) \le v_\Sigma(x_1) + v_\Sigma(x_2)$$

Hence also

$$v_{[u_0,u_1]}(x_i) \le L(P_0,P_1) \le v_{[u_0,u_1]}(x_1) + v_{[u_0,u_1]}(x_2)$$

Theorem 2-4. *A curve* $(x_1(u),x_2(u))$ *is rectifiable if and only if both* $x_1(u)$ *and* $x_2(u)$ *are of bounded variation.*

Exercise 2-1

1. Compute the arc length
   (*a*) Of the circle $x_1 = R\cos u$, $x_2 = R\sin u$, $0 \le u \le 2\pi$
   (*b*) Of the astroid $x_1 = a\cos^3 u$, $x_2 = a\sin^3 u$, $0 \le u \le 2\pi$
   (*c*) Of the logarithmic spiral $r = ce^{-m\phi}$, $0 \le \phi < \infty$
2. Show that in polar coordinates $ds = (\dot{r}^2 + r^2\dot{\phi}^2)^{\frac{1}{2}}\,dt$.
3. Compute ds for the ellipse $x_1 = a\cos u$, $x_2 = b\sin u$.
4. Show that the length of the graph of y_n in exercise 1-1, Prob. 11, is $>\frac{4}{3}$ times the length of the graph of y_{n-1}. Use this result to show that the C^0 curve, the graph of $y(t)$, is not rectifiable.
5. Show that a C^0 curve is rectifiable if and only if one of its representatives is rectifiable.
6. ♦ Give an example of a C^1 curve which is not K^0.
7. Give an example of a K^0 curve that is not a finite union of C^1 curves.
8. Prove that a monotone function has a right and a left limit at any point.
9. The left-hand derivative of a function $y(x)$ is defined as

$$D_l(x_0) = \lim_{x \uparrow x_0} [y(x) - y(x_0)]/(x - x_0)$$

The right-hand derivative is $D_r(x_0) = \lim_{x \downarrow x_0} [y(x) - y(x_0)]/(x - x_0)$.

Prove the following statements:

(*a*) For a *convex* function $y(x)$ and fixed x_0, the ratio $[y(x) - y(x_0)]/(x - x_0)$ is a monotone function of x.

(*b*) A convex function has a right-hand and a left-hand derivative at any point and for $x_1 > x_0$, $D_l y(x_1) \geq D_r y(x_0)$. [HINT: Use part (*a*) and Prob. 8.]

(*c*) A function is differentiable at x_0 if $D_l y(x_0) = D_r y(x_0)$. A convex function is differentiable except possibly in a countable set of points. The sum of the jumps $D_r(x_0) - D_l(x_0)$ is finite in any closed interval. [Use part (*b*).]

10. Assuming the validity of the statements of Prob. 9, define left and right tangents at the points of a K^0 curve, and discuss the existence of a tangent for K^0 curves.

2-2. The Moving Frame

A cartesian system of coordinates in a plane may be given by the two unit vectors on its $+x_1$, $+x_2$ axes. Conversely, any two mutually orthogonal unit vectors attached to some point in the plane determine a system of cartesian coordinates if we take them to define the axes with their positive orientations. The plane of euclidean geometry is quite distinct from a two-dimensional vector space. In a vector space all parallel vectors of the same length and same orientation are identified. In euclidean space the basic geometric object is a vector together with its starting point. The notion which will permit us to compute in the vector plane phenomena of the euclidean plane is that of a frame. For the moment, we shall use this notion in a more restricted sense than that introduced on page 8.

Definition 2-5. *A frame is a vector* $\{\mathbf{e}_1,\mathbf{e}_2\}$ *of mutually orthogonal unit vectors such that* $\mathbf{e}_2$ *is obtained from* $\mathbf{e}_1$ *by a rotation of* $+\pi/2$.

The coordinates in a plane are fixed by some frame $\{\mathbf{e}_1,\mathbf{e}_2\}$. Given a C^2 curve in terms of its arc length,

$$\mathbf{x}(s) = (x_1(s),x_2(s))\{\mathbf{e}_1,\mathbf{e}_2\} = x_1\mathbf{e}_1 + x_2\mathbf{e}_2$$

its *tangent* $\mathbf{t}(s) = x_1'\mathbf{e}_1 + x_2'\mathbf{e}_2$ is a unit vector. The *normal*

$$\mathbf{n}(s) = -x_2'\mathbf{e}_1 + x_1'\mathbf{e}_2$$

is the unit vector obtained from $\mathbf{t}(s)$ by a rotation of $+\pi/2$. $\{\mathbf{t}(s),\mathbf{n}(s)\}$ is the *moving frame* of the curve.

There are two alternative interpretations of this moving frame. In the euclidean plane it defines a cartesian system of coordinates for any point on the curve such that $\mathbf{x}(s)$ becomes the origin and the $+x_1$ axis will be tangent to the curve in the direction of increasing arc length

(Fig. 2-1). Otherwise, we may refer all frames $\{\mathbf{t}(s),\mathbf{n}(s)\}$ to the origin O of the fixed frame $\{\mathbf{e}_1,\mathbf{e}_2\}$ (Fig. 2-2). This means that now we consider the plane as a vector space of two dimensions. In this case the endpoint of $\mathbf{t}(s)$ moves on the unit circle about O; it describes the *tangent*

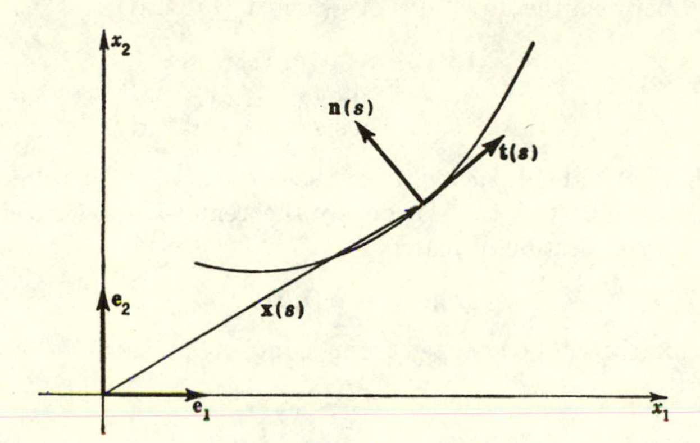

Fig. 2-1

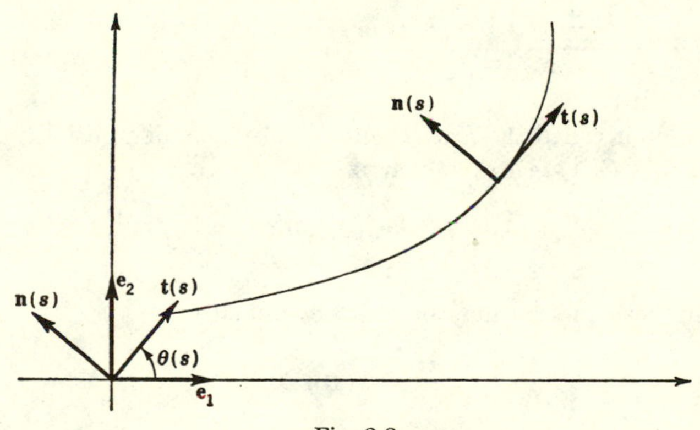

Fig. 2-2

image of the curve $\mathbf{x}(s)$. If the moving frame (or the tangent image) is given as a function of s in the vector plane, the curve in the euclidean plane may be found from

$$\mathbf{x}(s) = \mathbf{x}_0 + \int_0^s \mathbf{t}(\sigma)\, d\sigma \tag{2-1}$$

The operation $\mathbf{x} \to \mathbf{x} + \mathbf{x}_0$ is called a *translation* of the plane by $\mathbf{x}_0$.

Theorem 2-6. *The tangent image defines a curve up to a translation.*

If two curves $\mathbf{x}(s)$ and $\mathbf{y}(s)$ have identical tangent images,

$$\mathbf{x}(s) - \mathbf{y}(s) = \mathbf{x}_0 - \mathbf{y}_0$$

is constant, by Eq. (2-1). Hence $\mathbf{x}(s)$ is obtained from $\mathbf{y}(s)$ by a translation of the whole plane given in direction and length by the vector $\mathbf{x}_0 - \mathbf{y}_0$.

The moving frame is obtained from the fixed one by a rotation about the angle θ between the directions of $\mathbf{e}_1$ and $\mathbf{t}$ (Fig. 2-1),

$$\{\mathbf{t},\mathbf{n}\} = A(s)\{\mathbf{e}_1,\mathbf{e}_2\}$$

$$A(s) = \begin{pmatrix} \cos\theta & \sin\theta \\ -\sin\theta & \cos\theta \end{pmatrix} = \begin{pmatrix} x_1' & x_2' \\ -x_2' & x_1' \end{pmatrix} \tag{2-2}$$

The matrix $A(s)$ is the *frame matrix* of the curve $\mathbf{x}(s)$. The frame matrix defines the curve up to a translation, by theorem 2-6. If we change our fixed frame by a rotation of matrix B,

$$\{\mathbf{e}_1,\mathbf{e}_2\} = B\{\mathbf{i}_1,\mathbf{i}_2\}$$

the frame matrix will be changed from A into AB, since

$$\{\mathbf{t},\mathbf{n}\} = A(s)B\{\mathbf{i}_1,\mathbf{i}_2\}$$

The variation of the moving frame along the curve is determined by

$$\frac{d}{ds}\{\mathbf{t},\mathbf{n}\} = A'\{\mathbf{e}_1,\mathbf{e}_2\} = A'A^{-1}\{\mathbf{t},\mathbf{n}\}$$

that is,

$$\{\mathbf{t},\mathbf{n}\}' = C(A)\{\mathbf{t},\mathbf{n}\} \tag{2-3}$$

This equation is called the *Frenet equation* of plane differential geometry. By lemma 1-8, $C(A)$ is a skew matrix,

$$C(A) = \begin{pmatrix} 0 & k(s) \\ -k(s) & 0 \end{pmatrix} = k(s)J \tag{2-4}$$

and the Frenet equation may be written explicitly

$$\frac{d\mathbf{t}}{ds} = k(s)\mathbf{n} \qquad \frac{d\mathbf{n}}{ds} = -k(s)\mathbf{t} \tag{2-3a}$$

From Eq. (2-2) we have

$$k(s) = \frac{d\theta}{ds} = \mathbf{x}' \times \mathbf{x}'' \tag{2-5}$$

Definition 2-7. *$k(s)$ is the curvature of the curve $\mathbf{x}(s)$, and $1/k(s) = \rho(s)$ is the curvature radius of $\mathbf{x}(s)$.*

The curvature is defined by the unique parameter s; it is clear that it is really a function of the *curve* $\mathbf{x}(s)$ and does not depend on any special mapping function by which the curve might be represented. We go on to prove that two curves are congruent if and only if they have the same curvature. In elementary geometry, two figures are said to be congruent if it is possible to transform one into the other by a rotation and a translation.

Definition 2-8. *A map of vectors* $\mathbf{x} \to R\mathbf{x} + \mathbf{b}$ *is a euclidean motion if* R *is a constant rotation matrix and* $\mathbf{b}$ *a constant vector. Two curves are congruent if one is the image of the other in a euclidean motion.*

We may deal separately with translations and rotations. The curves $\mathbf{x}(s)$ and $\mathbf{x}(s) + \mathbf{b}$ evidently have an identical tangent vector $\mathbf{t}(s) = d\mathbf{x}/ds$; hence they have an identical arc length and frame matrix, as well as an identical Cartan matrix of the frame matrix.

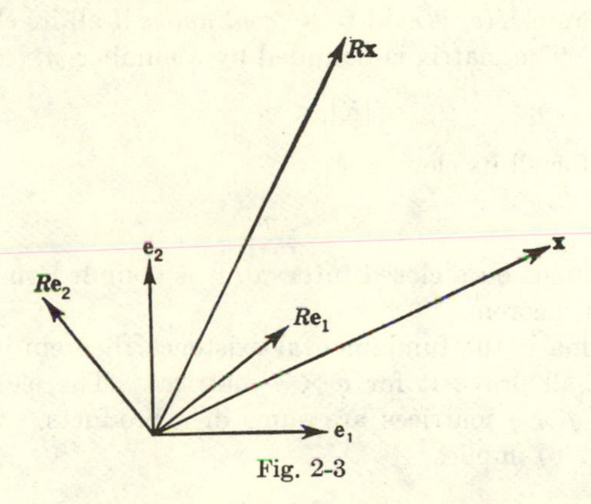

Fig. 2-3

A vector $\mathbf{x} = x_1\mathbf{e}_1 + x_2\mathbf{e}_2$ may be transformed into the rotated vector $R\mathbf{x}$ in two ways. We may leave the system of coordinates fixed and transform the coordinates as row vectors, $(x_1,x_2) \to (x_1,x_2)R$, or preferably we may leave rigid the whole configuration formed by $\mathbf{x}$ and the frame $\{\mathbf{e}_1,\mathbf{e}_2\}$ and rotate it as a whole. This means that we transform the frame into a new one $\{\mathbf{i}_1,\mathbf{i}_2\} = R\{\mathbf{e}_1,\mathbf{e}_2\}$. The coordinates of the rotated vector $R\mathbf{x}$ in the frame $R\{\mathbf{e}_1,\mathbf{e}_2\}$ are the coordinates of $\mathbf{x}$ in the frame $\{\mathbf{e}_1,\mathbf{e}_2\}$ (Fig. 2-3). It is quite important to distinguish between a rotation of the vector, $\mathbf{x} \to R\mathbf{x}$, and the corresponding transformation of its coordinates in a *fixed* system, $(x_1,x_2) \to (x_1,x_2)R$. If R is a rotation, $|dR\mathbf{x}/ds| = |R\,d\mathbf{x}/ds| = |\mathbf{t}(s)| = 1$; hence s is also the arc length for the rotated curve $R\mathbf{x}(s)$.

Theorem 2-9. *Congruent curves have equal arc lengths.*

In terms of the frame $\{\mathbf{e}_1,\mathbf{e}_2\}$, the coordinates of the tangent vector to the rotated curve are $(x_1',x_2')R$; the frame matrix of the rotated curve is $A(s)R$; its Cartan matrix, by Eq. (1-5), is $C(AR) = C(A)$; and the curvature is not changed in a rotation.

Theorem 2-10. *Two C^2 curves are congruent if and only if they have the same curvature as a function of the arc length.*

We have proved the "only if" part. We go on and prove that the curvature defines the frame matrix up to a rotation. The "if" part then follows immediately from theorem 2-6. It is much easier to work with the Cartan matrix (2-4) than with the curvature.

Lemma 2-11. *Two differentiable matrix functions $A(s), B(s)$ have equal Cartan matrices if and only if $A(s) = B(s)M$, M constant.*

If $A(s) = B(s)M$, $C(M) = 0$, and Eq. (1-5) shows that $C(A) = C(B)$ if one of the Cartan matrices exists. If $C(A) = C(B)$, let us write $A(s) = B(s)M(s)$. This is possible because the inverse of $B(s)$ exists. By Eq. (1-5), $BC(M)B^{-1} = 0$; hence $C(M) = 0$, $M = \text{const.}$

A matrix function $K(s)$ is said to be *continuous* if all its elements $k_{ij}(s)$ are continuous. The matrix is bounded by a number m,

$$|K| < m$$

if m is a bound for all its elements,

$$|k_{ij}| < m \tag{2-6}$$

If $K(s)$ is continuous on a closed interval, it is bounded on that interval by Weierstrass's theorem.

The next lemma is the fundamental existence theorem in differential geometry; we shall prove it for $q \times q$ matrices. The elements of the product of two $q \times q$ matrices are sums of q products. It follows by induction that (2-6) implies

$$|K^n| < (qm)^n \tag{2-7}$$

We are now ready to prove that any continuous matrix function is the Cartan matrix of some nonsingular differentiable matrix.

Lemma 2-12. *Let $K(s)$ be a continuous matrix function, $s_0 \leq s \leq s_1$. For any s^*, $s_0 \leq s^* \leq s_1$, there exists an interval about s^* and a nonsingular matrix function $A(s)$ defined in the interval such that $K(s) = C(A(s))$, $A(s^*) = U$.*

We have to solve the matrix differential equation

$$A'(s)A^{-1}(s) = K(s) \tag{2-8}$$

with the initial condition $A(s^*) = U$. We write this equation as $A' = KA$ and set up a process of successive approximations:

$$\begin{aligned} A_0 &= U \\ A_1(s) &= U + \int_{s^*}^{s} K(\sigma)A_0\, d\sigma \\ &\cdots\cdots\cdots\cdots \\ A_n(s) &= U + \int_{s^*}^{s} K(\sigma)A_{n-1}(\sigma)\, d\sigma \\ &\cdots\cdots\cdots\cdots\cdots \end{aligned}$$

The continuous $q \times q$ matrix $K(s)$ is bounded on any closed interval about s^*, $|K(s)| \leq m$. Then

$$|A_n(s) - A_{n-1}(s)| \leq q^{n-1}m^n \frac{|s - s^*|^n}{n!} \tag{2-9}$$

The proof is by induction. $|A_1(s) - A_0| \leq \int_{s^*}^{s} |K(\sigma)|\,|d\sigma| = m|s - s^*|$. Assume (2-9) to hold. For the next step,

$$\begin{aligned} |A_{n+1}(s) - A_n(s)| &= \left| \int_{s^*}^{s} K(\sigma)[A_n(\sigma) - A_{n-1}(\sigma)]\,d\sigma \right| \\ &\leq qmq^{n-1}m^n \int_{s^*}^{s} \frac{|\sigma - s^*|^n}{n!}\,|d\sigma| \\ &= q^n m^{n+1} \frac{|s - s^*|^{n+1}}{(n+1)!} \end{aligned}$$

Equation (2-9) is proved. The expression on the right-hand side of (2-9) is the $(n + 1)$-th term in the Taylor expansion of $e^{qm|s-s^*|}/q$; it tends to zero with increasing n. Therefore, $A(s) = \lim_{n\to\infty} A_n(s)$ exists and satisfies the integral equation

$$A(s) = U + \int_{s^*}^{s} K(\sigma)A(\sigma)\,d\sigma$$

All functions $A_n(s)$ are continuous functions, as integrals over continuous functions. The convergence to $A(s)$ is *uniform* since in Eq. (2-9) $|s - s^*|$ may be replaced by the constant $s_1 - s_0$. Hence $A(s)$ is continuous and, as an integral, differentiable: $A'(s) = K(s)A(s)$. By construction, $A(s^*) = U$. The determinant of $A(s^*)$ is 1. The determinant of a continuous function of s is a continuous function of s itself; hence there exists an interval about s^* in which this determinant is > 0. $A(s)$ has an inverse in that interval, and Eq. (2-8) holds.

It follows from lemma 2-11 that the *unique* solution of Eq. (2-8) for the initial condition $B(s^*) = M$ is given by $B(s) = A(s)M$.

In the case of plane geometry, $K(s) = k(s)J$, we may get an explicit solution of Eq. (2-8). For any matrix M define

$$e^M = U + M + \frac{1}{2!}M^2 + \frac{1}{3!}M^3 + \cdots + \frac{1}{n!}M^n + \cdots \tag{2-10}$$

If $|M| < m$, it follows from (2-7) that this series converges absolutely and that $|e^M| < e^{qm}$. If we put

$$L(s) = \int_{s^*}^{s} K(\sigma)\,d\sigma = \left(\int_{s^*}^{s} k(\sigma)\,d\sigma\right) J$$

the matrix $L(s)$ and its derivative $K(s)$ *commute:*

$$L(s)K(s) = K(s)L(s) = -k(s) \int_{s^*}^{s} k(\sigma)\, d\sigma\, U$$

In this case only it is true that $(L^n)' = nL^{n-1}K = nKL^{n-1}$; hence

$$(e^L)' = Ke^L \qquad e^{L(s^*)} = U$$

The unique solution of (2-8) with the given initial condition is

$$A(s) = e^{J \int_{s^*}^{s} k(\sigma)\, d\sigma} \tag{2-11}$$

The exponential matrix is rapidly computed from

$$J^2 = -U \qquad J^3 = -J \qquad J^4 = U \qquad \cdots$$

to be

$$A(s) = \begin{pmatrix} \cos \int_{s^*}^{s} k(\sigma)\, d\sigma & \sin \int_{s^*}^{s} k(\sigma)\, d\sigma \\ -\sin \int_{s^*}^{s} k(\sigma)\, d\sigma & \cos \int_{s^*}^{s} k(\sigma)\, d\sigma \end{pmatrix} \tag{2-12}$$

This is an orthogonal matrix. If we give an initial frame

$$\{\mathbf{t}(s^*),\mathbf{n}(s^*)\} = M\{\mathbf{e}_1,\mathbf{e}_2\}$$

the matrix M is also orthogonal; the unique curve with given curvature $k(s)$ and frame $\{\mathbf{t}(s^*),\mathbf{n}(s^*)\}$ at a point $\mathbf{x}(s^*)$ is obtained from the frame matrix $B(s) = A(s)M$. $A(s)$ always has the determinant $+1$; hence it exists in the whole domain of definition of $k(s)$, not only in some neighborhood of s^*. This proves not only theorem 2-10 but also the following theorem:

Theorem 2-13. *For any given continuous* $k(s)$ *and initial frame* $\{\mathbf{t}(s^*),\mathbf{n}(s^*)\}$ *at* $\mathbf{x}(s^*)$ *there exists a unique* C^2 *curve through* $\mathbf{x}(s^*)$, *having a tangent* $\mathbf{t}(s^*)$ *at* $\mathbf{x}(s^*)$ *and admitting* $k(s)$ *as its curvature at any point.*

We will not discuss the problem of whether a discontinuous function $k(s)$ might serve to define a curve. By Eq. (2-5), $k(s)$ is a derivative; derivatives behave almost like continuous functions. A frequent occurrence is C^1 curves that are piecewise C^2. The curvature exists and is continuous except at a finite number of points. There will be examples of this case later.

The equation $k = k(s)$ is the *natural equation* of the curves it defines, by theorem 2-13. All properties common to all curves congruent to a given one may be deduced from their common natural equation. Also, the natural equation does not depend on any special system of coordinates in the plane. In a sense, it brings back into analytical geometry the simplicity and the generality of synthetic euclidean geometry.

Exercise 2-2

1. If a C^2 curve $\mathbf{x}(u)$ is referred to a parameter other than the arc length, the frame matrix is

$$A(u) = [x_1^{\cdot}(u)^2 + x_2^{\cdot}(u)^2]^{-\frac{1}{2}} \begin{pmatrix} x_1^{\cdot}(u) & x_2^{\cdot}(u) \\ -x_2^{\cdot}(u) & x_1^{\cdot}(u) \end{pmatrix}$$

and the curvature is

$$k(u) = \frac{\mathbf{x}^{\cdot}(u) \times \mathbf{x}^{\cdot\cdot}(u)}{|\mathbf{x}^{\cdot}(u)|^3}$$

Especially, for $u = x_1$,

$$k(x_1) = \frac{d^2x_2/dx_1^2}{[(1 + (dx_2/dx_1)^2]^{\frac{3}{2}}}$$

Prove these formulas.

2. Prove that in polar coordinates

$$k(s) = \frac{r^2\phi^{\cdot 2} + r^{\cdot 2}(1 + \phi^{\cdot}) + rr^{\cdot}\phi^{\cdot\cdot} - rr^{\cdot\cdot}\phi^{\cdot}}{(r^{\cdot 2} + r^2\phi^{\cdot 2})^{\frac{3}{2}}}$$

[HINT: Use Eq. (2-5) and exercise 2-1, Prob. 2.] Write the expression of the curvature in the case that ϕ is the parameter.

◆ **3.** Compute the curvature of the following curves:

(*a*) A circle of radius R
(*b*) An ellipse of half axes a,b
(*c*) A cycloid $x_1 = u - \sin u$, $x_2 = 1 - \cos u$
(*d*) A spiral of Archimedes (exercise 1-3, Prob. 3)
(*e*) A logarithmic spiral (exercise 2-1, Prob. 1)
(*f*) A catenary $x_2 = a \operatorname{Cosh} x_1/a$
(*g*) A tractrix, the solution of $x_2 + (a^2 - x_2^2)^{\frac{1}{2}}(dx_2/dx_1) = 0$
(*h*) An astroid (exercise 2-1, Prob. 1)
(*i*) An involute of the circle, the solution of $dr/d\phi = ar/(r^2 - a^2)^{\frac{1}{2}}$

◆ **4.** Again let α denote the angle between the radius vector and the tangent of a curve. Find the curve whose curvature is

$$k(r) = \sin^3 \alpha/r$$

Compute its natural equation.

5. Find the natural equation of the catenary (Prob. 3*f*).

6. Show that the curvature of a spiral of Maclaurin of order n (exercise 1-3, Prob. 9) is

$$k(\phi) = \frac{n+1}{a} \sin^{(n-1)/n} n\phi$$

7. Show that for any curve which represents an integral of

$$\frac{dx_2}{dx_1} = \frac{[4a^2x_2^2 - (b^2 + x_2^2)^2]^{1/2}}{b^2 + x_2^2}$$

the difference between the curvature $k(x_1)$ and the reciprocal of the normal $|PN|$ is independent of b (Frenet).

8. Let $F(x_1,x_2)$ be a twice-differentiable function of both variables, and assume $F_{x_1}{}^2 + F_{x_2}{}^2 \neq 0$ (the indices denote partial derivatives). Show that the curvature of the level curves $F(x_1,x_2) = \text{const}$ is given by

$$k(x_1,x_2) = \frac{F_{x_1}{}^2F_{x_2x_2} - 2F_{x_1}F_{x_2}F_{x_1x_2} + F_{x_2}{}^2F_{x_1x_1}}{F_{x_1}{}^2 + F_{x_2}{}^2}$$

If $\mathbf{n} = (n_1,n_2)$ is the normal to the level curve, an alternative expression for the curvature is $k = \partial n_1/\partial x_1 + \partial n_2/\partial x_2$.†

9. Show that the Frenet equations may be transformed into

$$\frac{d\mathbf{t}}{d\theta} = \mathbf{n} \qquad \frac{d\mathbf{n}}{d\theta} = -\mathbf{t}$$

hence $d^2\mathbf{t}/d\theta^2 = -\mathbf{t}$. From this last equation obtain the curves of given curvature radius $\rho = \rho(\theta)$ as

$$x_1 = a_1 + \cos\gamma_1 \int_{\theta_0}^{\theta} \rho(\theta) \cos(\theta - \theta_0)\, d\theta$$

$$x_2 = a_2 + \sin\gamma_2 \int_{\theta_0}^{\theta} \rho(\theta) \sin(\theta - \theta_0)\, d\theta$$

a_1, a_2, γ_1, γ_2 are integration constants.

10. Obtain the formulas of Prob. 9 from Eq. (2-12).

11. A symmetry with respect to the x_1 axis is given by

$$S = \begin{pmatrix} 1 & 0 \\ 0 & -1 \end{pmatrix}$$

Show that $C(SA) = -C(A)$.

12. Any orthogonal matrix of determinant -1 may be written as the product of S (Prob. 11) and a rotation. Show that the sign of the curvature is changed under a euclidean motion followed by a symmetry.

13. If a C^3 curve is represented in a cartesian system of coordinates given by $e_1 = \mathbf{t}(s_0)$, $e_2 = \mathbf{n}(s_0)$, $\mathbf{x}(s_0) = 0$, show that

$$k(s_0) = 2 \lim_{x_1 \to 0} \frac{x_2}{x_1{}^2}$$

† R. Jerrard, *Bull. Am. Math. Soc.*, **67**:113 (1961).

[HINT: The Taylor expansion of x_2 as a function of x_1 in a neighborhood of $x_1 = 0$ is $x_2 = k(s_0)x_1^2/2 + R_3$.]

14. The "cocked-hat" curve

$$x_2 = \frac{a^2 - x_1^2}{2a \pm (a^2 - x_1^2)^{\frac{1}{2}}}$$

has two maxima at $x_1 = 0$. Use the result of Prob. 13 to compute the curvature at these two points, $(0,a/3)$ and $(0,a)$.

15. Solve the differential equation (2-8) for

$$K(s) = \begin{pmatrix} 0 & 1 \\ k(s) & 0 \end{pmatrix} \qquad A(0) = U$$

Compute the four first approximations.

◆**16.** Find a closed form for the solution of Eq. (2-8) for

$$K(s) = \begin{pmatrix} 0 & p & 0 \\ -p & 0 & q \\ 0 & -q & 0 \end{pmatrix} \qquad p,q \text{ const} \qquad A(0) = U$$

17. Show that Eq. (2-8) may also be solved for continuous $K(s)$ by the following process of successive approximations:

$$X_0 = U \qquad X_n = \int_{s^*}^{s} K(\sigma)X_{n-1}(\sigma)\,d\sigma \qquad A(s) = \sum_{0}^{\infty} X_n(s)$$

The convergence of the series is uniform and absolute in any closed interval of definition. What is the relation of the X_n and the A_n of the text?

***18.** The idea of the moving frame suggests a theory in which only increases of s are considered and a weakening of the differentiability assumptions. Let us define $A_l'(s) = \lim_{h\downarrow 0} (1/h)(A(s - h) - A(s))$, if the limit exists. Then $C_l(A) = A_l'A^{-1} = k_l(s)J$. This defines a left-hand curvature. Consider a curve for which k_l is defined, except possibly on a set of Riemann measure zero, and is bounded. Then prove the following statements: $C_l(AB) = C_l(A) + AC_l(B)A^{-1}$. The left-hand curvature is invariant under euclidean motions. Although no curve having a discontinuous $k_l(s)$ as its left-hand curvature may exist, for integrable k_l we obtain a C^1 curve from Eq. (2-12) which has $k_l(s)$ as its two-sided curvature almost everywhere.

***19.** Any derivative has the Bolzano property that it cannot change sign without becoming zero. Use this fact to prove that, if the curvature exists everywhere and is of bounded variation, it is continuous (for twice-differentiable curves, not necessarily C^2).

20. Define an appropriate notion of a "curve" such that a natural equation $k = k(s)$ defines exactly one "euclidean curve."

♦21. May a C^3 curve be given by a natural equation $k'(s) = F(k)$ and appropriate initial conditions?

22. If A and B commute, $AB = BA$, show that $e^A e^B = e^{A+B}$. Is this formula valid for 2×2 orthogonal matrices?

23. Deduce the addition formula for sin and cos from the matrix multiplication of two 2×2 orthogonal matrices.

2-3. The Circle of Curvature

The results of the preceding section have shown the importance of the notion of curvature in euclidean geometry. We go on to study properties of curves directly from the properties of the curvature, without integration of the natural equation. In this section, we shall not use continuity of the curvature; the results will apply to situations more general than those described by theorem 2-13.

Our first problem is that of the contact of curves. Assume that two curves $\mathbf{f}(s) = (x_{1f}(s), x_{2f}(s))$, $\mathbf{g}(s) = (x_{1g}(s), x_{2g}(s))$ have a point in common, the tangents coinciding at that point. As we are interested only in the behavior of the curves near the point of contact, we shall choose it as the point $s = 0$ on both $\mathbf{f}$ and $\mathbf{g}$. The fixed system of coordinates will be that defined by the moving frame common to both curves, $\mathbf{e}_1 = \mathbf{f}'(0) = \mathbf{g}'(0)$. θ_f and θ_g are the angles made by the x_1 axis with the tangents to $\mathbf{f}$ and $\mathbf{g}$, respectively, and the corresponding curvatures are $k_f(s)$ and $k_g(s)$. If the curvatures exist, the angles θ are continuous functions of s [see Eq. (2-5)]. As $\theta_f(0) = \theta_g(0) = 0$, there exists an interval about 0 on the x_1 axis on which for both curves the x_2 coordinate is a *function* of x_1. This interval is given by $|\theta_f|$, $|\theta_g| < \pi/2$. On it, $\mathbf{f}$ may be given as the graph of a function $x_2 = F(x_1)$, and $\mathbf{g}$ as the graph of $x_2 = G(x_1)$. If we say that $\mathbf{f}$ is *above* or *below* $\mathbf{g}$ in some x_1, it refers to the fixed system of coordinates and means that $F(x_1) > G(x_1)$ or $F(x_1) < G(x_1)$, $x_1 \neq 0$. The problem now is to obtain information from the curvature as to whether the two curves intersect at 0 or whether one touches the other from within.

Theorem 2-14. *Let* $\mathbf{f}$ *and* $\mathbf{g}$ *be two curves in contact at* 0. *If* $k_f(0) > k_g(0)$, $\mathbf{f}$ *is above* $\mathbf{g}$, *and the curves do not intersect. The same holds true if* $k_f(s) - k_g(s) > 0$ *in some interval about zero,* $s \neq 0$, *and if either* $k_f(0) = k_g(0) \neq 0$ *or if* $k_f(0) = k_g(0) = 0$ *is an isolated zero for both curvatures.*

Corollary 2-15. $\mathbf{f}$ *and* $\mathbf{g}$ *intersect at* 0 *if and only if* $k_f(s) - k_g(s)$ *changes sign at* 0.

The only case which is not covered by our theorem is that of curves whose curvature changes sign an infinity of times in the neighborhood of 0. The theorem then does not hold, as may be seen from a study of the relations of the curves $\mathbf{f} = (t, t^4 \sin 1/t)$, $\mathbf{g} = (t, t^4 \sin^2 1/t)$, $\mathbf{h} = (t,0)$.

If $k_f(0) > k_g(0)$,

$$k_f(0) - k_g(0) = \lim_{s \to 0} \frac{\theta_f(s) - \theta_g(s)}{s} > 0$$

Hence
$$\operatorname{sign}\, [\theta_f(s) - \theta_g(s)] = \operatorname{sign}\, s \tag{2-13}$$

for $|s|$ sufficiently small. Because of

$$\theta(s) = \int_0^s k(\sigma)\, d\sigma \tag{2-14}$$

Eq. (2-13) holds also if $k_f(0) = k_g(0)$, $k_f(s) > k_g(s)$, $s \neq 0$. The curves are then given by

$$\begin{aligned} x_1 &= \int_0^s \cos \theta(\sigma)\, d\sigma \\ x_2 &= \int_0^s \sin \theta(\sigma)\, d\sigma \end{aligned} \tag{2-15}$$

1. If either $k_f(0) > 0 > k_g(0)$ or $k_f(s) > 0 > k_g(s)$, $s \neq 0$,

$$\operatorname{sign}\, \theta_f(s) = \operatorname{sign}\, s \qquad \operatorname{sign}\, \theta_g(s) = -\operatorname{sign}\, s$$

Equations (2-15) show that **f** is above and **g** is below the x_1 axis; hence **f** is above **g**.

2. If $k_f(0) > k_g(0) = 0$, $\theta_f(s) = k_f(0)s + s\epsilon(s)$, $\theta_g(s) = s\delta(s)$,

$$\lim_{s \to 0} \epsilon(s) = \lim_{s \to 0} \delta(s) = 0$$

Hence there exists an interval about zero in which

$$|\theta_g(s)| < |\theta_f(s)| \qquad s \neq 0$$

If either $k_f(0) > k_g(0) > 0$ or if $k_f(0) = k_g(0)$, $k_f(s) > k_g(s) \geq 0$, Eq. (2-13) informs us that

$$|\theta_f(s)| > |\theta_g(s)| \geq 0$$

In both cases

$$\begin{aligned} |x_{1f}(s_1)| &= \left| \int_0^{s_1} \cos \theta_f(s)\, ds \right| < |x_{1g}(s_1)| = \left| \int_0^{s_1} \cos \theta_g(s)\, ds \right| \\ |x_{2f}(s_1)| &= \left| \int_0^{s_1} \sin \theta_f(s)\, ds \right| > |x_{2g}(s_1)| = \left| \int_0^{s_1} \sin \theta_g(s)\, ds \right| \end{aligned}$$

$\theta_g(s)$ being continuous, there exists an s_2 between 0 and s_1 such that $x_{1g}(s_2) = x_{1f}(s_1)$. $\theta_f(s)$ is not zero and of the sign of s; hence $|x_{2g}(s_2)| < |x_{2f}(s_2)| < |x_{2f}(s_1)|$. This means that the point $(x_1 = x_{1f}(s_1), x_2 = x_{2f}(s_1))$ is above the point $(x_1 = x_{1g}(s_2), x_2 = x_{2g}(s_2))$ and that **f** is above **g**.

3. The remaining case $0 \geq k_f(0) > k_g(0)$ or $0 \geq k_f(s) > k_g(s)$ may be treated exactly like case (2) to complete the proof of theorem 2-14.

As an application of the theorem, we study the contact of a curve $\mathbf{f}(s)$ with the circles tangent to it at $\mathbf{f}(s_0)$. The system of coordinates will again be given by $\mathbf{e}_1 = \mathbf{f}'(s_0)$. The circles tangent at $\mathbf{f}(s_0)$ have a parametric representation $x_1 = 2R \sin s/2R$, $x_2 = 2R(1 - \cos s/2R)$. The curvature of such a circle is $1/2R$. The sign of the radius indicates the half plane in which the circle is situated.

Let R_1 be the radius of any circle S_1 that is above $\mathbf{f}(s)$ in some interval about s_0. Also, we define the set of radii (with sign) R_2 that characterize the circles below $\mathbf{f}$ in some interval about s_0. Theorem 2-14 implies that

$$k_f(s_0) = \inf \{1/R_1\} = \sup \{1/R_2\} \tag{2-16}$$

for any curve with continuous tangents and a curvature at s_0. This statement admits a converse. If $\theta_f(s)$ is continuous and if

$$\inf \{1/R_1\} = \sup \{1/R_2\} = 1/R$$

then $\theta_f(s_0) = 0$ by definition and for any $\epsilon > 0$ the curve is in some interval between the circle of radius $R + \epsilon$ and that of radius $R - \epsilon$, that is,

$$\frac{s - s_0}{R + \epsilon} \leq \theta(s) \leq \frac{s - s_0}{R - \epsilon}$$

It follows that $1/R = d\theta/ds$ and that the curvature exists in s_0 and satisfies (2-16).

Definition 2-16. *The circle of curvature $S_f(s_0)$ of $\mathbf{f}$ at $\mathbf{f}(s_0)$ is the circle with center $\mathbf{f}(s_0) + [1/k_f(s_0)]\mathbf{n}(s_0)$ and radius $1/k_f(s_0)$. If $k_f(s_0) = 0$, the tangent line at $\mathbf{f}(s_0)$ is sometimes called the circle of curvature.*

A comparison of Defs. 2-7 and 2-16 shows that the radius of curvature is the radius of the circle of curvature. The circle of curvature is in some sense a best approximation of the arc $\mathbf{f}$ by circular arcs in a neighborhood of $\mathbf{f}(s_0)$. If $k(s)$ is strictly monotone in some neighborhood of s_0, the circle of curvature intersects the curve at $\mathbf{f}(s_0)$. In the next chapter, we study evolutes in order to obtain fuller information on the nature of this intersection. Here we note only an immediate corollary of theorem 2-14:

Theorem 2-17. *If the curvature of a C^2 curve $\mathbf{f}$ has a minimum or a maximum at $s = s_0$, $\mathbf{f}$ and $S_f(s_0)$ do not intersect at $\mathbf{f}(s_0)$.*

A *vertex* of a C^2 curve is a point at which the curvature is extremal. If the curve is closed, its graph is compact. A continuous function has a maximum and a minimum on a compact set. This shows that a C^2 curve has at least two vertices. One may prove much more:

Theorem 2-18 (Mukhopadhaya). *A closed Jordan C^2 curve has at least four vertices.*

The proof of this *four-vertices theorem* will be based on a theorem due to Axel Schur and Erhardt Schmidt:

Theorem 2-19. *Given two convex curves* **f**,**g** *with continuous tangents and piecewise continuous curvature, both of length L. If* $|k_f(s)| \geq |k_g(s)|$, $k_f(s) \not\equiv k_g(s)$, *the chord subtended by* **g** *is bigger than that subtended by* **f**.

We place both arcs in the lower half plane $x_2 \leq 0$ with endpoints on the x_1 axis, $x_{2f}(0) = x_{2f}(L) = x_{2g}(0) = x_{2g}(L) = 0$, and such that $x_{1f}(L) > x_{1f}(0)$, $x_{1g}(L) > x_{1g}(0)$. In this case, both curvatures are non-negative. The chords subtended are $d_f = x_{1f}(L) - x_{1f}(0) = \int_0^L \cos\theta_f(s)\,ds$, $d_g = x_{1g}(L) - x_{1g}(0) = \int_0^L \cos\theta_g(s)\,ds$.

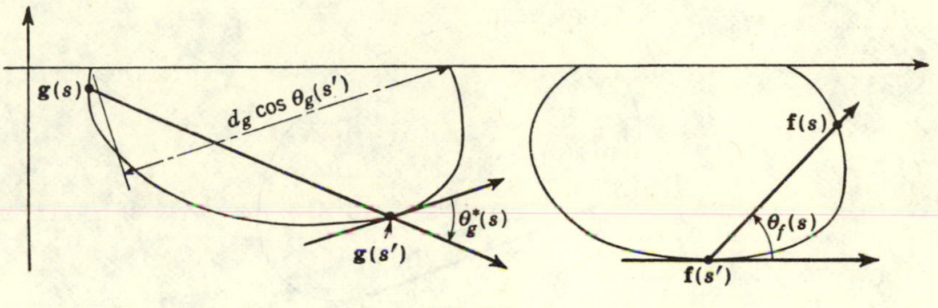

Fig. 2-4

Let s' be the value of the arc length for which the tangent to $\mathbf{f}(s)$ is parallel to the x_1 axis (Fig. 2-4). Then

$$\theta_f(s) = \int_{s'}^{s} k_f(\sigma)\,d\sigma$$

and, because of the convexity of the arc,

$$-\pi \leq \theta_f(s) \leq +\pi$$

The angle

$$\theta_g^*(s) = \theta_g(s) - \theta_g(s') = \int_{s'}^{s} k_g(\sigma)\,d\sigma \leq \theta_f(s)$$

Hence

$$d_f = \int_0^L \cos\theta_f(s)\,ds \leq \int_0^L \cos\theta_g^*(s)\,ds = d_g \cos\theta_g(s') \leq d_g$$

and $d_f = d_g$ only if $k_f(s) = k_g(s)$ for all s.

First we prove the four-vertices theorem for closed convex Jordan C^2 curves. Assume that the curvature has only one maximum at M and one minimum at m on the curve. M and m divide the curve into two arcs, for each of which the curvature is monotone and continuous. Because of the continuity of the curvature there exist two points P_1, P_2 which divide the curve into two arcs of equal length and for which the curvature takes the same values. The arcs P_1MP_2, P_1mP_2 have equal length, they subtend the same chord, and the curvature of P_1MP_2 is bigger than the corresponding curvature of P_1mP_2. This result contradicts theorem 2-19; therefore there must be more than two vertices.

If the closed Jordan C^2 curve has two inflection points, it has a double tangent. We consider the convex curve obtained by part of the arc and a segment T_1T_2 of the double tangent (Fig. 2-5). The curvature has a maximum at M. We may assume that the arc T_1M has a length not greater than that of the arc MT_2. Let P be the point which, together with T_1, divides the closed convex curve into two equal arcs. Then the curvature of T_1MP is greater than that of T_1T_2P. The previous argument shows that the convex curve has at least two maxima; hence the original curve has at least two maxima and two minima of the curvature.

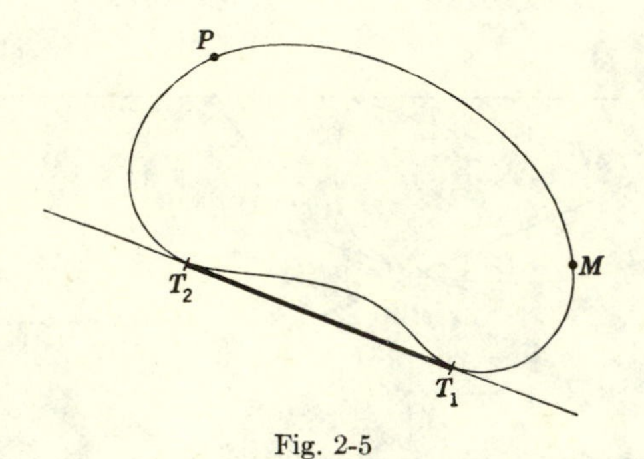

Fig. 2-5

A closed curve with more than two inflection points trivially has at least four vertices.

This proof is based on ideas due to D. Fog and E. Rodemich.

Exercise 2-3

1. Determine inf $1/R_1$ and sup $1/R_2$ for
   (*a*) $x_2 = x_1^2 \sin 1/x_1$
   (*b*) $x_2 = x_1^2 \sin^2 1/x_1$

2. Is it possible to have inf $1/R_1 = \sup 1/R_2$ for $\theta(s)$ discontinuous at s_0?

3. For C^3 curves, deduce theorem 2-14 from exercise 2-2, Prob. 13.

4. Verify that for an ellipse the vertices of differential geometry coincide with the vertices of analytical geometry and that the circles of curvature at the vertices are either completely inside or completely outside the ellipse. Which circles are inside the ellipse?

5. Verify that, for an ellipse $x_1 = a \cos u$, $x_2 = b \sin u$, the curvature is monotone in any one of the intervals $[0,\pi/2]$, $[\pi/2,\pi]$, $[\pi,3\pi/2]$, $[3\pi/2,2\pi]$. Use this result to prove that, if a circle is in contact with an ellipse and if it has exactly one point of intersection with the ellipse different from the point of contact, then it is the circle of curvature of the ellipse at the point of contact (J. P. Sydler).

6. Show that the radius of curvature of an ellipse is

$$\rho(u) = \frac{(a^2 \sin^2 u + b^2 \cos^2 u)^{3/2}}{ab}$$

the length of the normal $PN = b(a^2 \sin^2 u + b^2 \cos^2 u)^{1/2}/a$, and the cosine of the angle ψ between the normal and the focal radius FP is $\cos \psi = b(a^2 \sin^2 u + b^2 \cos^2 u)^{-1/2}$ (F a focus of the ellipse). Show that

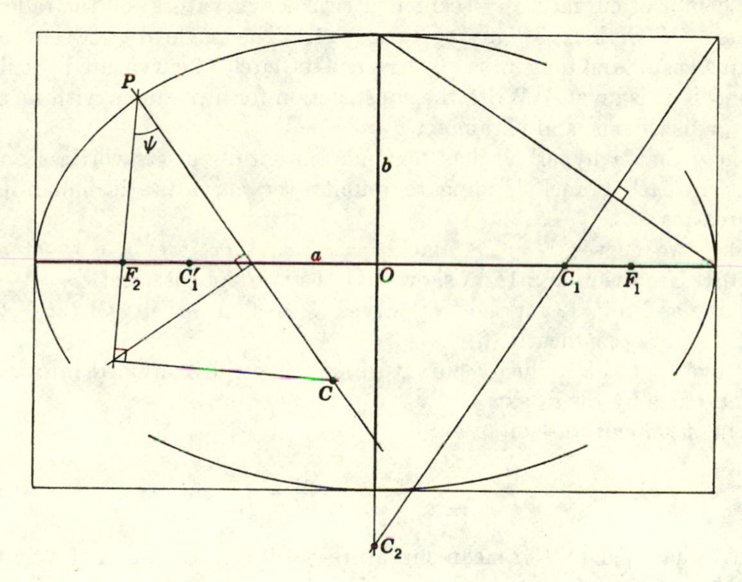

Fig. 2-6

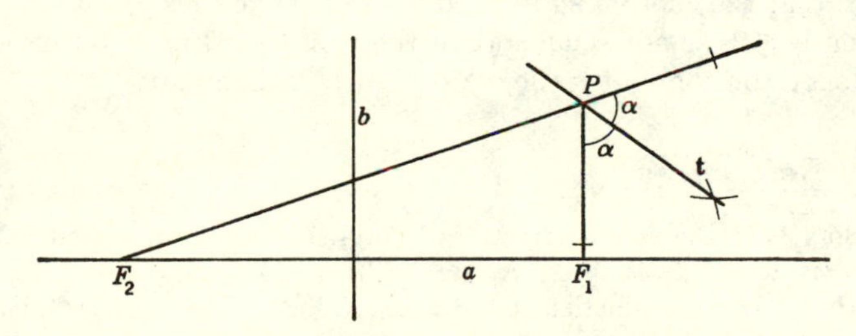

Fig. 2-7

the construction given in Fig. 2-6 for the center of curvature C of an ellipse at a point P is correct.

7. In the construction given in Fig. 2-6 we need the direction of the normal (or the tangent) at P. The construction of the tangent at a point P of an ellipse is given in Fig. 2-7. Justify the construction.

8. Show that the construction of Fig. 2-6 holds good also for the hyperbola and the parabola.

9. The radius of curvature of a logarithmic spiral is equal to the polar normal (exercise 2-2, Prob. 3*e*.) The angle is constant for a logarithmic spiral. Use these two facts to give a construction of the circle of curvature by ruler and compass.

10. For a spiral of Maclaurin (exercise 1-3, Prob. 9) the projection of the center of curvature (= center of circle of curvature) on the radius vector OP divides OP in the ratio $n:1$. The angle α between the radius vector and tangent is $n\phi$. Hence the circle of curvature is easily found if P is given. Write the construction for a parabola with focus 0, x_1 axis as axis, and parameter a $(n = -\frac{1}{2})$.

11. Show that a hyperbola has maximal curvature at its vertices and that on each branch tending to infinity the curvature is monotone decreasing.

12. The two curves $x_2 = x_1{}^2$ and $x_2 = x_1{}^5 \sin 1/x_1$ are in contact at (0,0). Use theorem 2-14 to show that they do not intersect.

13. Same as Prob. 12 for the two curves $(3t - t^2, 1 + t + 3t^2)$ $(3t - t^3, 1 + t + 3t^3)$, contact at (0,1).

14. Prove that the radius of curvature of a cycloid is divided into two equal parts by the x_1 axis.

15. The *mean curvature* of an arc is

$$K(s_0,s_1) = \frac{\theta(s_1) - \theta(s_0)}{s_1 - s_0} = \frac{1}{s_1 - s_0} \int_{s_0}^{s_1} k(\sigma)\, d\sigma$$

Prove that a curve is of mean curvature ≤ 1 if it is the union of a finite number of arcs of mean curvature ≤ 1 (Carathéodory).

16. Prove that the chord d subtended by a curve of mean curvature ≤ 1 and length $\leq 2\pi$ is not smaller than the chord subtended by an arc of equal length on the unit circle, $d \geq 2 \sin L/2$ (Schmidt).

References

Carathéodory, C.: Die Kurven mit beschränkten Biegungen, *Sitzber. Preuss. Akad. Wiss. Berlin, Phys. Math. Klasse,* 1933, pp. 102–125.

Coxeter, H. S. M.: "Introduction to Geometry," John Wiley & Sons, Inc., New York, 1961.

Fog, D.: Über den Vierscheitelsatz und seine Verallgemeinerungen, *Sitzber. Preuss. Akad. Wiss. Berlin, Phys. Math. Klasse,* 1933, pp. 251–254.

Graves, L. M.: "The Theory of Functions of Real Variables," 2d ed., McGraw-Hill Book Company, Inc., New York, 1946.

3

EVOLUTES AND INVOLUTES

3-1. The Riemann-Stieltjès Integral

The curvature and initial conditions define a curve completely. The differential geometry of curves therefore should use only second derivatives of the mapping functions; no higher derivatives are used in the computation of the curvature. It is possible to carry through this program by the use of the *Stieltjès integral*, a most important tool of modern analysis.

One of the most important techniques used in integration is the change of variables: If $x = g(u)$, $a = g(\alpha)$, $b = g(\beta)$, then under certain conditions on the functions involved

$$\int_a^b f(x)\,dx = \int_\alpha^\beta f(g(u))g'(u)\,du$$

If we write $f(g(u)) = h(u)$, $g'(u)\,du = dg(u)$, the formula becomes

$$\int_a^b f(x)\,dx = \int_\alpha^\beta h(u)\,dg(u) \tag{3-1}$$

The Stieltjès integral gives sense to Eq. (3-1) also if $g(u)$ is a nondifferentiable function!

Let $f(x)$ be a continuous function and $g(x)$ a function of bounded variation, in an interval $a \leq x \leq b$. Let Σ: $a = x_0 < x_1 < \cdots < x_N = b$ be a subdivision of the interval and Ξ: $\xi_0 \leq \xi_1 \leq \cdots \leq \xi_{N-1}$, $x_i \leq \xi_i \leq x_{i+1}$ a sequence of intermediate values taken from the intervals of Σ.

$$R_{\Sigma,\Xi}(f,g) = \sum_{i=0}^{N-1} f(\xi_i)(g(x_{i+1}) - g(x_i)) \tag{3-2}$$

is the *Riemann-Stieltjès* sum defined by Σ and Ξ. For $g(x) = x$ the Riemann-Stieltjès sum reduces to the Riemann sum used in the definition of the Riemann integral.

The *mesh size* of a subdivision Σ is $\mu(\Sigma) = \max (x_{i+1} - x_i)$.

Lemma 3-1. *Let $f(x)$ be continuous and $g(x)$ of bounded variation in $[a,b]$. For any sequence of subdivisions $\Sigma^{(k)}$ and intermediate values $\Xi^{(k)}$ with mesh size tending to zero, $\lim_{k\to\infty} \mu(\Sigma^{(k)}) = 0$, the limit $\lim_{k\to\infty} R_{\Sigma^{(k)},\Xi^{(k)}}(f,g)$ exists and is independent of the sequences $\Sigma^{(k)}$ and $\Xi^{(k)}$.*

By hypothesis, $f(x)$ is continuous on the closed interval $[a,b]$; hence it is bounded, $|f(x)| < M$. The Riemann-Stieltjès sums also are bounded:

$$R_{\Sigma,\Xi}(f,g) \leq M \sum_{i=0}^{N-1} |g(x_{i+1}) - g(x_i)| \leq M v_{[a,b]}(g) \tag{3-3}$$

For any sequence of subdivisions the set $\{R_{\Sigma^{(k)},\Xi^{(k)}}(f,g)\}$ is bounded; hence it has at least one cluster point. To prove that this cluster point is the unique limit point for all sequences of subdivisions with mesh size tending to zero, it suffices to show that there exists a $\delta(\epsilon) > 0$ such that

$$|R_{\Sigma,\Xi}(f,g) - R_{\Sigma^1,\Xi^1}(f,g)| < \epsilon \qquad \text{if } \mu(\Sigma) < \delta,\ \mu(\Sigma^1) < \delta \tag{3-4}$$

A subdivision Σ^1 is *finer* than Σ if all division points of Σ are also division points of Σ^1. One may write

$$\Sigma^1\colon a = x_0 < x_1 < \cdots < x_N = b \qquad \Sigma\colon a = x_0 < x_{i_1} < \cdots < x_{i_L} = b$$

The function $f(x)$ is uniformly continuous on $[a,b]$; hence there exists a $\Delta(\epsilon) > 0$ such that $|f(x) - f(x')| < \epsilon/v_{[a,b]}(g)$ if $|x' - x| < \Delta$. Choose Σ such that $\mu(\Sigma) < \Delta$, and let Σ^1 be finer than Σ. Ξ and Ξ^1 are any two sequences of intermediate values to Σ,Σ^1. Then

$$\begin{aligned} |R_{\Sigma^1,\Xi^1}(f,g) &- R_{\Sigma,\Xi}(f,g)| \\ &\leq \sum_{s=0}^{L-1} \sum_{j=0}^{i_{s+1}-i_s-1} |f(\xi'_{i_s+j}) - f(\xi_s)|\,|g(x_{i_s+j+1}) - g(x_{i_s+j})| \\ &< \frac{\epsilon}{v_{[a,b]}(g)} \sum_{i=0}^{N-1} |g(x_{i+1}) - g(x_i)| \leq \epsilon \end{aligned}$$

Inequality (3-4) holds for $\delta = \Delta(\epsilon/2)$. Given Σ and Σ^1, $\mu(\Sigma) < \Delta(\epsilon/2)$, $\mu(\Sigma^1) < \Delta(\epsilon/2)$, the subdivision Σ^* that contains all division points appearing in at least one of the subdivisions Σ and Σ^1 is finer than both Σ and Σ^1. Hence

$$\begin{aligned} |R_{\Sigma,\Xi}(f,g) - R_{\Sigma^1,\Xi^1}(f,g)| \leq{} & |R_{\Sigma,\Xi}(f,g) - R_{\Sigma^*,\Xi^*}(f,g)| \\ & + |R_{\Sigma^*,\Xi^*}(f,g) - R_{\Sigma^1,\Xi^1}(f,g)| < \epsilon \end{aligned}$$

Definition 3-2. *The Riemann-Stieltjès integral* $\int_a^b f(x)\, dg(x)$ *is the limit* $\lim_{k\to\infty} R_{\Sigma^{(k)},\Xi^{(k)}}(f,g)$, $(\lim_{k\to\infty} \mu(\Sigma^{(k)}) = 0)$ *if that limit exists independent of the sequences* $\Sigma^{(k)},\Xi^{(k)}$.

Theorem 3-3. *The Riemann-Stieltjès integral exists for $f(x)$ continuous and $g(x)$ of bounded variation.*

We have defined the integral only for $b > a$. In a more detailed treatment, we now should define the integral also for $b < a$ and prove the different linearity properties,

$$\int_a^b f(x)\, dg(x) + \int_b^c f(x)\, dg(x) = \int_a^c f(x)\, dg(x)$$

$$\int_a^b [f(x) + g(x)]\, dh(x) = \int_a^b f(x)\, dh(x) + \int_a^b g(x)\, dh(x)$$

The proofs are parallel to those of the corresponding formulas for Riemann integrals.† We shall not go into the details of integration theory, but we shall discuss three aspects of Riemann-Stieltjès integrals that will be needed later.

The expression (3-3) yields an estimate of Stieltjès integrals:

$$\left| \int_a^b f(x)\, dg(x) \right| \leq \max |f(x)| \cdot v_{[a,b]}(g) \tag{3-5}$$

Our next theorem gives the conditions for the validity of Eq. (3-1).

Theorem 3-4. *If $f(x)$ is continuous in $[a,b]$ and if $x = g(u)$ is a continuous piecewise monotone function of u, $a = g(\alpha)$, $b = g(\beta)$, then* $\int_a^b f(x)\, dx = \int_\alpha^\beta f(g(u))\, dg(u)$.

The conditions of the theorem are necessary. If $g(u)$ is not piecewise monotone, Eq. (3-1) is false even for Riemann integrals and differentiable $g(u)$. If $g(u)$ is monotone discontinuous, the integral $\int_\alpha^\beta f(g(u))\, dg(u)$ exists, but its value is not that of the corresponding Riemann integral $\int_a^b f(x)\, dx$. Simple examples may be constructed with step functions. Define

$$\delta(x) = \begin{cases} 0 & x < 0 \\ 1 & x \geq 0 \end{cases}$$

Then $\int_{-|a|}^{|a|} f(x)\, d\delta(x) = f(0)$. For the monotone substitution

$$x = g(u) = u/2 - \delta(u - \tfrac{1}{2})$$

one has

$$\int_0^2 f(g(u))\, dg(u) = 2 \int_0^1 f(x)\, dx - f(\tfrac{1}{2})$$

† See, for example, Tom M. Apostol, "Calculus," vol. I, Sec. 1.48, Blaisdell Publishing Company (Div. of Random House), New York, 1961.

The conditions of the theorem are sufficient. It needs proof only for monotone $g(u)$. The integral $\int_a^b f(x)\,dx$ may be approximated by a Riemann sum $\Sigma f(\xi_i)(x_{i+1} - x_i)$. The function $g(u)$ is monotone, hence one-to-one; let $h(x)$ be its inverse function. $h(x)$ is continuous. Put $u_i = h(x_i)$, $v_i = h(\xi_i)$; then

$$\Sigma f(\xi_i)(x_{i+1} - x_i) = \Sigma f(g(v_i))(g(u_{i+1}) - g(u_i))$$

The points u_i form a subdivision Σ of $[\alpha,\beta]$. The mesh size tends to zero with max $|x_{i+1} - x_i|$. The v_i are a sequence of intermediate values to Σ. The theorem follows from lemma 3-1.

The last of our theorems deals with integration by parts.

Theorem 3-5. *If $f(x)$ and $g(x)$ are defined on $[a,b]$, one function being continuous and the other of bounded variation,*

$$\int_a^b f(x)\,dg(x) + \int_a^b g(x)\,df(x) = f(b)g(b) - f(a)g(a)$$

Corollary 3-6. *The Riemann-Stieltjès integral $\int_a^b f(x)\,dg(x)$ exists also if $f(x)$ is of bounded variation and $g(x)$ is continuous.*

We may assume that $f(x)$ is continuous and that $g(x)$ is of bounded variation. The integral $\int_a^b g(x)\,df(x)$ is the limit of Riemann-Stieltjès sums $R_{\Sigma,\Xi}\ (g,f)$, if it exists. To given Σ,Ξ we take a sequence Ξ^* defined by $\xi_0^* = a$, $\xi_{N-1}^* = b$, $\xi_i^* = \xi_i$ for $1 \leq i \leq N - 2$. The difference

$$\begin{aligned}|R_{\Sigma,\Xi}(g,f) - R_{\Sigma,\Xi^*}(g,f)| &\leq |g(\xi_1) - g(a)|\,|f(x_1) - f(a)| \\ &\quad + |g(\xi_{N-1}) - g(b)|\,|f(b) - f(x_{N-1})| \leq v_{[a,b]}(g) \max |f(x_{i+1}) - f(x_i)|\end{aligned}$$

may be rendered arbitrarily small. Hence it is sufficient to prove theorem 3-5 for integrals defined by sequences Σ,Ξ^*. In this case

$$\begin{aligned}&\sum_{i=0}^{N-1} g(\xi_i^*)(f(x_{i+1}) - f(x_i)) \\ &\qquad = g(b)f(b) - g(a)f(a) - \sum_{i=1}^{N-1} f(x_i)[g(\xi_i^*) - g(\xi_{i-1}^*)]\end{aligned}$$

The Riemann-Stieltjès sum on the right-hand side of this equation converges to $\int_a^b f(x)\,dg(x)$ by lemma 3-1. Hence the sum on the left-hand side converges to a limit independent of the sequences Σ,Ξ chosen, and the formula of theorem 3-5 holds.

As a first application of theorem 3-5, we give a new proof for the four-vertices theorem. Let $\mathbf{x}(s)$ be a *convex* closed Jordan curve with curvature $k(s)$ of bounded variation, of length l. $\oint$ stands for $\int_0^l \cdots ds$.

Then $\oint dk(s) = 0$, and

$$\oint \mathbf{x}\, dk = -\oint k\mathbf{t}\, ds = \oint \mathbf{n}\, ds = \oint J\mathbf{t}\, ds = J\oint \mathbf{t}\, ds = J\oint d\mathbf{x} = 0$$

Hence $\oint x_1(s)\, dk(s) = \oint x_2(s)\, dk(s) = 0$, and, for any linear function, $L(x_1,x_2) = ax_1 + bx_2 + c$, $\oint L\, dk = 0$. Assume that the curvature has only one maximum at M and one minimum at m. The line Mm has a linear equation $L(x_1,x_2) = 0$. The linear function $L(x_1,x_2)$ is defined only up to a constant factor. It is possible to choose it so that the half plane $L(x_1,x_2) > 0$ contains the arc Mm with increasing curvature for increasing s, the arc Mm with decreasing curvature being in the half plane $L(x_1,x_2) < 0$. By construction, $\oint L\, dk > 0$, a contradiction to $\oint L\, dk = 0$; there must be more than one maximum and one minimum. This proof follows one given by G. Herglotz.

Exercise 3-1

1. A function is of bounded variation if and only if it is the difference of two monotone functions. Prove this statement in steps:
(i) If $g(x) = u(x) + v(x)$, $v_{[a,b]}(g) \leq v_{[a,b]}(u) + v_{[a,b]}(v)$. The "if" part follows.
(ii) Define $V(x) = v_{[a,x]}(g)$. $V(x)$ is monotone increasing.
(iii) For $x > x'$, $V(x) - V(x') \geq |g(x) - g(x')|$; hence

$$W(x) = V(x) - g(x)$$

is monotone. $g(x) = V(x) - W(x)$.

2. If $f(x) = x/|x|$, $x \neq 0$, $f(0) = 1$, the functions $f(x)$ and $\delta(x)$ both have a discontinuity at $x = 0$. Show that $\int_{-|a|}^{|a|} f(x)\, d\delta(x)$ does not exist.

3. For $g(x) = \delta(x) + \delta(x-1) + \cdots + \delta(x-n)$ compute $\int_0^n f(x)\, dg(x)$.

3-2. Involutes and Evolutes

Involutes and evolutes are important tools in the investigation of more hidden properties of curves.

The *evolute* $\mathbf{E}_f$ of a curve $\mathbf{f}(s)$ is the locus of the centers of curvature:

$$\mathbf{E}_f(s) = \mathbf{f}(s) + \frac{1}{k(s)}\mathbf{n}(s) \tag{3-6}$$

Here s is the arc length of $\mathbf{f}$; it is *not* also the arc length of $\mathbf{E}_f$. The evolute is defined for C^2 curves with curvature $\neq 0$.

The *involute* $\mathbf{I}_f$ of $\mathbf{f}(s)$ is the curve which cuts off the arc length s on any tangent to $\mathbf{f}(s)$,

$$\mathbf{I}_f(s) = \mathbf{f}(s) - s\mathbf{t}(s) \tag{3-7}$$

Again, s is the arc length of $\mathbf{f}$ only. The involute is defined for all C^1 curves. A change in the origin of the arc length, $s \to s - s_0$, will change the involute into $\mathbf{I}_f^* = \mathbf{f}(s) - (s - s_0)\mathbf{t}(s)$, as the involute is defined only up to some arbitrary constant s_0. This property suggests that involutes might be the result of some integration process [see Eq. (3-10)]. If this is true, we would expect the involute of a C^1 curve to be C^2. The evolute, like the derivative, is unique; we expect that the evolute of a C^2 will be only C^1. One of the first results in differential geometry published by its founder Christian Huygens in 1673 states that a curve is the evolute of its involutes and an involute of its evolute, just as a function is the derivative of its integrals and an integral of its derivative. The arc length of the evolute is measured by the radii of curvature of the original curve $\mathbf{f}$. We establish these results in two different ways, under more or less stringent differentiability conditions.

Theorem 3-7. *The involute* $\mathbf{I}_f$ *of a* C^2 *curve* $\mathbf{f}(s)$ *is a* C^3 *curve as a function of its arc length* s_I. *The curvature of the involute is* $k_I(s_I) = 1/s$, *and* $\mathbf{f}$ *is the evolute of its involute,* $\mathbf{f} = \mathbf{E}_{I_f}$.

The evolute $\mathbf{E}_f$ *of a* C^3 *curve* $\mathbf{f}(s)$ *with strictly monotone curvature is a* C^2 *curve as a function of its arc length* $s_E = \rho(s) + \text{const.}$ $\mathbf{f}$ *is the involute of its evolute* $\mathbf{f} = \mathbf{I}_{E_f}$.

By the defining equation (3-7),

$$\frac{d\mathbf{I}_f}{ds} = -sk(s)\mathbf{n}(s)$$

which shows the following:

1. *The involute of a* C^2 *curve is an orthogonal trajectory of the lines tangent to* $\mathbf{f}$.
2. $ds_I/ds = sk(s)$.

Hence

$$\frac{d\mathbf{I}_f}{ds_I} = -\mathbf{n}(s) \qquad \frac{d^2\mathbf{I}_f}{ds_I^2} = \frac{1}{s}\,\mathbf{t}(s) \qquad \frac{d^3\mathbf{I}_f}{ds_I^3} = -s^{-3}k^{-1}\mathbf{t}(s) + s^{-2}\mathbf{n} \tag{3-8}$$

It follows from Eq. (3-8) that the moving frame of the involute is $\mathbf{t}_I = -\mathbf{n}$, $\mathbf{n}_I = \mathbf{t}$, that $\mathbf{I}_f$ is C^3 except for $s = 0$, and that its curvature $k_I(s_I) = 1/s$. By (3-6), $\mathbf{E}_{I_f} = \mathbf{I}_f(s_I) + \rho(s_I)\mathbf{n}_I(s_I) = \mathbf{f}(s)$. This proves the first part of theorem 3-7.

For a proof of the second part of the theorem, we differentiate Eq. (3-6): $d\mathbf{E}_f/ds = \rho'(s)\mathbf{n}(s)$. Hence

1. *The normal to* $\mathbf{f}$ *is tangent to* $\mathbf{E}_f$.
2. $ds_E/ds = \pm(d\rho/ds)$, that is,

$$s_E(s_1) - s_E(s_0) = \pm(\rho(s_1) - \rho(s_0)) \tag{3-9}$$

The ambiguity of the sign is necessary to satisfy the condition

$$\frac{ds_E}{ds} > 0$$

which is needed to make the arc length a strictly monotone increasing function. The direction in which the arc length of the evolute measures changes in the curvature radius of **f** changes sign if ρ passes through a relative maximum or minimum. If an extremum of the curvature occurs at an interior point of the curve $\mathbf{f}(s)$, $dE_f/ds = 0$, the tangent is not defined. The evolute of an arc with nonmonotone curvature is not even C^1! Vertices of the curve appear as singular points (*cusps*) on the envelope. If the curvature is monotone and differentiable,

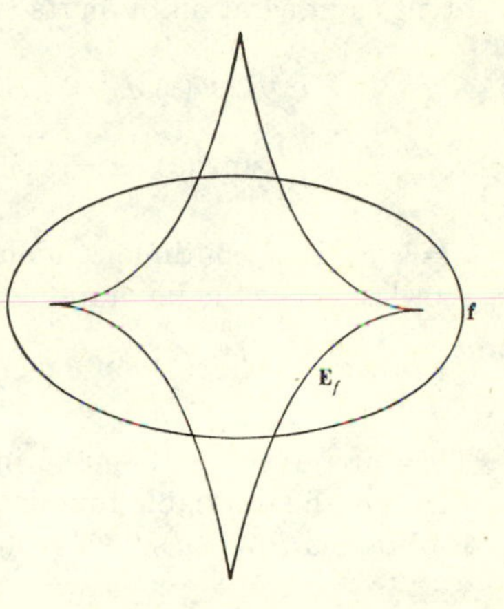

Fig. 3-1

$$\frac{d\mathbf{E}_f}{ds_E} = \frac{d\mathbf{E}_f}{d\rho} = \mathbf{n}(s) \qquad \frac{d^2\mathbf{E}_f}{ds_E{}^2} = -\frac{1}{\rho\rho'}\mathbf{t}(s)$$

$\mathbf{E}_f$ is C^2. It has an involute

$$\mathbf{I}_{E_f} = \mathbf{f}(s) + \rho(s)\mathbf{n}(s) - \rho(s)\mathbf{n}(s) = \mathbf{f}(s)$$

If the curvature is monotone, the curve is locally convex. A monotone function changes sign, at most, once. If $k(s)$ is of constant sign, so is $d\theta/ds$. Hence $\theta(s)$ is strictly monotone, and the arc from s_0 to s is convex as long as $|\theta(s) - \theta(s_0)| < \pi$. This means that a convexity condition appears naturally in theorem 3-7. It was shown in Sec. 2-1 that convexity may replace differentiability to some extent. The same phenomenon appears for the evolute-involute theorem.

Theorem 3-8. *The involute of a K^1 curve* **f** *is a K^2 curve as a function of its arc length s_I,* $\mathbf{f} = \mathbf{E}_{I_f}$. *The evolute of a K^2 curve* **f** *with strictly monotone nonzero curvature is K^1 as a function of the curvature,* $\mathbf{f} = \mathbf{I}_{E_f}$.

As in Sec. 2-3, the coordinates are measured in the fixed frame

$$\mathbf{e}_1 = \mathbf{t}(s_0) \qquad \mathbf{e}_2 = \mathbf{n}(s_0)$$

By hypothesis, we may choose s_0 such that **f** is convex in some interval about s_0; $\theta(s_0) = 0$, and the continuous function

$$\theta(s) = \arctan\frac{dx_2}{dx_1}$$

is a monotone function of s in that interval. Hence $\theta(s)$ is locally a one-to-one continuous function of s and may serve as a parameter in some interval about s_0. The moving frame is obtained from Eq. (2-2). The coordinates of the involute $\mathbf{I}_f = (x_{1I}, x_{2I})$ are

$$x_{1I} = \int_{s_0}^{s} \cos \theta(\sigma)\, d\sigma - s \cos \theta(s)$$

$$x_{2I} = \int_{s_0}^{s} \sin \theta(\sigma)\, d\sigma - s \sin \theta(s)$$

Stieltjès integration by parts (theorem 3-5) gives

$$\int_{s_0}^{s} \cos \theta(\sigma)\, d\sigma = s \cos \theta(s) - s_0 + \int_{s_0}^{s} \sigma \sin \theta(\sigma)\, d\theta(\sigma)$$

$$\int_{s_0}^{s} \sin \theta(\sigma)\, d\sigma = s \sin \theta(s) - \int_{s_0}^{s} \sigma \cos \theta(\sigma)\, d\theta(\sigma)$$

$s = s(\theta)$ is a continuous monotone function of the parameter θ. The coordinates of the involute may be written

$$x_{1I} = \int_0^{\theta} s(\vartheta) \sin \vartheta\, d\vartheta - s_0 \qquad x_{2I} = -\int_0^{\theta} s(\vartheta) \cos \vartheta\, d\vartheta \tag{3-10}$$

They are given by Riemann integrals over continuous integrands; hence they are differentiable functions of the upper limit θ. The arc length s_I is obtained from $(ds_I/d\theta)^2 = (dx_{1I}/d\theta)^2 + (dx_{2I}/d\theta)^2 = s^2(\theta)$; hence

$$\frac{d\mathbf{I}_f}{ds_I} = \frac{d\mathbf{I}_f}{d\theta}\frac{d\theta}{ds_I} = (\sin \theta, \; -\cos \theta)\{\mathbf{e}_1, \mathbf{e}_2\} = -\mathbf{n}$$

$$\frac{d^2\mathbf{I}_f}{ds_I{}^2} = -\frac{d\mathbf{n}}{d\theta}\frac{d\theta}{ds_I} = \frac{1}{s}\,\mathbf{t}$$

The angle $\theta_I(s) = \theta(s) - \pi/2$ of the tangent to $\mathbf{I}_f$ with the $+x_1$ axis is a monotone function of s; hence $\mathbf{I}_f$ is K^2. $\mathbf{f} = \mathbf{E}_{I_f}$ follows as before.

For the evolute theorem it is assumed that the curvature of $\mathbf{f}$ is strictly monotone; hence $\rho(s)$ is a continuous function [Eq. (2-5) and exercise 2-2, Prob. 19]. ρ may serve as a parameter for $\mathbf{f}$ and also for the evolute $\mathbf{E}_f = (x_{1E}, x_{2E})$,

$$x_{1E} = \int_{\rho_0}^{\rho} \cos \theta(r)\, ds(r) - \rho \sin \theta(\rho)$$

$$x_{2E} = \int_{\rho_0}^{\rho} \sin \theta(r)\, ds(r) + \rho \cos \theta(\rho)$$

Here we use the same fixed frame as for the involute. Stieltjès integration by parts (and $ds = \rho\, d\theta$) gives

$$x_{1E} = -\int_{\rho_0}^{\rho} \sin \theta(r)\, dr \qquad x_{2E} = \int_{\rho_0}^{\rho} \cos \theta(r)\, dr + \rho_0 \tag{3-11}$$

Here again the coordinates are Riemann integrals over continuous integrands; hence they are differentiable functions of $\rho = \pm s_E$, and $d\mathbf{E}_f/d\rho = \mathbf{n}(s)$. The remaining statements of the theorem are easily verified.

The normals to $\mathbf{f}$ are tangents to $\mathbf{E}_f$, and *the evolute* $\mathbf{E}_f$ *is the envelope of the normal lines to* $\mathbf{f}$. We show that this statement is true under the conditions of theorem 3-8.

A C^n *family of curves* is a C^n map of the unit square I^2 $(0 \leq t \leq 1, 0 \leq c \leq 1)$ into the plane:

$$\mathbf{F} = (x_1(t,c),x_2(t,c)) : I^2 \to R^2$$

For constant $c = c_0$, $\mathbf{F}(t,c_0)$ is a C^n curve. An *envelope* $\mathcal{E}_F$ of the family $\mathbf{F}$ is any C^1 curve that has the following three properties:

1. *For each* $u \in I$ *there is* $(t_0,c_0) \in I^2$ *such that* $\mathcal{E}_F(u) = \mathbf{F}(t_0,c_0)$; that is, every point of the envelope is on some curve of the family $\mathbf{F}$.

2. If $\mathcal{E}_F(u) = \mathbf{F}(t,c_0)$, *the tangents* $\mathbf{t}_\mathcal{E}$ *of the envelope and* $\mathbf{t}_F$ *of the curve* $\mathbf{F}(t,c_0)$ *coincide.*

3. $\mathcal{E}_F(u)$ *and* $\mathbf{F}(t,c_0)$ *have only isolated points in common.* The envelope may not contain an arc of a curve of the family.

By condition 1, $\mathcal{E}_F(u) = \mathbf{F}(u,c(u))$, the function $c(u)$ being differentiable. $\mathbf{t}_F$ is the unit vector in the direction of $((\partial x_1/\partial u)(u,c_0),\ (\partial x_2/\partial u)(u,c_0))$, and $\mathbf{t}_\mathcal{E}$ is the unit vector in the direction of

$$\left(\frac{\partial x_1}{\partial u}(u,c_0) + \frac{\partial x_1}{\partial c}(u,c_0)\frac{dc}{du}_{c=c_0},\ \frac{\partial x_2}{\partial u} + \frac{\partial x_2}{\partial c}\frac{dc}{du}_{c=c_0}\right)$$

The two directions are identical if the vectors are proportional, i.e., if

$$\left(\frac{\partial x_1}{\partial c},\frac{\partial x_2}{\partial c}\right)\frac{dc}{du} = a(u,c)\left(\frac{\partial x_1}{\partial u},\frac{\partial x_2}{\partial u}\right) \tag{3-12}$$

Condition 3 implies $dc/du \neq 0$. The system of equations (3-12) has *in general* only the trivial solution

$$\begin{aligned} a(u,c) &= 0 \\ \frac{\partial x_1}{\partial c} = \frac{\partial x_2}{\partial c} &= 0 \end{aligned} \tag{3-13}$$

The function $c(u)$ may be obtained from (3-13). *In general* the resulting curve will contain the envelopes. Although Eq. (3-13) seems to be necessary, it is by no means a sufficient condition to ensure the existence of an envelope. The same condition is obtained for a great number of other singular loci of families of curves.

***Example* 3-1.** The *lemniscate* $r = a \cos^{\frac{1}{2}} 2\phi$ is the locus

$$(x_1^2 + x_2^2)^2 = a^2(x_1^2 - x_2^2)$$

The family of curves $(x_1^2 + (x_2 - c)^2)^2 = a^2(x_1^2 - (x_2 - c)^2)$ is composed of lemniscates translated along the x axis. Equation (3-13) implies $2(x_1^2 + (x_2 - c)^2)(x_2 - c) = a^2(x_2 - c)$. A solution of this equation is $x_2 = c$. The equation of the family then becomes $x_1^4 = a^2x_1^2$, or $x_1 = 0$, $+a$, $-a$. Nothing new is obtained for $x_2 \neq c$. The two lines $x_1 = \pm a$ are envelopes, but $x_1 = 0$ is a line of double points (Fig. 3-2).

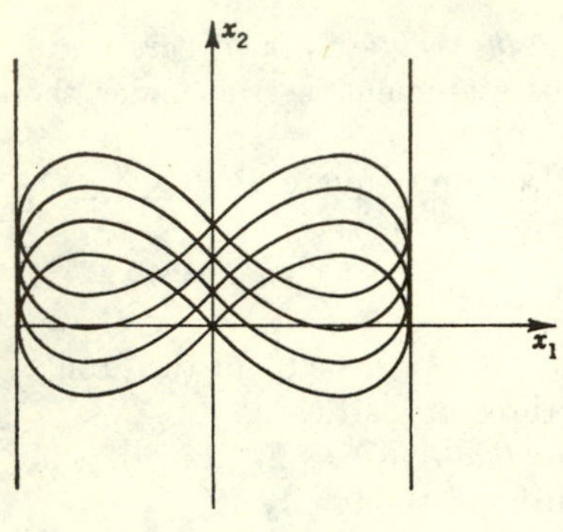

Fig. 3-2

The general theory of envelopes is not yet in a completely satisfactory state, but we will not discuss it further except for the case of a family of straight lines where there exists a simple and exhaustive theory.

Given a curve $\mathbf{f}(s)$, let (X_1,X_2) be a point on the tangent line to the curve in $\mathbf{f}(s_0) = (x_1,x_2)$. If we write $p = dx_2/dx_1$, the slope of the tangent line is p, and its equation is

$$X_2 = pX_1 + (x_2 - px_1)$$

If we assume that p is a strictly monotone function of s in some interval, we may use it as a parameter for $\mathbf{f}$ and write $x_2(s) - px_1(s) = H(p)$. The family of lines tangent to $\mathbf{f}$ is given by the *Clairaut* differential equation

$$X_2 = pX_1 + H(p) \qquad p = \frac{dX_2}{dX_1} \tag{3-14}$$

If $\mathbf{f}$ is a C^2 curve, $dH/dp = -x_1$. θ is a monotone function of s and of p; hence it is possible to choose the system of coordinates so as to make x_1 a monotone function of s in some interval about s_0.

Theorem 3-9. *The differential equation of the family of tangents to a C^1 curve with monotone slope may always be brought into the form of a Clairaut differential equation. If the curve is C^2, $H(p)$ is differentiable and dH/dp is monotone.*

This theorem has a converse:

Theorem 3-10. *A Clairaut differential equation* (3-14) *with continuous monotone dH/dp represents the family of tangents to a K^1 arc.*

Differentiate Eq. (3-14):

$$\left(X_1 + \frac{dH}{dp}\right) dp = 0 \tag{3-15}$$

A first solution of this equation is given by $dp = 0$, $p = c$; this yields the family of straight lines

$$X_2 = cX_1 + H(c) \tag{3-16}$$

If $dp \neq 0$, Eqs. (3-14) and (3-15) are satisfied by

$$x_1 = -\frac{dH}{dp} \qquad x_2 = -p\frac{dH}{dp} + H(p) \tag{3-17}$$

We have to show that the curve (3-17) is K^1 and that the family of lines (3-15) is that of the tangents to (3-17) (dH/dp need not be differentiable). Since dH/dp is continuous monotone, the first equation (3-17) may be inverted,

$$p = G(x_1) = \frac{dx_2}{dx_1}$$

$G(x_1)$ continuous monotone. Hence

$$x_2 = x_2(x_{10}) + \int_{x_{10}}^{x_1} G(x)\,dx$$

is a differentiable function of x_1 and Eq. (3-16) describes the tangent lines $X_2 - x_2 = p(X_1 - x_1)$. $\theta = \arctan p$ is a monotone function of x_1; hence the curve (3-17) is K^1.

The *normal* lines of a K^2 curve $\mathbf{f}$ with monotone curvature have an envelope $\mathcal{E}_n$. The solution of the Clairaut equation is

$$(p,-1)\cdot(\mathcal{E}_n - \mathbf{f}) = 0 \qquad p = -\frac{dx_1}{dx_2} \tag{3-18}$$

The vector $(p,-1)$ has the direction of $\mathbf{t}$; $\mathcal{E}_n - \mathbf{f}$ then has the direction of $\mathbf{n}$. By theorem 3-10, there is only one envelope, given by (3-17). We know that the normals are tangents to the evolute; hence

$$\mathbf{E}_f = \mathcal{E}_n \tag{3-19}$$

as stated.

***Example* 3-2.** Formulas (3-18) and (3-17) are very convenient for the computation of evolutes. In the case of the ellipse,

$$\mathbf{f} = (a\cos t, b\sin t) \qquad p = (a/b)\tan t$$
$$H(p) = x_2 - x_1p = b\sin t - pa\cos t = (b^2 - a^2)p(a^2 + b^2p^2)^{-\frac{1}{2}}$$

By Eq. (3-17) the evolute is given in parametric representation by

$$x_{1E} = \frac{(a^2 - b^2)a^2}{(a^2 + b^2p^2)^{\frac{3}{2}}} \qquad x_{2E} = \frac{(a^2 - b^2)b^2p^3}{(a^2 + b^2p^2)^{\frac{3}{2}}}$$

This is the astroid $(ax_{1E})^{\frac{2}{3}} + (bx_{2E})^{\frac{2}{3}} = (a^2 - b^2)^{\frac{2}{3}}$ (Fig. 3-1). For a circle, $a = b$, the evolute reduces to the center (0,0).

Exercise 3-2

1. All involutes of a circle are congruent. Find a parametric representation for them.

◆ **2.** Determine the evolute

(*a*) Of the parabola $x_2^2 = 2ax_1$

(*b*) Of the equilateral astroid $x_1 = a\cos^3 t$, $x_2 = a\sin^3 t$

(*c*) Of the logarithmic spiral $r = ce^{m\phi}$

(*d*) Of the cycloid $(t - \sin t, 1 - \cos t)$

(*e*) Of the catenary $x_2 = a \operatorname{Cosh} x_1/a$

◆ **3.** Find the envelope of the family of lines

$$X_1 \sin t - X_2 \cos t = h(t)$$

and its evolute. What are the minimal conditions of differentiability to be imposed on $h(t)$ for the problem to make sense?

4. Find the Clairaut equation of the tangents of a circle.

5. Involutes are orthogonal trajectories of the tangents to $\mathbf{f}$. Prove that $\mathbf{I}_f = \mathbf{f}(s_0) + \int_{s_0}^{s} \mathbf{n}(\sigma)\,d\sigma$.

◆ **6.** Find the envelope of the family of straight lines for which the sum of the intercepts is constant.

◆ **7.** Find the envelope of the family of lines for which the product of the intercepts is constant.

◆ **8.** The *evolutoid* of angle α of a curve $\mathbf{f}(s)$ is the envelope of the lines making a fixed angle α with $\mathbf{n}(s)$ at $\mathbf{f}(s)$ (Fig. 3-3.) In the frame $\{\mathbf{a},\mathbf{b}\} = \begin{pmatrix} \sin\alpha & \cos\alpha \\ \cos\alpha & -\sin\alpha \end{pmatrix} \{\mathbf{t},\mathbf{n}\}$ the lines are $(\mathbf{X} - \mathbf{f}(s)) \cdot \mathbf{b} = 0$.

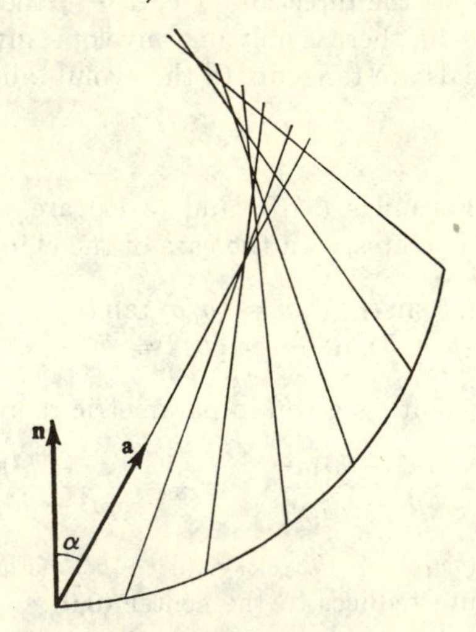

Fig. 3-3

(*a*) Show that the evolutoid is the curve $\mathbf{R} = \mathbf{f}(s) + \rho(s)\mathbf{a}(s)$.

(*b*) Find the radius of curvature of the evolutoid.

(*c*) Let $A(s)$ be the frame matrix of $\mathbf{f}$, $B(\rho)$ that of $\mathbf{E}_f$. Find the frame matrix of the evolutoid. State your differentiability assumptions.

(*d*) What are the evolutoids of a circle (Réaumur)?

9. Show that the evolute may still be defined in a reasonable way if the curvature exists except for a finite number of values of s, if it is strictly monotone, and if at all points $\lim_{s \downarrow s_0} k(s) - \lim_{s \uparrow s_0} k(s)$ has a constant sign. If $k(s)$ is discontinuous, one has to insert in the evolute a straight segment whose length is equal to the jump of the radius of curvature at the corresponding point of the curve. With this definition the evolute is still a C^1 curve.

10. Show that the evolute of a curve with continuous nonzero curvature cannot contain a straight segment.

11. A *cusp* of a C^2 curve is a point at which the curvature radius changes sign, passing through zero. Prove that the evolute of a closed C^3 curve has at least four cusps.

12. Show without computation that an ellipse is not the evolute of a closed curve. (Use Prob. 11.)

13. For a logarithmic spiral $r = ae^{m\phi}$, show that the endpoint of the polar subnormal describes the evolute, the endpoint of the polar normal describes the involute, and both evolute and involute are again logarithmic spirals.

14. Given a quarter of an ellipse, a convex arc joining a maximum and a minimum of the curvature, divide the plane into regions according to the number of normals one may draw from a given point to the arc of ellipse (Appolonius). (Hint: A normal to the ellipse is tangent to its evolute.)

15. Condition (3-13) for the envelope is also the condition for the *nodal locus*, or set of double points of the curves of the family, the *cusp locus*, and the *tac locus*, the set of points of contact of two curves of the family that belong to parameter values c_1, c_2, $|c_1 - c_2| > \epsilon$ (at least for curves given by analytic functions). In the following problems, draw a sketch of the curves of the family and identify the different loci.

(*a*) $(x_1 - c)^2 + x_2^2 = c^2 \sin^2 \alpha$ — two envelopes, one tac locus
(*b*) $x_1^3 = (x_2 + c)^2$ — one cusp locus, no envelope
(*c*) $(x_2 + c)^2 = x_1(x_1 - a)^2$ — one envelope, one nodal locus, one tac locus

(Cayley).

16. Solve the differential equation $x_2 = px_1 + (1 + p^2)^{\frac{1}{2}}$.

3-3. Spiral Arcs

Definition 3-11. *A spiral arc is a curve with monotone curvature of constant sign.*

A sufficiently small subarc of a spiral arc is always a convex curve. All spiral arcs are K^2's. If a spiral arc $\mathbf{r}$ contains a subarc of constant curvature, that subarc belongs to its own circle of curvature. The evolute of a spiral arc $\mathbf{r}$ will not contain any straight segment since the direction of $\mathbf{n}$ is not constant. This means that the length of the arc connecting two points of $\mathbf{E}_r$ is always greater than the distance between

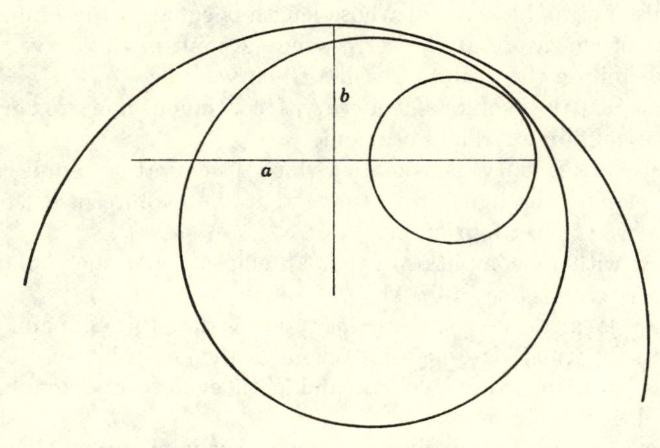

Fig. 3-4

the two points, or that the distance between two centers of curvature of $\mathbf{r}$ is less than the difference between the corresponding curvature radii. The smaller circle of curvature will be completely in the interior of the bigger one.

Theorem 3-12 (Kneser). *Any circle of curvature of a spiral arc contains every smaller circle of curvature of the arc in its interior and in its turn is contained in the interior of every circle of curvature of greater radius.*

Corollary 3-13. *Two distinct circles of curvature of a spiral arc never intersect. Such an arc is not the envelope of its circles of curvature.*

Figure 3-4 shows some circles of curvature of a spiral arc of an ellipse.

A point $P = \mathbf{r}(s_0)$ of a spiral arc decomposes the arc into two branches, one of increasing curvature, $k(s) > k(s_0)$, and one of decreasing curvature, $k(s) < k(s_0)$. Kneser's theorem assigns distinct regions in the plane to the two branches:

Theorem 3-14. *If the curvature of a spiral arc is not constant in a neighborhood of s_0, the branch of increasing curvature is completely in the* *interior of the circle of curvature* $S_r(s_0)$, *and the branch of decreasing curvature is in the exterior of* $S_r(s_0)$.

No point can be in the interior and in the exterior of a circle:

Corollary 3-15. *If a spiral arc does not contain the circumference of a circle, it has no double point.*

From a point inside a circle of curvature there is no tangent to the circle:

Corollary 3-16. *A spiral arc has no double tangent.*

The next theorems deal with *convex spiral arcs.*

Theorem 3-17 (Vogt). *Let A and B be the endpoints of a spiral arc, the curvature nondecreasing from A to B. The angle* β *of the tangent to the arc at B with the chord AB is not less than the angle* α *of the tangent at A with AB.* $\alpha = \beta$ *only if the curvature is constant.*

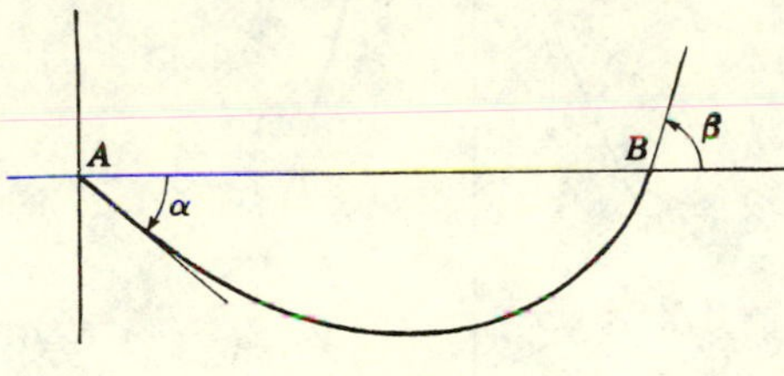

Fig. 3-5

Proof (Fig. 3-5): We choose AB as the x_1 axis such that $x_2(s) < 0$ for the inner points of the arc AB, $x_2(0) = x_2(l) = 0$. Here l is the length of the arc AB. We have to show that $\theta(l) = \beta \geq 2\pi - \theta(0) = \alpha$, or that $\cos\theta(l) \leq \cos\theta(0)$. By hypothesis, $k(s_{i+1}) - k(s_i) \geq 0$ for any subdivision of the interval $[0,l]$. The theorem follows by Stieltjès integration by parts:

$$\begin{aligned}\cos\theta(0) - \cos\theta(l) &= \int_{\theta(0)}^{\theta(l)} \sin\theta\, d\theta = \int_0^l x_2'(s)k(s)\,ds \\ &= -\int_0^l x_2(s)\,dk(s) \geq 0\end{aligned}$$

Equality can hold only if k is constant on $[0,l]$. This proof is due to A. Ostrowski.

A *line element* is a pair $(P,\mathbf{t})$ of a point and a direction (unit vector). A C^1 curve $\mathbf{f}(s)$ *defines its line elements* $(\mathbf{f}(s),\mathbf{f}'(s))$. Vogt's theorem may be reformulated in terms of line elements: *For any two line elements* $(A,\mathbf{r}'(0))$, $(B,\mathbf{r}'(l))$ *defined by a convex spiral arc of increasing curvature,* $\angle(\mathbf{r}'(0),AB) \leq \angle(\mathbf{r}'(l),AB)$. The question now arises whether any two given line elements may be joined by a convex spiral arc which defines the given elements at its endpoints. The answer will be a corollary to a stronger theorem on curvature elements.

A *curvature element* $(P,\mathbf{t},\rho\mathbf{n})$ is a triple of a point, a direction, and a directed segment perpendicular to the given direction. A C^2 curve defines its curvature elements as triples of a point on the curve, the tangent, and the curvature radius at the point. We want to find necessary and sufficient conditions for the existence of a convex spiral arc *joining* two given curvature elements $(A,\mathbf{t}_1,AP_1)$, $B,\mathbf{t}_2,BP_2)$, that is, a convex spiral arc with endpoints A,B, tangents $\mathbf{t}_1$ at A and $\mathbf{t}_2$ at B, and centers of curvature P_1 for A and P_2 for B. We may assume (Fig. 3-6) that the

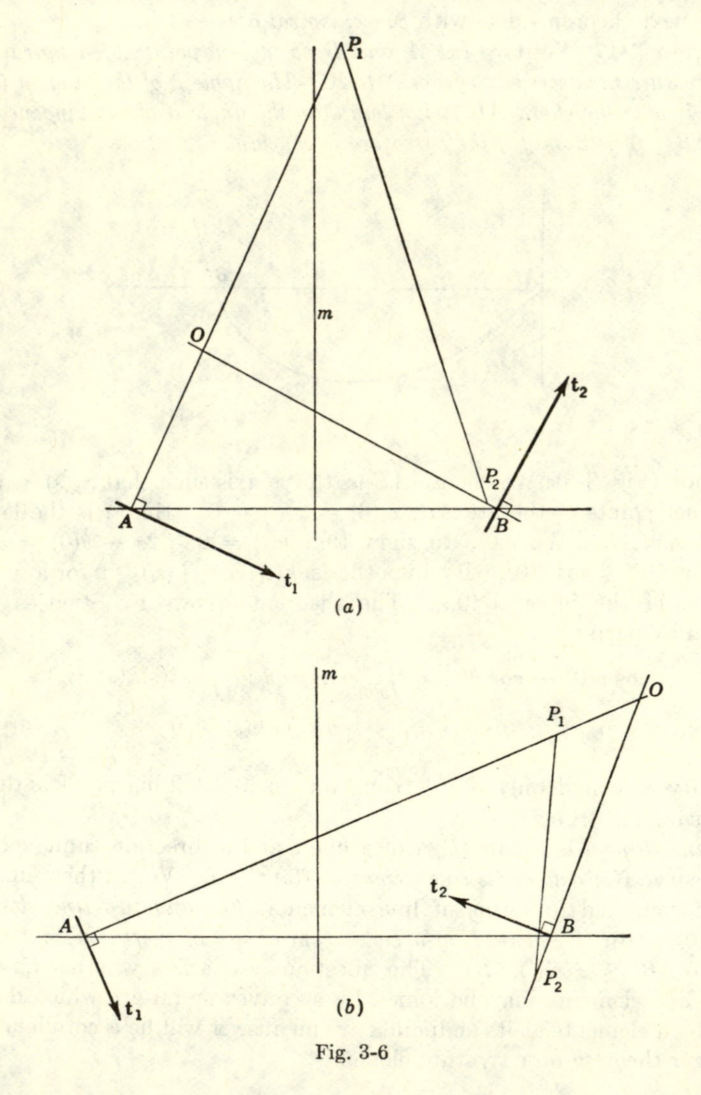

Fig. 3-6

orientation of the tangents has been chosen so as to make the curvature positive and that the (unoriented) angle between $\mathbf{t}_1$ and the line AB is smaller than that between $\mathbf{t}_2$ and AB:

$$\angle(\mathbf{t}_1,AB) \leq \angle(\mathbf{t}_2,AB) \tag{3-20}$$

If a convex spiral arc exists joining the two curvature elements, its curvature must increase from A to B, by Vogt's theorem:

$$|AP_1| \geq |BP_2|$$

The perpendicular bisector m of the segment AB divides the plane into two half planes. The point P_1 must be situated in the half plane that contains B, since by Kneser's theorem B is in the circle with center P_1 and radius $|AP_1|$. By the same theorem, we even have

$$|P_1P_2| \leq |AP_1| - |BP_2| \tag{3-21}$$

Two cases are now to be distinguished.

(*a*) $$\angle(\mathbf{t}_2,AB) - \angle(\mathbf{t}_1,AB) < \pi \tag{3-20a}$$

Let O be the point of intersection of the lines AP_1 and BP_2.

$$\angle OAB = \pi/2 - \angle(\mathbf{t}_1,AB) \geq \pi/2 - \angle(\mathbf{t}_2,AB) = \angle OBA$$

Hence $|OB| \geq |OA|$, and O is in the half plane of A with respect to m. It follows that the evolute of the spiral arc AB must be contained in the interior of the triangle $\triangle P_1OP_2$; it is tangent at P_1 to AP_1 and at P_2 to BP_2. The polygonal line P_1OP_2 is circumscribed to the (convex) evolute. It is longer than the length of the evolute, the difference of the curvature radii:

$$|AP_1| - |BP_2| \leq |P_1O| + |P_2O| \tag{3-22}$$

Given any number l,

$$|P_1P_2| < l < |P_1O| + |P_2O|$$

it is possible to find a convex curve of length l from P_1 to P_2, tangent at P_1 to P_1O and at P_2 to P_2O, and inscribed at P_1OP_2. Such a curve, for example, may be composed of circular arcs smoothly pieced together. Its involute through A is a convex spiral arc solving our problem, for $l = |AP_1| - |BP_2|$.

(*b*) If $$\angle(\mathbf{t}_2,AB) - \angle(\mathbf{t}_1,AB) > \pi \tag{3-20b}$$

(Fig. 3-6*b*) the evolute must be tangent at P_1 to P_1A and at P_2 to BP_2. Its length is subject only to condition (3-21). The construction of the arc AB follows as before. (See also exercise 3-3, Prob. 21.)

Theorem 3-18. *Two curvature elements* $(A,\mathbf{t}_1,AP_1)$ *and* $(B,\mathbf{t}_2,BP_2)$ *may be joined by a convex spiral arc with curvature increasing from A to B if and only if conditions* (3-21) *and either* (3-20*a*) *and* (3-22) *or* (3-20*b*) *hold.*

If we are given two *line elements* such that condition (3-20) holds, any choice of P_1 in the half plane of B with respect to m allows the construction of a P_2 to satisfy the conditions of theorem 3-18:

Corollary 3-19 (Ostrowski). *Two line elements* $(A,\mathbf{t}_1)$ *and* $(B,\mathbf{t}_2)$ *may be joined by a convex spiral arc with curvature increasing from A to B if and only if condition* (3-20) *holds.*

Two line elements are *concircular* if they may be joined by an arc of a circle, i.e., if $\angle(\mathbf{t}_1,AB) = \angle(\mathbf{t}_2,AB)$. By Vogt's theorem, *two concircular line elements may not be joined by any convex spiral arc with variable curvature.*

Any K^2 curve is the union of a finite number of convex spiral arcs. The four-vertices theorem suggests that a closed K^2 curve cannot be the union of too few convex spiral arcs.

Theorem 3-20. *A closed K^2 curve which is not a circle is the union of at least three convex spiral arcs.*

A closed curve which is not *simple* is not the union of less than three convex arcs. Here the theorem is trivial. (For an example of a closed curve with a double point, the union of three convex spiral arcs, see exercise 3-3, Prob. 19.)

A simple closed curve with nowhere constant curvature is the union of at least *four* spiral arcs (four-vertices theorem). If a simple closed curve contains a circular arc, either it is a circle or the arc joining its endpoints contains at least two vertices. Since it joins two concircular line elements, it cannot be a spiral arc. If it contains only one vertex, say a maximum of the curvature at M, any point of the circular arc may serve as the locus of the minimum m, and the contradictions that prove the four-vertices theorem apply in this case also, since the curvature is continuous on the closed curve. The arc joining the two concircular line elements is the union of at least three convex spiral arcs; the closed curve is not the union of a smaller number of such arcs.

Exercise 3-3

1. The notion of a spiral arc may be generalized to allow for a finite number of points at which the curvature is not defined under the condition that the one-sided curvatures exist and are monotone. Show that Kneser's theorem remains true with the addition that at points with unequal one-sided curvatures the smaller one-sided circle of curvature touches the bigger one from within. (See exercise 3-2, Prob. 9.)

2. Show that any compact subarc of the spiral of Archimedes (exercise 1-3, Prob. 3), of the hyperbolic spiral (exercise 1-3, Prob. 5), and of the logarithmic spiral (exercise 2-2, Prob. 3*e*) is a spiral arc.

3. Derive the four-vertices theorem from Vogt's theorem.

4. Prove that the four-vertices theorem remains true for discontinuous curvature if a one-sided extremum is counted as a full vertex (i.e., one point might represent two vertices).

5. Prove that no line through a point of a spiral arc is tangent to that arc at a point of smaller curvature (Vogt).

6. The tangent line at a point A of a spiral arc divides the plane into two half planes. Show that the branch of increasing curvature defined by A is completely in one of the half planes (Vogt).

7. Let $\mathbf{r}(s)$ be a spiral arc. Show that the rotation of the vector $\mathbf{r}(s) - \mathbf{r}(s_0)$ about the point $A = \mathbf{r}(s_0)$ never changes sense for increasing s (Vogt).

8. Let $k(s)$ be a decreasing function of the arc length on a spiral arc $\mathbf{r}(s)$. Let $s_1 < s_2 < s_3 < \cdots$ be the parameter values of the intersections of the spiral arc with a fixed straight line. Show that the segment of that line bounded by $\mathbf{r}(s_i)$ and $\mathbf{r}(s_{i+1})$ will contain *all* $\mathbf{r}(s_j)$, $j < i$, but *no* $\mathbf{r}(s_k)$, $k > i + 1$ (Vogt).

9. Let $s_1 < s_2 < s_3 < \cdots$ be the parameter values of the points of intersection of a spiral arc with a circle. Two successive points of intersection, $\mathbf{r}(s_i)$ and $\mathbf{r}(s_{i+1})$, divide the circle into two arcs. Show that one of them contains all $\mathbf{r}(s_j)$, $j < i$; the other one, all points $\mathbf{r}(s_k), k > i + 1$ (Vogt).

10. A spiral arc has, at most, two points of contact with a given circle. If it has two, show that the given circle contains one of the circles of curvature at the points of contact and is contained in the other one (Vogt).

11. Prove that a convex spiral arc and a circle have, at most, three points of intersection without contact or one point of contact and one noncontact intersection (Vogt).

12. If a spiral arc has two points of contact with some circle, show that it is not convex (Vogt).

13. If a closed convex curve with continuous curvature has $2n$ points of intersection with some circle, prove that it has at least $2n$ vertices (Vogt). HINT: Use Prob. 11.

14. Prove that no circle intersects an ellipse in more than four points. HINT: Use Prob. 13.

15. Let AB be a convex arc (C^2) such that one circle S is the circle of curvature both at A and at B. If S is tangent to the arc either both times from the outside or both times from the inside, show that the arc has at least three vertices and that otherwise it has at least four vertices (Fabricius-Bjerre). HINT: The given arc and an arc of S define a C^2 simple closed curve.

16. (Sequel to Prob. 15.) Given a simple closed C^2 curve such that some circle S is the circle of curvature at two distinct points A,B. If A and B are ordinary points on the curve, show that there exist on it at least six vertices. If A and B are vertices and S is tangent to the curve both times from the same side, show that there are at least eight vertices. If A and B are both vertices and S is tangent from different sides, show that there are at least 10 vertices (Fabricius-Bjerre).

17. Show that no circle tangent to an ellipse at a vertex has exactly one other point of intersection with the ellipse.

18. Show that the property of the circle of curvature of the ellipse stated in exercise 2-3, Prob. 5, holds (a) for all conics other than the circle and (b) for all closed convex curves with four vertices.

19. A closed convex curve, the union of three spiral arcs, may be constructed as follows: By exercise 3-2, Prob. 13, one may find an arc of logarithmic spiral which cuts the polar axis once and for which the curvature centers at the endpoints are on the polar axis. The arc subtends a polar angle of $3\pi/2$. This arc and the symmetric one with respect to the polar axis are then joined by arcs of the circles of curvature at the endpoints. Draw the curve.

20. If a closed convex curve with one double point is the union of three convex spiral arcs, show that a tangent from a vertex of maximal curvature passes through a vertex of minimal curvature.

21. Find an example showing that in case (3-20b) the length of the evolute may be arbitrarily great. (HINT: Study, for example, the hyperbolic spiral, exercise 1-3, Prob. 5.)

22. By Kneser's theorem, it is impossible to compose an arc of nonconstant curvature from arcs of curvature circles. But for practical use it is possible to keep the arcs of curvature circles small enough so that the deviation from the actual shape of the given curve is less than the thickness of the graphite band used to represent it. Obtain an estimate for the maximal aperture to be used in drawing an ellipse with a lead of about $\frac{1}{100}$ in. thickness (half axes: a few inches). Draw an ellipse composed of arcs of curvature circles, using the construction method of exercise 2-3, Prob. 6. Draw also the evolute of the ellipse by the same construction.

A convex C^2 arc may be given as a function of the tangent angle θ, or also of the angle $\psi = \theta - \pi/2$ of the normal with the $+x_1$ axis. The convex arc is the envelope of its tangents

$$X_1 \cos \psi + X_2 \sin \psi = h(\psi) \tag{3-23}$$

$h(\psi)$, the distance of the tangent from the origin, is the *support function* of the convex curve. The derived equation

$$-X_1 \sin \psi + X_2 \cos \psi = h^{\cdot}(\psi) \tag{3-24}$$

shows that the normal at the point of contact of the tangent (3-23) is a distance $dh/d\psi$ from the origin. A parametric representation of the envelope of the lines (3-23) is

$$x_1 = h(\psi) \cos \psi - h^{\cdot}(\psi) \sin \psi \qquad x_2 = h(\psi) \sin \psi + h^{\cdot}(\psi) \cos \psi$$

(theorem 3-10). The radius of curvature is

$$\rho(\psi) = h(\psi) + h^{\cdot\cdot}(\psi) \tag{3-25}$$

If the support function of a convex C^2 arc is given, the curve itself may be

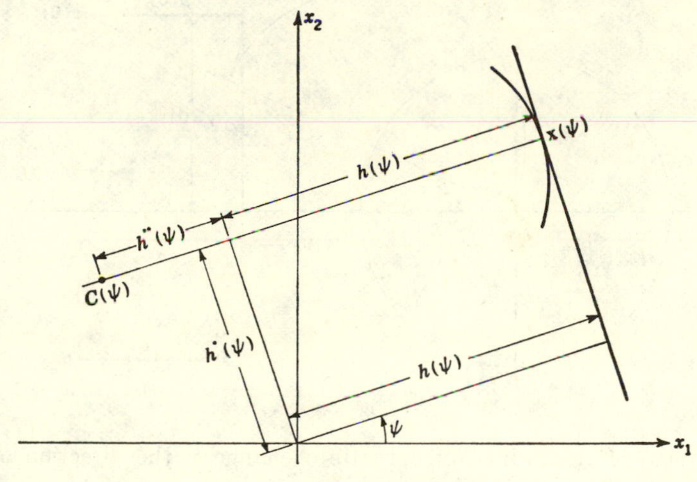

Fig. 3-7

found analytically as the envelope of its tangents, or graphically from the known center of curvature $C(\psi)$ and radius of curvature (3-25) (Fig. 3-8). The precautions explained in Prob. 22 should be observed.

The preceding formulas have been used by E. Meissner to give a simple method of integration of nonlinear equations:

$$y'' + g(y',y) = F(y',y) \tag{3-26}$$

The independent variable is taken to be the angle ψ, and the dependent variable y is the support function of a curve, the *curve representation* (c.r.) of the function y. If $y(0)$, $y'(0)$ are given, it is possible to draw a small arc of the circle of curvature in $x_1(0) = y(0)$, $x_2(0) = y'(0)$. At the endpoint of that arc the new circle of curvature may be found by Eq. (3-25). The center of curvature naturally is on the normal (Fig. 3-7). The following problems present a discussion of the method.

23. Discuss the relation of $\mathbf{x} = x_1\mathbf{e}_1 + x_2\mathbf{e}_2$ and the support vector $\mathbf{h} = h\mathbf{e}_1 + h^{\cdot}\mathbf{e}_2$.

24. A differential equation $ay'' + by' + cy = F(y,y')$, a,b,c constant, $a \neq 0$, may be transformed into $z'' + z = G(z,z')$ by a substitution $y = e^k z$. Find k.

25. The differential equation

$$y'' + y = \begin{cases} +1 & y' < 0 \\ -1 & y' > 0 \end{cases} \tag{3-27}$$

appears in the mechanics of dry friction and in push-pull control processes. Its c.r. consists of an arc of circles (Fig. 3-8). Why does

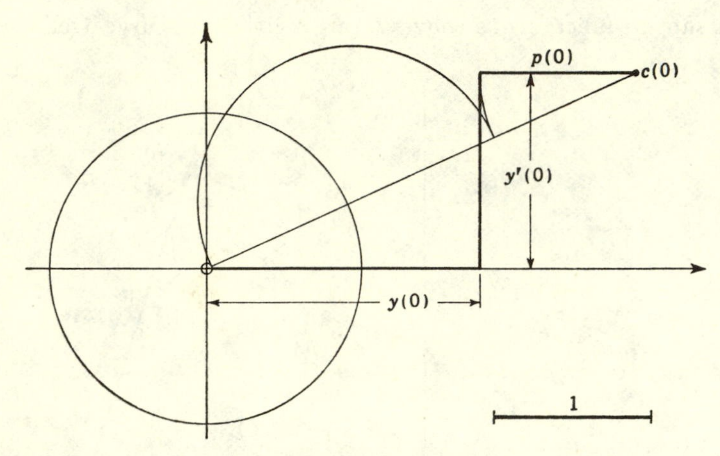

Fig. 3-8

the sign of the radius (of curvature) change if the direction of the radius points to the origin?

26. In the mechanical interpretation of Eq. (3-27), movement of a mass point under the influence of dry friction and a system's force proportional to the elongation, the point will come to a standstill if the force $|y|$ is less than the friction ± 1 at a point of velocity 0 ($y' = 0$). Show that the c.r. of a constant $c = c(\psi)$ is a circle about the origin and that every system (3-27) will come to a standstill, for arbitrary initial conditions.

27. Solve

$$y'' + y = \begin{cases} -1 & y' > 0 \\ +1 & y' < 0 \end{cases} \qquad y(0) = 2 \qquad y'(0) = 1$$

28. Fnid the c.r. for

$$\begin{aligned} y'' + \tfrac{1}{2}y &= 0 \qquad y < 0 \\ y'' - \tfrac{1}{2}y &= 0 \qquad y > 0 \end{aligned}$$

$y(-\pi/2) = 0$, $y'(-\pi/2) = 1$, y and y' continuous at 0.

29. If Eq. (3-27) is changed into

$$y'' + y = \begin{cases} -1 & y' > -k^2 \\ +1 & y' < -k^2 \end{cases}$$

we have a system *with retarded switching.* Let ψ measure the time. The time needed to bring the system to a standstill is a monotone decreasing function of k for $0 \leq k < k_{\min}$ (Fig. 3-9). Compute $k^2_{\min}$ from the condition that for it the c.r. is tangent to the standstill

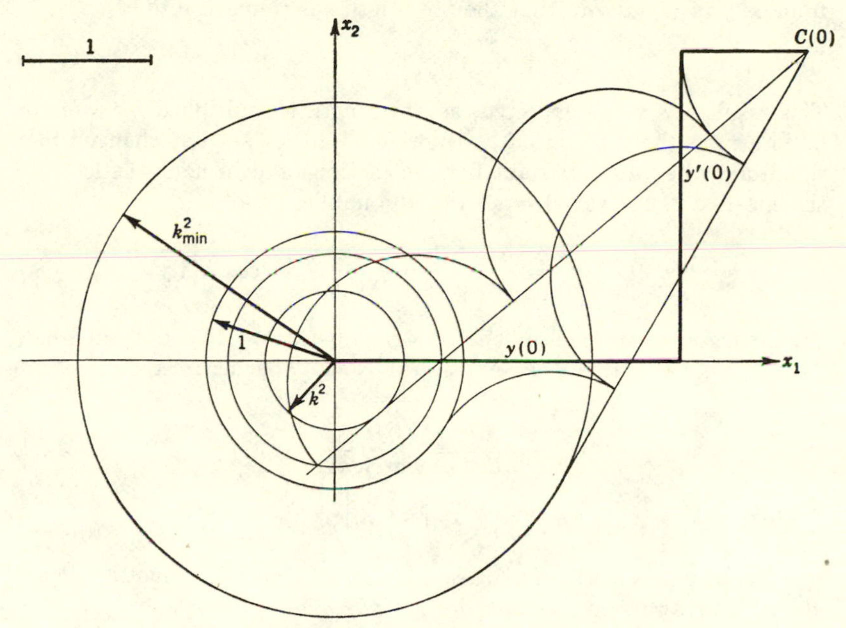

Fig. 3-9

circle of radius 1 (Prob. 25) about the origin. Show that the "optimal" solution so obtained is unstable in the sense that the system will not come to a standstill for any $k^2 > k^2_{\min}$.

30. Control systems may be built in which not only the control force is a discontinuous function of y' but in which under certain conditions an *instantaneous impulsion* $\Delta y'(\psi) = y'(\psi + 0) - y'(\psi - 0)$ is imparted to the system from an outside source. By Fig. 3-7, this means that we have to insert in the c.r. a straight segment of length $\Delta y'$ in a tangential direction at stated places. Solve graphically the following system:

$$y'' + y = \begin{cases} -1 & y' > 0 \\ +1 & y' < 0 \end{cases} \qquad \Delta y' = \begin{cases} -1 & \text{if } y'(\psi - 0) = +1 \\ +1 & \text{if } y'(\psi - 0) = -1 \end{cases}$$

with initial conditions $y(0) = 5$, $y'(0) = 3$.

3-4. Congruence and Homothety†

Definition 3-21. *A homothety of ratio λ and with its center the origin is a map $\mathbf{x} \to \lambda\mathbf{x}$ (λ a real number $\neq 0$). A similitude of ratio λ is any map, a composition of a homothety of ratio λ and a euclidean motion.*

A geometric figure and its image under a similitude are said to be *similar* (of ratio λ).

A curve $\mathbf{x}(u)$ is changed into $\mathbf{x}^*(u) = \lambda\mathbf{x}(u)$ in a homothety. It follows from $d\mathbf{x}^*/du = \lambda\, d\mathbf{x}/du$ that the arc length s is changed into s^*,

$$s^* = |\lambda| s \tag{3-28}$$

This result was to be expected, as any length is multiplied by a factor $|\lambda|$ in a homothety of ratio λ. Unoriented angles are not changed in a similitude; they are invariant both in motions and in homotheties. In the latter case, $\cos \angle(\mathbf{x},\mathbf{y}) = \mathbf{x} \cdot \mathbf{y}/|\mathbf{x}|\,|\mathbf{y}|$ implies

$$\cos \angle(\mathbf{x}^*,\mathbf{y}^*) = \frac{\mathbf{x}^* \cdot \mathbf{y}^*}{|\mathbf{x}^*|\,|\mathbf{y}^*|} = \frac{\lambda\mathbf{x} \cdot \lambda\mathbf{y}}{|\lambda\mathbf{x}|\,|\lambda\mathbf{y}|} = \cos \angle(\mathbf{x},\mathbf{y})$$

Oriented angles are multiplied by sign $\lambda = |\lambda|/\lambda$. The tangent angle θ^* of $\mathbf{x}^*$ is equal to that θ of $\mathbf{x}$ multiplied by sign λ; hence for the curvature radius,

$$\rho^*(s^*) = \frac{ds^*}{d\theta^*} = \frac{|\lambda|}{\operatorname{sign}\lambda}\frac{ds}{d\theta} = \lambda\rho(s) \tag{3-29}$$

Finally,

$$\left|\frac{d\rho^*}{ds^*}\right| = \left|\lambda \frac{d\rho}{ds}\frac{ds}{ds^*}\right| = \left|\frac{d\rho}{ds}\right|$$

in a homothety. Both ρ and s are invariant in a euclidean motion; hence $|d\rho/ds|$ is invariant in a similitude.

Theorem 3-22. *$|d\rho/ds|$ is an invariant in any similitude.*

In the applications, it is much better to work with Eqs. (3-28) and (3-29) than with theorem 3-22. The theorem makes sense only for C^3 curves, and if $d\rho^*/ds^* = d\rho/ds$, $ds^*/ds = \lambda$, one may infer only

$$\rho^*(s^*) = \lambda\rho(s) + c$$

This does not imply similarity of the curves unless $c = 0$.

Theorem 3-23. *If $\rho_x(|\lambda|s) = \lambda\rho_y(s)$ holds for two C^2 curves $\mathbf{x}(s_x)$, $\mathbf{y}(s_y)$ with curvature radii ρ_x, ρ_y, the two curves are similar, of ratio λ.*

PROOF: A curve is defined up to a congruence by its curvature radius as a function of the arc length. By Eqs. (3-28) and (3-29) the curvature

† For a discussion of homothety ("dilatation") and similitude within the framework of elementary geometry, see, for example, H. S. M. Coxeter, "Introduction to Geometry," John Wiley & Sons, Inc., New York, 1961.

of $\mathbf{x}$ and that of a homothetic image $\mathbf{y}^*$ of $\mathbf{y}$ under a homothety of ratio λ are identical. Hence $\mathbf{x}$ and $\mathbf{y}^*$ are congruent.

The differential geometry of homothetic curves was first discovered by Leonhard Euler. Some of his problems are discussed here.

***Example* 3-3.** *Find all curves for which similarity implies congruence.*

The problem is impossible for *compact* curves, since the distance between the endpoints of such a curve is multiplied by $|\lambda|$ in a homothety. A similar argument shows that a closed curve also cannot remain congruent with itself in all similitudes. Therefore the search is restricted to unbounded C^2 curves.

Let us take $\lambda > 0$ for the moment. Equation (3-29) shows that a curve solving the problem has a curvature radius which satisfies the functional equation

$$\rho(\lambda s) = \lambda \rho(s)$$

It follows that ρ is a differentiable function of s and that

$$\frac{d\rho}{ds} = \lim_{h \to 0} \frac{1}{h} (\rho(s + h) - \rho(s)) = \lim_{h \to 0} \left(1 + \frac{h}{s} - 1\right) \frac{\rho(s)}{h} = \frac{\rho(s)}{s}$$

Hence $\rho(s) = cs$. This is the natural equation of the *logarithmic spirals.* A more familiar form of this equation is obtained from

$$k(s) = \frac{d\theta}{ds} = \frac{1}{cs}$$

Hence $s = e^{c(\theta - \theta_0)}$, $\rho(s) = ce^{c(\theta-\theta_0)}$ (see exercise 2-2, Prob. 3*e*). Another solution of our problem which escapes our method is given by the *straight lines* $[k(s) = 0]$.

***Example* 3-4.** *Find the C^3 curves similar to their evolutes.*

The moving frame of the evolute is $\{\mathbf{n}, -\mathbf{t}\}$. From

$$\frac{d\mathbf{n}}{d\rho} = \left(\frac{d\mathbf{n}}{ds}\right)\left(\frac{ds}{d\rho}\right) = -\frac{\mathbf{t}}{\rho\rho'}$$

the curvature radius of the evolute is

$$\rho_E = \rho\rho' = \frac{d\rho}{ds}\frac{ds}{d\theta} = \frac{d\rho}{d\theta} = \frac{d^2 s}{d\theta^2}.$$

The condition of similitude (3-29) becomes

$$\rho_E = \frac{d\rho}{d\theta} = \lambda\rho$$

Hence $\rho = ce^{\lambda(\theta-\theta_0)}$. The solution again is the *logarithmic spirals.* It is interesting to note that a curve similar to its evolute is always congruent with its evolute.

The problem may be generalized in various directions. The *second evolute* $\mathbf{E}_f^{(2)}$ of a curve of high order of differentiability is the evolute of the evolute: $\mathbf{E}_f^{(2)} = \mathbf{E}_{E_f}$. In the same way, the nth *evolute* $\mathbf{E}_f^{(n)}$ is the evolute of the $(n-1)$-th one, if it exists. The curvature radius $\rho^{(n)}$ of $\mathbf{E}_f^{(n)}$ and the curvature radius ρ of $\mathbf{f}$ are related by

$$\rho^{(n)} = \frac{d^n\rho}{d\theta^n} \tag{3-30}$$

The formula, or its equivalent $\rho^{(n)} = d\rho^{(n-1)}/d\theta$, is proved by induction. A curve is similar to its nth evolute if the curvature radius is an integral of $\lambda\rho = d^n\rho/d\theta^n$.

As we have remarked before, we consider curves as *maps* and not merely as *point sets*. Similitude of curves therefore implies not only

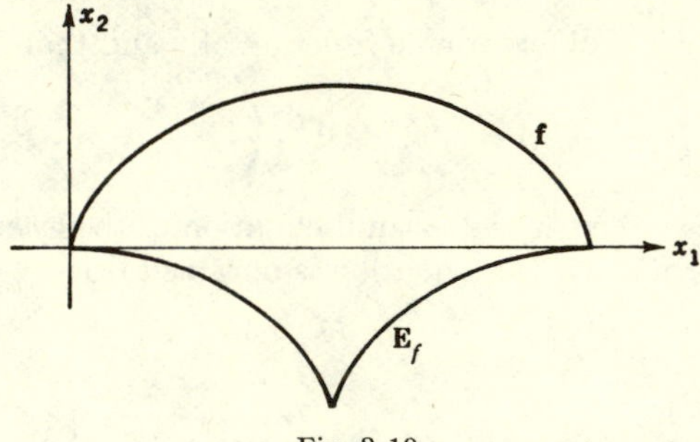

Fig. 3-10

similitude of the point sets in the sense of elementary geometry but also correspondence of the arc lengths. A complete analytical discussion of similitudes therefore must include the study of *indirect* similitude. Two curves $\mathbf{f}(s)$, $\mathbf{g}(s)$ are said to be *indirectly similar* if $\mathbf{f}(s)$ is similar to $\mathbf{g}(s_0 - s)$. If a C^4 curve is indirectly similar to its evolute,

$$\frac{d\rho}{d\theta}(\theta) = \lambda\rho(\theta_0 - \theta)$$

then by another differentiation

$$\frac{d^2\rho}{d\theta^2}(\theta) = \lambda\frac{d\rho(\theta_0 - \theta)}{d\theta} = -\lambda^2\rho(\theta)$$

we see that the curve is directly similar to its second evolute, in a ratio $-\lambda^2$. For $\lambda = \pm 1$, the curves indirectly congruent with their evolutes have curvature radii $d^2\rho/d\theta^2 = -\rho$, $\rho(\theta) = c\sin(\theta + \theta_0)$. The particular solution $\theta_0 = 0$ gives $s = c(1 - \cos\theta)$, and, by the formulas of exercise 2-2, Prob. 9,

$$x_1 = -\frac{c}{4}\sin 2\theta + \frac{c}{2}\theta \qquad x_2 = -\frac{c}{4}\cos 2\theta + \frac{c}{4}$$

a *cycloid* (Fig. 3-10).

Exercise 3-4

1. Show that a closed curve cannot be congruent with itself in a similitude of ratio $\neq \pm 1$.

2. Obtain Eq. (3-29) directly, without computation, from the geometric definition of the curvature radius given in Sec. 3-2.

3. Find the general form of the curves that are indirectly congruent with their evolutes (epicycloids and hypocycloids).

4. If $\mathbf{f}$ is similar to $\mathbf{E}_f^{(2)}$, with ratio $+\lambda^2$, show that $\mathbf{f}$ belongs to the family of curves

$$x_1 = \frac{c}{\lambda^2 + 1}(\lambda \sin\theta \operatorname{Sinh}\lambda\theta - \cos\theta \operatorname{Cosh}\lambda\theta + 1)$$
$$x_2 = \frac{c}{\lambda^2 + 1}(\lambda \cos\theta \operatorname{Sinh}\lambda\theta + \sin\theta \operatorname{Cosh}\lambda\theta)$$

5. Find the natural equation of a curve indirectly similar to its nth evolute.

6. The radius of curvature of a parabola $x_2 = 2px_1^2$ is $\rho = \cos^{-3}\theta/4p$. Show that all parabolas are similar to one another. How does p change in a homothety?

7. The natural equation of a catenary is $\rho = a + s^2/a$ (exercise 2-2, Prob. 3*f*). Show that all catenaries are similar to one another. How is the parameter a changed in a homothety of ratio λ?

8. Use Prob. 7 to show that all curves with the natural equation $\rho = a \operatorname{Cosh} s/a$ are similar to one another.

9. Show that similar curves have similar evolutes.

10. Two ellipses $x_1^2/a^2 + x_2^2/b^2 = 1$ and $x_1^2/A^2 + x_2^2/B^2 = 1$ are homothetic if and only if $a/b = A/B$. Prove this statement and deduce from it and Prob. 9 an analogous statement on astroids $(Ax_1)^{\frac{2}{3}} + (Bx_2)^{\frac{2}{3}} = 1$ (example 3-2).

11. Find all curves similar to their involutes.

12. Define the nth involute of a curve and discuss its relation to the given curve.

13. Show that all hyperbolic spirals are pairwise similar (exercise 2-2, Prob. 4).

3-5. The Moving Plane

We have studied curves by means of the frame $\{\mathbf{t},\mathbf{n}\} = A(s)\{\mathbf{i},\mathbf{j}\}$ moving along it. We assume now that a plane Π_f is rigidly connected with the moving frame $\{\mathbf{t},\mathbf{n}\}$ and that it glides over the fixed plane Π of the fixed frame $\{\mathbf{i},\mathbf{j}\}$. We may imagine that a point in Π moves along a curve $\mathbf{g}$. If that point is represented by a graphite pencil and if Π_f is a sheet of paper, the moving point will trace a curve $\boldsymbol{\gamma}$ in Π_f. We want to

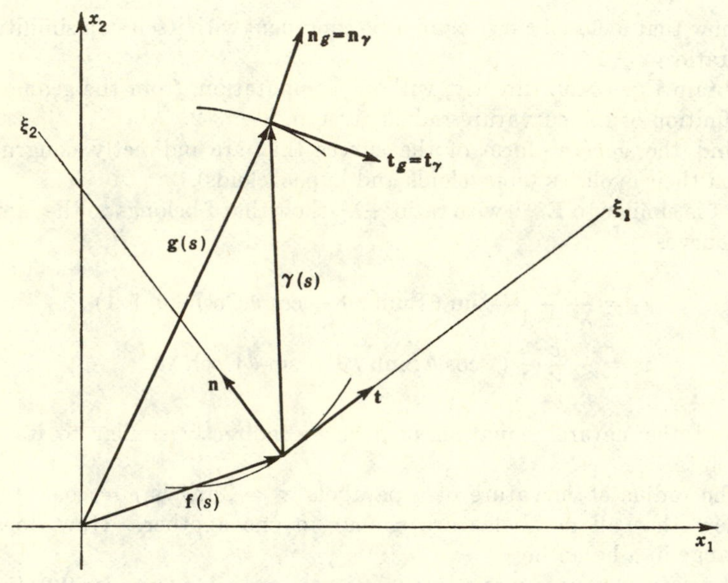

Fig. 3-11

find $\boldsymbol{\gamma}$, given $\mathbf{f}$ and $\mathbf{g}$. This problem appears in the theory of dented gears. There we have two planes fixed to the axes of the gears. If both planes rotate with appropriate angular velocities, the curves bounding the two gears are traces of one another. The kinematic problem is the same if we keep one of the planes at rest and rotate the second plane about it.

We select our notations. Roman letters will be used for all coordinates in Π, and Greek ones for coordinates in Π_f. In order to avoid cumbersome indices, the symbol of a curve will also be used to denote its coordinates, $\mathbf{g} = (g_1,g_2)$, $\boldsymbol{\gamma} = (\gamma_1,\gamma_2)$. The curve $\mathbf{f}$ which defines the moving plane has coordinates (f_1,f_2) in Π and $(0,0)$ in Π_f. The arc length s_f of $\mathbf{f}$ will serve as a parameter also for $\mathbf{g}$; it is not, in general, the arc length of $\mathbf{g}$. The arc length and curvature of $\boldsymbol{\gamma}$ will be denoted by s_γ and k_γ, as usual. The frame matrices of $\mathbf{f}$ and $\mathbf{g}$ with respect to $\{\mathbf{i},\mathbf{j}\}$ are $F(s_f)$ and $G(s_g)$;

the frame matrix of $\boldsymbol{\gamma}$ with respect to $\{\mathbf{t},\mathbf{n}\}$ is $\Gamma(s_\gamma)$. The moving frames of $\Gamma(s_f)$ and $G(s_f)$ coincide for every value of s_f (Fig. 3-11); hence

$$G(s_f) = \Gamma(s_f)F(s_f) \tag{3-31}$$

and, by Eq. (1-5),

$$C(G(s_g))\,\frac{ds_g}{ds_f} = C(\Gamma(s_\gamma))\,\frac{ds_\gamma}{ds_f} + \Gamma(s_f)C(F(s_f))\Gamma(s_f)^{-1}$$

All matrices in Eq. (3-31) are rotations; hence, by the statement preceding Def. 1-7,

$$k_g\,\frac{ds_g}{ds_f} = k_\gamma\,\frac{ds_\gamma}{ds_f} + k_f \tag{3-32}$$

The vector $\boldsymbol{\gamma}$ is given in the plane Π_f. Its trace in the plane Π is the vector $\boldsymbol{\gamma}F = (\gamma_1,\gamma_2)F\{\mathbf{i},\mathbf{j}\}$. Then (Fig. 3-11)

$$\mathbf{g}(s_f) = \mathbf{f}(s_f) + \boldsymbol{\gamma}(s_f)F(s_f) \tag{3-33}$$

Hence

$$\frac{d\mathbf{g}}{ds_f} = \frac{d\mathbf{f}}{ds_f} + \boldsymbol{\gamma}(s_f)C(F(s_f))F(s_f) + \frac{d\boldsymbol{\gamma}}{ds_f}F(s_f) \tag{3-34}$$

Let us now choose the moving frame at a point of $\mathbf{f}$ as the fixed frame $\{\mathbf{i},\mathbf{j}\} = \{\mathbf{t}(s_0),\mathbf{n}(s_0)\}$. Equation (3-34) becomes for $s_f \to s_0$

$$\frac{d\mathbf{g}}{ds_f} = \frac{d\boldsymbol{\gamma}}{ds_f} + k_f(s_f)\boldsymbol{\gamma}(s_f)J + \mathbf{i} \tag{3-35a}$$

or, in coordinates,

$$\frac{dg_1}{ds_f} = \frac{d\gamma_1}{ds_f} - k_f\gamma_2 + 1 \qquad \frac{dg_2}{ds_f} = \frac{d\gamma_2}{ds_f} + k_f\gamma_1 \tag{3-35b}$$

The form of Eqs. (3-35) is not changed in a euclidean motion. Hence it represents the differential equation of the transition from $\boldsymbol{\gamma}$ to $\mathbf{g}$. By construction, $\mathbf{g}(s_0) = \boldsymbol{\gamma}(s_0)$; hence also

$$\frac{d\boldsymbol{\gamma}}{ds_f} = \frac{d\mathbf{g}}{ds_f} - k_f\mathbf{g}J - \mathbf{i} \tag{3-36}$$

By Eqs. (3-35), the arc lengths are related by

$$\left(\frac{ds_g}{ds_f}\right)^2 = \left(\frac{d\gamma_1}{ds_f} - k_f\gamma_2 + 1\right)^2 + \left(\frac{d\gamma_2}{ds_f} + k_f\gamma_1\right)^2$$

or, in differential form,

$$ds_g{}^2 = ds_\gamma{}^2 + 2\left[k_f\begin{Vmatrix}\gamma_1 & \gamma_2\\ d\gamma_1 & d\gamma_2\end{Vmatrix} + d\gamma_1\right]ds_f + (1 + k_f{}^2|\boldsymbol{\gamma}|^2)\,ds_f{}^2 \tag{3-37}$$

$\|\ \|$ denotes the determinant. Equations (3-32) and (3-37) determine the natural equation of one of the three curves **f**, **g**, $\boldsymbol{\gamma}$ if the two others are given.

***Example* 3-5.** Find the trajectories in Π of the points *fixed* in Π_f.

By hypothesis, the curve $\boldsymbol{\gamma}$ reduces to a point, $d\boldsymbol{\gamma} = 0$. Hence, by (3-36),

$$\frac{dg_1}{ds_f} = 1 - k_f g_2 \qquad \frac{dg_2}{ds_f} = k_f g_1 \tag{3-38}$$

are the differential equations of the trajectories. Is it possible for a point to be fixed simultaneously in both planes? By Eq. (3-38), the fixed point must be $(0,1/k_f)$. This implies that k_f is constant; the only case in which a point of Π_f always covers the same point of Π is that of the center of a circle **f**, the plane Π_f moving along the circumference of the circle.

***Example* 3-6.** A point in the Π_f plane moves along the ξ_1 axis with velocity -1. Find its trace in the Π plane.

By hypothesis, $\boldsymbol{\gamma} = (-s_f,0)\{\mathbf{t},\mathbf{n}\}$. By Eqs. (3-32) and (3-37),

$$ds_g = |k_f| s_f \, ds_f \qquad k_g = \frac{1}{s_f}$$

that is, $\mathbf{g} = \mathbf{I}_f$.

***Example* 3-7.** Find the natural equation of a conic.

The equation of a conic in any one of its moving frames (Fig. 3-12) is of the form

$$\xi_2 = \tfrac{1}{2}(a\xi_1^2 + 2b\xi_1\xi_2 + c\xi_2^2) \tag{3-39}$$

since it must be given by a quadratic equation, and for $\xi_1 = 0$

$$\xi_2 = \frac{d\xi_2}{d\xi_1} = 0$$

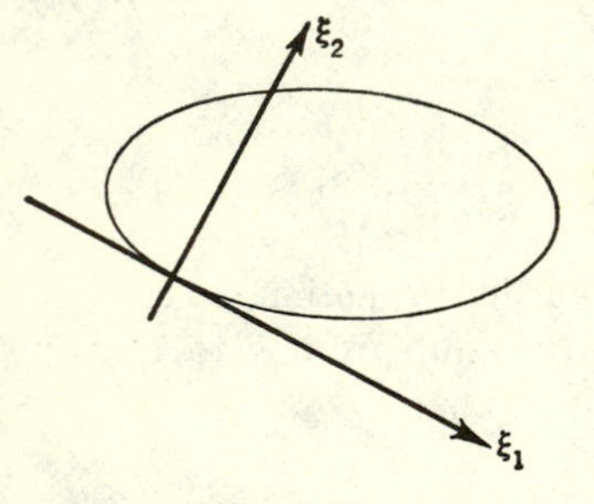

Fig. 3-12

The coefficients a, b, c are functions of the arc length $s = s_f$ of the conic. The ellipse is the trace in Π_f of the origin of the coordinates in Π if for some value $s = s_0$ the frames in both planes coincide. (ξ_1,ξ_2) satisfy Eq. (3-36) for $dg/ds = 0$:

$$\frac{d\xi_1}{ds} = k\xi_2 - 1 \qquad \frac{d\xi_2}{ds} = -k\xi_1$$

We differentiate Eq. (3-39) and use the preceding equations to eliminate ξ_1' and ξ_2':

$$(a - k)\xi_1 + b\xi_2 = \tfrac{1}{2}[(a' - 2bk)\xi_1^2 + 2(ak + b' - ck)\xi_1\xi_2 + (c' + 2bk)\xi_2^2]$$

Since we may use any value of s for s_0, this equation again must represent the same conic as does Eq. (3-39). Hence

$$\begin{aligned} a &= k \\ a' - 2bk &= ab \\ b' + ak - ck &= b^2 \\ c' + 2bk &= cb \end{aligned}$$

From the second equation we obtain $b = k'/3k = (\log k^{\frac{1}{3}})'$. The two *invariants* of the quadratic form on the right-hand side of Eq. (3-39) are

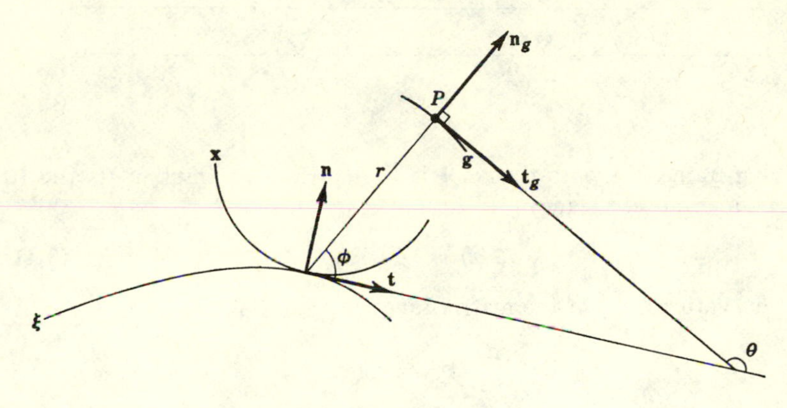

Fig. 3-13

the *trace* $T = a + c$ and the *determinant* $D = ac - b^2$. For these we have $(\log T)' = b = (\log k^{\frac{1}{3}})'$, $(\log D)' = 2b = (\log k^{\frac{2}{3}})'$. Hence $T = Ak^{\frac{1}{3}}$, $D = Bk^{\frac{2}{3}}$ with *constant* factors A, B. The algebraic identity $b^2 = -D + aT - a^2$ finally yields the differential equation

$$\frac{k'^2}{9k^2} = Ak^{\frac{4}{3}} - Bk^{\frac{2}{3}} - k^2$$

or, in terms of the curvature radius ρ,

$$\frac{1}{9}\left(\frac{d\rho}{ds}\right)^2 = A\rho^{\frac{2}{3}} - B\rho^{\frac{4}{3}} - 1 \tag{3-40}$$

An important field of application of the theory of the moving plane is the study of roulettes. There we have two C^2 curves, $\mathbf{x}$ and $\boldsymbol{\xi}$, each in its own plane, Π_x and Π_ξ. Π_x will be kept at rest, but Π_ξ will be moved in such a way that the moving frames of $\mathbf{x}$ and $\boldsymbol{\xi}$ coincide for every value of $s = s_x = s_\xi$ (Fig. 3-13).

Definition 3-24. *The roulettes of* $\boldsymbol{\xi}$ *on* $\mathbf{x}$ *are the traces* $\mathbf{g}$ *in* Π_x *of the points* P in Π_ξ.

A familiar example of roulettes is that of the different kinds of cycloids generated by a circle rolling on a line (Fig. 3-14).

We fix some cartesian system of coordinates both in Π_x and in Π_ξ. Let $X(s)$, $\Xi(s)$, $k_x(s)$, $k_\xi(s)$ be the frame matrices and the curvatures of $\mathbf{x}$ and $\boldsymbol{\xi}$ in their respective planes. The rotation of Π_x which brings the

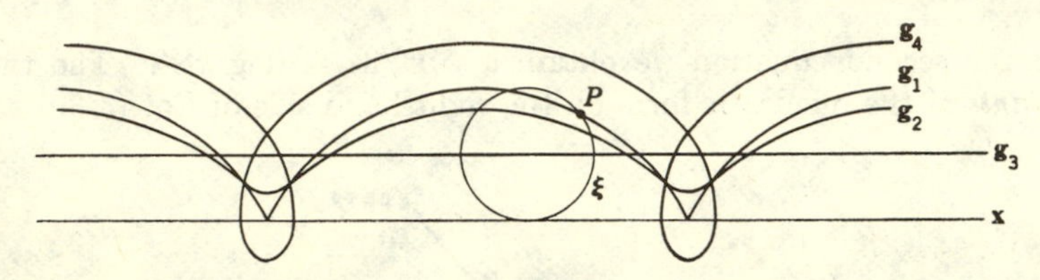

Fig. 3-14

moving frame of $\mathbf{x}$ onto that of $\boldsymbol{\xi}$ is $Q(s)$. Here we think of Π_x and Π_ξ as of superimposed planes. In formulas,

$$\Xi(s) = Q(s)X(s) \tag{3-41}$$

The curvature $k_Q(s)$ of the matrix function $Q(s)$, defined by

$$\frac{dQ}{ds}Q^{-1} = k_Q(s)J$$

is

$$k_Q(s) = k_\xi(s) - k_x(s) \tag{3-42}$$

The coordinates of $\mathbf{g}$ are γ_1, γ_2 in the moving frame common to $\mathbf{x}$ and $\boldsymbol{\xi}$, and g_1, g_2 in the fixed frame of Π_x. The latter frame will coincide with the moving one for $s = s_0$. Again, we compute the differential equations for $s = s_0$, using $\gamma_i(s_0) = g_i(s_0)$, $i = 1, 2$. The point P is fixed in Π_ξ; hence (see example 3-7)

$$\frac{d\gamma_1}{ds} = k_\xi\gamma_2 - 1 \qquad \frac{d\gamma_2}{ds} = -k_\xi\gamma_1$$

and, by Eq. (3-35*b*),

$$\frac{dg_1}{ds} = k_Q g_2 \qquad \frac{dg_2}{ds} = -k_Q g_1 \tag{3-43}$$

We also use polar coordinates (Fig. 3-13):

$$r(s) = (\gamma_1^2 + \gamma_2^2)^{\frac{1}{2}} \qquad \phi = \arctan\frac{\gamma_2}{\gamma_1}$$

From Eq. (3-43) we see that

$$ds_g = |k_Q(s)|r(s)\,ds \tag{3-44}$$

and that the *normal to the roulette at P is the unit vector* $\mathbf{g}/r = (\gamma_1/r, \gamma_2/r)$. For the curvature of g we obtain from Eq. (3-32)

$$k_g \frac{ds_g}{ds} = k_x + \frac{d\theta}{ds}$$

θ is the tangent angle of $\mathbf{g}$ in the moving frame,

$$\frac{d\theta}{ds} = \frac{d(\phi + \pi/2)}{ds} = \frac{\gamma_2 - k_\xi(\gamma_1^2 + \gamma_2^2)}{\gamma_1^2 + \gamma_2^2} = -k_\xi + \frac{\sin\phi}{r}$$

Hence

$$k_g |k_Q| r = -k_Q + \frac{\sin\phi}{r}$$

If we put $\varepsilon = \operatorname{sign} k_Q$, the curvature of the roulette is

$$k_g = -\frac{\varepsilon}{r} + \frac{\sin\phi}{r^2 k_Q} \tag{3-45a}$$

and its curvature radius satisfies

$$\frac{1}{r} - \frac{\varepsilon}{\rho_g + \varepsilon r} = \frac{k_Q}{\sin\phi} \tag{3-45b}$$

This formula, due to L. Euler, leads to a simple construction of ρ_g. We formulate it in the case $\operatorname{sign} k_x \neq \operatorname{sign} k_\xi$ shown in Fig. 3-13.

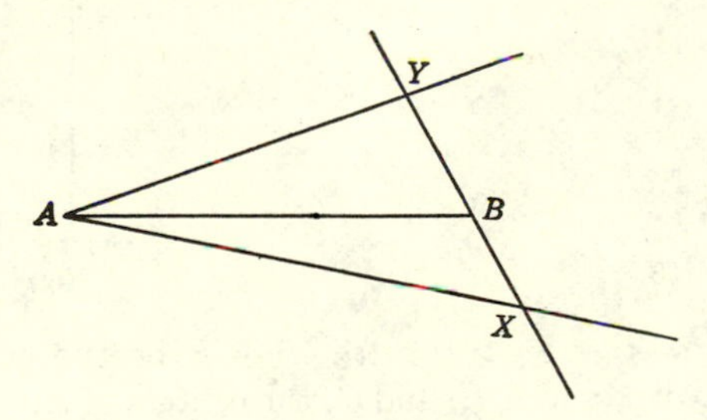

Fig. 3-15

Lemma 3-25. *Let X and Y be the points of intersection with the legs of an angle A of a transversal of that angle through a fixed point B. The quantity*

$$\left(\frac{1}{|XB|} + \frac{1}{|YB|}\right) \csc \angle ABX$$

is independent of the transversal (Fig. 3-15).

By the sine theorem,

$$\frac{\sin \angle XAB}{|XB|} = \frac{\sin (\angle ABX + \angle BAX)}{|AB|}$$

$$\frac{\sin \angle BAY}{|BY|} = \frac{\sin (\angle ABX - \angle BAY)}{|AB|}$$

or
$$\left(\frac{1}{|XB|} + \frac{1}{|BY|}\right) \csc \angle ABX = \frac{1}{|AB|} \frac{\sin (\angle XAB + \angle BAY)}{\sin \angle XAB \sin \angle BAY}$$

The right-hand side of this equation depends only on the angle A and the point B.

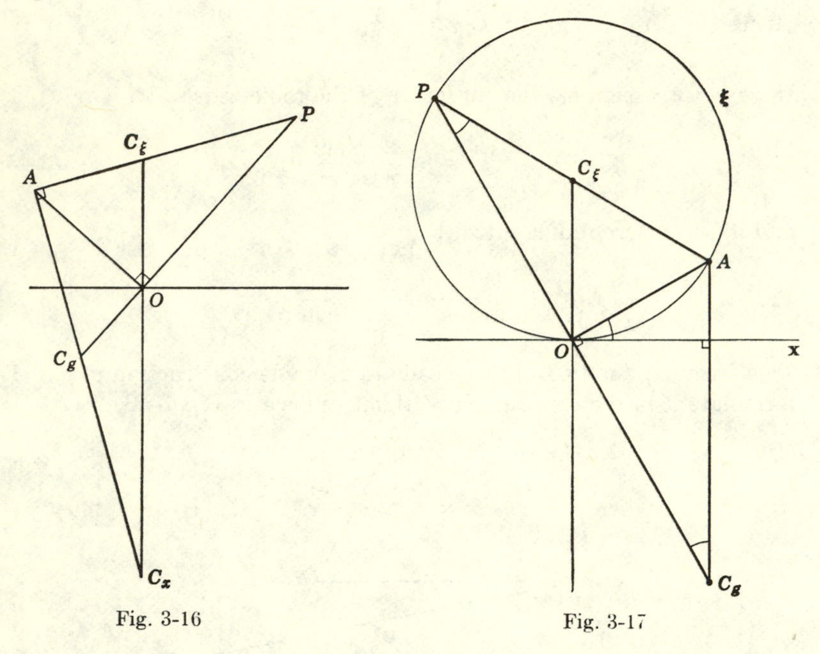

Fig. 3-16

Fig. 3-17

We assume now that $k_\xi > 0 > k_x$. Let P be the point on **g**, O the point of contact of **ξ** and **x**, C_ξ and C_x the centers of curvature of **x** and **ξ**. The perpendicular in O to OP meets PC_ξ in A. *PO meets AC_x in the center of curvature C_g of* **g** (Fig. 3-16). In fact, we know that C_g must be on the line PO. By lemma 3-25, applied to $\angle PAC_x$ and point O,

$$\left(\frac{1}{|PO|} + \frac{1}{|OC_g|}\right) \csc \angle AOP = \left(\frac{1}{|C_xO|} + \frac{1}{|OC_\xi|}\right) \csc \angle AOC$$

Now in our case, $\angle AOP = \pi/2$, $\angle AOC = \phi$, $|PC_g| = -\rho_g$, $|OC_x| = -\rho_x$, $|OC| = \rho_\xi$; the formula reduces to (3-45*b*).

***Example* 3-8.** The construction of the center of curvature of a cycloid is shown in Fig. 3-17. The cycloid is generated by the movement of the point P on the circle $\boldsymbol{\xi}$ which rolls on the line $\mathbf{x}$. Since

$$\angle \mathbf{x}OA = \angle APO = \angle AC_gO$$

the right triangles ΔAPO and ΔAC_gO are congruent: The radius of curvature of a cycloid is twice the distance of P from the point of contact O of circumference and line.

The mathematical interest of the theory of roulettes lies in the fact that formula (3-45*b*) leads in many cases to a simple discussion of their curvature properties. *Every C^2 curve can be obtained as a roulette;* in fact, Eqs. (3-44) and (3-45) show that, if two of the three curves $\mathbf{x}$, $\boldsymbol{\xi}$, $\mathbf{g}$ are given, the third may be found. In the theory of dented gears this means that one profile may be given arbitrarily.

Formula (3-45) gives information on singular points of roulettes.

In a cusp, $\rho_g = 0$ by definition. If $r \neq 0$, this implies $k_Q = 0$.

Theorem 3-26 (La Hire). *If $k_Q \neq 0$, only the roulettes of points of $\boldsymbol{\xi}$ may have cusps and only where the roulette meets $\mathbf{x}$.*

The cycloids are a nice example for this theorem.

At an inflection point, $k_g = 0$; hence $r/\sin \phi = \varepsilon/k_Q$. If $k_Q \neq 0$, the locus of the points (r,ϕ) is a circle with its center on the common normal to $\mathbf{x}$ and $\boldsymbol{\xi}$ and diameter ε/k_Q, that is, in the upper half plane if $\varepsilon > 0$ and in the lower one for $\varepsilon < 0$. If we rewrite Eqs. (3-45) as

$$k_g = \frac{1}{rk_Q}\left(\frac{\sin \phi}{r} - \varepsilon k_Q\right)$$

we see that k_g changes sign if the roulette crosses the circle $r/\sin \phi = \varepsilon/k_Q$. This circle, *La Hire's circle,* is therefore the locus of inflections for $s = s_0$.

Theorem 3-27 (La Hire). *In the movement of $\boldsymbol{\xi}$ on $\mathbf{x}$, the locus of inflections of all roulettes $\mathbf{g}(s)$ for $s = s_0$ is a circle with diameter $1/k_Q(s_0)$ and tangent to the common tangent of $\mathbf{x}(s_0)$ and $\boldsymbol{\xi}(s_0)$, with the possible exception of the point of contact.*

Exercise 3-5

1. Find the trajectories in Π_f for all points of Π in case $\mathbf{f}$ is a circle (see example 3-5).

2. A line in Π, $x_1 \cos \omega - x_2 \sin \omega + p = 0$, has as its trace in Π_f a line for all values of s, $\xi_1 \cos \alpha(s) - \xi_2 \sin \alpha(s) + \pi(s) = 0$. Show that $\alpha' = -k_f(s)$, $\pi' = \cos \alpha(s)$. [HINT: All points (ξ_1,ξ_2) are images of points at rest in Π. Use the equations from example 3-7.]

◆ **3.** The trace in Π of a fixed line in Π_f is a line

$$x_1 \cos \omega(s) - x_2 \sin \omega(s) + p(s) = 0$$

Compute ω' and p'. (See Prob. 2.)

4. Show that the envelope of the family of lines discussed in Prob. 3 is obtained by the elimination of s from the equation $L(x_1,x_2) = 0$ of the line and $(1 - k_f x_2)(\partial L/\partial x_1) + k_f x_1(\partial L/\partial x_2) = 0$.

5. Use Prob. 4 to show that the envelope of the lines, traces of the normal to $\mathbf{f}$ ($\xi_1 = 0$), is $\mathbf{E}_f$.

6. Prove that

$$ds_\gamma{}^2 = ds_g{}^2 - 2\left(k_f \left\| \begin{matrix} g_1 & g_2 \\ dg_1 & dg_2 \end{matrix} \right\| + dg_1\right) ds_f + (1 + k_f{}^2|g|^2)\, ds_f{}^2$$

7. Prove that, in polar coordinates, (3-38) becomes

$$\frac{dr_g}{ds} = -\cos\phi_g \qquad \frac{d\phi_g}{ds} = k_f + \frac{\sin\phi_g}{r_g}$$

8. If $\mathbf{f}$ is a straight line, show that the trajectories in Π_f of points in Π are all straight lines.

9. Use example 3-7, Eq. (3-40), to show that the radius of curvature of the evolute of a conic is given by $\rho_E = -3b\rho^2$.

10. Find the equation of a circle in any one of its moving frames.

11. Let $(r,\phi),(r^*,\phi^*)$ be the polar coordinates of the foci of an ellipse in one of its moving frames. In elementary geometry it is shown that $\phi + \phi^* = \pi$. Use Prob. 7 to show that

$$2k = \left(\frac{1}{r} + \frac{1}{r^*}\right)\sin\phi$$

For a hyperbola, $\phi + \phi^* = 0$. Find the corresponding formula for $2k$ (L'Hospital).

12. The *inverse movement* to that of $\boldsymbol{\xi}$ on $\mathbf{x}$ is obtained if the plane of $\boldsymbol{\xi}$ is kept unmoved and $\mathbf{x}$ is moved along $\boldsymbol{\xi}$. This movement is characterized by the matrix $Q^{-1}(s)$. Prove that, if C_g is the center of curvature of a roulette $\mathbf{g}$ at a point P, then P is the center of curvature of the roulette image of C_g in the inverse movement.

13. Discuss the validity of the construction of C_g in the case that $\boldsymbol{\xi}$ and $\mathbf{x}$ are both convex in the same direction.

14. A fixed line in Π_ξ defines the envelope $\mathcal{E}$ of its traces in Π_x. Use Prob. 2 to obtain for the envelope

$$\frac{ds_{\mathcal{E}}}{ds} = \sin\phi$$

$$\frac{d}{ds}(x_{1\mathcal{E}},x_{2\mathcal{E}}) = \left(\frac{1}{r}\sin\phi + k_Q\right)(x_{2\mathcal{E}},-x_{1\mathcal{E}})$$

$$\rho_{\mathcal{E}} = r + \frac{\sin\phi}{k_Q}$$

Use the last equation to show that the locus of all points at which the envelope may have cusps is a circle, either identical or symmetric to La Hire's circle with respect to the common tangent of **x** and **ξ** (Aronhold).

15. If a point P is on La Hire's circle for all s, show that its roulette is a straight line (La Hire).

16. If a circle of radius r is rolling on a straight line, La Hire's circle has radius $r/2$. The roulettes of the points not on the circumference are called *trochoidal* curves (hypotrochoid: the roulette of a point inside the circumference; epitrochoid: outside the circumference). Which kind of trochoids may have inflections?

17. A *cycloidal* curve is defined by the movement of a circle of radius r on a circle of radius R. Show that its natural equation is

$$\frac{s^2}{[4r(R+r)/R]^2} + \frac{\rho^2}{[4r(R+r)/(R+2r)]^2} = 1$$

for points on the moving circle. Which cycloidal curves may have inflections?

18. The natural equation of a cycloidal curve in the general case is $s^2/a^2 + \rho^2/b^2 = 1$. Check this statement by integration.

19. In the problem of dented gears, two circles (M_1) and (M_2) turn with *constant* angular velocities ω_1 and ω_2 (Fig. 3-18). This movement cannot be used directly for the transmission of forces, since the smooth surfaces would start to glide if some resistance occurs to the movement of one of the circles. Therefore we fix a curve **x** in the plane Π_x of the circle (M_1), and we look for a curve **ξ** in the plane Π_ξ of (M_2) such that the movement of **ξ** on **x** should induce the movement of (M_2) on (M_1). If the radius of (M_i) is r_i, $\omega_1 r_1 = \omega_2 r_2$. The rotations of the circles are given by

$$X = \begin{pmatrix} \cos\omega_1 t & \sin\omega_1 t \\ -\sin\omega_1 t & \cos\omega_1 t \end{pmatrix}$$

$$\Xi = \begin{pmatrix} \cos\omega_2 t & \sin\omega_2 t \\ -\sin\omega_2 t & \cos\omega_2 t \end{pmatrix}$$

$Q(t)$ may be computed. (M_2) will be the roulette of (M_1); hence the

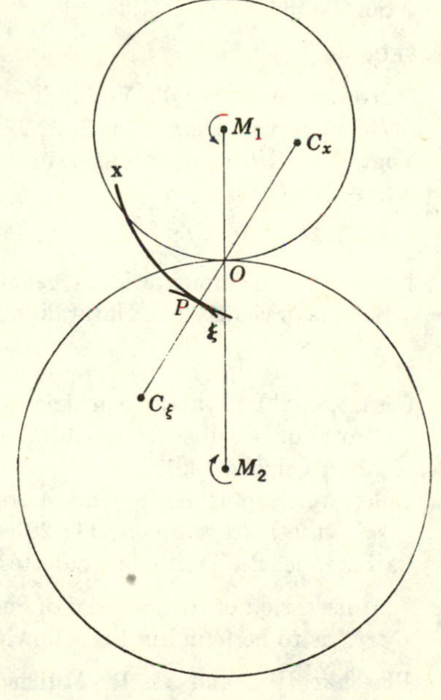

Fig. 3-18

normal at the point of contact of **x** and **ξ** must pass through the point of contact O of the two circles at the same moment. Show that **ξ** is uniquely determined.

20. Given C_x, C_ξ, P, C_g, construct the direction of the tangent common to **x** and **ξ** (Bobilier).

21. Find the natural equation of the roulettes of a logarithmic spiral on a line.

22. State explicitly the differentiability conditions under which the formulas of the preceding section are valid.

References

SEC. 3-1

A detailed discussion of the Stieltjès integral is found in the following:

Widder, D. V.: "Advanced Calculus," 2d ed., Prentice-Hall, Inc., Englewood Cliffs, N.J., 1961.

SEC. 3-2

Huygens, C.: "Horologium oscillatorium," "Oeuvres complètes," vol. 18, Martinus Nijhoff, La Haye, 1934.

Ostrowski, A.: Un applicazione dell'integrale di Stieltjès alla teoria elementare delle curve piane, *Atti Accad. Naz. Lincei, Rend., Classe Sci. Fis. Mat. Nat.*, Ser. 8, **18**: 373–375 (1955).

SEC. 3-3

Ostrowski, A.: Über die Verbindbarkeit von Linien- und Krümmungselementen, *Enseignement Math.*, Ser. 2, **2**: 277–292 (1956).

Vogt, W.: Über monotongekrümmte Kurven, *J. Reine Angew. Math.*, **144**: 239–248 (1914).

SEC. 3-4

Euler, L.: "Commentationes geometricae," "Opera Omnia," Ser. I., vol. 27, Societas Scientiarum Naturalium Helvetica, Lousannae et Turici, 1954.

SEC. 3-5

Cesàro, E.: "Lezioni di geometria intrinseca," Naples, 1896; German translation: "Vorlesungen über natürliche Geometrie," B. G. Teubner Verlagsgesellschaft, mbH, Leipzig, 1901.

Euler, L.: Supplementum de figura dentium rotarum, *Novi commentarii accad. sci. imp. Petropolitanae*, **11**: 207–231 (1767).

La Hire, de, P.: Traité des roulettes, *Mém. Acad. Royale Sci.*, 1706, pp. 439–499.

A discussion of the material of Sec. 3-5 from the viewpoint of "integral geometry" is to be found in the following:

Blaschke, W., and H. R. Müller: "Ebene Kinematik," R. Oldenbourg KG, Munich, 1956.

4

CALCULUS OF VARIATIONS

4-1. Euler Equations

Many interesting problems in differential geometry deal with curves, surfaces, etc., characterized by extremal properties.

***Example* 4-1.** The segment between two points $A = (a_1,a_2)$, $B = (b_1,b_2)$ is the C^1 curve $\mathbf{x} = (x_1(t),x_2(t))$, $x_i(0) = a_i$, $x_i(1) = b_i$, $i = 1, 2$, making a minimum the integral

$$I = \int_0^1 ds(t) = \int_0^1 (\dot{x}_1(t)^2 + \dot{x}_2(t)^2)^{\frac{1}{2}}\, dt$$

***Example* 4-2.** A famous problem in mechanics is that of the *brachistochrone* (*βραχύς* = short, *χρόνος* = time). Two points A, B are given in a vertical plane (Fig. 4-1). We are asked to find a curve such that a

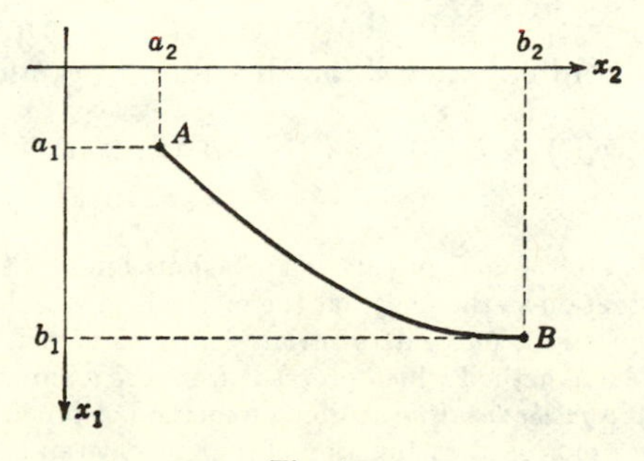

Fig. 4-1

material point will reach B from A in minimum time if it falls along the curve under the influence of a constant gravitational force acting in the direction of the $+x_1$ axis. Mathematically speaking, we look for a C^1 curve with boundary conditions, as in the preceding example, that will make

$$\int_0^1 \left[\frac{\dot{x}_1^2 + \dot{x}_2^2}{2g(x_1 - a_1)}\right]^{\frac{1}{2}} dt$$

a minimum.

***Example* 4-3.** In hyperbolic (non-euclidean) geometry, the role of straight lines is played by the curves for which

$$\int_{t_0}^{t_1} \frac{(\dot{x}_1^2 + \dot{x}_2^2)^{\frac{1}{2}}}{x_2}\, dt$$

is minimal.

Most problems in physics and many problems in differential geometry may be formulated in terms of such *variational* problems in which a curve is sought that makes an integral $\int_{t_0}^{t_1} F(x_1(t),x_2(t),\dot{x}_1(t),\dot{x}_2(t))\, dt$ a maximum or a minimum within a certain class of *admissible* curves $\mathbf{x}(t)$. The function $F(a,b,c,d)$ which appears under the integral must have continuous second derivatives in all four variables.

We shall reduce the problem of the calculus of variations to the solution of a certain system of differential equations. In order to derive the equations, we shall assume that the proposed problem has a solution. If no solution exists, our hypothesis is false, and from a wrong premise no valid conclusion may be deduced. Therefore no solution of a variational problem is valid without a proof of the existence of such a solution.†

We assume that there exists a C^1 curve $\mathbf{x}(t) = (x_1(t),x_2(t))$, $\mathbf{x}(0) = A$, $\mathbf{x}(1) = B$, for which the integral

$$I = \int_0^1 F(t,x_1(t),x_2(t),\dot{x}_1(t),\dot{x}_2(t))\, dt \tag{4-1}$$

is extremal. If we take any two C^1 functions $\xi_i(t)(i = 1,2)$ that vanish in 0 and 1, $\xi_i(0) = \xi_i(1) = 0$, the curves $x^*(t) = (x_1 + \epsilon_1\xi_1,\ x_2 + \epsilon_2\xi_2)$ will still have endpoints A and B for all values of ϵ_i, and the integral

$$I(\epsilon_1,\epsilon_2) = \int_0^1 F(t,\ x_1(t) + \epsilon_1\xi_1(t),\ x_2(t) + \epsilon_2\xi_2(t),\ \dot{x}_1(t) + \epsilon_1\dot{\xi}_1(t),\ \dot{x}_2(t) + \epsilon_2\dot{\xi}_2(t))\, dt$$

† This situation is often found in calculus. Leibniz himself, the principal founder of calculus, drew attention to the fact that the methods of calculus are excellent tools for a *methodus inveniendi*, a method to invent solutions of mathematical problems. He stated that any such method which proceeds from the assumption of the existence of a solution is never a *methodus demonstrandi*, a method of proof. This explains why the ϵ,δ technique of proof used in function theory belongs to a world of ideas completely different from that of the computational aspect of calculus. It is not surprising to see the same situation also in the calculus of variations.

by hypothesis, is extremal for $\epsilon_1 = \epsilon_2 = 0$. Hence

$$\left.\frac{\partial I(\epsilon_1,\epsilon_2)}{\partial \epsilon_1}\right|_{\epsilon_1=\epsilon_2=0} = \int_0^1 \xi_1(t) \frac{\partial F}{\partial x_1} (t,x_1(t),x_2(t),\dot{x}_1(t),\dot{x}_2(t))\, dt + \int_0^1 \dot{\xi}_1(t) \frac{\partial F}{\partial \dot{x}_1} (t,x_1(t),x_2(t),\dot{x}_1(t),\dot{x}_2(t))\, dt = 0$$

The second integral is transformed through integration by parts:

$$\int_0^1 \dot{\xi} \frac{\partial F}{\partial \dot{x}}\, dt = -\int_0^1 \xi \frac{d}{dt}\frac{\partial F}{\partial \dot{x}}\, dt$$

The condition for the existence of an extremum is then

$$\int_0^1 \xi_1(t) \left[\frac{\partial F}{\partial x_1} - \frac{d}{dt}\frac{\partial F}{\partial \dot{x}_1}\right] dt = 0$$

which must hold for *any* C^1 function $\xi(t)$. The quantity in brackets must vanish, by the following lemma:

Lemma 4-1. *If* $\int_0^1 \xi(t)f(t)\, dt = 0$ *for continuous* $f(t)$ *and all differentiable (or continuous)* $\xi(t)$, *then* $f(t) = 0$.

PROOF: If there is a τ, $0 < \tau < 1$, such that $f(\tau) \neq 0$, say $f(\tau) > 0$, then by continuity there exists a whole interval $t_0 < \tau < t_1$ in which $f(t)$ is positive. Choose now

$$\begin{aligned} \xi(t) &= 0 & t < t_0 \quad &\text{or} \quad t_1 < t \\ \xi(t) &= (t - t_0)^2(t_1 - t)^2 & t_0 \leq t &\leq t_1 \end{aligned}$$

Then clearly $\int_0^1 \xi(t)f(t)\, dt > 0$. Hence no such τ may exist.

We have shown that an extremal curve of (4-1) satisfies

$$\frac{\partial F}{\partial x_1} - \frac{d}{dt}\frac{\partial F}{\partial \dot{x}_1} = 0 \tag{4-2a}$$

In the same way, $\partial I(\epsilon_1,\epsilon_2)/\partial \epsilon_2 = 0$ yields

$$\frac{\partial F}{\partial x_2} - \frac{d}{dt}\frac{\partial F}{\partial \dot{x}_2} = 0 \tag{4-2b}$$

Equations (4-2) are the *Euler equations* of the calculus of variations.

***Example* 4-1 (*continued*).** $F = (\dot{x}_1^2 + \dot{x}_2^2)^{\frac{1}{2}}$ does not depend explicitly on x_1 or x_2. In this case it is of advantage to take one of them, for example, x_1, as an independent variable. Equation (4-2*a*) becomes trivial ($\dot{x}_1 = dx_1/dx_1 = 1$), and (4-2*b*) is

$$\frac{d}{dx_1}\frac{\dot{x}_2}{(1 + \dot{x}_2^2)^{\frac{1}{2}}} = 0$$

or $x_2 = Cx_1 + C^*$. If there is a shortest curve between two points, it must be a straight segment.

***Example* 4-2 (*continued*).** The extremal property does not depend on the value of the gravitational constant $2g$. We take the origin in A, and x_1 as the independent variable. The Euler equation (4-2*b*) of this problem is

$$\frac{d}{dx_1} \frac{\dot{x}_2}{[x_1(1 + \dot{x}_2^2)]^{\frac{1}{2}}} = 0$$

The second factor on the left-hand side of the integrated equation

$$\frac{1}{\sqrt{x_1}} \frac{\dot{x}_2}{(1 + \dot{x}_2^2)^{\frac{1}{2}}} = C$$

is $\sin \theta$, where, as always, θ is the tangent angle. If we put $a = \frac{1}{2}C^{-2}$, the integral curve is given by

$$x_1 = 2a \sin^2 \theta = a(1 - \cos 2\theta) \tag{4-3a}$$

and

$$\frac{dx_2}{d\theta} = \frac{dx_2}{dx_1}\frac{dx_1}{d\theta} = \frac{dx_1}{d\theta} \tan \theta = 4a \sin^2 \theta = 2a(1 - \cos 2\theta)$$

Hence

$$x_2 = a(2\theta - \sin 2\theta) \tag{4-3b}$$

The brachistochrones are cycloids.

It remains to show that these cycloids really give a minimum time for the fall from A to B. Euler's teacher John Bernoulli gave a proof for the brachistochrone in 1718 (the Euler equations date from 1741). The same idea in an analytical setting was used 150 years later by Weierstrass to give existence proofs for a wide class of variational problems.

John Bernoulli started out to look for the *curves of equal duration of fall* from A. For all cycloids starting from A with $\theta = 0$, the time T needed to fall to a point with tangent angle θ is

$$T = 2^{\frac{1}{2}}\theta a^{\frac{1}{2}} = \int_0^\theta \left[\frac{(\dot{x}_1^2(\theta) + \dot{x}_2^2(\theta))}{x_1(\theta)} \right]^{\frac{1}{2}} d\theta$$

Here the units of measurement have been chosen to make $2g = 1$. An *isochrone* is a curve $T =$ const. Its differential equation is $dT/da = 0$ or

$$\frac{d\theta}{da} = -\frac{\theta}{2a}$$

A brachistochrone (cycloid) is determined by the value of its parameter a. Points on the brachistochrone are then fixed by the time T. An isochrone, on the other hand, is given by the value of T, and points on it are singled out by the different values of a, showing the brachistochrone along which the point A must fall to arrive at the isochrone in time T.

Therefore we may use $T = u$ and $\sqrt{2a} = v$ as new coordinates in the domain $|x_2| \leq \pi|x_1|/2$ in which there pass only one brachistochrone and one isochrone through each point. From

$$2x_1 = v^2\left(1 - \cos\frac{u}{v}\right) \qquad 2x_2 = uv - v^2 \sin\frac{u}{v}$$

a straightforward computation shows that

$$\int_0^\theta \left[\frac{dx_1^2 + dx_2^2}{x_1}\right]^{\frac{1}{2}}$$
$$= \int_0^\theta \left(du^2 + \frac{1}{2v^2[1 - \cos(u/v)]}\left\{\left[2v\left(1 - \cos\frac{u}{v}\right) - u\sin\frac{u}{v}\right]^2 + \left[2u - 2v\sin\frac{u}{v} - u\left(1 - \cos\frac{u}{v}\right)\right]^2 + 4u^2\right\} dv^2\right)^{\frac{1}{2}}$$

The factor multiplying dv^2 in this integral is positive. Hence for all curves joining A to a point within the domain the curve $dv = 0$ (i.e., a brachistochrone) will give a minimal value to the integral: $\int_0^\theta du = T$.

One should not think that the above method is always successful. The first example of a variational problem in which the integral curves of the Euler equations give neither a minimum nor a maximum is due to Euler himself.

***Example* 4-4.** The integral curves of the Euler equations of

$$\int[x_1(dx_1^2 + dx_2^2)]^{\frac{1}{2}} = \text{extremum}$$

are the parabolas $x_2 = 2(ax_1 - a^2)^{\frac{1}{2}} + b$. Let us take $b = 0$. The parabola of parameter $a = fh/(f + h)$ passes through two points $A(f,d)$, $B(h,k)$ chosen so that $(k - d)^2 = 4fh$. The value of the integral along the parabola from A to B is $\frac{2}{3}(f + h)^{\frac{3}{2}}$, whereas the integral along the broken line AA_0B_0B (Fig. 4-2) is $\frac{2}{3}(f^{\frac{3}{2}} + h^{\frac{3}{2}})$ which for positive f,h is always less than the first expression. It is easily seen that the parabolas do not yield a maximum; it is possible to approximate the parabola $\mathbf{x}$ by nearby curves of arbitrary length, for example, $\mathbf{x} + (1/n)\sin ns \cdot \mathbf{e}_1$. For such a curve, the integral can be made arbitrarily large.

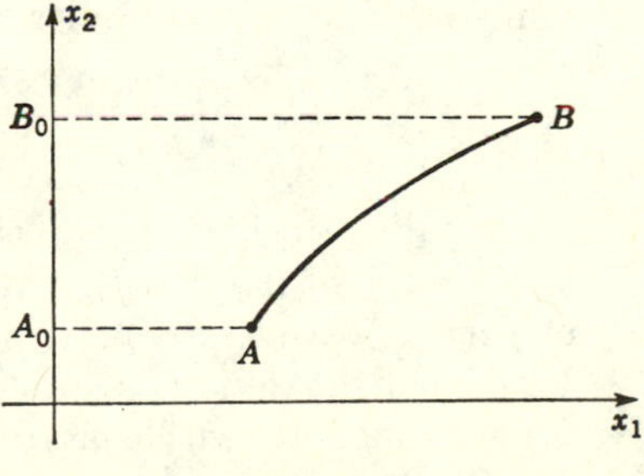

Fig. 4-2

The problem of the existence of a minimum is treated in a systematic way in Sec. 11-1.

Exercise 4-1

1. Check the computations of Sec. 3-4 for their validity. Is one entitled to accept the results of an integration of Eq. (3-30) as a solution of the similitude problem without check that the curve obtained satisfies the conditions asked for?

2. A variational problem that has geometric meaning should depend only on $\mathbf{x}$ and on $d\mathbf{x}/ds$ but not on the parameter t chosen to represent the problem. If $c = dt/dt^*$, a change of parameter $t \to t^*$ will transform dt into $c\,dt^*$ and $d\mathbf{x}/dt$ into $(1/c)\,d\mathbf{x}/dt^*$. Show that a variational problem on $I = \int F(x_1,x_2,\dot{x}_1,\dot{x}_2)\,dt$ has geometric meaning if and only if $F(x_1,x_2,c\dot{x}_1,c\dot{x}_2) = |c|F(x_1,x_2,\dot{x}_1,\dot{x}_2)$. Check the conditions on the examples given in the text.

3. Show that the isochrones are the orthogonal trajectories of the brachistochrones. Find the finite equations of the isochrones.

4. Show that a straight segment is actually the shortest curve connecting two points. Which system of coordinates corresponds to the (u,v) system of example 4-2?

5. In example 4-3, show that both the extremals and the lines of equal "distance" from a fixed point are circles. Characterize the two families of circles. Are they mutually orthogonal?

◆ **6.** Find the Euler equations for the curves making

$$\int[E(x_1,x_2)\,dx_1^2 + 2F(x_1,x_2)\,dx_1\,dx_2 + G(x_1,x_2)\,dx_2^2]^{\frac{1}{2}}$$

a minimum.

7. An n-dimensional problem of variations asks for functions $x_i(t)$, $i = 1, 2, \ldots, n$, making $\int F(x_1, \ldots, x_n, \dot{x}_1, \ldots, \dot{x}_n)\,dt$ an extremum. Prove that the Euler equations of this problem are

$$\frac{\partial F}{\partial x_i} - \left(\frac{\partial F}{\partial \dot{x}_i}\right)^{\cdot} = 0, \qquad i = 1, \ldots, n.$$

◆ **8.** Fermat's principle states that in a medium with index of refraction $n(x_1,x_2)$ a light ray describes an integral curve of the Euler equations of $\int n(x_1,x_2)(dx_1^2 + dx_2^2)^{\frac{1}{2}}$.

Find the light rays

(*a*) In a medium with constant index of refraction

(*b*) For $n = 1 + cx_2$, c constant

(*c*) For $n = cx_2$

9. The Euler equations are local conditions; they indicate that a short arc of the curve has an extremal property. Even if they have locally unique solutions, there may be many extremals connecting two points at a finite distance. If we look for the shortest curve connecting two points in the upper half plane $x_2 \geq 0$, we have as a solution not only the absolute minimum, the straight segment connecting the two points, but also a relative minimum, the line reflected at the x_1 axis. This example shows that a solution of an extremal problem may have cusps. If the cusp occurs at $t = t_0$, the partial integration used to establish the Euler equations must be broken up into two integrals, from 0 to t_0 and from t_0 to 1. Show that the condition for the existence of an extremum is that

$$-\xi(t_0+0)\frac{\partial F}{\partial \dot{x}_1}(x_1(t_0),x_2(t_0),\dot{x}_1(t_0+0),\dot{x}_2(t_0+0))$$
$$+\ \xi(t_0-0)\frac{\partial F}{\partial \dot{x}_1}(x_1(t_0),x_2(t_0),\dot{x}_1(t_0-0),\dot{x}_2(t_0-0))$$
$$+\int_0^1 \xi(t)\left[\frac{\partial F}{\partial x_1}-\frac{d}{dt}\frac{\partial F}{\partial \dot{x}_1}\right]dt = 0$$

(and the corresponding equation for $\dot{x}_2$) for all C^1 functions $\xi(t)$, $\xi(0) = \xi(1) = 0$. Deduce from this that in a cusp

$$\frac{\partial F}{\partial \dot{x}_i}(x_1(t_0),x_2(t_0),\dot{x}_1(t_0+0),\dot{x}_2(t_0+0))$$
$$= \frac{\partial F}{\partial \dot{x}_i}(x_1(t_0),x_2(t_0),\dot{x}_1(t_0-0),\dot{x}_2(t_0-0)) \qquad i = 1, 2$$

is a necessary condition for an extremal curve (*Erdmann's* condition).

10. Check that Erdmann's condition holds for the reflection of light rays at a line and for the refraction of light rays at the straight boundary of two domains with indices of refraction n_1,n_2, respectively, if Snell's law holds: $n_1 \cos\theta(t_0 - 0) = n_2 \cos\theta(t_0 + 0)$ (see Prob. 8).

11. Check that Erdmann's condition holds along the broken line AA_0B_0B of example 4-4.

12. Complete the argument of the proof of lemma 4-1 to infer $f(t) = 0$ from $\int_0^1 \xi(t)f(t)\,dt = 0$ in the cases $\tau = 0$ and $\tau = 1$.

4-2. The Isoperimetric Problem

One of the famous problems of Greek geometry is a variational problem with side conditions: *Among all closed curves of given length L, find the one with maximal area.*

A K^0 curve has a tangent everywhere, except possibly in the complement of a countable union of open sets (exercise 2-1, Prob. 9). Its arc length may be computed by a Stieltjès integral, which we write for the perimeter of a closed K^0 curve

$$L = \oint ds = \oint \mathbf{t}\cdot\mathbf{t}\,ds = \oint \mathbf{t}\cdot d\mathbf{x} = -\oint \mathbf{x}\cdot d\mathbf{t} \tag{4-4}$$

The constant term in the integration by parts vanishes if we take the point $\mathbf{x}(0) = \mathbf{x}(L)$ to be one at which a tangent exists. As an example, we find the perimeter of a square (Fig. 4-3). $d\mathbf{t} = 0$ except in the four vertices of the square which correspond to values u_i ($i = 1,2,3,4$) of the

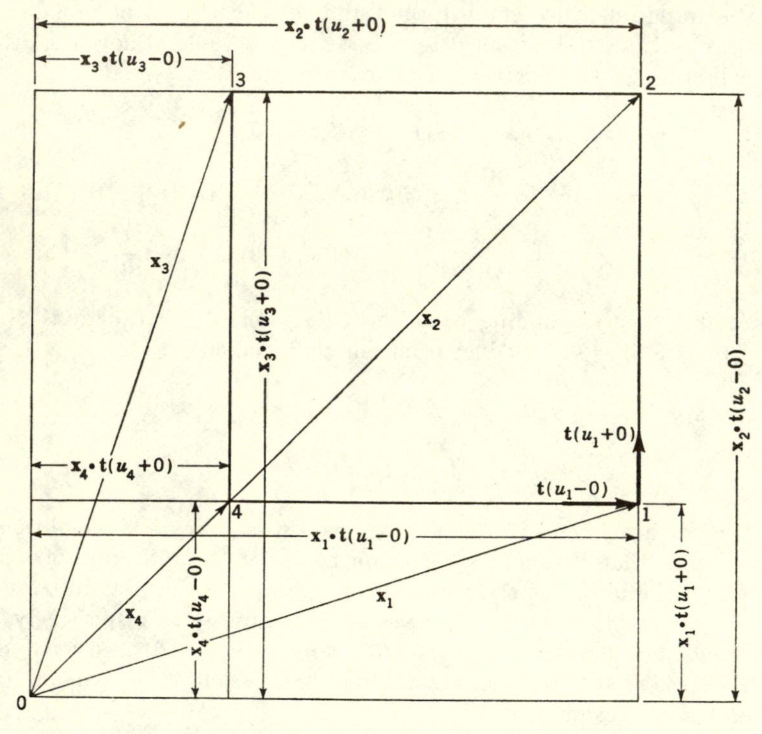

Fig. 4-3

parameter. The integral (4-4) then becomes

$$L = -\sum_{i=1}^{4} \mathbf{x}(u_i) \cdot [\mathbf{t}(u_i + 0) - \mathbf{t}(u_i - 0)]$$

The vector $\mathbf{t}$ is a unit vector, and $\mathbf{x} \cdot \mathbf{t}$ is the length of the projection of $\mathbf{x}$ onto the direction of $\mathbf{t}$, with the appropriate sign. The formula is easily verified in Fig. 4-3.

Definition 4-2. *The parallel set S_p of distance p of a set S is the union of all closed circular disks of radius p, the centers of which are points of S. If S is a domain bounded by a closed Jordan curve $\mathbf{c}$, the boundary of S_p is the parallel curve $\mathbf{c}_p$ of $\mathbf{c}$.*

If S is a single point P, the parallel set P_p is the circle of center P and radius p. Figure 4-4 shows the parallel sets of a segment, a square, and a circle.

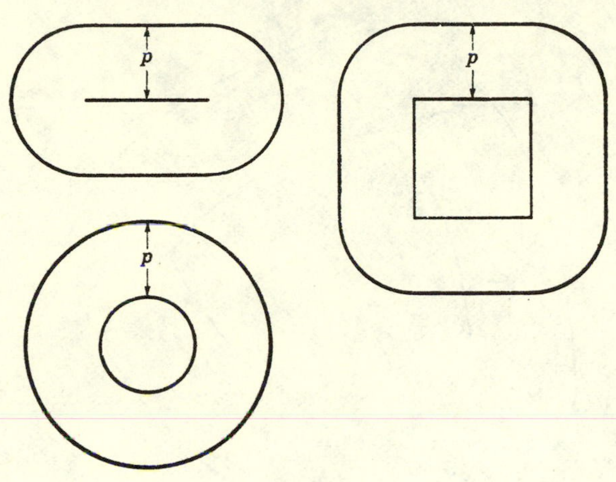

Fig. 4-4

Theorem 4-3 (Steiner). *Let L be the perimeter of a simple closed rectifiable curve* $\mathbf{c}$ *and A the area of its interior, and let L_p, A_p denote the corresponding quantities for the parallel curve* $\mathbf{c}_p$. *Then*

$$L_p \leq L + 2\pi p \tag{4-5}$$
$$A_p \leq A + Lp + \pi p^2 \tag{4-6}$$

In both relations, equality holds for convex $\mathbf{c}$.

We assume that the direction of increasing s has been chosen so that $\mathbf{n}$ is the inner normal, at points where it exists.

If $\mathbf{c}$ is a convex C^1 curve, $\mathbf{c}_p(s) = \mathbf{c}(s) - p\mathbf{n}(s)$, $\mathbf{t}_p(s) = \mathbf{t}(s)$, $d\mathbf{t} = \mathbf{n}\,d\theta$; hence $L_p = -\oint(\mathbf{c} - p\mathbf{n}) \cdot \mathbf{n}\,d\theta = L + p2\pi$.

If $\mathbf{c}$ is a convex C^0 curve, it has, at most, a countable set of cusps, corresponding to values of the arc length s_i, $i = 1, \ldots$. Let

$$\Delta_i = \theta(s_i + 0) - \theta(s_i - 0)$$

be the angle of the one-sided tangents in a cusp, $\sum_{i=0}^{\infty} \Delta_i \leq 2\pi$. The parallel set of the interior of $\mathbf{c}$ is bounded by arcs $\mathbf{c}(s) - p\mathbf{n}(s)$, $s_i < s < s_{i+1}$, $i = 1, \ldots,$ and by circular arcs of radius p, center $\mathbf{c}(s_i)$, and aperture Δ_i (Fig. 4-5). The curve $\mathbf{c}$ is closed convex; hence

$$\Sigma_i\Delta_i + \Sigma_i[\theta(s_{i+1} - 0) - \theta(s_i + 0)] = 2\pi$$

and

$$\begin{aligned} L_p &= L + p\Sigma_i \int_{s_i+0}^{s_{i+1}-0} d\theta + p\Sigma_i\Delta_i \\ &= L + 2\pi p \end{aligned}$$

If **c** is not convex, parts of the curve $\mathbf{c} - p\mathbf{n}$ or of the circular arcs connecting such parallel arcs may be submerged in a parallel set (Fig. 4-6). Hence in this case $L_p \leq L + 2p\pi$.

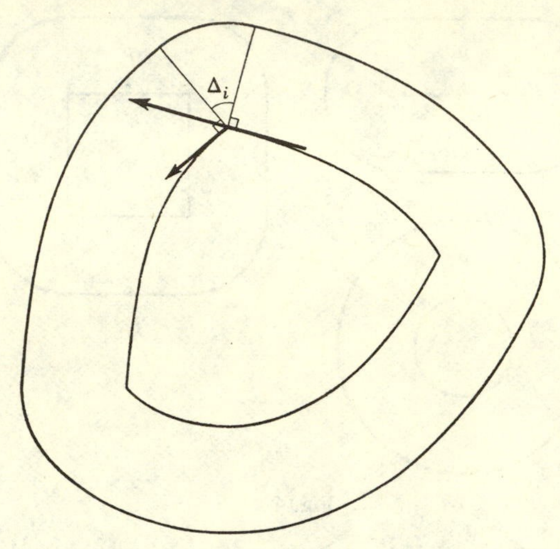

Fig. 4-5

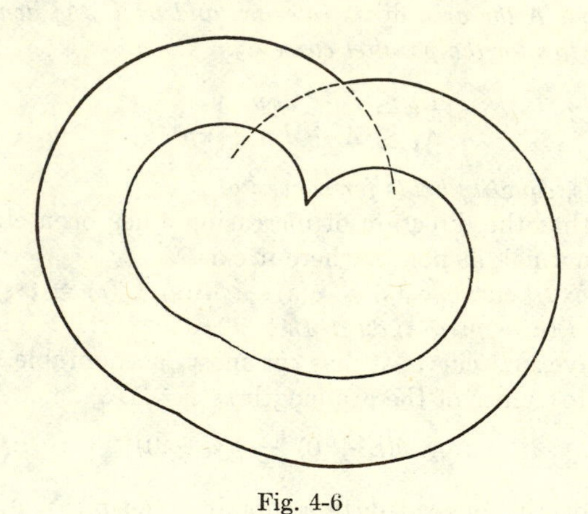

Fig. 4-6

Equation (4-5) is now completely proved, and (4-6) follows by integration:

$$A_p = \int_0^p L_q \, dq + A$$

for convex curves.

Definition 4-4. *An incircle of a simple closed curve is an inscribed circle of maximal radius.*

The incircle may not be uniquely defined, but the *inradius* r is. The center of an incircle is an incenter of the curve. Incenter and inradius appear in the following *isoperimetric inequality:*

Theorem 4-5. *Let r be the inradius of a rectifiable closed Jordan curve* **c**, *and a the length of any chord of* **c** *passing through an incenter. Then*

$$L^2 - 4\pi A \geq \frac{\pi^2}{4}(a - 2r)^2 \tag{4-7}$$

It follows from (4-7) that $L^2 = 4\pi A$ only if every secant through the incenter is a diameter of the incircle, i.e., that **c** coincides with its incircle.

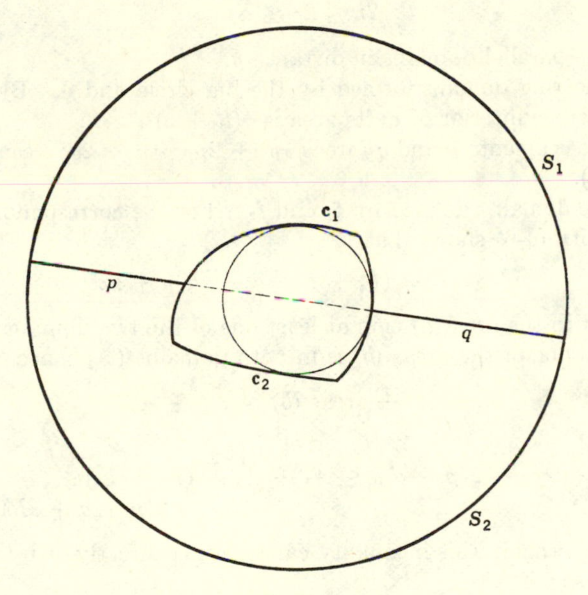

Fig. 4-7

Corollary 4-6. *Among all closed Jordan curves of perimeter L, the circle of radius $L/2\pi$ defines the largest area.*

Inequality (4-7) must be proved only for convex curves. If **c** is not convex, it is contained in a smallest convex domain (see Fig. 2-5). The boundary of that domain is convex; it has a smaller perimeter and defines a bigger area than the nonconvex curve **c**.

For the proof we enclose **c** in a big circle of radius R whose center is an incenter of **c** (Fig. 4-7). A line drawn through the incenter cuts off a secant of length $a \geq 2r$ in **c** and segments of lengths p,q in the ring domain between **c** and the big circle. The line divides both **c** and the circle into two arcs, $\mathbf{c}_1$ and $\mathbf{c}_2$, S_1 and S_2, respectively. The ring domain is divided into two domains whose boundaries are two closed Jordan curves which we may indicate symbolically by

$$\mathbf{f}_1 = \{S_1 q \mathbf{c}_1 p\} \qquad \mathbf{f}_2 = \{S_2 p \mathbf{c}_2 q\}$$

Let L_i be the length of $\mathbf{c}_i$ ($i = 1,2$) and A_i the area enclosed by $\mathbf{f}_i$. A and L refer to $\mathbf{c}$ itself. Then

$$L = L_1 + L_2 \tag{4-8a}$$

$$\pi R^2 = A + A_1 + A_2 \tag{4-8b}$$

Let ρ be a real number such that

$$2r \leq 2\rho \leq a \tag{4-9}$$

We form the parallel domains in distance ρ:

(*a*) Of the ring domain formed by the big circle and $\mathbf{c}$. By (4-9) it covers the whole interior of $\mathbf{c}$; its area is $\pi(R + \rho)^2$.

(*b*) Of the segments p and q (areas $\pi\rho^2 + 2p\rho$, $\pi\rho^2 + 2q\rho$, respectively; see Fig. 4-4).

(*c*) Of the domains defined by $\mathbf{f}_1$ and $\mathbf{f}_2$. For the corresponding areas A_i^*, inequality (4-6) shows that

$$A_i^* \leq A_i + (L_i + p + q + \pi R)\rho + \pi\rho^2$$

Any point in the area (*a*) is in at least one of the two domains (*c*), and any point in one of the areas (*b*) is in both domains (*c*); hence

$$\text{area } (a) + \text{areas } (b) \leq A_1^* + A_2^*$$

or

$$\pi(R + \rho)^2 + 2\pi\rho^2 + 2(p + q)\rho \leq A_1 + A_2 + (L_1 + L_2)\rho + 2(p + q + \pi R)\rho + 2\pi\rho^2$$

Most of the terms in this inequality cancel either directly or by (4-8); all that remains is

$$A \leq L\rho - \pi\rho^2$$

This relation is equivalent to

$$L^2 - 4\pi A \geq (L - 2\pi\rho)^2 \tag{4-10}$$

an alternative form of the isoperimetric inequality.

In (4-10) we first put $\rho = r$, then $\rho = a/2$, and add the two inequalities. Formula (4-7) then follows from the algebraic inequality

$$2(x^2 + y^2) \geq (x - y)^2$$

Exercise 4-2

1. Check formula (4-4) for a triangle. Choose the origin as an interior point of the triangle.

2. Determine the area of the parallel set of a segment (Fig. 4-4). Why has the length of the segment to be counted twice in formulas (4-5) and (4-6)?
3. Prove that the inradius of a closed Jordan curve exists and is finite.
4. Give an example of a closed convex curve with an infinity of incircles.
5. Prove that $L = \oint \mathbf{x} \times d\mathbf{n}$.
6. A closed convex curve may be given by its support function $h(\psi)$ [see Eq. (3-23)]. Prove that

$$L = \int_0^{2\pi} h(\psi)\, d\psi \qquad A = \tfrac{1}{2} \int_0^{2\pi} |h(\psi)|\, ds(\psi)$$

7. Formulate the isoperimetric problem as a variational problem of the type considered in Sec. 4-1, under the side condition that a certain integral be kept constant.
8. Given the support function of $\mathbf{c}$, find that of $\mathbf{c}_p$. Then use Prob. 6 to derive Steiner's formulas (4-5) and (4-6).
9. Does there exist a real number x, the root of $\pi x^2 + Lx + A = 0$, L and A being the perimeter and area of a closed curve?
10. Does the following problem have a solution? Among all regular polygons of perimeter L, find the polygon of maximal area.
11. For the regular n-gon, compute max $(\pi^2/2)(a - 2r)^2$, the *isoperimetric deficit.*
12. Prove that among all n-gons of perimeter L, the regular n-gon has maximal area. Here n is a fixed number (theorem of Zenodorus).

References

SEC. 4-1

A classic on the calculus of variations is the following:

Bliss, G. A.: "Calculus of Variations," University of Chicago Press, Chicago, 1924

Other references are:

Bernoulli, Jean: Problème de la plus vîte descente resolu d'une manière directe et extraordinaire, *Mém. Acad. Royale Sci.*, 1718, pp. 172–175.

Euler, L.: De insigni paradoxo quod in analysi maximorum et minimorum occurrit (E 735), "Opera Omnia," Ser. I, vol. 25, pp. 286–292, Societas Scientiarum Naturalium Helvetica, Lousannae et Turici, 1952.

SEC. 4-2

Hadwiger, H.: Eine elementare Ableitung der isoperimetrischen Ungleichung für Polygone, *Comment. Math. Helv.*, **16:** 305–309 (1943).

Compare also:

Bonnesen, T., and W. Fenchel: Theorie der konvexen Körper, *Ergeb. Math. Grenzg.*, **3:** 1–172 (1934).

Yaglom, I. M., and V. G. Boltyanskii: "Convex Figures," translated by Paul J. Kelly and Lewis F. Walton, Holt, Rinehart and Winston, Inc., New York, 1961.

5
INTRODUCTION TO TRANSFORMATION GROUPS

5-1. Translations and Rotations

The theory of transformation groups is one of the strongest tools available in differential geometry. Actually, much of our work in the first three chapters dealt with phenomena of group theory. In order to get acquainted with the subject, we study in this chapter some fundamental notions and a number of examples. A formal treatment will start in Chap. 6.

In algebra, a group is a set G for which a composition (usually called *multiplication*) is defined that associates with each pair (a,b) of elements of G a *unique* element $c = ab \in G$. This map $(a,b) \to ab$ must satisfy the axioms

i. $x \cdot yz = xy \cdot z$ for all $x,y,z \in G$.

ii. For all $x,y \in G$ there exist *unique* z_1 and z_2 such that $xz_1 = y$, $z_2x = y$.

From these axioms it may be shown that there exists in G a unique *unit* e ($xe = ex = x$ for all $x \in G$) and a unique *inverse* x^{-1} for every $x \in G$ ($x^{-1}x = xx^{-1} = e$). The group is *commutative* or *abelian* if

iii. $xy = yx$ for all $x,y \in G$.

In this case, the composition is usually written as addition $x + y$.

If the set G is finite, the composition law may be made explicit in a multiplication table (see exercise 5-1, Prob. 1). If the set G is not finite, some other way has to be found to express the group law.

***Example* 5-1.** The real numbers form an abelian group R under addition.

***Example* 5-2.** The 2×2 rotation matrices

$$A(\theta) = \begin{pmatrix} \cos\theta & \sin\theta \\ -\sin\theta & \cos\theta \end{pmatrix}$$

form an abelian group O_2 under matrix multiplication, since

$$A(\theta_1)A(\theta_2) = A(\theta_1 + \theta_2)$$

The map

$$\theta \rightarrow A(\theta)$$

defines a matrix for each real number θ. The composition of two real numbers is mapped into the composition of the corresponding matrices. Such a map, which is compatible with the group structure of its domain (R) and its range (O_2), is called a *homomorphism*, or, in the special case of a homomorphism into a matrix group, a *representation*. O_2 is the range of a representation of R, but there is a fundamental difference between the two groups. For $\theta_1 = \theta + 2k\pi$, k integer, $A(\theta_1) = A(\theta)$. The map of R onto O_2 is not one-to-one; an element of O_2 corresponds to all members of the class of real numbers differing from a given number by an integer multiple of 2π. The elements of the group R may be represented by the points on the real-number line, which is not a compact set. The elements of O_2 correspond to the points $\mathbf{x} = (\cos\theta, \sin\theta)$ on the (compact) unit circle S^1. $A(\theta)$ is the rotation that brings (1,0) into $\mathbf{x}$. The map $R \rightarrow O_2$ may be visualized by taking a real-number line, gluing the zero of R onto (1,0) of the unit circle, and then winding the real-number line on the circle in a two-sided infinite staircase. The real-number line is a "universal covering" of the unit circle.

Both O_2 and R are examples of *continuous* groups. They are characterized not only by the algebraic law of composition but also by the analytic properties of the point sets which may represent the groups. The "multiplication" in both groups is continuous in the sense that, if x is "near" x', yx will be near yx' for all y. In R, nearness is measured by distance, $|x' - x| < \epsilon$; in O_2 nearness may be measured, for example, by the distance on the unit circle S^1.

An *isomorphism* of two continuous groups is a homomorphism which is one-to-one and both ways continuous. O_2 is not isomorphic to R since there exists no one-to-one continuous map of S^1 into R. But the neighborhood of the unit element of O_2 given by $-\pi/3 < \theta < +\pi/3$ is isomorphic to the neighborhood $-\pi/3 < r < +\pi/3$ of the unit 0 of R.

This example may have made it clear that a treatment of continuous groups must be based on an analysis of the spaces that may be given a group structure. Such an analysis needs the tools of algebraic topology; it is outside the scope of the methods used in this book. We will deal only with that part of the group structure that can be studied in a (cartesian) neighborhood of the unit element. Such a neighborhood is called a *group germ*. O_2 and R have isomorphic group germs. We shall see later that all differentiable abelian groups of the same dimension have isomorphic group germs. They are all equivalent from a local point of view.

The elements of O_2 may be represented by the points (x_1,x_2) on the unit circle $x_1^2 + x_2^2 = 1$. The multiplication formula of O_2,

$$\begin{pmatrix} \cos\theta_1 & \sin\theta_1 \\ -\sin\theta_1 & \cos\theta_1 \end{pmatrix}\begin{pmatrix} \cos\theta_2 & \sin\theta_2 \\ -\sin\theta_2 & \cos\theta_2 \end{pmatrix}$$
$$= \begin{pmatrix} \cos\theta_1\cos\theta_2 - \sin\theta_1\sin\theta_2 & \sin\theta_1\cos\theta_2 + \cos\theta_1\sin\theta_2 \\ -\sin\theta_1\cos\theta_2 - \cos\theta_1\sin\theta_2 & \cos\theta_1\cos\theta_2 - \sin\theta_1\sin\theta_2 \end{pmatrix}$$

is equivalent to the algebraic identity

$$(x_1^2 + x_2^2)(\xi_1^2 + \xi_2^2) = (x_1\xi_1 - x_2\xi_2)^2 + (x_1\xi_2 + x_2\xi_1)^2 \qquad (5\text{-}1)$$

If (x_1,x_2) and (ξ_1,ξ_2) are points on the unit circle, so is (X_1,X_2):

$$X_1 = x_1\xi_1 - x_2\xi_2 \qquad X_2 = x_1\xi_2 + x_2\xi_1$$

The composition formula $(x_1,x_2) \cdot (\xi_1,\xi_2) = (X_1,X_2)$ defines a group structure on S^1; the trigonometric addition formulas show that this group is O_2. Equation (5-1) may also serve as a basis for the theory of complex numbers (see exercise 5-1, Prob. 5).

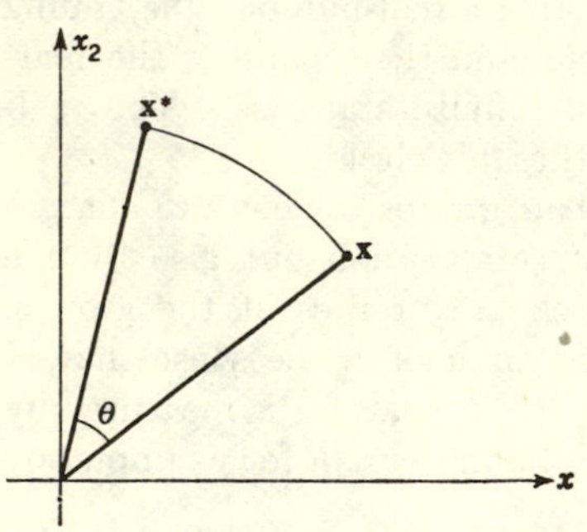

***Example* 5-3.** The rotations about the origin (Fig. 5-1)

$$\begin{aligned} x_1^* &= x_1\cos\theta - x_2\sin\theta \\ x_2^* &= x_1\sin\theta + x_2\cos\theta \end{aligned} \qquad (5\text{-}2)$$

describe the *action* of O_2 on the plane R^2. To each value of θ there corresponds a homeomorphism of R^2 onto itself,

Fig. 5-1

$$\mathbf{x}^* = F(\mathbf{x},\theta)$$

such that $F(F(\mathbf{x},\theta_1),\theta_2) = F(\mathbf{x}, \theta_1 + \theta_2)$. This means that a composition of homeomorphisms of parameters θ_1,θ_2 results in the homeomorphism belonging to the composition $\theta_1 + \theta_2$ of the parameters. $F(\ ,\theta)$ is in some sense a realization of O_2 by operators. It is a *pseudo group of transformations* (Chap. 7).

O_2 may be realized as a pseudo group of transformations in many other ways. A possible procedure is the following: Let θ measure the center angle in a circle S^1 of radius $\frac{1}{2}$. From the point $\theta = \pi$ the circle is projected on a line R^1 tangent to S^1 in $\theta = 0$ in "stereographic projection" (Fig. 5-2). The point of parameter θ is mapped into $x = \tan \theta/2$. The point of parameter $(\theta_1 + \theta_2)$ is mapped into

$$\tan \frac{\theta_1 + \theta_2}{2} = \frac{x_1 + x_2}{1 - x_1 x_2}$$

Every group acts as a transformation group on itself. If one factor in the multiplication is fixed, e.g., the left one, the operation $x \to ax$ is

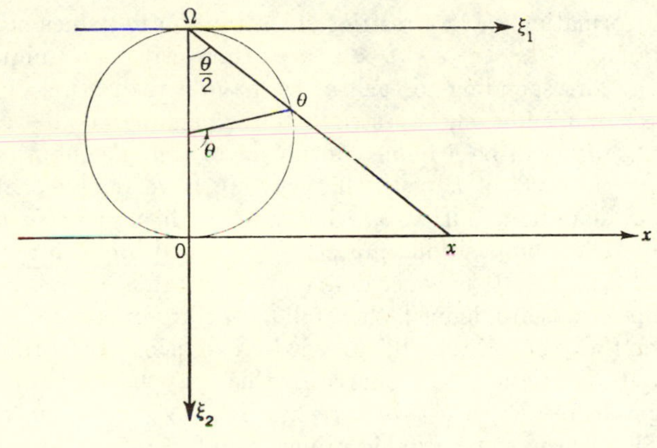

Fig. 5-2

the transformation by a *left translation* with a, $ax = L_a(x)$. Here again, the composition of translations is the translation that belongs to the product of the parameters:

$$L_b[L_a(x)] = L_{ba}(x)$$

In the case of the group O_2, $L_\alpha(\theta) = \alpha + \theta$. Put $a = \tan \alpha/2$. The image of L_α on the line R^1 is a transformation

$$y_a(x) = \frac{a + x}{1 - ax} \tag{5-3}$$

Since

$$y_b(y_a(x)) = \frac{(a + b)/(1 - ab) + x}{1 - x(a + b)/(1 - ab)} = y_{b \cdot a}(x)$$

the addition of parameters in O_2 is transformed into the composition according to the addition formula of the tan function,

$$b \cdot a = \frac{a + b}{1 - ab} \tag{5-4}$$

The composition formula (5-4) is much more complicated than the simple addition of angles in O_2; a good choice of the parameters will do much to simplify computations with continuous groups.

The transformation described by (5-3) is neither a realization of the full group O_2, nor is it a homeomorphism. No transformation corresponds to $\theta = \pi$, and, for $a \neq 0$, the point $x = 1/a$ is not in the domain of definition of y_a. There are two ways out of this difficulty. We might accept the fact that there are certain cases excluded from the definition of the transformation. If we restrict our attention to values of parameters $a = \tan \alpha/2$, $-\pi < -\epsilon < \alpha < \epsilon < \pi$, there will be a unique transformation y_a corresponding to each a; we have a realization of a group germ in O_2. If, in addition, we restrict x to the neighborhood $-\cot \epsilon/2 < x < \cot \epsilon/2$, $y_a(x)$ will be a homeomorphism of that neighborhood onto another open interval of R^1. In this way we have removed all exceptions to the definition, but we are left only with a *groupoid* of transformations: The composition $y_b[y_a(x)]$ is defined only for $|y_a(x)| < \cot \epsilon/2$, and $y_c\{y_b[y_a(x)]\}$ is never defined for $|\arctan a + \arctan b| > 2\epsilon$. *If* the compositions are defined, they follow the group laws.

The second way out of the difficulty is by extension of the function y_a, instead of its restriction. In our case, this may be accomplished by imbedding the line R^1 in a *projective line* P^1. A projective line is an image of S^1. In the stereographic projection of Fig. 5-2, there is a one-to-one correspondence between the points on the x axis and the lines through the origin Ω of the (ξ_1,ξ_2) system, except for the parallel to x through Ω. On any line through Ω, the ratio $\xi_1:\xi_2$ is constant, if we assume the usual rules for proportions involving a zero. The points x correspond to the lines (cx,c), $c \neq 0$, in the (ξ_1,ξ_2) plane. The projective line P^1 is the set of lines through Ω, each line representing one point of P^1, with the neighborhood properties induced by those of R^1. In this way, a single *point at infinity* $(c,0)$ is added to R^1. On R^1 proper we have introduced homogeneous coordinates $x = \xi_1:\xi_2$. The transformation (5-3) may be made homogeneous, $y = y_1:y_2$. If we write the parameter a also as a proportion $a:1 = a_1:a_2$, then

$$\begin{aligned} y_1 &= a_1\xi_2 + a_2\xi_1 \\ y_2 &= a_2\xi_2 - a_1\xi_1 \end{aligned} \tag{5-5}$$

Formula (5-5) is just a new edition of (5-2). [Note the orientation of the coordinate system (ξ_1,ξ_2)!]

In this book we shall restrict ourselves to methods dealing with groupoids of transformations defined on a group germ. The problems of extension are of a topological nature; they cannot be treated with the analytical tools we have at our disposal.

***Example* 5-4.** The euclidean motions in a plane are the transformations

$$\begin{aligned} x_1^* &= x_1 \cos\theta - x_2 \sin\theta + a_1 \\ x_2^* &= x_1 \sin\theta + x_2 \cos\theta + a_2 \end{aligned} \tag{5-6}$$

or, in vector form, in a fixed frame $\mathbf{e}_1,\mathbf{e}_2$,

$$\mathbf{x}^* = \mathbf{x}A(\theta) + (a_1,a_2) = F(\mathbf{x};\theta,a_1,a_2) \tag{5-6a}$$

A euclidean motion may be split up into a rotation,

$$\mathbf{x}_1 = \mathbf{x}A(\theta) \tag{5-7a}$$

followed by two translations,

$$\mathbf{x}_2 = \mathbf{x}_1 + (a_1,0) \tag{5-7b}$$
$$\mathbf{x}^* = \mathbf{x}_2 + (0,a_2) \tag{5-7c}$$

Any plane vector (a_1,a_2) represents a translation $\mathbf{x} \to \mathbf{x} + (a_1,a_2)$. The composition of two translations (a_1,a_2), (b_1,b_2) is the translation of the vector $(b_1 + a_1,\ b_2 + a_2)$ by the parallelogram law of vector addition. This fits into a more general frame of group composition. Given two groups G_1 and G_2 with multiplications $\cdot_1$ and $\cdot_2$, their *direct product* $G_1 \times G_2$ is the group defined on the set of ordered pairs (g_1,g_2), $g_1 \in G_1$, $g_2 \in G_2$, by the multiplication law $(g_1,g_2) \cdot (h_1,h_2) = (g_1 \cdot_1 h_1,\ g_2 \cdot_2 h_2)$. The group of plane vectors under vector addition is the direct product $R \times R$ of the group of real numbers with itself. Equations (5-7*b*) and (5-7*c*) describe the action of $R \times R$ as a transformation group in the plane. The action of the rotation group (5-7*a*) is not that of a factor in a direct product. The motion (5-6) depends on parameters θ, $\mathbf{a} = (a_1,a_2)$. The composition of two motions of parameters $(\theta,\mathbf{a})$ and $(\phi,\mathbf{b})$ is the motion belonging to $(\theta + \phi,\ \mathbf{a}A(\phi) + \mathbf{b})$. In the multiplication formula there occurs mixing of elements belonging to different constituent groups. Note also that the group of euclidean motions is *not* commutative, although it is generated by the plane rotation group and the plane translation group, both being abelian. This situation is impossible for a direct product.

The same situation is found on P^1 if we combine the action of the rotation group according to (5-3) with that of R as a group of translations on R^1,

$$y = x + b \tag{5-8a}$$

or in homogeneous coordinates $x = \xi_1 : \xi_2$, $b = b_1 : b_2$,

$$\begin{aligned} y_1 &= b_2\xi_1 + b_1\xi_2 \\ y_2 &= b_2\xi_2 \end{aligned} \tag{5-8b}$$

On P^1 this is a group of transformations leaving invariant the point at infinity $\xi_2 = 0$. The combination of (5-8*a*) and (5-3) generates a group of transformations

$$y = \frac{a + bx}{1 - cx} \tag{5-9a}$$

with the nonvanishing determinant

$$\begin{Vmatrix} a & b \\ 1 & -c \end{Vmatrix} \neq 0$$

In homogeneous form the same group appears as

$$\begin{aligned} y_1 &= \xi_1 A_1 + \xi_2 A_2 \\ y_2 &= \xi_1 A_3 + \xi_2 A_4 \end{aligned} \tag{5-9b}$$

or in matrix form

$$(y_1, y_2) = (\xi_1, \xi_2) \begin{pmatrix} A_1 & A_3 \\ A_2 & A_4 \end{pmatrix} \qquad \|A_i\| \neq 0$$

The last equation shows that the transformation group (5-9) represents the action of the group of 2×2 matrices with a nonzero determinant on P^1. This group of all nonsingular 2×2 matrices is not the direct product $O_2 \times R$, and it is not abelian, since, in general, matrix multiplication is not commutative.

Exercise 5-1

1. In the following multiplication table, an entry is the product of the element appearing at the head of its row times the element at the head of its column. Check that the table defines a group. Is the group abelian?

	e	a	b	c	d	f
e	e	a	b	c	d	f
a	a	b	e	d	f	c
b	b	e	a	f	c	d
c	c	f	d	e	b	a
d	d	c	f	a	e	b
f	f	d	c	b	a	e

◆ **2.** Find all groups of two and of three elements.

3. Check the group axioms for R and for O_2.

◆ **4.** It has been shown in the text that a neighborhood $-\epsilon < \theta < \epsilon$ of O_2 is isomorphic to the corresponding neighborhood $-\epsilon < r < \epsilon$ of R for small ϵ. Find sup ϵ for all admissible ϵ.

5. Define $(1,0) = \mathbf{1}$, $(0,1) = \mathbf{i}$. A point (x,y) in the plane may be written as $x\mathbf{1} + y\mathbf{i}$. Define the product of two points by

$$(x,y) \cdot (a,b) = (xa - yb, xb + ya)$$

Show that this definition is that of the multiplication of complex numbers: It is distributive and $\mathbf{1} \cdot \mathbf{1} = \mathbf{1}, \mathbf{1} \cdot \mathbf{i} = \mathbf{i} \cdot \mathbf{1} = \mathbf{i}, \mathbf{i} \cdot \mathbf{i} = -\mathbf{1}$.

6. Explain the parallelism of formulas (5-2) and (5-5). (Use $\tan\theta = \sin\theta : \cos\theta$.)

7. The euclidean motions in a plane may be referred to three parameters s_1,s_2,s_3 by

$$x_1^* = x_1 s_1 - x_2(1 - s_1^2)^{\frac{1}{2}} + s_2$$
$$x_2^* = x_1(1 - s_1^2)^{\frac{1}{2}} + x_2 s_1 + s_3$$

Find the law of composition of the parameters.

8. Show that P^1 is homeomorphic to S^1.

9. Describe the action of R on P^1 in the groupoid of transformations $y_1 : y_2 = a_1\xi_1 : a_2\xi_2$.

10. The map $r \to e^r$ brings R into the multiplicative group of the positive real numbers. Is the map an isomorphism?

11. Find the multiplication law of the groups

(*a*) $R_2 = R \times R$

(*b*) $R_3 = R \times R \times R$

(*c*) $O_2 \times O_2$

(*d*) $R \times O_2$

Determine the spaces underlying each group.

12. Determine the coefficients A_i in formula (5-9*b*) as a function of the a_i [Eqs. (5-5)] and the b_i [Eqs. (5-8*b*)].

13. Verify the group axioms for the direct product of two groups.

14. Show that the direct product of two abelian groups is abelian.

15. Verify the associative law for the transformations (5-4).

◆**16.** Show that the matrices

$$\begin{pmatrix} 1 & x \\ 0 & 1 \end{pmatrix}$$

form a group under matrix multiplication. This group is isomorphic to one discussed in the text. Which one?

◆**17.** Show that the matrices

$$\begin{pmatrix} 1 & a \\ 0 & p \end{pmatrix} \qquad a \text{ real}, p > 0$$

form a multiplicative group. Indicate a domain in the plane that may serve as a space of the group.

5-2. Affine Transformations

***Example* 5-5.** The groups discussed in Sec. 5-1 may be generalized to higher dimensions in various ways. In his proof that every positive integer is the sum of, at most, four integer squares, Euler uses the formula

$$(a^2 + b^2 + c^2 + d^2)(x^2 + y^2 + z^2 + u^2) = X^2 + Y^2 + Z^2 + U^2 \tag{5-10}$$

where

$$\begin{aligned} X &= ax - by - cz - du \\ Y &= bx + ay + dz - cu \\ Z &= cx - dy + az + bu \\ U &= dx + cy - bz + au \end{aligned} \tag{5-11}$$

As before, this shows that the three-dimensional sphere S^3 in four-dimensional space

$$x_1^2 + x_2^2 + x_3^2 + x_4^2 = 1$$

may be given the structure of a continuous group. The product

$$(x,y,z,u) \cdot (a,b,c,d)$$

of two points on the sphere

$$a^2 + b^2 + c^2 + d^2 = 1 \qquad x^2 + y^2 + z^2 + u^2 = 1$$

is the point (X,Y,Z,U) defined by (5-11). It is on the sphere, by Eq. (5-10). The unit element is $e = (1,0,0,0)$. The composition formula (5-11) is known as the *quaternion multiplication*.†

***Example* 5-6.** The most general linear transformation of an n-dimensional cartesian space R^n onto itself,

$$x_i^* = \sum_{k=1}^{n} x_k a_{ki} + b_i \qquad \|a_{ki}\| \neq 0$$

or, in vector notation,

$$\mathbf{x}^* = \mathbf{x}A + \mathbf{b} \qquad \|A\| \neq 0 \tag{5-12}$$

† With topological methods it is possible to show that no n-dimensional sphere $\sum_{i=1}^{n+1} x_i^2 = 1$ other than S^0 ($x^2 = 1$), S^1, and S^3 may be a continuous group. It follows that a formula

$$\left(\sum_{i=1}^{n} a_i^2\right)\left(\sum_{i=1}^{n} x_i^2\right) = \sum_{i=1}^{n} X_i^2$$

is possible only for $n = 1, 2, 4$ if the X_i are continuous (e.g., algebraic) functions of $(a_1, \ldots, a_n, x_1, \ldots, x_n)$.

is a composition of the action of the multiplicative group GL_n of all nonsingular $n \times n$ matrices with the translation group

$$R_n = R \times R \times \cdots \times R$$

Such a linear transformation is also called an *affine transformation* or an *affinity*. For the study of the geometric properties of affine transformations we may restrict our attention to *central affinities* or *linear maps*

$$\mathbf{y} = \mathbf{x}A \tag{5-13}$$

since the remaining translation $\mathbf{x}^* = \mathbf{y} + \mathbf{b}$ is a well-known euclidean motion.

The row vectors $\mathbf{a}_{i\cdot} = (a_{i1}, \ldots, a_{in})$ are the images in the transformation of the unit vectors $\mathbf{e}_i$ whose coordinates are all zero except 1 at place i. Since $\|A\| \neq 0$, the $\mathbf{a}_{i\cdot}$ are linearly independent. We can obtain an orthogonal basis by a *Gram-Schmidt process*

$$\mathbf{b}_1 = \mathbf{a}_{1\cdot}$$

$$\mathbf{b}_2 = \mathbf{a}_{2\cdot} - \frac{\mathbf{a}_{2\cdot}\mathbf{b}_1{}^t}{\mathbf{b}_1\mathbf{b}_1{}^t}\,\mathbf{b}_1$$

$$\cdots$$

$$\mathbf{b}_{k+1} = \mathbf{a}_{k+1} - \sum_{j=1}^{k} \frac{\mathbf{a}_{k+1\cdot}\mathbf{b}_j{}^t}{\mathbf{b}_j\mathbf{b}_j{}^t}\,\mathbf{b}_j$$

$$\cdots$$

The formulas show that the matrix B whose rows are the $\mathbf{b}_j$ is obtained from A by multiplication with a *triangular* matrix Δ in which the diagonal entries are 1 and all entries below the main diagonal are 0. The unit vectors $\mathbf{c}_j = \mathbf{b}_j/|\mathbf{b}_j|$ form an orthonormal basis which is obtained from the $\mathbf{e}_i$ by an orthogonal matrix O. B is obtained from the matrix of the $\mathbf{c}_j$ by the *diagonal* matrix $D = (d_{ij})$ where $d_{ii} = |\mathbf{b}_i|$, $d_{ij} = 0$ for $i \neq j$. From $B = OD = A\Delta$ we get $A = OD\Delta^{-1}$ where Δ^{-1} is again triangular: *Every nonsingular matrix is the product of an orthogonal matrix, a diagonal matrix and a triangular matrix.* The affine map given by D is a *pure deformation* (change of scale) and that by a triangular matrix is a *shear*: *Every linear map is the product of an orthogonal transformation, a pure deformation, and a shear.* A function is an affine invariant if it is invariant under O, D, and Δ.

If a function can be defined using only the notion of length $|\mathbf{x}| = (\mathbf{x}\mathbf{x}^t)^{1/2}$ it is an affine invariant if it is invariant for the symmetric map given by $\mathbf{y}\mathbf{y}^t = \mathbf{x}AA^t\mathbf{x}^t$. Since AA^t is the matrix of a positive definite form there exist O and D such that $AA^t = (OD)(OD)^t$: *A metric function is an affine invariant if it is invariant in orthogonal transformations and pure deformations.* A transformation of matrix OD is a *normal* affinity.

From elementary geometry, the use of euclidean motions and similitudes is well known. We give here some applications of plane affinities to show the use of transformation groups even in elementary problems.

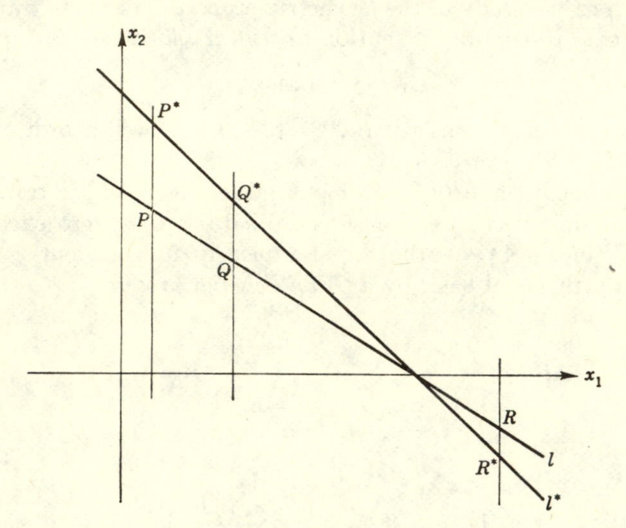

Fig. 5-3

As a prototype of a plane normal affinity we study (Fig. 5-3).

$$(x_1^*,x_2^*) = (x_1,x_2)\begin{pmatrix}1 & 0\\ 0 & a\end{pmatrix} \qquad a > 0 \tag{5-15}$$

Since $x_1 = x_1^*$, corresponding points P and P^* are on parallels to the x_2 axis. The image of a line $Ax_1 + Bx_2 + C = 0$ is a line

$$Ax_1^* + (B/a)x_2^* + C = 0$$

Parallel lines are mapped into parallel lines. The x_1 axis is invariant in (5-15); hence a line l not parallel to the x_1 axis will intersect its image l^* on the x_1 axis. It can be checked immediately in Fig. 5-3 that *ratios of segments on one line are invariant* ($|PQ|:|PR| = |P^*Q^*|:|P^*R^*|$).

A polygon may be divided into triangles with bases parallel to the x_1 axis. Such a triangle is transformed into one with the same base, the height being multiplied by a. The area of the triangle is multiplied by $a = \left\|\begin{matrix}1 & 0\\ 0 & a\end{matrix}\right\|$. The determinant of a product of matrices is equal to the product of the determinants; hence *the area of any polygon is multiplied by the determinant of the transformation in a plane affinity.* (A similar result holds for n-dimensional volumes in n-dimensional affinities.) All properties stated in italics hold in all affine transformations, since they hold in normal ones and in euclidean motions.

The image of a circle $x_1^2 + x_2^2 = r^2$ in a pure deformation $\begin{pmatrix} a & 0 \\ 0 & b \end{pmatrix}$ is the ellipse $x_1^2/a^2 + x_2^2/b^2 = r^2$. The image of an ellipse in a euclidean

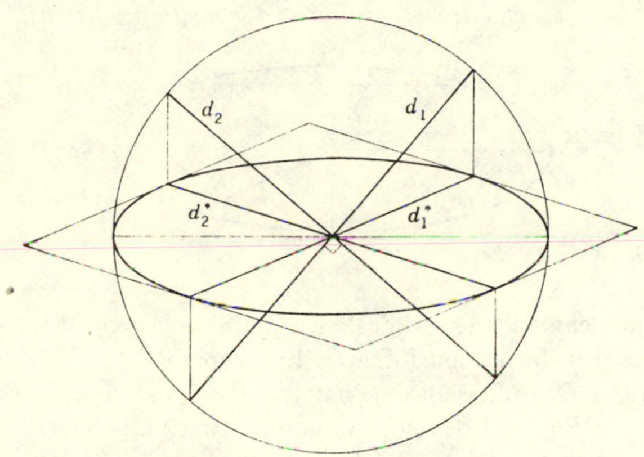

Fig. 5-4

motion is again an ellipse. Hence, *the image of a circle in an affine transformation is an ellipse.* If a theorem holds for the circle and if it expresses a property invariant under affine transformations, it is automatically true for ellipses. Some applications of this principle of proof will now be given.

Conjugate diameters of the ellipse are affine images of a pair of perpendicular diameters of the circle (Fig. 5-4). The tangents to the circle at the endpoints of one diameter are parallel to the perpendicular diameter; hence *the tangents to the ellipse at the endpoints of a diameter are parallel, and their direction is that of the conjugate diameter.* The four tangents at the endpoints of a pair of perpendicular diameters of a circle of radius a determine a square of area $4a^2$. Hence *the parallelogram defined by the tangents at the endpoints of a pair of conjugate diameters of an ellipse of half axes a,b has an area 4ab.* A diameter of a circle subtends a right angle at any point of the circumference. Hence *the two legs of the angle subtended by a diameter at any point of an ellipse give the directions of a pair of con-*

jugate diameters (Fig. 5-5). More applications of affine transformations are given in the exercises.

Of special interest are the affine transformations in R^n that conserve n-dimensional volumes

$$\mathbf{x}^* = \mathbf{x}A + \mathbf{b} \qquad \|A\| = 1$$

The unit matrix U has determinant $\|U\| = 1$, and the product of two matrices of determinant 1 again has determinant 1. Hence the matrices of determinant 1 form a group SL_n with the matrix multiplication as composition (the *special linear group* or *unimodular group*). An affine

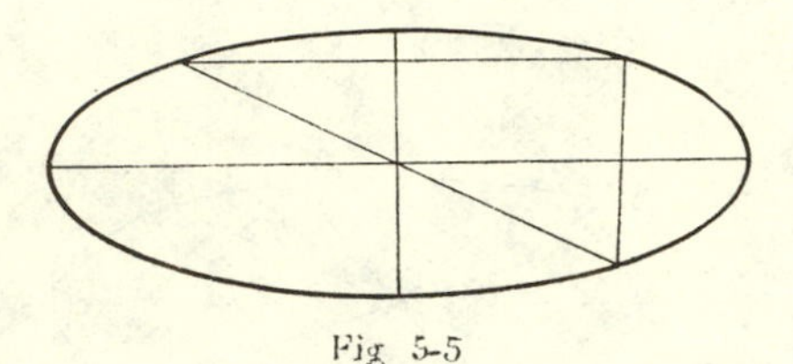

Fig. 5-5

transformation can always be broken up into a "special" volume-conserving affine transformation followed by a homothety. This approach is better suited for differential geometry than the one based on pure deformations. We shall develop affine differential geometry mostly for transformations in the *subgroup* SL_n of GL_n.

***Example* 5-7.** A very important subgroup of GL_n is the group O_n of $n \times n$ orthogonal matrices ($AA^t = U$). Affinities with orthogonal matrices are euclidean motions. For $n > 2$ the group O_n is not abelian; even for group germs, O_n is quite distinct from the group R_{n-1} of $(n-1)$-dimensional vectors under vector addition, in contrast to the situation for $n = 2$ studied in example 5-2. It may be shown also that the space of O_n is not a sphere for $n \neq 2, 3$. The orthogonal group acts by $\mathbf{x}^* = \mathbf{x}A$ on the $(n-1)$-dimensional sphere S^{n-1}; $\mathbf{x}\mathbf{x}^t = 1$ implies $\mathbf{x}^*\mathbf{x}^{*t} = 1$. The *antipodal point* of $\mathbf{x} \in S^{n-1}$ is $-\mathbf{x}$. Pairs of antipodal points are mapped into antipodal points by the action of an orthogonal matrix. The lines through the origin intersect S^{n-1} in pairs of antipodal points. The *projective space* P^{n-1} is the space whose points are the lines through the origin in R^n, the distance of lines being measured by the minimal distance of their respective points of intersection with S^{n-1} (see example 5-4). O_n then acts also as a transformation group on P^{n-1}. It may be shown that P^{n-1} is *not* homeomorphic to S^{n-1} for $n > 2$, in contrast to the case $n = 2$.

For the study of solid geometry, we need more information on the group O_3. We may restrict our attention to the group O_3^* of *rotations*, i.e., of orthogonal matrices with determinant $+1$. Every orthogonal matrix of determinant -1 is the product of $-U$ and a rotation. The rotation

$$A = \begin{pmatrix} a_{11} & a_{12} & a_{13} \\ a_{21} & a_{22} & a_{23} \\ a_{31} & a_{32} & a_{33} \end{pmatrix}$$

satisfies the conditions

$$\sum_k a_{ki}^2 = 1 \qquad i = 1, 2, 3$$
$$\sum_k a_{ki}a_{kj} = 0 \qquad i \neq j$$

This shows that the columns of A are composed of the direction cosines of three mutually orthogonal lines through the origin,

$$A = \begin{pmatrix} \cos\alpha_1 & \cos\beta_1 & \cos\gamma_1 \\ \cos\alpha_2 & \cos\beta_2 & \cos\gamma_2 \\ \cos\alpha_3 & \cos\beta_3 & \cos\gamma_3 \end{pmatrix}$$

If we apply A to the unit vectors on the coordinate axes, $\mathbf{e}_1 = \{1,0,0\}$, $\mathbf{e}_2 = \{0,1,0\}$, $\mathbf{e}_3 = \{0,0,1\}$, we see that the cosines are the direction cosines of the coordinate axes x_i^* $(i = 1,2,3)$ after the rotation.

The *Euler angles* give a description of any rotation by three independent parameters (Fig. 5-6). All we have to do is to describe a rotation that

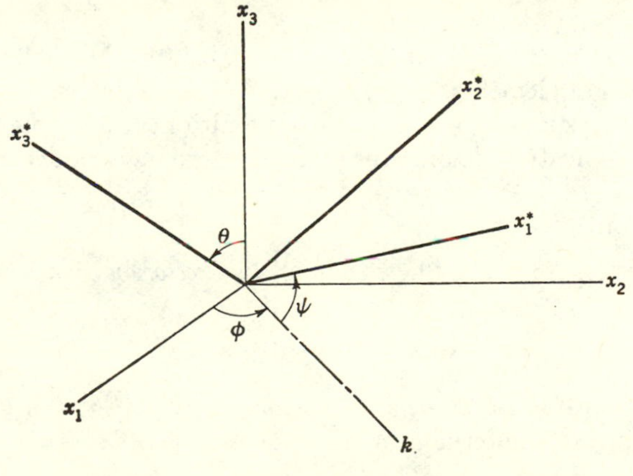

Fig. 5-6

brings three coordinate axes x_1, x_2, x_3 onto three preassigned mutually orthogonal lines x_1^*, x_2^*, x_3^*. Note that we are going to rotate the system of coordinates. We apply the matrices to column vectors. Let k be the line of intersection of the x_1x_2 and the $x_1^*x_2^*$ planes, and ϕ the angle of the rotation A_ϕ about x_3 that brings the x_1 axis into k,

$$A_\phi = \begin{pmatrix} \cos\phi & \sin\phi & 0 \\ -\sin\phi & \cos\phi & 0 \\ 0 & 0 & 1 \end{pmatrix}$$

The rotation

$$A_\theta = \begin{pmatrix} 1 & 0 & 0 \\ 0 & \cos\theta & \sin\theta \\ 0 & -\sin\theta & \cos\theta \end{pmatrix}$$

about k brings x_3 onto x_3^*, and also x_2 into the $x_1^*x_2^*$ plane. A new rotation

$$A_\psi = \begin{pmatrix} \cos\psi & \sin\psi & 0 \\ -\sin\psi & \cos\psi & 0 \\ 0 & 0 & 1 \end{pmatrix}$$

about the x_3^* axis brings k into x_1^*, and x_2 into x_2^*. This shows that any rotation may be represented by its Euler angles as

$$A = A_\psi A_\theta A_\phi \tag{5-16}$$

Exercise 5-2

1. Write Eq. (5-11) as a matrix formula $\mathbf{X} = \mathbf{x}A$, $\mathbf{X} = (X,Y,Z,U)$, $\mathbf{x} = (x,y,z,u)$. Find the matrix A.
2. Let $1 = (1,0,0,0)$, $i = (0,1,0,0)$, $j = (0,0,1,0)$, $k = (0,0,0,1)$. Show that the eight elements $1, i, j, k, -1, -i, -j, -k$ form a group under the composition law (5-11). Find its multiplication table.
3. The multiplication defined by (5-11) for two *quaternions*

   $$A = a + bi + cj + dk \qquad X = x + yi + zj + uk$$

   will be written $A \circ X$. Show that for two *vectors* $\mathbf{a} = a_1 i + a_2 j + a_3 k$, $\mathbf{x} = x_1 i + x_2 j + x_3 k$,

   $$\mathbf{a} \circ \mathbf{x} = -\mathbf{a} \cdot \mathbf{x} + \mathbf{a} \times \mathbf{x}$$

4. The length $|A|$ of a quaternion A is defined by $|A|^2 = a^2 + b^2 + c^2 + d^2$. The conjugate quaternion $\bar{A}$ of A is $\bar{A} = a - bi - cj - dk$. Show that
   (*a*) $|A|^2 = A \circ \bar{A}$.
   (*b*) $|A \circ B| = |A|\,|B|$.
   (*c*) $\overline{A \circ B} = \bar{B} \circ \bar{A}$.
   (*d*) $|A| = |\bar{A}|$.

(*e*) If $|A| = 1$, A may be written $A = \cos \phi + \mathbf{n} \sin \phi$, $\mathbf{n}$ a unit *vector* (see Prob. 3).

5. If $\mathbf{n}$ is a unit *vector*, show that $\mathbf{x}' = \mathbf{n} \circ \mathbf{x} \circ \mathbf{n}$ is the vector symmetric to $\mathbf{x}$ with respect to the plane through the origin and normal to $\mathbf{n}$.

6. Let i, j, k be the unit vectors on the coordinate axes in R^3. If a vector $\mathbf{x}$ (see Prob. 3) is rotated about an axis through the origin ($\mathbf{n}$ a unit vector on the axis of rotation) and if the angle of rotation is θ, show that the result of the rotation is the vector $\mathbf{x}^* = D \circ \mathbf{x} \circ \bar{D}$, where D is the unit quaternion $D = \cos \theta/2 + \mathbf{n} \sin \theta/2$ (Euler). [HINT: Split the rotation into two reflections (symmetries at a plane) and use Prob. 5.]

7. The composition of two rotations belonging to unit quaternions D_1, D_2 (see Prob. 6) is the rotation given by $D_2 \circ D_1$. Show that the multiplicative group of unit quaternions is isomorphic to O_3^*.

8. Use Probs. 6 and 1 to compute the rotation matrix A to a rotation of axis $\mathbf{n} = (n_1,n_2,n_3)$ and angle θ.

9. If D is not a unit quaternion, $\mathbf{x}^* = D \circ \mathbf{x} \circ \bar{D}$ is a rotation followed by a homothety. What is the ratio of the homothety?

10. With a quaternion $A = a_0 + a_1 i + a_2 j + a_3 k$ we associate a matrix

$$\mathfrak{A} = \begin{pmatrix} a_0 + \sqrt{-1}\, a_3 & -a_2 + \sqrt{-1}\, a_1 \\ a_2 + \sqrt{-1}\, a_1 & a_0 - \sqrt{-1}\, a_3 \end{pmatrix}$$

Show that the matrix of $A \circ B$ is $\mathfrak{A}\mathfrak{B}$, that of $A + B$ is $\mathfrak{A} + \mathfrak{B}$. Find the matrices that correspond to the base vectors $1, i, j, k$ of the quaternions (Moebius).

11. (Sequel to Probs. 7 and 10.) There exists a representation of O_3^* as a group of complex-valued 2×2 matrices. Characterize that group as a subgroup of the group $GL_2(C)$ of nonsingular 2×2 matrices with complex elements.

12. In Prob. 1, determine the matrix $A(t)$ that belongs to the action of the quaternion $\cos t + k \sin t$. If $x{:}y{:}z{:}u$ are homogeneous coordinates in R^3, show that $\mathbf{x}(t) = \mathbf{x}_0 A(t)$ describes a straight line as t varies. Different orbits under $A(t)$ are *skew;* they are neither parallel nor intersecting (Clifford).

13. Show that the matrix described in Prob. 1 maps $\mathbf{x}\mathbf{x}^t = 0$ into $\mathbf{X}\mathbf{X}^t = 0$.

14. In an affine transformation, the midpoint of a segment is mapped into the midpoint of the image. Show that the centroid of a triangle is mapped into the centroid of the image of the triangle in any affinity.

15. Show that the locus of the midpoints of the chords of an ellipse parallel to a diameter is the diameter conjugate to the given one.

16. Prove that the product of the segments defined on a tangent to an ellipse by two conjugate diameters is equal to the square of the half diameter parallel to the tangent (Fig. 5-7).

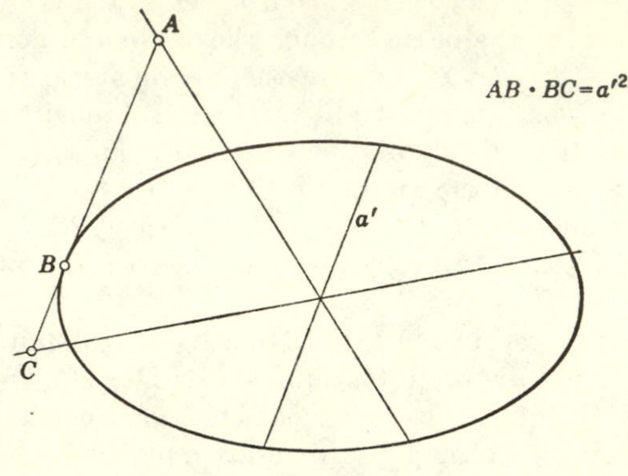

Fig. 5-7

17. Discuss the definition and properties of triples of conjugate diameters of an ellipsoid in R^3. Prove that the parallelepipedon determined by the tangent planes at the endpoints of a triple of conjugate diameters has the volume $8abc$ (a,b,c half axes of the ellipsoid).

18. The Moebius pseudo group is generated by the combination of the group of similitudes with $\mathbf{x}^* = \mathbf{x}/|\mathbf{x}|^2$. Show that a transformation from this pseudo group maps any arc of a circle into an arc of a circle or into a straight segment.

19. Show that an orthogonal matrix εO_3 but $\not\varepsilon O_3^*$ is the product of a rotation and $-U$. Such a matrix A, $\|A\| = -1$, is called a *reflection*. Show also that any rotation is the product of two reflections.

20. Use Eq. (5-16) to find the explicit formula of a rotation in terms of its Euler angles.

21. If the rotation A has Euler angles ϕ, θ, ψ, find the Euler angles of A^{-1}.

References

Freudenthal, H.: Lie Groups, Lecture Notes, Yale University, 1961.

Klein, Felix: "Elementary Mathematics from an Advanced Standpoint: Geometry," vol. II, 1914 (reprint), Dover Publications, Inc., New York.

Lie, S.: "Vorlesungen über kontinuierliche Gruppen," edited and published by G. Scheffers, B. G. Teubner Verlagsgesellschaft, mbH, Leipzig, 1893.

Rothe, H.: Systeme geometrischer Analyse, "Encyclopädie der mathematischen Wissenschaften," vol. III, pp. 1277–1423, especially secs. 12 and 13, B. G. Teubner Verlagsgesellschaft, mbH, Leipzig, 1921.

Stéphanos, C.: Mémoire sur la représentation des homographies binaires par des points de l'espace avec applications à l'étude des rotations sphériques, *Math. Ann.*, **22:** 299–367 (1883).

6

LIE GROUP GERMS

6.1. Lie Group Germs and Lie Algebras

The contents of this and the following chapter are mainly elaborations of the work of Sophus Lie (1842–1899). This classical part of Lie group theory uses differentiability assumptions and analytical methods. It does *not* consider topological questions.

Definition 6-1. *A Lie group germ G in a neighborhood V of the origin* $\mathbf{e}$ *in an n-dimensional cartesian space R^n is defined by a function*

$$(\mathbf{x},\mathbf{y}) \to \mathbf{f}(\mathbf{x},\mathbf{y}) \in R^n$$

subject to the following axioms:

($L1$). $\mathbf{f}(\mathbf{x},\mathbf{y})$ *is defined for all* $\mathbf{x} \in V$, $\mathbf{y} \in V$.

($L2$). $\mathbf{f}(\mathbf{x},\mathbf{y})$ *is twice continuously differentiable in all its $2n$ arguments.*

($L3$). *If* $\mathbf{f}(\mathbf{x},\mathbf{y}) \in V$, $\mathbf{f}(\mathbf{y},\mathbf{z}) \in V$, *then* $\mathbf{f}(\mathbf{f}(\mathbf{x},\mathbf{y}),\mathbf{z}) = \mathbf{f}(\mathbf{x},\mathbf{f}(\mathbf{y},\mathbf{z}))$.

($L4$). $\mathbf{f}(\mathbf{e},\mathbf{y}) = \mathbf{y}$, $\mathbf{f}(\mathbf{x},\mathbf{e}) = \mathbf{x}$.

We shall write $\mathbf{f}(\mathbf{x},\mathbf{y}) = \mathbf{xy}$ (or $= \mathbf{x} \cdot \mathbf{y}$) if there can be no misunderstanding. In this group multiplication, ($L3$) is the associative law $\mathbf{xy} \cdot \mathbf{z} = \mathbf{x} \cdot \mathbf{yz}$ if all the compositions involved are defined. ($L4$) shows that the origin $\mathbf{e} = (0, \ldots, 0)$ is a unit of the multiplication.

The notion of a Lie group germ is somewhat hazy, since any open subset $V' \subset V$ that contains the origin gives rise to another Lie group germ defined on the same group structure. It is possible to get rid of this indetermination by the construction of an *algebra* that uniquely characterizes all germs derived from the same group structure.

Definition 6-2. *A (real) algebra of dimension n is a (real) vector space of dimension n in which is defined a distributive multiplication* $(\mathbf{a},\mathbf{b}) \to [\mathbf{a},\mathbf{b}]$.

The only laws of multiplication prescribed for an algebra are (a,b

real numbers)

$$[a\mathbf{x} + b\mathbf{y}, \mathbf{z}] = a[\mathbf{x},\mathbf{z}] + b[\mathbf{y},\mathbf{z}] \qquad [\mathbf{x}, a\mathbf{y} + b\mathbf{z}] = a[\mathbf{x},\mathbf{y}] + b[\mathbf{x},\mathbf{z}] \tag{6-1}$$

Neither the associative nor the commutative laws need hold. Addition is the usual vector addition in R^n.

In the whole discussion the coordinate system in V will not be changed. Therefore, vectors $\mathbf{x} \in R^n$ may be given by their coordinate vectors $(x_1, \ldots, x_n)$.

Let $\mathbf{a}(t) = (a_1(t), \ldots, a_n(t))$, $\mathbf{a}(0) = \mathbf{e}$, be a differentiable curve in V. The tangent vector to $\mathbf{a}(t)$ in $\mathbf{e}$ is $\alpha = (d\mathbf{a}/dt)_{t=0}$. α is called the *infinitesimal vector* of $\mathbf{a}(t)$. All vectors $\alpha \in R^n$ are infinitesimal vectors of some differentiable curves in V, for example, of $\mathbf{a}(t) = \alpha t$. The set of infinitesimal vectors therefore is an n-dimensional vector space. It is possible to connect vector addition of infinitesimal vectors with the group multiplication in V.

Theorem 6-3. *Let* $\mathbf{a}(t),\mathbf{b}(t)$ $(\mathbf{a}(0) = \mathbf{b}(0) = \mathbf{e})$ *be* C^n $(n \geq 1)$ *curves in a Lie group germ, with infinitesimal vectors* α, β. *The curve* $\mathbf{a}(t)\mathbf{b}(t)$ *is differentiable, and its infinitesimal vector is* $\alpha + \beta$.

Let $f_k(a_1, \ldots, a_n, b_1, \ldots, b_n) = (\mathbf{ab})_k$ be the kth coordinate of the product $\mathbf{ab} = \mathbf{f}(\mathbf{a},\mathbf{b})$. Then

$$\frac{d(\mathbf{ab})_k}{dt} = \sum_{i=1}^{n} \frac{\partial f_k}{\partial a_i}\frac{da_i}{dt} + \sum_{i=1}^{n} \frac{\partial f_k}{\partial b_i}\frac{db_i}{dt} \tag{6-2}$$

Hence $\mathbf{a}(t)\mathbf{b}(t)$ is a differentiable curve, $\mathbf{a}(0)\mathbf{b}(0) = \mathbf{e}$. The second part of the theorem follows from the next lemma.

Lemma 6-4. *At the origin, the jacobians* $J_{\mathbf{x}}\mathbf{f} = (\partial f_k(\mathbf{x},\mathbf{e})/\partial x_j)$ *and* $J_{\mathbf{y}}\mathbf{f} = (\partial f_k(\mathbf{e},\mathbf{y})/\partial y_j)$ *both are the unit matrix.*

$\mathbf{h}_j = (0, \ldots, 0, h, 0, \ldots, 0)$ is the vector of length $|h|$ on the jth coordinate axis; all components of $\mathbf{h}_j$ are zero except the jth. By axiom $(L4)$,

$$\left.\frac{\partial \mathbf{f}(\mathbf{x},\mathbf{e})}{\partial x_j}\right|_{\mathbf{x}=\mathbf{e}} = \lim_{h\to 0} \frac{1}{h}\left(\mathbf{f}(\mathbf{h}_j,\mathbf{e}) - \mathbf{f}(\mathbf{e},\mathbf{e})\right) = \mathbf{1}_j$$

or

$$J_{\mathbf{x}}\mathbf{f}\Big|_{\mathbf{x}=\mathbf{y}=\mathbf{e}} = \left(\frac{\partial f_k}{\partial x_j}\right)_{\mathbf{x}=\mathbf{y}=\mathbf{e}} = (\delta_{kj}) = U \tag{6-3}$$

δ_{kj} is the *Kronecker symbol* of the unit matrix [see Eq. (1-4)]. An analogous computation shows that $J_{\mathbf{y}}\mathbf{f}\Big|_{\mathbf{x}=\mathbf{y}=\mathbf{e}} = U$.

Equation (6-2) may be written in matrix form

$$\frac{d(\mathbf{ab})}{dt} = \frac{d\mathbf{a}}{dt} J_{\mathbf{a}}\mathbf{f} + \frac{d\mathbf{b}}{dt} J_{\mathbf{b}}\mathbf{f} \tag{6-2a}$$

Hence, by lemma 6-4,

$$\left.\frac{d(\mathbf{ab})}{dt}\right|_{t=0} = \alpha U + \beta U = \alpha + \beta$$

This completes the proof of theorem 6-3.

Our axioms of a Lie group germ differ in their algebraic aspect from the axioms of a group given in Sec. 5-1. The existence of a unit does not, in general, imply the group axiom ii. This axiom is equivalent to the existence and unicity of both unit and inverse. But for differentiable groups, the existence and uniqueness of an inverse follows from lemma 6-4. The right inverse $\mathbf{a}_r^{-1}$ of an $\mathbf{a} \varepsilon V$ is a vector $\mathbf{y}$ such that $\mathbf{f}(\mathbf{a},\mathbf{y}) = \mathbf{e}$. Since $\|J_\mathbf{y}\mathbf{f}\| \neq 0$ at $\mathbf{e}$, the inverse-function theorem† implies that there exists a unique differentiable $\mathbf{y}(\mathbf{a})$ in some neighborhood of the origin. By the same reason there exists a unique differentiable left inverse $\mathbf{a}_l^{-1} = \mathbf{x}$, $\mathbf{f}(\mathbf{x},\mathbf{a}) = \mathbf{e}$. The intersection of the domains of existence of $\mathbf{a}_r^{-1}$ and $\mathbf{a}_l^{-1}$ is a neighborhood V' of $\mathbf{e}$. In V', axiom ($L3$) holds for the inverses; hence

$$\mathbf{a}_l^{-1} = \mathbf{a}_l^{-1} \cdot \mathbf{a}\mathbf{a}_r^{-1} = \mathbf{a}_l^{-1}\mathbf{a} \cdot \mathbf{a}_r^{-1} = \mathbf{a}_r^{-1}$$

The left inverse is the right inverse; thus there exists a unique two-sided inverse.

Theorem 6-5. *In any Lie group germ G there is a neighborhood V' of the origin such that for any $\mathbf{a} \varepsilon V'$ there exists a unique $\mathbf{a}^{-1} \varepsilon V$. If $\mathbf{a}(t)$ is a differentiable curve, $\mathbf{a}(0) = \mathbf{e}$, the curve $\mathbf{a}^{-1}(t)$ is differentiable in V'.*

The infinitesimal vector of the curve $\mathbf{e}(t) = \mathbf{e}$ is the zero vector. If the infinitesimal vector of $\mathbf{a}(t)$ is α, that of $\mathbf{a}^{-1}(t)$ is $-\alpha$, by theorem 6-3. The curves $\mathbf{a}^{-1}(t)$ and $-\mathbf{a}(t)$ have identical infinitesimal vectors.

Definition 6-6. *The Lie product $[\alpha,\beta]$ of two infinitesimal vectors α, β belonging to curves $\mathbf{a}(t)$, $\mathbf{b}(t)$ is the infinitesimal vector of $(\mathbf{ab} - \mathbf{ba})$ (τ), where $\tau = t^2$. The subtraction is to be understood in the sense of the vector addition in R^n.*

A vector α is the infinitesimal vector of an infinity of different curves $\mathbf{a}(t)$. In order to show that the Lie product is independent of the chosen curves $\mathbf{a}(t)$ and $\mathbf{b}(t)$ and to explain the change of parameters $t \rightarrow \tau$, we write the Taylor expansion of $\mathbf{f}(\mathbf{x},\mathbf{y})$ in $(\mathbf{e},\mathbf{e})$. It is possible to compute a second-order expansion by axiom ($L2$). The constant term vanishes [axiom ($L4$)] and the linear term is $\mathbf{x} + \mathbf{y}$ [lemma 6-4 and Eq. (6-2)]:

$$f_k(x_1, \ldots ,x_n,y_1, \ldots ,y_n) = x_k + y_k + \sum_{i,j} a_{k,ij}x_iy_j + \epsilon_k(\mathbf{x},\mathbf{y}) \qquad (6\text{-}4)$$

† See, for example, R. C. Buck, "Advanced Calculus," Sec. 5.8, McGraw-Hill Book Company, Inc., New York, 1956.

The infinitesimal vectors of $\mathbf{a}(t)$ and $\mathbf{b}(t)$ are the coefficients of the linear terms in an expansion

$$a_i(t) = \alpha_i t + \cdots \qquad b_i(t) = \beta_i t + \cdots$$

Hence

$$(ab)_k = (\alpha_k + \beta_k)t + \sum_{i,j} a_{k.ij}\alpha_i\beta_j t^2 + \epsilon(t)$$

or

$$(ab - ba)_k = \sum_{i,j} (a_{k,ij} - a_{k,ji})\alpha_i\beta_j t^2 + \epsilon(t) \tag{6-5}$$

These equations show that the product is commutative in the first approximation near the unit of a Lie group. The

$$c_{k,ij} = a_{k,ij} - a_{k,ji} \tag{6-6}$$

are the *structure constants* of the group germ G; they vanish only for abelian groups. In Eq. (6-4)

$$\lim_{|\mathbf{x}|+|\mathbf{y}|\to 0} \frac{\epsilon(\mathbf{x},\mathbf{y})}{|\mathbf{x}|^2 + |\mathbf{y}|^2} = 0$$

Hence $\lim_{t\to 0} \epsilon(t)/t^2 = 0$ in Eq. (6-5). By Def. 6-6, the kth coordinate of the Lie product is

$$[\alpha,\beta]_k = \sum_{i,j} c_{k,ij}\alpha_i\beta_j \tag{6-7}$$

By theorem 6-3, the infinitesimal vector of $-\mathbf{ba}$ is the same as that of $(\mathbf{ba})^{-1} = \mathbf{a}^{-1}\mathbf{b}^{-1}$. A direct verification on Eq. (6-5) shows the following:

Theorem 6-7. *$[\alpha,\beta]$ is the infinitesimal vector of the commutator curve* $\mathbf{aba}^{-1}\mathbf{b}^{-1}(\tau)$.

We have said before that we shall keep the coordinate system fixed throughout Secs. 6-1 and 6-2, and we have made extensive use of this fixed coordinate system. Nevertheless, it is important to know which properties are independent of any system of coordinates. Addition of infinitesimal vectors has this property as vector addition and also by theorem 6-3. Theorem 6-7 shows that the Lie product can be defined by reference to the product in the Lie group germ only. On the other hand, the structure constants make sense only if referred to the fixed system of coordinates.

Equation (6-7) is linear both in α and in β, and the distributive law (6-1) holds for the Lie product. The Lie product turns the vector space of infinitesimal vectors into an algebra with *anticommutative* multiplication

$$[\alpha,\beta] = -[\beta,\alpha] \tag{6-8}$$

Definition 6-8. *The Lie algebra $\mathfrak{L}(G)$ of a Lie group germ is the algebra of infinitesimal vectors defined by the Lie product.*

Before we proceed with the theory, we determine the Lie algebras of some groups discussed in Chap. 5.

***Example* 6-1.** The elements of the general linear group GL_n and of any multiplicative matrix group will be written

$$X = \begin{pmatrix} 1 + x_{11} & x_{12} & \cdots & x_{1n} \\ x_{21} & 1 + x_{22} & \cdots & x_{2n} \\ \cdots & \cdots & \cdots & \cdots \\ x_{n1} & x_{n2} & \cdots & 1 + x_{nn} \end{pmatrix} = U + (x_{ik}) \qquad (6\text{-}9)$$

It is necessary to put in evidence the 1's in the principal diagonal, because the unit element U of the group must be identified with the origin $\mathbf{e}$ ($x_{ik} = 0$) of the space of coordinates R^{n^2}. Equation (6-9) makes sense only for matrices in some neighborhood of the unit matrix. If the x_{ik}'s become too big, the determinant $\|X\|$ may vanish and X no longer belongs to GL_n. A differentiable curve in the group is given by $x_{ik} = x_{ik}(t)$, $(dx_{ik}/dt)_{t=0} = a_{ik}$. The Lie algebra $\mathfrak{L}(GL_n)$ is the space of *all* $n \times n$ matrices $\alpha = (\alpha_{ik})$ (determinant vanishing or not). The product of two matrices $X = U + (x_{ik})$, $Y = U + (y_{ik})$ is

$$XY = U + (x_{ik}) + (y_{ik}) + \Big(\sum_j x_{ij}y_{jk}\Big) \qquad (6\text{-}10)$$

If X and Y represent curves with infinitesimal vectors $\alpha = (\alpha_{ik})$, $\beta = (\beta_{ik})$, the Lie product becomes [Eq. (6-5)]

$$[\alpha,\beta] = \Big(\sum_j \alpha_{ij}\beta_{jk} - \beta_{ij}\alpha_{jk}\Big) = \alpha\beta - \beta\alpha \qquad (6\text{-}11)$$

The products on the right-hand side of this equation are matrix products.

Theorem 6-9. *$\mathfrak{L}(GL_n)$ is the algebra of all $n \times n$ matrices with the product $[\alpha,\beta] = \alpha\beta - \beta\alpha$.*

***Example* 6-2.** The orthogonal group O_n is a subgroup of GL_n. Its multiplication is the matrix multiplication. By Eqs. (6-10) and (6-11) the multiplication in $\mathfrak{L}(O_n)$ is again $\alpha\beta - \beta\alpha$. Let $X(t)$ be a curve in O_n, $X(t)X(t)^t = U$, $X(0) = U$, $X^{\cdot}(0) = \alpha$. By differentiation,

$$X^{\cdot}X^t + XX^{t\cdot} = 0$$

Hence for $t = 0$

$$\alpha^t + \alpha = 0 \qquad (6\text{-}12)$$

(See also exercise 6-1, Prob. 6.)

Theorem 6-10. *$\mathfrak{L}(O_n)$ is the algebra of skew-symmetric $n \times n$ matrices with the product $[\alpha,\beta] = \alpha\beta - \beta\alpha$.*

The Cartan matrices $k(s)J$ used in plane differential geometry are curves in $\mathfrak{L}(O_2)$. The curve is given by its frame matrix $A(s)$. For each s_0, the curve $X(t) = A(s_0 + t)A^{-1}(s_0)$ is in O_2. It starts from the origin, $X(0) = U$, with infinitesimal vector

$$X'(0) = A'(s_0)A^{-1}(s_0) = C(A)(s_0)$$

This reasoning is valid in any Lie group germ.

Theorem 6-11. *The Cartan map*

$$\mathbf{a}(t) \to \frac{d\mathbf{a}(t)}{dt}\,\mathbf{a}^{-1}(t)$$

is a map from a Lie group (germ) into its Lie algebra.

***Example* 6-3.** SL_n is the group of matrices of determinant 1. In the coordinates (6-9),

$$1 = \|X\| = 1 + \sum_i x_{ii} + \text{higher-order terms in } x_{ik} \tag{6-13}$$

For a curve from the origin, $x_{ik}(0) = 0$, $(dx_{ik}/dt)_{t=0} = \alpha_{ik}$. As always, we differentiate (6-13) and put $t = 0$, to obtain

$$\sum_i \alpha_{ii} = 0 \tag{6-14}$$

The expression on the left-hand side of Eq. (6-14) is the sum of the diagonal elements of $\alpha = (\alpha_{ik})$, the *trace* of α.

Theorem 6-12. $\mathfrak{L}(SL_n)$ *is the algebra of matrices of trace zero, with the product* $[\alpha,\beta] = \alpha\beta - \beta\alpha$.

***Example* 6-4.** In an abelian group, $\mathbf{ab} - \mathbf{ba} = 0$.

Theorem 6-13. *In the Lie algebra of any commutative Lie group germ all Lie products are zero.*

The group O_2 is commutative (example 5-2). $\mathfrak{L}(O_2)$ is the algebra of matrices kJ, k a real number. The Lie product in $\mathfrak{L}(O_2)$ is

$$[k_1J,k_2J] = k_1k_2(JJ - JJ) = 0$$

as it must be by theorem 6-13.

Exercise 6-1

◆ **1.** Find the Lie algebra of the additive group of vectors in the plane R_2.

◆ **2.** Find the Lie algebra of the multiplicative group of matrices λA, λ a real number $\neq 0$, $A \in O_n$.

3. Determine the Lie algebra of the multiplicative group of nonzero quaternions. [Hint: Use (5-11), but note that in that formula the unit has coordinates (1,0,0,0).]

4. Same question as Prob. 3 for the unit quaternions D, $|D| = 1$.

◆ **5.** Show that the matrices

$$\begin{pmatrix} 1 & x & z \\ 0 & 1 & y \\ 0 & 0 & 1 \end{pmatrix}$$

form a group under matrix multiplication, and find its Lie algebra. If the matrix is represented by a vector $(x,y,z) \in R^3$, give an explicit formula for the Lie multiplication of two infinitesimal vectors $(\alpha_1,\alpha_2,\alpha_3)$ and $(\beta_1,\beta_2,\beta_3)$.

6. In the proof of theorem 6-10 we have shown only that $\mathfrak{L}(O_n)$ consists of skew matrices. It remains to be shown that *all* skew $n \times n$ matrices appear as infinitesimal vectors. The proof is best broken up into steps.
(*a*) Show that the matrices $\alpha = (\alpha_{ij})$, $\alpha_{i_0j_0} = -\alpha_{j_0i_0} = 1$, $i_0 < j_0$, all other $\alpha_{ij} = 0$, form a basis of the vector space of $n \times n$ skew-symmetric matrices. (Check dimensions and linear independence.)
(*b*) Show that the matrix (α_{ij}) is the infinitesimal vector of $X(t)$,

$$\begin{aligned} x_{ii}(t) &= 1 \qquad i \neq i_0,j_0 \\ x_{i_0i_0}(t) &= x_{j_0j_0}(t) = \cos t \\ x_{i_0j_0}(t) &= -x_{j_0i_0}(t) = \sin t \\ x_{ij}(t) &= 0 \qquad \text{otherwise} \end{aligned}$$

◆ **7.** Find the Lie algebra of the group defined in exercise 5-1, Prob. 17. Give its multiplication table in terms of a basis of the algebra. What is the dimension of the algebra? (Dimension of an algebra = dimension of its vector space.)

8. The elements of $\mathfrak{L}(O_3)$ are the matrices $\alpha = \begin{pmatrix} 0 & p & -q \\ -p & 0 & r \\ q & -r & 0 \end{pmatrix}$.
Such a matrix may be represented by a vector $\boldsymbol{\alpha} = (p,q,r)$. Find the vector which belongs to $[\alpha,\beta]$ as a function of vectors $\boldsymbol{\alpha},\boldsymbol{\beta}$.

9. Show that the vectors $(\lambda x_0,\lambda y_0)$, $\lambda \in R$, x_0,y_0 fixed real numbers, form a subgroup H_1 of R_2. Find its Lie algebra.

10. Prove that the set of vectors (m,n), m,n integers, is a subgroup H_2 of R_2. Is it a Lie group?

11. Let a be a fixed irrational number. The set of vectors $(m + c, n + ca)$, m,n integers, c real, defines a subgroup H_3 of R_2. Show that no neighborhood of the origin in H_3 is an open set in R_2 but that the point set of H_3 is dense in the plane. H_3 is not a Lie group.

12. For a matrix $X = (x_{ij})$ with *complex* coefficients, the transpose of its complex conjugate $X^h = (\bar{x}_{ji})$ is important. The matrix is *hermitian* if $X = X^h$, *skew-hermitian* if $X = -X^h$, and *unitary* if

$XX^h = U$. Show that all unitary $n \times n$ matrices form a Lie group and find its Lie algebra.

13. The groups O_3 and O_3^* (example 5-7) have identical Lie algebras. Explain this fact.

14. If α,β are skew-symmetric matrices, show that $\alpha\beta - \beta\alpha$ is skew-symmetric.

15. Same as Prob. 14 for matrices of trace zero.

16.

$$I = \begin{pmatrix} 0 & 0 & 1 & 0 \\ 0 & 0 & 0 & 1 \\ -1 & 0 & 0 & 0 \\ 0 & -1 & 0 & 0 \end{pmatrix}$$

All 4×4 matrices A which satisfy $AI^t + IA = 0$ form a Lie group Sp_2. Find $\mathfrak{L}(Sp_2)$.

◆**17.** The group Sp_n is defined for $2n \times 2n$ matrices analogous to Sp_2, starting with $I_n = \begin{pmatrix} 0 & U \\ -U & 0 \end{pmatrix}$, where U and 0 are $n \times n$ unit and zero matrices. Find $\mathfrak{L}(Sp_n)$.

6-2. The Adjoint Representation

A homomorphism of a structure G into a structure G' of the same kind is a map that preserves the structure (Sec. 5-1).† If G and G_1 are groups with multiplications $\cdot$ and $\cdot_1$, a homomorphism is a map

$$h\colon G \to G_1$$

such that

$$h(x \cdot y) = h(x) \cdot_1 h(y) \tag{6-15}$$

If G and G_1 are Lie groups, the homomorphism must conserve the differentiable structure; h is a homomorphism of Lie groups only if it is a C^2 function and satisfies Eq. (6-15).

The structure of an algebra is given by its addition $+$ and its multiplication $[\ ,\ ]$. A homomorphism of an algebra $\mathfrak{L}(+,[\ ,\ ])$ into an algebra $\mathfrak{L}_1(+_1,[\ ,\ ]_1)$ is a map $h\colon \mathfrak{L} \to \mathfrak{L}_1$ such that

$$\begin{aligned} h(\alpha + \beta) &= h(\alpha) +_1 h(\beta) \\ h([\alpha,\beta]) &= [h(\alpha), h(\beta)]_1 \end{aligned} \tag{6-16}$$

It follows from Eq. (6-15) that a group homomorphism maps the unit $e \in G$ into the unit e_1 of G_1. e is the unique element such that

† A systematic theory of structures and their morphisms is given in N. Bourbaki, "Eléments de Mathématique," part 1, book 1, chap. 4, Hermann & Cie, Paris, 1957.

$x \cdot e = x$; hence $h(x) \cdot_1 h(e) = h(x)$, that is, $h(e) = e_1$. Since

$$h(x \cdot x^{-1}) = h(x) \cdot_1 h(x)^{-1} = e_1$$

it also follows that $h(x^{-1}) = h(x)^{-1}$. Finally, h maps the commutator $x \cdot y \cdot x^{-1} \cdot y^{-1}$ into the commutator $h(x) \cdot_1 h(y) \cdot_1 h(x)^{-1} \cdot_1 h(y)^{-1}$. If G and G_1 are Lie group germs, a homomorphism maps a product curve $a(t) \cdot b(t)$ onto the product curves of the images $h(a(t)) \cdot_1 h(b(t))$ and the commutator curve into the commutator curve of the images. Since h is differentiable, the image curves also have infinitesimal vectors. Theorems 6-3 and 6-7 show that the map defined by the correspondence of the infinitesimal vectors is an algebra homomorphism.

Theorem 6-14. *A homomorphism h of Lie group germs G,G_1 defines an induced algebra homomorphism $h^*\colon \mathfrak{L}(G) \to \mathfrak{L}(G_1)$ by $h^*(\alpha) = \lim_{t \to 0} dh(\alpha t)/dt$.*

A homomorphism which is a one-to-one map is an *isomorphism*. An isomorphism of a set onto itself is an *automorphism*. An element g of a group G defines an automorphism

$$h_g(x) = gxg^{-1} \qquad x \in G \tag{6-17}$$

This map of the group into itself is a homomorphism since

$$h_g(xy) = gxyg^{-1} = gxg^{-1}gyg^{-1} = h_g(x)\,h_g(y)$$

It is an isomorphism because $gxg^{-1} = gyg^{-1}$ implies $g^{-1}gxg^{-1}g = g^{-1}gyg^{-1}g$; hence $x = y$. It is onto, because every element of G is an image under h_g, $x = h_g(g^{-1}xg)$. Any automorphism that is of the form (6-17) is an *inner* automorphism. If h_g is defined on a Lie group germ, it is still called an inner automorphism, although it will not, in general, be a map of V onto V. In this case, theorem 6-7 and Eq. (6-7) show that in coordinates

$$(h_{\mathbf{g}}(\mathbf{x})\mathbf{x}^{-1})_k = \sum_{i,j} c_{k,ij} g_i x_j + \epsilon$$

Hence, by theorem 6-5,

$$h_{\mathbf{g}}(\mathbf{x})_k = x_k + \sum_{i,j} c_{k,ij} g_i x_j + \epsilon \tag{6-18}$$

The matrix $C_{\mathbf{g}}$,

$$C_{\mathbf{g}} = (\delta_{jk} + \sum_i c_{k,ij} g_i) \tag{6-19}$$

allows the expression of $h_{\mathbf{g}}$ approximately by a linear transformation

$$h_{\mathbf{g}}(\mathbf{x}) = \mathbf{x} C_{\mathbf{g}} + \varepsilon \tag{6-20}$$

Since $h_{\mathbf{m}}(h_{\mathbf{g}}(\mathbf{x})) = \mathbf{x} C_{\mathbf{g}} C_{\mathbf{m}} + \varepsilon$, the map

$$\mathbf{g} \to C_{\mathbf{g}}{}^t$$

is a homomorphism of G into GL_n (for $\mathbf{g}$ near $\mathbf{e}$, $C_{\mathbf{g}}$ is nonsingular). Note that we have to use the transpose matrix to ensure that right-hand multiplication is mapped into right-hand multiplication. By theorem 6-14, the induced homomorphism of $\mathfrak{L}(G)$ into $\mathfrak{L}(GL_n)$ is

$$\alpha \to \text{ad } \alpha = \Big(\sum_s c_{i,sj}\alpha_s\Big) \tag{6-21}$$

The mapping "ad" is the *adjoint representation* of $\mathfrak{L}(G)$. It is an algebra homomorphism,

$$\text{ad } [\alpha,\beta] = \text{ad } \alpha \text{ ad } \beta - \text{ad } \beta \text{ ad } \alpha \tag{6-22}$$

This means for the operation of the adjoint matrices on an arbitrary vector γ that

$$[[\alpha,\beta],\gamma] = [\alpha,[\beta,\gamma]] - [\beta,[\alpha,\gamma]]$$

or

$$[[\alpha,\beta],\gamma] + [[\beta,\gamma],\alpha] + [[\gamma,\alpha],\beta] = 0 \tag{6-23}$$

holds. This is an identity for arbitrary vectors and is equivalent to the relation

$$\sum_s (c_{i,sj}c_{s,kl} + c_{i,sk}c_{s,lj} + c_{i,sl}c_{s,jk}) = 0 \tag{6-24}$$

for the structure constants. Formulas (6-23) and (6-24) are referred to as the *Jacobi identities*. They replace the associative law for Lie algebras. It is an interesting problem to find all algebras for which Eqs. (6-8) and (6-23) hold. This problem has grown out of the study of Lie groups.

Definition 6-15. *An abstract Lie algebra is an algebra in which the multiplication is anticommutative and satisfies the Jacobi identity.*

We have proved the following theorem:

Theorem 6-16 (Lie's Third Fundamental Theorem). *The Lie algebra of a Lie group germ is a real abstract Lie algebra.*

Abstract Lie algebras are of the utmost importance in Lie group theory, because they contain all possible information about Lie group germs. The next section will be devoted to the proof of this fact, more specifically of the following theorem:

Theorem 6-17 (Converse of the Third Fundamental Theorem). *Any real abstract Lie algebra is the Lie algebra of a Lie group germ. It is always possible to find coordinates in which the multiplication is given by analytic functions. Two Lie group germs with identical algebras are isomorphic in neighborhoods of their unit elements.*

This theorem shows that for algebras in real vector spaces there is no need to distinguish between Lie algebras and abstract Lie algebras. It also shows that a complete study of Lie groups must be based on the solution of two classification problems. First, one needs an enumeration of all possible abstract Lie algebras. This problem was solved by W. Killing and E. Cartan in the last decades of the nineteenth century for *complex* Lie algebras and later by E. Cartan and others for real Lie algebras. Then one needs a classification of all Lie groups that may be derived from one Lie group germ, a problem solved by O. Schreier in 1926. Another important theorem, due to I. Ado (1934), says that any Lie algebra (hence also any Lie group germ) has an *isomorphic* representation: the study of Lie groups is the study of multiplicative matrix groups. All these theorems need very advanced tools in algebra or topology, and so they will not be treated in this text.

***Example* 6-5.** To illustrate the stringency of the conditions (6-8) and (6-23) imposed on Lie algebras, we determine all Lie algebras in one and two dimensions.

In dimension one, there is only one basis vector α, and the product is trivial, by Eq. (6-8), $[\alpha,\alpha] = 0$.

In dimension two, the vector space underlying the algebra has a base of two linearly independent vectors α,β.

i. A possible algebra is always the trivial one in which all products are zero,

$$[\alpha,\alpha] = [\alpha,\beta] = [\beta,\beta] = 0$$

By theorem 6-13, this is the algebra of any two-dimensional abelian Lie group, e.g., the group of vector addition in the plane.

ii. If the algebra is not trivial, $[\alpha,\beta] = a\alpha + b\beta$, we may assume that $b \neq 0$. The vectors

$$\alpha_1 = \frac{1}{b}\alpha \qquad \alpha_2 = \frac{a}{b}\alpha + \beta$$

are a new base, for which the multiplication table is

$$[\alpha_1,\alpha_1] = [\alpha_2,\alpha_2] = 0 \qquad [\alpha_1,\alpha_2] = \alpha_2$$

This discussion shows that there are only two nonisomorphic Lie algebras in dimension two.

Exercise 6-2

1. A subset H of a group G is *invariant* in G if $h_x(a) \in H$ for all $a \in H$, $x \in G$ [see Eq. (6-17)]. Show that the set of vectors $\mathbf{k} = (0,k_1,k_2,k_3)$ is invariant in the group of quaternions. What is the geometric interpretation of $h_D(\mathbf{k})$ in this case? (See exercise 5-2, Prob. 6.)

2. An *ideal* I in an algebra $\mathfrak{L}$ is a subgroup of the additive group of the algebra such that $[\xi,\alpha] \in I$ for all $\xi \in \mathfrak{L}$, $\alpha \in I$. Let H be an invariant (see Prob. 1) subgroup of a Lie group G, and suppose that H is itself a Lie group. Show that $\mathfrak{L}(H)$ is an ideal in $\mathfrak{L}(G)$.

3. A Lie algebra $\mathfrak{L}$ is *semisimple* if its only commutative ideal (see Prob. 2) different from $\mathfrak{L}$ is $\{0\}$. Show that there are no semisimple Lie algebras in dimensions one and two.

4. Prove that the action of the matrix ad α on the vector group of $\mathfrak{L}(G)$ is given by $\xi(\text{ad }\alpha)^t = [\alpha,\xi]$ (matrix multiplication on the left-hand side).

5. The *kernel* of a homomorphism is the set of all elements mapped into the unit element of G_1. Prove that the kernel of the adjoint map (6-21) is an ideal (see Prob. 2) in $\mathfrak{L}(G)$. This ideal is the *center* of $\mathfrak{L}(G)$. Show that a vector is in the center if and only if its Lie product with all other vectors in the algebra vanishes. (Use Prob. 4.)

6. Show by direct computation that the matrix commutator

$$[A,B] = AB - BA$$

satisfies the Jacobi identity.

7. The following construction is of importance in quantum physics: On the set of C^∞ functions $f(x_1, \ldots, x_n)$ defined in a region of a cartesian space R^n, define operations q_i: $q_i f = x_i f$, p_i: $p_i f = \partial f/\partial x_i$, $i = 1, \ldots, n$. Prove that these operations form a Lie algebra with the product $[p_i,q_j] = p_i q_j - q_j p_i$, and find the multiplication table for the products $[p_i,p_j]$, $[q_i,q_j]$, $[p_i,q_j]$.

8. Prove that, for any four elements of a Lie algebra,

$$[[[\alpha,\beta],\gamma],\delta] + [[[\beta,\alpha],\delta],\gamma] + [[[\gamma,\delta],\alpha],\beta] + [[[\delta,\gamma],\beta],\alpha] = 0$$

* **9.** Determine all Lie algebras in dimension three. For any Lie algebra $\mathfrak{L}$, the *derived algebra* $\mathfrak{L}'$ is the smallest subalgebra of $\mathfrak{L}$ that contains *all* products $[\alpha,\beta]$.

(*a*) If $\mathfrak{L}' = 0$, prove that $\mathfrak{L}$ is the trivial algebra in which all products vanish, e.g., the Lie algebra of the three-dimensional abelian Lie groups.

(*b*) If the dimension of $\mathfrak{L}'$ is *one*, prove that there are two possibilities (up to isomorphisms and changes of name) (basis vectors α_i):

I. $\quad [\alpha_1,\alpha_2] = [\alpha_1,\alpha_3] = 0 \qquad [\alpha_2,\alpha_3] = \alpha_1$

II. $\quad [\alpha_1,\alpha_2] = [\alpha_2,\alpha_3] = 0 \qquad [\alpha_1,\alpha_3] = \alpha_1$

(For the proof, use the methods of example 6-5.)

(*c*) If $\mathfrak{L}'$ is generated by two elements, say α_1,α_2, put

$$[\alpha_1,\alpha_3] = a\alpha_1 + b\alpha_2 \qquad [\alpha_2,\alpha_3] = c\alpha_1 + d\alpha_2$$

1. Show that the Jacobi identity implies $[\alpha_1,\alpha_2] = 0$.
2. One may try to find new basis vectors β_1, β_2 in $\mathfrak{L}'$, such that $[\beta_1,\alpha_3] = a^*\beta_1$. Show that a^* must be a root of

$$a^{*2} - (a + d)a^* + ad - bc = 0 \qquad (6\text{-}25)$$

3. If Eq. (6-25) has real roots, show that possible Lie algebras are

I. $[\beta_1,\beta_2] = 0 \quad [\beta_1,\alpha_3] = \beta_1 \quad [\beta_2,\alpha_3] = \beta_1 + \beta_2$
II. $[\beta_1,\beta_2] = 0 \quad [\beta_1,\alpha_3] = \beta_1 \quad [\beta_2,\alpha_3] = k\beta_2 \quad k \neq 0$

4. If Eq. (6-25) has complex roots, show that a base may be found such that the multiplication table of the algebra becomes

III. $[\beta_1,\beta_2] = 0 \quad [\beta_1,\alpha_3] = \beta_2 \quad [\beta_2,\alpha_3] = -\beta_1 + h\beta_2 \quad h^2 < 4$

(*d*) If $\mathfrak{L}' = \mathfrak{L}$, the determinant

$$\|C\| = \begin{Vmatrix} c_{1,23} & c_{2,23} & c_{3,23} \\ c_{1,31} & c_{2,31} & c_{3,31} \\ c_{1,12} & c_{2,12} & c_{3,12} \end{Vmatrix}$$

is different from zero. The Jacobi identities imply that $C^t = C$. The matrix C defines a nondegenerate conic in homogeneous coordinates $\mathbf{x} = (x_1,x_2,x_3)$ by $\mathbf{x}C\mathbf{x}^t = 0$.

1. If the conic is real, show that by a change of coordinates it may be brought into $x_1x_3 - x_2^2 = 0$, giving the algebra

I. $[\alpha_1,\alpha_2] = \alpha_1 \quad [\alpha_1,\alpha_3] = 2\alpha_2 \quad [\alpha_2,\alpha_3] = \alpha_3$

2. If the conic has no real points, it may be given the normal form $x_1^2 + x_2^2 + x_3^2 = 0$. Show that the corresponding algebra is

II. $[\alpha_1,\alpha_2] = \alpha_3 \quad [\alpha_1,\alpha_3] = -\alpha_2 \quad [\alpha_2,\alpha_3] = \alpha_1$

10. (See Prob. 9.) Show that, for the algebras of type (*c*)II in Prob. 9, no algebra with $k = 1$ is isomorphic to one with $k \neq 1$.

11. Check that the following multiplication table defines a Lie algebra of dimension five $(i < j)$: $[\alpha_i,\alpha_j] = 0 (i,j = 1, \ldots ,5)$ except for $[\alpha_1,\alpha_2] = \alpha_5$, $[\alpha_1,\alpha_3] = \alpha_3$, $[\alpha_2,\alpha_4] = \alpha_4$.
Find the algebras $\mathfrak{L}'$ and $\mathfrak{L}'' = (\mathfrak{L}')'$ (see Prob. 9). Find also the subspace S of $\mathfrak{L}$ such that $[\sigma,\alpha] = 0$, $\sigma \in S$, $\alpha \in \mathfrak{L}$.

12. Show that the following spaces of matrices form Lie algebras under the matrix commutator product (6-11):
(*a*) The space of triangular $n \times n$ matrices $A = (a_{ij})$, $a_{ij} = 0$ for $i < j$
(*b*) The space of triangular matrices of diagonal zero, $a_{ij} = 0$ for $i \leq j$
(*c*) The space of triangular matrices of traces zero
Identify these abstract Lie algebras with algebras determined in exercise 6-1.

13. Verify that a Lie algebra is defined by $[\alpha_1,\alpha_2] = \alpha_3$, $[\alpha_1,\alpha_3] = \alpha_4$, all other $[\alpha_i,\alpha_j] = 0$, $1 \leq i < j \leq 4$.

14. An algebra $\mathcal{L}$ is *simple* if its only ideals (Prob. 2) are the trivial ones $\mathcal{L}$ and $\{0\}$. Show that in the list of Prob. 9 the only simple algebras are those of types (*d*)I and II.

15. Use the Jacobi identity to prove the following theorem: If a four-dimensional Lie algebra $\mathcal{L}$ has a simple (Prob. 14) three-dimensional subalgebra $\mathcal{L}_1$, then
(*a*) $\mathcal{L}_1 = \mathcal{L}' = [\mathcal{L},\mathcal{L}]$ is the derived algebra (Prob. 9).
(*b*) It is possible to find a basis α_i ($i = 1,2,3,4$) of $\mathcal{L}$ such that $\mathcal{L}'$ is spanned by $\alpha_1,\alpha_2,\alpha_3$ and $[\alpha_4,\xi] = 0$, $\xi \in \mathcal{L}$.

6-3. One-parameter Subgroups

For the proof of theorem 6-17 we have to find the multiplication formulas of a Lie group germ

$$(\mathbf{x} \cdot \mathbf{y})_k = f_k(x_1, \ldots ,x_n,y_1, \ldots ,y_n)$$

if an abstract Lie algebra is given by its multiplication table

$$[\alpha,\beta]_k = \sum c_{k,ij}\alpha_i\beta_j$$

The multiplication in the Lie group depends on a *vector function of* $2n$ *variables* $\mathbf{f}(\mathbf{x},\mathbf{y})$; the multiplication in the algebra is given by *constants*. The main idea of the proof is to show that $\mathbf{f}(\mathbf{x},\mathbf{y})$ may be found if the *matrix function of n variables*

$$P(\mathbf{y}) = \left[\frac{\partial f_k(\mathbf{x},\mathbf{y})}{\partial x_j}\right]_{\mathbf{x}=\mathbf{e}} = J_{\mathbf{x}}\mathbf{f}(\mathbf{x},\mathbf{y})\Big|_{\mathbf{x}=\mathbf{e}} \tag{6-26}$$

is known and that $P(\mathbf{y})$ is the solution of a simple matrix differential equation with constant coefficients $c_{k,ij}$.

The matrix P appears in the Taylor expansion of $\mathbf{x} \cdot \mathbf{y}$ in $\mathbf{x} = \mathbf{e}$ if $\mathbf{y}$ is taken as a vector of parameters,

$$\mathbf{x} \cdot \mathbf{y} = \mathbf{y} + \mathbf{x}P(\mathbf{y}) + \cdots \tag{6-27}$$

The associative law $\mathbf{z} \cdot \mathbf{xy} = \mathbf{zx} \cdot \mathbf{y}$ becomes

$$\begin{aligned} \mathbf{xy} + \mathbf{z}P(\mathbf{xy}) + \cdots &= f(\mathbf{x} + \mathbf{z}P(\mathbf{x}) + \cdots ,\mathbf{y}) \\ &= \mathbf{xy} + \mathbf{z}P(\mathbf{x})J_{\mathbf{x}}\mathbf{f}(\mathbf{x},\mathbf{y}) + \cdots \end{aligned}$$

This can be an identity in $\mathbf{x},\mathbf{y},\mathbf{z}$ only if the *Lie differential equation*

$$P(\mathbf{f}(\mathbf{x},\mathbf{y})) = P(\mathbf{x})J_{\mathbf{x}}\mathbf{f}(\mathbf{x},\mathbf{y}) \tag{6-28}$$

holds. This matrix equation

$$p_{jk}(\mathbf{f}(\mathbf{x},\mathbf{y})) = \sum_s p_{js}(\mathbf{x}) \frac{\partial f_k(\mathbf{x},\mathbf{y})}{\partial x_s}$$

may be given another form in which its meaning is more easily seen. By lemma 6-4, $P(\mathbf{e}) = U$, and the matrix $P(\mathbf{x})$ has an inverse

$$Q(\mathbf{x}) = P(\mathbf{x})^{-1}$$

in some neighborhood of **e**. Lie's equation for the inverse is

$$Q(\mathbf{x}) = J_{\mathbf{x}}\mathbf{f}(\mathbf{x},\mathbf{y})Q(\mathbf{f}(\mathbf{x},\mathbf{y}))$$

This equation we multiply by $d\mathbf{x}$ on the left and use the formula for the total differential (**y** is a vector of parameters!)

$$d\mathbf{f}(\mathbf{x},\mathbf{y}) = d\mathbf{x}J_{\mathbf{x}}\mathbf{f}(\mathbf{x},\mathbf{y}) = \left\{\sum_s dx_s \left(\frac{\partial f_k}{\partial x_s}\right)\right\}$$

to obtain

$$d\mathbf{x}Q(\mathbf{x}) = d(\mathbf{xy})Q(\mathbf{xy}) \tag{6-29}$$

Definition 6-18. *$F(\mathbf{x})$ is right invariant on the group G if $F(\mathbf{x}) = F(\mathbf{xa})$ for all $\mathbf{a} \in G$.*

Equation (6-29) means therefore:

Theorem 6-19. *The "differential vector" $\omega(\mathbf{x}) = d\mathbf{x}Q(\mathbf{x})$ is right invariant.*

If α is a vector in a Lie algebra $\mathfrak{L}(G)$, the line $\{r\alpha\}$, r real, is a subalgebra $\mathfrak{L}_\alpha$ since it contains the origin and is closed under addition and multiplication. The multiplication is trivial, $[r\alpha,s\alpha] = 0$. We ask if it is possible to find a Lie subgroup G_α of G such that $\mathfrak{L}(G_\alpha) = \mathfrak{L}_\alpha$. G_α must be one-dimensional, and it must contain the origin **e**; hence it is natural to try to get G_α as a curve $\mathbf{x}(t)$, $\mathbf{x}(0) = 0$, $\mathbf{x}'(0) = \alpha$. By a change of parameter, all vectors of $\mathfrak{L}_\alpha$ appear as infinitesimal vectors of G_α since $d\mathbf{x}/d\tau(0) = r\alpha$ for $t = r\tau$. More specifically, we will try to find G_α such that in it the group multiplication will be the addition of parameters

$$\mathbf{x}(t_1)\mathbf{x}(t_2) = \mathbf{x}(t_1 + t_2) \tag{6-30}$$

This condition is suggested by theorem 6-17 which would imply that all groups with trivial Lie algebra (for example, all groups of algebra $\mathfrak{L}_\alpha$) are abelian. Equation (6-27) becomes in this case

$$\begin{aligned}\mathbf{x}(h + t) &= \mathbf{x}(t) + \mathbf{x}(h)P(\mathbf{x}(t)) + \cdots \\ &= \mathbf{x}(t) + \alpha P(\mathbf{x}(t))h + \cdots\end{aligned}$$

Hence
$$\frac{d\mathbf{x}}{dt} = \alpha P(\mathbf{x}) \tag{6-31}$$

The next theorems state how the group multiplication is obtained from Eq. (6-31).

Theorem 6-20. *Let $\mathbf{x}(t)$ be the integral curve of Eq. (6-31) with initial condition $\mathbf{x}(0) = \mathbf{e}$, and $\hat{\mathbf{x}}(t)$ the integral curve to the initial condition $\hat{\mathbf{x}}(0) = \mathbf{a}$. Then $\hat{\mathbf{x}}(t) = \mathbf{x}(t) \cdot \mathbf{a}$.*

Equation (6-31) is a system of first-order differential equations with differentiable coefficients $p_{jk}(\mathbf{x})$. By the standard existence and unicity theorems,† the initial condition determines a *unique* solution. If $\mathbf{x}(t)\mathbf{a}$ is a solution of (6-31), it must be identical to $\hat{\mathbf{x}}(t)$, since $\mathbf{x}(t)\mathbf{a}_{t=0} = \mathbf{a}$. By Eqs. (6-28) and (6-31),

$$\begin{aligned} \frac{d}{dt}\mathbf{x}(t)\mathbf{a} = \frac{d}{dt}\mathbf{f}(\mathbf{x},\mathbf{a}) &= \frac{d\mathbf{x}}{dt}J_{\mathbf{x}}\mathbf{f}(\mathbf{x},\mathbf{a}) = \alpha P(\mathbf{x})J_{\mathbf{x}}\mathbf{f}(\mathbf{x},\mathbf{a}) \\ &= \alpha P(\mathbf{x}(t)\mathbf{a}) \end{aligned}$$

$\mathbf{x}(t)\mathbf{a}$ is a solution of (6-31).

Corollary 6-21. *Let G be a Lie group germ and $\mathfrak{L}(G)$ its Lie algebra. For any $\alpha \in \mathfrak{L}(G)$, the integral curve $\mathbf{x}(t)$ of Eq.* (6-31) *with initial condition $\mathbf{x}(0) = \mathbf{e}$ is an abelian subgroup germ G_α of G such that $\mathbf{x}(t_1 + t_2) = \mathbf{x}(t_1)\mathbf{x}(t_2)$ for small t_1,t_2. $\mathfrak{L}(G_\alpha) = \mathfrak{L}_\alpha$.*

PROOF: $\mathbf{x}(t + t_2)$ is an integral curve of (6-31). Its initial point is $\mathbf{x}(0 + t_2) = \mathbf{x}(t_2)$. Hence $\mathbf{x}(t_1 + t_2) = \mathbf{x}(t_1)\mathbf{x}(t_2)$, by theorem 6-20. G_α contains $\mathbf{e}$. It is closed under multiplication as long as the product exists; hence it is a subgroup germ of G.

Definition 6-22. *A right coset of a subgroup H in a group G is the set $Hg_0 = \{hg_0\}$, h any element in H and g_0 a fixed element in G.*

It follows from theorem 6-20 and corollary 6-21 that we may completely characterize the integral curves of Eq. (6-31) in a group-theoretic way:

Corollary 6-23. *The integral curves of Eq.* (6-31) *are the one-parameter group G_α and its right cosets.*

Having found a unique additive subgroup G_α to each $\mathfrak{L}_\alpha$, we try to find coordinates in a neighborhood of $\mathbf{e}$ such that *every* G_α becomes the line αt and that on it the group multiplication is the vector addition

$$(\alpha t_1)(\alpha t_2) = \alpha(t_1 + t_2)$$

This will be possible because of the interesting property of one-parameter subgroups

$$\mathbf{x}(ct,\alpha) = \mathbf{x}(t,c\alpha) \tag{6-32}$$

The proof of (6-32) again is based on the unicity theorem for differential equations. If $\mathbf{x}(t,\alpha)$ is a solution of (6-31), $\mathbf{x}(0,\alpha) = \mathbf{e}$, then

$$\frac{d\mathbf{x}(ct,\alpha)}{dt} = c\alpha P(\mathbf{x}) \qquad \mathbf{x}(0,\alpha) = \mathbf{e}$$
$$\frac{d\mathbf{x}(t,c\alpha)}{dt} = c\alpha P(\mathbf{x}) \qquad \mathbf{x}(0,c\alpha) = \mathbf{e}$$

† See, for example, E. A. Coddington and Norman Levinson, "Theory of Ordinary Differential Equations," chap. 1, McGraw-Hill Book Company, Inc., New York, 1955, or any other standard text on differential equations.

Both sides of Eq. (6-32) solve the same differential equation with identical initial conditions; they are equal.

The algebra $\mathfrak{L}(G)$ may be defined in the same cartesian space R^n in which the group germ G is defined. In this case, the function

$$\mathbf{F}(\xi) = \mathbf{x}(1,\xi) \tag{6-33}$$

$\mathbf{x}(t,\xi)$ a one-parameter subgroup, is a map $R^n \to R^n$. New coordinates $\mathbf{x}^*$ in R^n to which the group germ will be referred may now be defined by

$$\mathbf{x} = \mathbf{F}(\mathbf{x}^*) \tag{6-34}$$

The correspondence $\mathbf{x} \to \mathbf{x}^*$ is a differentiable coordinate transformation:

1. $\mathbf{F}(\mathbf{x}^*)$ exists for all $\mathbf{x}^* \in R^n$, since $\mathbf{F}(\mathbf{x}^*) = \mathbf{x}(1,\mathbf{x}^*) = \mathbf{x}(1/c,c\mathbf{x}^*)$. Since equation (6-31) has a unique solution for all α and $|t| < \epsilon$, we may take $|1/c| < \epsilon$.

2. For any solution of Eq. (6-31), $J_\alpha(d\mathbf{x}/dt) = (d/dt)J_\alpha\mathbf{x} = P(\mathbf{x})$. Hence $\mathbf{x}(t,\alpha)$ is differentiable in α; that is, $\mathbf{F}$ is a differentiable function of $\mathbf{x}^*$.

3. By (6-31), $[d\mathbf{x}(t,\alpha)/dt]_{t=0} = \alpha P(\mathbf{e}) = \alpha$. This means that, for a vector $\boldsymbol{\varepsilon}_i = (0, \ldots, 0, \epsilon, 0, \ldots, 0)$, $\mathbf{F}(\boldsymbol{\varepsilon}_i) = \boldsymbol{\varepsilon}_i U +$ higher-order terms. The jacobian $(J_{\mathbf{x}^*}\mathbf{F}(\mathbf{x}^*))_{\mathbf{x}^*=\mathbf{e}} = U$ has a nonzero determinant. The inverse-function theorem now guarantees that (6-34) defines a one-to-one differentiable transformation of coordinates in some neighborhood of $\mathbf{e}$.

Definition 6-24. *The coordinates* $\mathbf{x}^*$ *defined by* (6-34) *are the canonical coordinates of* G.

Theorem 6-25. *In canonical coordinates, the one-parameter subgroups* G_α *are the lines* αt *with the vector addition.*

PROOF: $\mathbf{F}(\alpha t) = \mathbf{x}(1,t\alpha) = \mathbf{x}(t,\alpha)$.

From now on we shall omit the asterisk for canonical coordinates.

Theorem 6-26. *The point* $\mathbf{ab}$ *in* G *is the point* $\mathbf{x}(1,\mathbf{a})$ *on the coset through* $\mathbf{b}$ *to the one-parameter subgroup* $\mathbf{a}t$, *in canonical coordinates. Hence* $P(\mathbf{x})$ *defines* $\mathbf{f}(\mathbf{x},\mathbf{y})$.

PROOF: Theorems 6-20 and 6-25.

By theorem 6-25, Eq. (6-31) becomes, in canonical coordinates,

$$\alpha = \alpha P(\mathbf{x}) \tag{6-35}$$

and, upon multiplication by t,

$$\mathbf{x} = \mathbf{x}P(\mathbf{x}) \tag{6-36}$$

or its equivalent

$$\mathbf{x} = \mathbf{x}Q(\mathbf{x})$$

It is very easy to find $P(\mathbf{x})$ for groups of matrices. The product of two matrices $X = U + (x_{ij})$, $Y = U + (y_{ij})$ is [see Eq. (6-10)]

$$XY = Y + X(y_{ij}) + (x_{ij})$$

Hence
$$P(Y) = J_X(XY)_{X=U} = Y \tag{6-37}$$

The Lie algebra in this case is a vector space of matrices $\mathrm{A} = (\alpha_{ij})$, and (6-31) becomes

$$\frac{dX}{dt} = \mathrm{A}X \qquad X(0) = U \tag{6-38}$$

that is
$$C(X) = \mathrm{A} \qquad X(0) = U \tag{6-38a}$$

The solution is

$$X(t) = e^{\mathrm{A}t} \tag{6-39}$$

since A and $\mathrm{A}t = \int_0^t \mathrm{A}\,dt$ commute [see lemma 2-12 and Eq. (2-10)].

Theorem 6-27. *Any Lie algebra of matrices $\mathfrak{L}$ is the Lie algebra of the multiplicative group of matrices $e^{\mathrm{A}t}$, $\mathrm{A} \in \mathfrak{L}$.*

This result, together with the Ado theorem quoted on page 113 proves theorem 6-17. But since we shall not prove Ado's theorem, and also because it may be very difficult to find an isomorphic representation of a given Lie algebra, we give an elementary proof of theorem 6-17, due to F. Schur and H. Freudenthal, which in itself yields other useful information.

The adjoint representation is a map of the element $\alpha \in \mathfrak{L}$ onto the linear operator $[\alpha,\ \ ]$. The representation is not isomorphic if and only if there exists an $\alpha \neq 0$ which is mapped onto the zero matrix, i.e., if $[\alpha,\xi] = 0$ for *some* α and *all* ξ in $\mathfrak{L}$. The set of all such vectors α is the *center* of $\mathfrak{L}$; it is a linear subspace of $\mathfrak{L}$ (see also exercise 6-2, Prob. 5). The center is reduced to the zero element if in a base $\epsilon_1, \ldots, \epsilon_n$ of the vector space $\mathfrak{L}$ there is no ϵ_j such that $[\epsilon_j,\epsilon_i] = 0$ for all $i = 1, \ldots, n$. In this case $\mathfrak{L}$ is isomorphic to its adjoint representation which is an algebra of matrices.

Corollary 6-28. *An (abstract) Lie algebra $\mathfrak{L}$ whose center is the zero element is the Lie algebra of the group of matrices $X = e^{t\,\mathrm{ad}\,\alpha}$, $\alpha \in \mathfrak{L}$.*

***Example* 6-6.** The nontrivial Lie algebra of dimension two (example 6-5, ii) has the multiplication table

	ϵ_1	ϵ_2
ϵ_1	0	ϵ_2
ϵ_2	$-\epsilon_2$	0

for a suitable base ϵ_1,ϵ_2 of R^2. The center of the algebra is 0. A vector $\alpha = a_1\epsilon_1 + a_2\epsilon_2$ acts on any other vector $\xi = x_1\epsilon_1 + x_2\epsilon_2$ by

$$(\mathrm{ad}\ \alpha)\xi^t = [\alpha,\xi] = (a_1x_2 - a_2x_1)\epsilon_2$$

Hence ad α is the matrix

$$\text{ad } \alpha = \begin{pmatrix} 0 & 0 \\ -a_2 & a_1 \end{pmatrix}$$

The corresponding element of the Lie group is

$$X = e^{t \text{ ad } \alpha} = \begin{pmatrix} 1 & 0 \\ \frac{a_2}{a_1}(1 - e^{a_1 t}) & e^{a_1 t} \end{pmatrix}$$

or, alternatively, $X = \begin{pmatrix} 1 & 0 \\ a & p \end{pmatrix}$, a real, $p > 0$. This example also shows that canonical coordinates $(a_1 t, a_2 t)$ may lead to more complicated expressions than do other types of coordinates like $(p - 1, a)$.

If the algebra $\mathfrak{L}$ has a nontrivial center, the matrix P must be obtained from a differential equation. The usual definition of the directional derivative in the direction of a unit vector β,

$$D_\beta f = \beta \text{ grad } f$$

immediately leads to the definition of the directional derivative of a vector, a function of a vector

$$D_\beta \mathbf{f}(\alpha) = \beta J_\alpha \mathbf{f}(\alpha) \tag{6-40}$$

and of a matrix, $D_\beta P(\alpha) = \Sigma \beta_s (\partial p_{ij} / \partial \alpha_s)$. Equation (6-35) is now differentiated in $\mathfrak{L}$ with respect to a vector β:

$$\beta P(\alpha u) + \alpha D_\beta P(\alpha u) = \beta \tag{6-41}$$

By defining equations (6-26) and (6-4), $P(\alpha u) = U + (\Sigma a_{j,is} \alpha_s u)$; hence $\alpha D_\beta P(\alpha u) = (\Sigma a_{j,is} \alpha_i \beta_s u)$ and

$$[\alpha u, \beta] = \alpha D_\beta P(\alpha u) - \beta D_\alpha P(\alpha u) \tag{6-42}$$

We insert (6-42) into (6-41) and use also $[\alpha u, \beta] = \beta(\text{ad } \alpha u)^t$ to obtain

$$\beta(D_\alpha P(\alpha u) + (\text{ad } \alpha u)^t + P(\alpha u) - U) = 0$$

This equation must hold for all $\beta \in \mathfrak{L}$; the parenthesis vanishes:

$$D_\alpha P(\alpha u) + P(\alpha u) + (\text{ad } \alpha u)^t - U = 0 \tag{6-43}$$

This equation may be simplified by the introduction of *Schur's matrix*

$$S(\alpha, u) = u P(\alpha u)$$

for which
$$\frac{dS}{du} = P(\alpha u) + D_\alpha P(\alpha u)$$

It follows also from Eq. (6-35) that

$$(\text{ad } \alpha u)^t = [\alpha u, \quad] = [\alpha u, \quad] P(\alpha u) = (\text{ad } \alpha)^t S(\alpha, u)$$

Equation (6-43) implies now that S solves the *Schur differential equation*

$$\frac{dS}{du} = U - (\text{ad } \alpha)^t S \qquad S(\alpha,0) = 0 \tag{6-44}$$

Theorem 6-29. *A solution $S(\alpha,u)$ of the Schur differential equation is an analytic function of u for all real u and all $\alpha \in R^n$. The matrix*

$$P(\alpha) = S(\alpha,1)$$

satisfies $\alpha P(\alpha) = \alpha$ for all α.

The matrix equation (6-44) has a unique solution. With the methods of Sec. 2-2, one sees that it is

$$S(\alpha,u) = Uu - \frac{u^2}{2!}(\text{ad } \alpha)^t + \frac{u^3}{3!}(\text{ad } \alpha)^{t2} + \cdots + (-1)^{n+1}\frac{u^n}{n!}(\text{ad } \alpha)^{tn} + \cdots$$

The series converges absolutely and uniformly for all values of u and α, $|S(\alpha,u)| < |e^{u(\text{ad } \alpha)^t}|$; it may be derived term by term. Then

$$P(\alpha) = U - \frac{1}{2!}(\text{ad } \alpha)^t + \frac{1}{3!}(\text{ad } \alpha)^{t2} - \cdots$$

Now $\alpha(\text{ad } \alpha)^t = [\alpha,\alpha] = 0$; hence $\alpha P(\alpha) = \alpha$. $P(\alpha)$ is unique.

The straight line $\mathbf{x} = \alpha t$ is a solution of $d\mathbf{x}/dt = \alpha P(\mathbf{x})$, $\mathbf{x}(0) = \mathbf{e}$. Theorem 6-20 and its consequence theorem 6-26 are based only on this equation and on Eq. (6-28) which is equivalent to (6-29). By corollary 6-23, the multiplication law $\mathbf{f}(\mathbf{a},\mathbf{b})$ is obtained as $\mathbf{F}(1)$, where $\mathbf{F}(t)$ is a solution of $d\mathbf{F}/dt = \alpha P(\mathbf{F})$, $\mathbf{F}(0) = \mathbf{b}$. This means that

$$d\mathbf{F}\, P(\mathbf{F})^{-1} = \alpha\, dt = dt\, \alpha P(\mathbf{a})^{-1} = d\mathbf{a}\, P(\mathbf{a})^{-1}$$

For $t = 1$, this equation reduces to $d(\mathbf{ab})Q(\mathbf{ab}) = d\mathbf{a}Q(\mathbf{a})$, and Eq. (6-29) holds. $P(\mathbf{x})$ defines a group germ by theorems 6-20 and 6-26.

It follows from theorem 10-25 that **ab** is well defined. This completes the proof of theorem 6-17.

Exercise 6-3

1. If $\mathbf{x}(t,\alpha)$ is a solution of Eq. (6-31), $\mathbf{x}(0,\alpha) = \mathbf{e}$, prove that $(d\mathbf{x}/dt)_{t=0} = \alpha$ and that $\mathbf{x}(t)^{-1} = \mathbf{x}(-t)$.
2. Show that a homomorphism h: $G \to G'$ of two Lie group germs given in canonical coordinates is always a *linear* map. (HINT: A homomorphism must map any one-parameter subgroup $G_\alpha \subset G$ on a one-parameter group G'_α. G'_α may be $\{\mathbf{e}\}$.)

3. Use Prob. 2 to show that the induced mapping $\mathfrak{L}(G) \to \mathfrak{L}(G')$ is a linear map.
4. Given a homomorphism of Lie algebras $h^*\colon \mathfrak{L} \to \mathfrak{L}'$, show that it is always possible to find a homomorphism $h\colon G \to G'$ of the respective groups such that h^* is the algebra homomorphism induced by h.
5. Use Prob. 4 to show that the homomorphic image of a Lie group germ is again a Lie group germ or $\{\mathbf{e}\}$.

◆ 6. Do there exist subgroups G' of a Lie group G that are not homomorphic images of G? Use Prob. 5.

7. State and prove the converse of theorem 6-13.
8. The *center* of a *group* G is the set C of all elements c that commute with all elements of G: $cg = gc$ for all $g \in G$. Prove that C is a subgroup of G and that $\mathfrak{L}(G)$ is of zero center if and only if C does not contain any one-parameter subgroup of G.
9. Prove that a semisimple (exercise 6-2, Prob. 3) Lie algebra has zero center. (Use exercise 6-2, Prob. 5.)
10. Check that, by our hypotheses, $P(\mathbf{x})$ has continuous partial derivatives. Show that this implies

$$||P(\mathbf{x}')| - |P(\mathbf{x})|| < K|\mathbf{x}' - \mathbf{x}|$$

for $|\mathbf{x}' - \mathbf{x}| < \epsilon(\mathbf{x}',\mathbf{x})$.

11. Define $\mathbf{x}_0 = \alpha t$, $\mathbf{x}_n = \int_0^t \alpha P(\mathbf{x}_{n-1})\,dt$. Use Prob. 10 to show that $|\mathbf{x}_{n+1} - \mathbf{x}_n| < (K|\alpha|T)^n/n!$ for $t < T$. Then show that $\mathbf{x} = \lim_{n\to\infty} \mathbf{x}_n$ exists and that $d\mathbf{x}/dt = \alpha P(\mathbf{x})$. (See Sec. 2-2.)
12. Find all algebras with zero center and all semisimple algebras in the list of exercise 6-2, Prob. 9.
13. Write explicitly the differential equations (6-38) for the group discussed in example 6-6. Integrate the equations componentwise and compare with the results of the text.
14. Find the groups to the algebras of exercise 6-2, Prob. 9. For which algebras does the method of corollary 6-28 break down?
15. Find the six-dimensional group to the algebra of base $\alpha_1, \ldots, \alpha_6$ and multiplication table

	α_1	α_2	α_3	α_4	α_5	α_6
α_1	0	α_3	$-\alpha_2$	0	α_6	$-\alpha_5$
α_2	$-\alpha_3$	0	α_1	$-\alpha_6$	0	α_3
α_3	α_2	$-\alpha_1$	0	α_5	$-\alpha_4$	0
α_4	0	α_6	$-\alpha_5$	0	$-\alpha_3$	α_2
α_5	$-\alpha_6$	0	α_4	α_3	0	$-\alpha_1$
α_6	α_5	$-\alpha_3$	0	$-\alpha_2$	α_1	0

References

Freudenthal, H.: Ein Aufbau der Lie'schen Gruppentheorie, *Jahresber. Deutsch. Math. Verein.*, **43**: 26–39 (1934).

An excellent modern treatment of Lie group theory is the following:

Freudenthal, H.: "Lie Groups," Lecture Notes, Yale University, 1961, 320 pp. (mimeographed).

The topological theory of global Lie groups may be found in the following:

Chevalley, C.: "Theory of Lie Groups I," Princeton University Press, Princeton, N.J., 1946.

Pontrjagin, L.: "Topological Groups," translated by Emma Lehmer, Princeton University Press, Princeton, N.J., 1939.

A proof of Ado's theorem is given in

Bourbaki, N.: "Groupes et algèbres de Lie," chap. I, Algèbres de Lie, Actualités Scientifiques et Industrielles, 1285, Hermann & Cie, Paris, 1960.

For Lie algebras, see

Jacobson, N.: "Lie Algebras," Interscience Publishers, Inc., New York, 1962.

7
TRANSFORMATION GROUPS

7-1. Transformation Groups

In Chap. 5 we discussed some examples of Lie group germs G realized by a pseudo group of transformations in some "homogeneous" space, such that the composition of two transformations corresponded to the multiplication in G. A transformation group involves two spaces in its description: the homogeneous space in which it acts and the group space. To avoid ambiguities, only vectors in the homogeneous space will be indicated by boldface, and elements of the group G will be denoted by lightface, $a = (a_1, \ldots, a_n)$. Again, we shall work with a fixed coordinate system in each one of the spaces; our vectors in this discussion will always be coordinate (row) vectors.

Definition 7-1.† *An r-parameter (Lie) pseudo group of transformations in an open set $V \varepsilon R^m$ is given by an r-dimensional Lie group germ G and a family of maps $\mathbf{y} = \mathbf{F}(\mathbf{x},a)$, $\mathbf{x} \varepsilon V$, $\mathbf{y} \varepsilon R^m$, $a \varepsilon G$, such that*

(*T*1). *$\mathbf{F}$ is C^2 in all its arguments.*

(*T*2). $\mathbf{F}(\mathbf{x},e) = \mathbf{x}$.

(*T*3). *If $\mathbf{F}(\mathbf{x},a) \varepsilon V$, then $\mathbf{F}(\mathbf{F}(\mathbf{x},a),b) = \mathbf{F}(\mathbf{x},ba)$.*

By (*T*1), $\mathbf{F}(\mathbf{x},a)$ is a continuous function of $a \varepsilon G$. For a near e, the set $\mathbf{F}(V,a) = \{\mathbf{F}(\mathbf{x},a), \mathbf{x} \varepsilon V\}$ has a nonvoid intersection with V, by (*T*2). For all $\mathbf{x}$ such that $\mathbf{F}(\mathbf{x},a) \varepsilon V$, $\mathbf{F}(\mathbf{F}(\mathbf{x},a),a^{-1}) = \mathbf{x}$. This shows that $\mathbf{F}(\mathbf{y},a^{-1})$ is the inverse function of $\mathbf{F}(\mathbf{x},a)$ for fixed a. The map $\mathbf{F}(\mathbf{x},a)$

† A more general (non-Lie) notion of a pseudo group of transformations is needed in differential topology. For a satisfactory definition, see the papers by Ehresmann cited in the references at the end of this chapter.

must be one-to-one for constant a; otherwise it could not have an inverse.

Theorem 7-2. *The maps of a pseudo group of transformations are one-to-one.*

By an abuse of language, we shall in the future speak of "transformation groups" instead of "Lie pseudo groups of transformations."

Even more than for Lie groups, one-parameter subgroups are important in the study of transformation groups. Let $a(t) = \alpha t + \cdots$ be a one-parameter group G_α in G (not necessarily in canonical coordinates). $\mathbf{y}(t) = \mathbf{F}(\mathbf{x},a(t))$ is the *trajectory* of the fixed point $\mathbf{x}$ under the one-parameter group $a(t)$. $\mathbf{y}(t)$ has a Taylor expansion, by ($T1$) and ($T2$),

$$\mathbf{y}(t) = \mathbf{x} + a(t)W(\mathbf{x}) + \cdots = \mathbf{x} + \alpha W(\mathbf{x})t + \cdots \tag{7-1}$$

where the matrix $W(\mathbf{x})$ is the jacobian

$$W(\mathbf{x}) = J_a\mathbf{F}(\mathbf{x},a)_{a=e} = \left[\frac{\partial F_j(\mathbf{x},a)}{\partial a_i}\right]_{a=e} \tag{7-2}$$

It is an $r \times m$ matrix independent of α. The vector

$$\Xi_\alpha = \alpha W(\mathbf{x})$$

is the tangent at $\mathbf{x}$ to the trajectory of G_α through $\mathbf{x}$. On this trajectory, $\mathbf{x}$ belongs to the parameter value $t = 0$. By Eq. (7-1), the differential equation of the trajectories is

$$\frac{d\mathbf{x}}{dt} = \Xi_\alpha = \alpha W(\mathbf{x}) \tag{7-3}$$

analogous to Eq. (6-31). We shall prove later that all integral curves of (7-3) are trajectories of one-parameter groups.

There is another way of looking at the trajectories of one-parameter groups. We may compare the flow of points in their trajectories ("streamlines") to the flow of molecules in a river. Each molecule takes with it its characteristics, and if we fix an observatory midstream, we may measure the change in time of temperature, density, etc. By analogy, the map $\mathbf{x} \to \mathbf{F}(\mathbf{x},a)$ defines not only a flow of points but also a flow of functions. For any function $f(\mathbf{x})$ we may attach its value to $\mathbf{F}(\mathbf{x},a(t))$ at time t. This means that we define a family of functions

$$f^{a(t)}(\mathbf{x}) = f(\mathbf{F}(\mathbf{x},a(t)^{-1})) \tag{7-4}$$

The value of f^a at $\mathbf{x}$ is the value of f at the point which the action of a brings into $\mathbf{x}$; it is the value of f that flows over $\mathbf{x}$ at the moment of the measurement. We say that f^a results from f by *dragging along a trajectory* of G_α. The change of f in the flow is the *Lie derivative.*

Definition 7-3. *The Lie derivative* $\pounds_\alpha f$ *of a function* $f(\mathbf{x})$ *under a one-parameter group* G_α *is* $\pounds_\alpha f = -[df^{a(t)}\mathbf{x}/dt]_{t=0}$.

For numerical functions the Lie derivative is simply the directional derivative in the direction tangent to the trajectory at x. The definition is nevertheless important, because it may be generalized to more complicated geometric objects that are not easily described in terms of directional derivatives (Secs. 11-2 and 13-4). By definition,

$$\begin{aligned}\pounds_\alpha f(\mathbf{x}) &= -\frac{d}{dt} f(\mathbf{F}(\mathbf{x},a(t)^{-1})_{t=0} \\ &= -\frac{d}{dt} f(\mathbf{x} - \alpha t + \cdots)_{t=0} \\ &= \sum_{i=1}^{m} \Xi_{\alpha i} \frac{\partial f(x)}{\partial x_i} \\ &= \Xi_\alpha \operatorname{grad} f = \alpha W(\mathbf{x}) \operatorname{grad} f \end{aligned} \tag{7-5}$$

grad f is the vector $\{\partial f/\partial x_1, \ldots, \partial f/\partial x_m\}$, sometimes written ∇f. The columns in a jacobian matrix are gradients.

***Example* 7-1.** The group of euclidean motions in a plane is

$$\begin{aligned} y_1 &= x_1 \cos a_1 + x_2 \sin a_1 + a_2 \\ y_2 &= -x_1 \sin a_1 + x_2 \cos a_1 + a_3 \end{aligned}$$

It is a three-parameter group in two space. Putting $a_i = \alpha_i t + \cdots$ and using the Taylor expansion of the trigonometric functions, we obtain

$$\begin{aligned} y_1 &= x_1 + \alpha_1 x_2 t + \alpha_2 t + \cdots \\ y_2 &= x_2 - \alpha_1 x_1 t + \alpha_3 t + \cdots \end{aligned}$$

Hence
$$W(\mathbf{x}) = \begin{pmatrix} x_2 & -x_1 \\ 1 & 0 \\ 0 & 1 \end{pmatrix}$$

In $\mathfrak{L}(O_2 \times R_2)$ we take the basis vectors $\epsilon_1 = (1,0,0)$, $\epsilon_2 = (0,1,0)$, and $\epsilon_3 = (0,0,1)$. The vector fields to the groups $G_i = G_{\epsilon_i}$ are

$$\Xi_1 = (x_2, -x_1) \qquad \Xi_2 = (1,0) \qquad \Xi_3 = (0,1)$$

They are the row vectors of $W(\mathbf{x})$. The corresponding Lie derivatives are

$$\pounds_1 = x_2 \frac{\partial}{\partial x_1} - x_1 \frac{\partial}{\partial x_2} \qquad \pounds_2 = \frac{\partial}{\partial x_1} \qquad \pounds_3 = \frac{\partial}{\partial x_2}$$

and the Lie derivative of the infinitesimal vector $\beta = b_1\epsilon_1 + b_2\epsilon_2 + b_3\epsilon_3$ becomes $\pounds_\beta = b_1\pounds_1 + b_2\pounds_2 + b_3\pounds_3$.

***Example* 7-2.** Any rotation in R^3 is the product of rotations about the coordinate axes (example 5-7). The one-parameter groups

$$\mathbf{y}_1(t) = \mathbf{x}\begin{pmatrix} 1 & 0 & 0 \\ 0 & \cos\alpha t & \sin\alpha t \\ 0 & -\sin\alpha t & \cos\alpha t \end{pmatrix} = \mathbf{x} + \alpha(0,-x_3,x_2)t + \cdots$$

$$\mathbf{y}_2(t) = \mathbf{x}\begin{pmatrix} \cos\beta t & 0 & \sin\beta t \\ 0 & 1 & 0 \\ -\sin\beta t & 0 & \cos\beta t \end{pmatrix} = \mathbf{x} + \beta(-x_3,0,x_1)t + \cdots$$

$$\mathbf{y}_3(t) = \mathbf{x}\begin{pmatrix} \cos\gamma t & \sin\gamma t & 0 \\ -\sin\gamma t & \cos\gamma t & 0 \\ 0 & 0 & 1 \end{pmatrix} = \mathbf{x} + \gamma(-x_2,x_1,0)t + \cdots$$

give rise to the differential operators

$$\pounds_1 = -x_3\frac{\partial}{\partial x_2} + x_2\frac{\partial}{\partial x_3} \qquad \pounds_2 = -x_3\frac{\partial}{\partial x_1} + x_1\frac{\partial}{\partial x_3}$$
$$\pounds_3 = -x_2\frac{\partial}{\partial x_1} + x_1\frac{\partial}{\partial x_2}$$

The properties of $\pounds_\alpha$ closely resemble those of ad α. In order to make $\pounds_\alpha$ a linear mapping in a space of functions and to obtain simple statements, we assume from now on that the functions on which $\pounds_\alpha$ operates possess derivatives of all orders, $f \in C^\infty$. If f were only of class C^r, it would not be possible to compare $(\pounds_\alpha\pounds_\beta - \pounds_\beta\pounds_\alpha)f$ (of class C^{r-2}) with $\pounds_{[\alpha,\beta]}f$ (of class C^{r-1}). The set of Lie derivatives of a transformation group acting on C^∞ functions is a linear space.

Theorem 7-4 (Second Fundamental Theorem of Lie). *The Lie derivatives of a transformation group form an algebra under addition and commutation* $[\pounds_\beta,\pounds_\alpha] = \pounds_\alpha\pounds_\beta - \pounds_\beta\pounds_\alpha$. *This algebra is homomorphic to the Lie algebra of the group of parameters:*

$$\pounds_{\alpha+\beta} = \pounds_\alpha + \pounds_\beta \qquad \pounds_{[\beta,\alpha]} = [\pounds_\beta,\pounds_\alpha]$$

The Lie derivative is a linear operator. It follows immediately from the defining equation (7-5) that $\pounds_{\alpha+\beta} = \pounds_\alpha + \pounds_\beta$. Therefore we may compute the vector $\Xi_{[\beta,\alpha]}$ as the difference of the vectors Ξ belonging to $(ba)(t^2)$ and to $-(ba)(t^2)$, respectively. The vectors are obtained from the Taylor expansion for $a(t) = \alpha t + \cdots$, $b(t) = \beta t + \cdots$,

$$\begin{aligned}\mathbf{F}(\mathbf{x},ba) &= \mathbf{F}(\mathbf{F}(\mathbf{x},a),b) = \mathbf{F}(\mathbf{x},a) + \beta W(\mathbf{F}(\mathbf{x},a))t + \cdots \\ &= \mathbf{x} + \alpha W(\mathbf{x})t + \beta W(\mathbf{x} + \alpha W(\mathbf{x})t + \cdots)t + \cdots \\ &= \mathbf{x} + (\alpha+\beta)W(\mathbf{x})t + \alpha W(\mathbf{x})J_{\mathbf{x}}(\beta W(\mathbf{x}))t^2 + \cdots\end{aligned}$$

The third vector in the last expression is in components:

$$(\alpha_i)(w_{ij})\left(\frac{\partial}{\partial x_j}\{(\beta_s)(w_{sk})\}\right) = \left(\sum_{i,j,s} \alpha_i\beta_s w_{ij}\frac{\partial w_{sk}}{\partial x_j}\right)$$

The tangent vector to the trajectory of the commutator curve is therefore

$$\begin{aligned}\Xi_{[\beta,\alpha]} &= \alpha W(\mathbf{x})J_{\mathbf{x}}(\beta W(\mathbf{x})) - \beta W(\mathbf{x})J_{\mathbf{x}}(\alpha W(\mathbf{x}))\\ &= \pounds_\alpha \Xi_\beta - \pounds_\beta \Xi_\alpha\end{aligned}$$

On the other hand, the second derivatives of f in $[\pounds_\beta,\pounds_\alpha]f$ are

$$\sum_{i,j,k,l}\left\{\alpha_i\beta_j w_{ik}w_{jl}\frac{\partial^2 f}{\partial x_k \partial x_l} - \beta_i\alpha_j w_{ik}w_{jl}\frac{\partial^2 f}{\partial x_k \partial x_l}\right\} = 0$$

because of the symmetry of the second derivatives. Hence

$$[\pounds_\beta,\pounds_\alpha]f = (\pounds_\alpha\Xi_\beta - \pounds_\beta\Xi_\alpha)\,\mathrm{grad}\,f = \pounds_{[\beta,\alpha]}f$$

by direct computation. As a homomorphic image of $\mathfrak{L}(G)$ the set of Lie derivatives is a Lie algebra.

The vector fields $\Xi_1(\mathbf{x}), \ldots, \Xi_r(\mathbf{x})$ corresponding to a basis $\epsilon_1, \ldots, \epsilon_r$ of $\mathfrak{L}(G)$ are linearly independent if no relation $\sum_{i=1}^{r} c_i\Xi_i(\mathbf{x}) = 0$ with constant c_i exists in some neighborhood V, unless $c_i = 0$, $i = 1, \ldots, r$. If the Ξ_i are linearly independent, so are the operators $\pounds_1, \ldots, \pounds_r$. The vector space of all linear combinations $\Sigma c_i\pounds_i$ has dimension r, and it is isomorphic to the vector space underlying $\mathfrak{L}(G)$. In this case the algebra of Lie derivatives is *isomorphic* to $\mathfrak{L}(G)$.

Definition 7-5. *A Lie group G acts effectively in a transformation group if its Lie algebra is isomorphic to the algebra of Lie derivatives.*

If the action of G in F is not effective, there is an $\alpha \in \mathfrak{L}(G)$ which is mapped into the zero derivative, $\pounds_\alpha = 0$. All vectors mapped into the zero derivative form a linear subspace $\mathfrak{L}'$ of $\mathfrak{L}(G)$, and, since

$$\pounds_\alpha\pounds_\beta - \pounds_\beta\pounds_\alpha = 0 \qquad \text{if } \pounds_\alpha = \pounds_\beta = 0$$

it follows that $[\beta,\alpha] \in \mathfrak{L}'$ if $\alpha \in \mathfrak{L}'$, $\beta \in \mathfrak{L}'$. $\mathfrak{L}'$ is a *subalgebra* of $\mathfrak{L}(G)$. It is possible to find a basis $\epsilon_1, \ldots, \epsilon_s, \epsilon_{s+1}, \ldots, \epsilon_r$ of $\mathfrak{L}(G)$ such that $\epsilon_{s+1}, \ldots, \epsilon_r$ is a basis of $\mathfrak{L}'$. Let M be the subspace of $\mathfrak{L}(G)$ spanned by $\epsilon_1, \ldots, \epsilon_s$. M is not, in general, a subalgebra of $\mathfrak{L}(G)$. For $\mu \in M$, $\lambda \in \mathfrak{L}'$, $[\pounds_\mu,\pounds_\lambda] = 0$; hence $[\mu,\lambda] \in \mathfrak{L}'$. If $[A,B]$ denotes the set of all products $[\alpha,\beta]$, $\alpha \in A$, $\beta \in B$, the last result may be written $[\mathfrak{L}',M] \subset \mathfrak{L}'$.

Definition 7-6. *A Lie algebra $\mathfrak{L}$ is reductive if it is the (direct) sum of two vector spaces $\mathfrak{L} = \mathfrak{L}' + M$, $\mathfrak{L}' \cap M = 0$, such that $\mathfrak{L}'$ is a subalgebra of $\mathfrak{L}$, $[\mathfrak{L}',\mathfrak{L}'] \subset \mathfrak{L}'$, and $[\mathfrak{L}',M] = (\mathrm{ad}\,\mathfrak{L}')M \subset \mathfrak{L}'$.*

Theorem 7-7. *If a Lie group G does not act effectively in a transformation group, its Lie algebra is reductive.*

This result shows that a Lie group may act as a transformation group only in a few dimensions that are determined by the structure of its algebra.

A one-parameter group $G_\alpha \subset G$ defines a flow of the points of R^m in trajectories. The field of tangent vectors to the trajectories of G_α are [Eq. (7-3)]

$$\Xi_\alpha = \pounds_\alpha \mathbf{x} \tag{7-6}$$

The next three theorems show that one-parameter groups are as important for transformation groups as for Lie groups.

Theorem 7-8. *The integral curves of*

$$\frac{d\mathbf{x}}{dt} = \pounds_\alpha \mathbf{x} \tag{7-7}$$

are the trajectories of the one-parameter group G_α.

If the Lie group G is given in canonical coordinates, the theorem states that a solution of Eq. (7-7) with $\mathbf{x}(0) = \mathbf{x}_0$ satisfies $\mathbf{x}(t) = \mathbf{F}(\mathbf{x}_0,\alpha t)$. The method of proof is the same as for Lie groups: The function

$$\mathbf{y}(t) = \mathbf{F}(\mathbf{x}_0,\alpha t)$$

satisfies (7-7) and the initial condition [see Eq. (7-1)]. The solution is unique because the vector function $\Xi(\mathbf{x})$ is differentiable.

There is no element in G that does not belong to a one-parameter group G_α. By theorem 7-8, any transformation $\mathbf{F}(\mathbf{x}_0,a)$ may be obtained by the integration of a system (7-7):

Corollary 7-9 (Lie's First Fundamental Theorem). *A transformation group is completely determined by its Lie derivatives.*

The main theorem on transformation groups states that a Lie algebra of differentiation operators $\pounds = \sum_{i=1}^{m} \xi_i(\mathbf{x})(\partial/\partial x_i) = \Xi(\mathbf{x})\,\text{grad}$ is always the algebra of Lie derivatives of a Lie pseudo group of transformations. Theorem 7-10 is the analogon of theorem 6-17.

Theorem 7-10 (Converse of Lie's Second Fundamental Theorem). *Given a set of linearly independent linear differentiation operators*

$$\pounds_i(\mathbf{x}) = \Xi_i(\mathbf{x})\,\text{grad}$$

$i = 1, \ldots, r$, *defined in an open set $V \subset R^m$ and closed under the bracket operation* $[\pounds_i,\pounds_j] = \pounds_j\pounds_i - \pounds_i\pounds_j = \sum_{k=1}^{r} c_{k,ij}\pounds_k$ *(constant $c_{k,ij}$), then the linear combinations of the $\pounds_i$'s form an algebra which is the algebra of Lie* *derivatives of a transformation pseudo group. The general transformation of the pseudo group is obtained by the integration of* $d\mathbf{x}/dt = (\Sigma a_i \mathcal{L}_i)\mathbf{x}$, a_i *constant.*

It follows from the definition of the bracket operator that the $\mathcal{L}_i$'s generate an abstract Lie algebra (see theorem 6-9). The algebra defines a unique parameter group germ G, in canonical coordinates. It is possible to choose a base in $\mathcal{L}(G)$ such that the product of the basis vectors $\epsilon_1, \ldots, \epsilon_r$ is given by the structure constants $c_{k,ij}$ defined in theorem 7-10. Any element $a \in G$ may be written as $a = \Sigma a_i \epsilon_i t_0$ if the spaces of G and $\mathcal{L}(G)$ are identified. We have to show that the function $\mathbf{F}(\mathbf{x},a)$ defined as the value for $t = t_0$ of

$$\frac{d\mathbf{F}}{dt} = \left(\sum_{i=1}^{r} a_i \mathcal{L}_i\right)\mathbf{F} \qquad \mathbf{F}(0) = \mathbf{x}$$

is a transformation group in which G acts on R^m, that is, that

$$\mathbf{F}(\mathbf{F}(\mathbf{x},a),b) = \mathbf{F}(\mathbf{x},ba)$$

By corollary 7-9, the transformation group is unique if it exists. The idea of the proof is the same as that for theorem 6-25. The matrix $W(\mathbf{x})$ whose row vectors are $\Xi_1, \ldots, \Xi_r$ will play the role of the matrix appearing in Eq. (7-3). A vector $\alpha = (a_1, \ldots, a_r)$ defines a one-parameter group G_α and trajectories $d\mathbf{x}/dt = \alpha W(\mathbf{x})$. For $q = \alpha t$, the differential equation of the trajectory may be written

$$d\mathbf{x} = \alpha\, dt\, W(\mathbf{x}) = dq\, Q(q) W(\mathbf{x})$$

[see Eq. (6-35)]. For a,b constant, define

$$\mathbf{z}(t) = \mathbf{F}(\mathbf{x}, bt \cdot a) \qquad \mathbf{z}^*(t) = \mathbf{F}(\mathbf{F}(\mathbf{x},a),bt)$$

We have to show that $\mathbf{z}(1) = \mathbf{z}^*(1)$. The differential equation for $\mathbf{z}^*$ is

$$\frac{d\mathbf{z}^*}{dt} = bQ(bt)W(\mathbf{z}^*) \qquad \mathbf{z}^*(0) = \mathbf{F}(\mathbf{x},a)$$

By Eq. (6-29), the equation for $\mathbf{z}$ is

$$d\mathbf{z} = d(bt \cdot a)Q(bt \cdot a)W(\mathbf{z}) = b\, dt\, Q(bt)W(\mathbf{z})$$

Hence
$$\frac{d\mathbf{z}}{dt} = bQ(bt)W(\mathbf{z}) \qquad \mathbf{z}(0) = \mathbf{F}(\mathbf{x},ea)$$

Since $\mathbf{z}$ and $\mathbf{z}^*$ satisfy the same differential equation with identical initial conditions, they coincide.

***Example* 7-1 (*continued*).** The matrix of the action of the group of euclidean motions in the plane is

$$W(\mathbf{x}) = \begin{pmatrix} x_2 & -x_1 \\ 1 & 0 \\ 0 & 1 \end{pmatrix}$$

By theorem 7-8, we have three one-parameter families of trajectories $\mathbf{x}^{(i)} = (x_1^{(i)}(t), x_2^{(i)}(t))$, $i = 1, 2, 3$, solutions of

$$\begin{aligned} \frac{dx_1^{(1)}}{dt} &= x_2^{(1)} & \frac{dx_2^{(1)}}{dt} &= -x_1^{(1)} \\ \frac{dx_1^{(2)}}{dt} &= 1 & \frac{dx_2^{(2)}}{dt} &= 0 \\ \frac{dx_1^{(3)}}{dt} &= 0 & \frac{dx_2^{(3)}}{dt} &= 1 \\ x_1^{(i)}(0) &= x_1 & x_2^{(i)}(0) &= x_2 \end{aligned}$$

or

$$\begin{aligned} x_1^{(1)} &= x_1 \cos t + x_2 \sin t & x_2^{(1)} &= -x_1 \sin t + x_2 \cos t \\ x_1^{(2)} &= x_1 + t & x_2^{(2)} &= x_2 \\ x_1^{(3)} &= x_1 & x_2^{(3)} &= x_2 + t \end{aligned}$$

Any plane euclidean motion is a linear combination of these three particular types of transformations, i.e., rotation and translations parallel to the coordinate axes, as asserted by theorem 7-10.

***Example* 7-3.** We want to find all "groups" of analytic transformations acting on some interval of the real-number line R^1. By theorem 7-10, any differentiation operator generates trajectories of a one-parameter group. Hence it is convenient to start with the simplest possible operator $\mathfrak{L}_1 = d/dx$. (Since we have only one variable, the derivative is a total one.) The trajectory is the integral curve of $dx/dt = \mathfrak{L}_1 x = 1$, $x(0) = x_0$,

$$x^{(1)} = x_0 + t$$

This is the one-parameter group of translations.

A second operator suggests itself: $\mathfrak{L}_2 = x(d/dx)$. Since $[\mathfrak{L}_2, \mathfrak{L}_1] = \mathfrak{L}_1$, $\mathfrak{L}_2$ is admissible, by theorem 7-10. The group is obtained by integration of $dx/dt = \mathfrak{L}_2 x = x$; it is

$$x^{(2)} = x_0 e^t$$

the group of homotheties on the line. $\mathfrak{L}_1$ and $\mathfrak{L}_2$ together generate the group of similitudes on R^1.

For $\mathfrak{L}_3 = x^2(d/dx)$ we have $[\mathfrak{L}_3,\mathfrak{L}_1] = 2\mathfrak{L}_2$, $[\mathfrak{L}_3,\mathfrak{L}_2] = \mathfrak{L}_3$; $\mathfrak{L}_3$ is also admissible. The corresponding transformation $dx/dt = x^2$, $x(0) = x_0$, is

$$x^{(3)} = \frac{x_0}{x_0 t - 1}$$

(This last transformation is defined only on a neighborhood of R^1 and only for t in a group germ of R. See example 5-3 for details.) $x^{(1)}$, $x^{(2)}$, and $x^{(3)}$ together generate the transformation group of projectivities

$$y = \frac{ax + b}{cx + d} \qquad ad - bc \neq 0$$

It turns out that in some sense this is the most general analytic transformation group on the line. Let us try another operator $\mathfrak{L}_k = x^{k-1}(d/dx)$. Then $[\mathfrak{L}_k,\mathfrak{L}_3] = (k - 3)\mathfrak{L}_{k+1}$. If a transformation group contains any operator $\mathfrak{L}_k$, $k > 3$, it must contain *all* operators $\mathfrak{L}_i$, $i = k, k + 1, k + 2, \ldots$. The infinite set of functions $\{x,x^2, \ldots, x^k,x^{k+1}, \ldots\}$ contains only elements linearly independent over R (see exercise 7-1, prob. 5). The algebra of Lie derivatives therefore has infinite dimension; it is not the Lie algebra of a group. This shows that no $\mathfrak{L}_k(k > 3)$ can be a generator in a transformation group of more than one parameter. The same argument applies to other functions given by a Taylor series.

On the other hand, any analytic coordinate transformation $\xi = \xi(x)$ with nonvanishing derivative gives rise to a new group

$$\eta = \frac{a\xi + b}{c\xi + d}$$

which is isomorphic to, but not identical with, the projective group in the original coordinates.

Exercise 7-1

1. Prove by direct computation that constants $c_{k,ij}$ defined by $\mathfrak{L}_i\mathfrak{L}_j - \mathfrak{L}_j\mathfrak{L}_i = \Sigma c_{k,ij}\mathfrak{L}_k$ are the structure constants of a Lie algebra.

2. Given a Lie group G, the multiplication $a * b = ba$ defines a Lie group G^*. Use this fact to show that a transformation group may also be defined if condition ($T3$) is replaced by $\mathbf{F}(\mathbf{F}(\mathbf{x},a),b) = \mathbf{F}(\mathbf{x},ab)$.

3. Let M be a subspace of the vector space in which a Lie algebra $\mathfrak{L}$ is defined. Take a base $\epsilon_1, \ldots, \epsilon_r$ of $\mathfrak{L}$ such that $\epsilon_1, \ldots, \epsilon_s$ is a base of M. The Lie product is given by $[\epsilon_i,\epsilon_j] = \Sigma c_{k,ij}\epsilon_k$. Put

$$c^*_{k,ij} = \begin{cases} c_{k,ij} & \text{for } 1 \leq k,i,j \leq s \\ 0 & \text{for } s < k \leq r, 1 \leq i,j \leq s \end{cases}$$

Show that $[\epsilon_i,\epsilon_j] = \Sigma c^*_{k,ij}\epsilon_k$ defines a Lie algebra $\mathfrak{M}$ on the space M.

4. Use Prob. 3 or theorem 7-10 to prove that, if a Lie group germ G does not act effectively in a transformation group, it has a biggest subgroup germ H which acts effectively.

5. From the fundamental theorem of algebra follows: if a relation $\sum_{i=0}^{N} a_i x^i = 0$ is satisfied by at least $N + 1$ distinct values of x, then all $a_i = 0$. Use this fact to prove that no finite set of powers of x may identically satisfy a linear relation with constant coefficients.

◆ **6.** The group of unimodular affine transformations in the plane is

$$\begin{aligned} x_1^* &= a_1x_1 + a_2x_2 + a_3 \\ x_2^* &= a_4x_1 + a_5x_2 + a_6 \end{aligned} \qquad a_1a_5 - a_2a_4 = 1$$

Find its matrix $W(\mathbf{x})$ and its Lie derivatives $\pounds_1, \ldots, \pounds_5$. (HINT: Use theorem 6-12.)

7. Find the Lie derivatives $\pounds_1, \ldots, \pounds_6$ of the euclidean motions

$$\mathbf{x}^* = \mathbf{x}A + \mathbf{b} \qquad A \in O_3 \qquad \mathbf{b} \in R_3$$

in R^3.

8. Example 7-3 shows that the maximal dimension of a group acting on R^1 is three. The dimension of the groups acting on R^2 is *not* bounded. To prove this statement, show that any two Lie derivatives $\pounds_1 = g(x_1)(\partial/\partial x_2)$, $\pounds_2 = h(x_1)(\partial/\partial x_2)$ with C^∞ functions g,h define a transformation group on R^2 (Lie). Is this group acting effectively on R^2? What are the trajectories?

◆ **9.** Find the transformation group in the plane given by

$$W(\mathbf{x}) = \begin{pmatrix} x_2 & x_1 \\ 1 & 0 \\ 0 & 1 \end{pmatrix}$$

Also find its parameter group.

10. Integrate the group defined in the plane by

$$\pounds_1 = \frac{\partial}{\partial x_1} \qquad \pounds_2 = x_1\frac{\partial}{\partial x_1} + x_2\frac{\partial}{\partial x_2} \qquad \pounds_3 = (x_1{}^2 - x_2{}^2)\frac{\partial}{\partial x_1} + 2x_1x_2\frac{\partial}{\partial x_2}$$

Find the parameter group in the list of groups of dimension three (exercise 6-2, Prob. 9).

11. Same question as Prob. 10 for

$$\pounds_1 = \frac{\partial}{\partial x_2} \qquad \pounds_2 = \sin x_2\frac{\partial}{\partial x_1} + \cot x_1 \cos x_2\frac{\partial}{\partial x_2}$$

$$\pounds_3 = \cos x_2\frac{\partial}{\partial x_1} - \cot x_1 \sin x_2\frac{\partial}{\partial x_2}$$

12. Show that the following operations define a transformation group in R^3. Find the multiplication table of the Lie algebra.

$$\mathcal{L}_1 = \frac{\partial}{\partial x_2}$$

$$\mathcal{L}_2 = \cos x_2 \frac{\partial}{\partial x_1} - \cot x_1 \sin x_2 \frac{\partial}{\partial x_2} + \frac{n \sin x_2}{\sin x_1} \frac{\partial}{\partial x_3}$$

$$\mathcal{L}_3 = -\sin x_2 \frac{\partial}{\partial x_1} - \cot x_1 \cos x_2 \frac{\partial}{\partial x_2} + \frac{n \cos x_2}{\sin x_1} \frac{\partial}{\partial x_3}$$

$$\mathcal{L}_4 = \frac{\partial}{\partial x_3}$$

13. Find a reductive decomposition of the algebra of exercise 6-3, Prob. 15.

14. Find the commutation rules for the operators of example 7-3, and show that the algebra of the projective group on R^1 is isomorphic to $\mathcal{L}(O_3)$.

15. In analytical mechanics, the bracket $[p_i,p_j]$ of the operators defined in exercise 6-2, Prob. 2, is called the *Poisson* bracket. Compute explicitly the Poisson bracket.

7-2. Invariants

The study of invariants is the central topic of differential geometry.

Definition 7-11. *A function $f(\mathbf{x})$ is an invariant of a transformation group $\mathbf{F}(\mathbf{x},a)$ if $f(\mathbf{x}) = f(\mathbf{F}(\mathbf{x},a))$.*

An invariant is constant on any trajectory of a one-parameter group:

Lemma 7-12. *A differentiable function $f(\mathbf{x})$ is an invariant if and only if $\mathcal{L}f = 0$ for all Lie derivatives of the transformation group.*

The most important transformation groups are those which can transform any given point $\mathbf{x}$ into any other point $\mathbf{y}$, that is, the equation $\mathbf{y} = \mathbf{F}(\mathbf{x},a)$ has a solution a for all $\mathbf{x},\mathbf{y}$. For our local theory, we need this property only locally:

Definition 7-13. *A transformation group $\mathbf{F}(\mathbf{x},a)$ is locally transitive at $\mathbf{x}$ if for all $\mathbf{y}$ in some neighborhood of $\mathbf{x}$ there is an a such that $\mathbf{y} = \mathbf{F}(\mathbf{x},a)$.*

A transformation group is transitive if the neighborhood in which it is locally transitive is the whole space in which it acts. A transformation group in a locally compact, connected space is transitive if it is locally transitive at any point. We will not use the notion of (global) transitivity, which is of a topological character.

If a group is locally transitive in $\mathbf{x}$, every vector is the tangent vector of some trajectory through $\mathbf{x}$ of a one-parameter subgroup. By lemma 7-12, all directional derivatives vanish.

Theorem 7-14. *A locally transitive transformation group has no nonconstant differentiable invariant.*

The locally transitive transformation groups are the important ones in geometry. Theorem 7-14 shows that Def. 7-11 is inadequate. We shall see better how to arrive at an interesting definition of an invariant if we study first some algebraic implications of lemma 7-12.

A Lie derivative of the transformation group $\mathbf{F}(\mathbf{x},a)$ is a linear combination of the components of the vector $W(\mathbf{x})$ grad. Let $V(\mathbf{x})$ be a matrix formed by a maximal set of row vectors of $W(\mathbf{x})$, linearly independent in a neighborhood of $\mathbf{x}$, and let $r' \leq r$ be the number of its rows. We put $\mathbf{u} = \text{grad } f$. f is an invariant if and only if

$$V(\mathbf{x})\mathbf{u} = 0 \tag{7-8}$$

For a locally transitive group $r' \geq m$, the rank of $V(\mathbf{x})$ is m and the homogeneous linear system (7-8) has only the trivial solution $\mathbf{u} = 0$. We may obtain nontrivial invariants only if we are able to create a situation such that the rank of $V(\mathbf{x})$ is less than the dimension of the space in which it acts. By a fundamental existence theorem on partial differential equations (exercise 10-4, Prob. 11), the solutions of $V(\mathbf{x}) \text{ grad } f = 0$ in some neighborhood of $\mathbf{x}$ form a linear space of the same dimension (m − rank V) as do the solutions of the algebraic system $V(\mathbf{x})\mathbf{u} = 0$ at any point of the neighborhood, if $V(\mathbf{x})$ is of constant rank. This theorem, which we shall assume here, reduces the discussion to an algebraic one.

A first method for the generation of invariants is based on the fact that we may have $\mathbf{F}$ act not only on the space R^m of points but also on the space $R^m \times R^m$ of ordered couples of points $(\mathbf{x}^{(1)},\mathbf{x}^{(2)})$, by the definition

$$\mathbf{F}^{[2]}((\mathbf{x}^{(1)},\mathbf{x}^{(2)}),a) = (\mathbf{F}(\mathbf{x}^{(1)},a),\mathbf{F}(\mathbf{x}^{(2)},a)) \tag{7-9}$$

In the same way, $\mathbf{F}$ may be *prolonged* to an operation on k-tuples of points, i.e., a transformation group on $R^m \times \cdots \times R^m = (R^m)^k$, by

$$\mathbf{F}^{[k]}((\mathbf{x}^{(1)}, \ldots ,\mathbf{x}^{(k)}),a) = (\mathbf{F}(\mathbf{x}^{(1)},a), \ldots ,\mathbf{F}(\mathbf{x}^{(k)},a))$$

If the space $(R^m)^k$ is written as an mk-dimensional space R^{mk} with coordinates $x_i^{(j)} = x_{(j-1)m+i}$, the Lie derivatives of $\mathbf{F}^{[k]}$ are obtained from those of $\mathbf{F} = \mathbf{F}^{[1]}$ by the matrix

$$V^{[k]}(\mathbf{x}^{(1)}, \ldots ,\mathbf{x}^{(k)}) = (V(\mathbf{x}^{(1)}) \vdots V(\mathbf{x}^{(2)}) \vdots \cdots \vdots V(\mathbf{x}^{(k)})) \tag{7-10}$$

In this matrix $V(\mathbf{x})$ is written k times, with different arguments. $V^{[k]}$ acts on the *prolonged* gradient operator $\text{grad}^{[k]} = \{\partial/\partial x_s\}$, $s = 1, \ldots, mk$, and each one of the $V(\mathbf{x}^{(j)})$ acts on the gradient operator to $\mathbf{x}^{(j)}$. The rank of $V^{[k]}$ is r' since linear dependence in $V(\mathbf{x})$ implies linear dependence in $V^{[k]}(\mathbf{x})$ and vice versa. The dimension of $(R^m)^k$ is mk;

hence there is an infinite set of integers $k \geq k_0$ for which the homogeneous system corresponding to (7-8)

$$V^{[k]}(\mathbf{x}^{(1)}, \ldots, \mathbf{x}^{(k)})\mathbf{u} = 0$$

has nontrivial solutions. This will be the case for $k_0 >$ the greatest integer $[r'/m]$ contained in r'/m.

Definition 7-15. *A differentiable k-point invariant of a transformation group is a differentiable function* $f(\mathbf{x}^{(1)}, \ldots, \mathbf{x}^{(k)})$ *invariant under* $\mathbf{F}^{[k]}$.

The existence theorem on partial differential equations quoted above then assures the existence of k-point invariants.

Theorem 7-16. *Every transformation group has differentiable k-point invariants for* $k > [r'/m]$.

A comparison of theorems 7-16 and 7-14 shows that $\mathbf{F}^{[k]}$ cannot be locally transitive on $(R^m)^k$ for large k; that is, no transitive group may map any k-tuple of points onto any other k-tuple. This fact is illustrated in the next example.

***Example* 7-4.** The (transitive) group of euclidean motions in a plane has been discussed in example 7-1. The row vectors of $W(\mathbf{x})$ are linearly independent; hence $W(\mathbf{x}) = V(\mathbf{x})$, $r = r' = 3$, $m = 2$. The rank of $V^{[2]}$ is three, the matrix $V^{[2]}$ acts in a four-dimensional space, $4 - 3 = 1$, and there exists *one* two-point invariant, the solution of

$$V^{[2]} \operatorname{grad}^{[2]} f = \begin{pmatrix} x_2^{(1)} & -x_1^{(1)} & x_2^{(2)} & -x_1^{(2)} \\ 1 & 0 & 1 & 0 \\ 0 & 1 & 0 & 1 \end{pmatrix} \begin{pmatrix} \dfrac{\partial f}{\partial x_1^{(1)}} \\ \dfrac{\partial f}{\partial x_2^{(1)}} \\ \dfrac{\partial f}{\partial x_1^{(2)}} \\ \dfrac{\partial f}{\partial x_2^{(2)}} \end{pmatrix} = 0$$

or

$$\begin{aligned} x_2^{(1)} \frac{\partial f}{\partial x_1^{(1)}} - x_1^{(1)} \frac{\partial f}{\partial x_2^{(1)}} + x_2^{(2)} \frac{\partial f}{\partial x_1^{(2)}} - x_1^{(2)} \frac{\partial f}{\partial x_2^{(2)}} &= 0 \\ \frac{\partial f}{\partial x_1^{(1)}} + \frac{\partial f}{\partial x_1^{(2)}} &= 0 \\ \frac{\partial f}{\partial x_2^{(1)}} + \frac{\partial f}{\partial x_2^{(2)}} &= 0 \end{aligned}$$

The last two equations show that f is a function only of $u = x_1^{(1)} - x_1^{(2)}$, $v = x_2^{(1)} - x_2^{(2)}$. The first one reduces to $vf_u - uf_v = 0$ (here lower indices indicate partial derivatives). A *first integral* of this equation is a curve $f(u,v) = \text{const}$ in the (u,v) plane, or $df = f_u\,du + f_v\,dv = 0$. The first integrals therefore are given by $du/v = -dv/u$, or $u\,du + v\,dv = 0$,

$u^2 + v^2 = \text{const.}$ It is easily checked that

$$f(u,v) = (x_1^{(1)} - x_1^{(2)})^2 + (x_2^{(1)} - x_2^{(2)})^2$$

is a solution of the system of partial differential equations. $\sqrt{f}$ is the distance between the two points $x^{(1)}$ and $x^{(2)}$; clearly it is a euclidean invariant. Two couples of points may be brought one into the other by a euclidean motion only if they give the same value to the two-point invariant, i.e., if their distances are equal.

We may expect $mk - r = 2 \cdot 3 - 3 = 3$ independent three-point invariants. But three distinct points define three distances between pairs of points, and so any three-point invariant is a function of two-point invariants. The same reasoning applies to k-point invariants (see exercise 7-2, Prob. 3).

Theorem 7-17. *In plane euclidean geometry, any differentiable k-point invariant is a function of distances of couples of points.*

Plane euclidean geometry is a *two-point geometry;* there is no need to consider invariants of more than two points. Every transformation group is a k^*-point geometry; i.e., every invariant of $k > k^*$ points is a function of invariants of $\leq k^*$ points. In fact, for $k > m$, the number $k + 1$ of k-tuples formed of $k + 1$ points is greater than the number m of new invariants introduced by the passage from k to $k + 1$ [a k-point invariant is trivially a $(k + 1)$-point invariant].

The theory of invariants as given here has one serious drawback from the geometric point of view. The transformation group is given in a special system of coordinates that is kept fixed during the whole discussion. The invariants obtained may depend on the special system of coordinates. This is not the case in euclidean geometry; the invariant $f = |\mathbf{x}^{(1)} - \mathbf{x}^{(2)}|^2$ is unchanged in any transformation of coordinates. An invariant that does not have this property obviously is of no intrinsic geometric significance. All differentiable coordinate transformations form a pseudo group in a certain sense but definitely not a Lie pseudo group. The methods of this section do not furnish an a priori method of singling out geometric invariants among all invariants of a transformation group (see exercise 7-2, Prob. 1).

Definition 7-18. *A geometric invariant is an invariant of a transformation group which is not changed in any (sufficiently-many-times) differentiable coordinate transformation.*†

The problem of the geometric significance of an analytic development is not restricted to the theory of invariants. (Compare exercise 4-1, Prob. 2.) The admissible changes of coordinates and parameters may change with the geometry (see Sec. 9-4).

† These invariants are sometimes called "absolute geometric invariants."

We are now ready to explain the meaning of the word "geometry" in modern mathematics.

Definition 7-19. *A Klein geometry is the theory of geometric invariants of a transitive transformation group* ("Erlangen Program," 1872).

Several Klein geometries will be studied in the following sections. Non-Klein geometries are introduced in Chaps. 11 and 13.

The prolongation of a transformation group to k-point groups is not the only way to obtain invariants. The remainder of this section is devoted to the theory of differential invariants due to S. Lie. In the rest of the book, we shall expound and use the theory of these invariants due to Elie Cartan.

The importance of Lie's method lies in the fact that it easily leads to general theorems, especially on the "natural" differentiability assumption of a Klein geometry, and that the technique of prolongation of Lie derivatives has important applications later on in the computation of Lie derivatives of tensor fields. Its drawbacks are that invariants are found only by integration of systems of partial differential equations and that geometric invariants must be constructed in *Grassmann manifolds* which are not linear spaces. Elie Cartan's method needs only differentiations—no integrations—to find the invariants; automatically it yields only geometric invariants. It does not lead to general theorems of Lie's kind, but it is very well adapted to the discrimination of special cases that escape the general theorems (see, for example, Chap. 12).

As the name indicates, differential invariants are functions of points $\mathbf{x}$ and their derivatives. Since the definition of the derivatives depends on the choice of the independent variables, there will be different differential invariants for different geometric beings within the framework of one geometry.

A curve in R^m is a map $\mathbf{f}: I \rightarrow R^m$ (see Chap. 1) locally one-to-one and several times differentiable. (We shall come back later to the question of the "natural" condition of differentiability to be imposed on $\mathbf{f}$.) The parameter $\tau \in I$ will be identified with the coordinate x_1 in R^m. This is always possible in some neighborhood of any given value τ_0 through a good choice of the coordinate axes. $x_2, \ldots, x_m$ are functions of x_1 on the curve.

A one-parameter group of transformations determines its Lie derivative

$$\pounds^{(0)} = \sum_{i=1}^{m} \xi_i \frac{\partial}{\partial x_i} \tag{7-11}$$

We write $\pounds^{(0)}$ instead of $\pounds$ to indicate that the operator corresponds to Def. 7-3 of the Lie derivative only if applied to a function of the variables $x_1, \ldots, x_m$ alone. One may ask what the Lie derivative is for a function $\psi(x_1, \ldots, x_m; dx_2/dx_1, \ldots, dx_m/dx_1)$ which depends on the variables and certain first derivatives. We may formally introduce a new series of variables by

$$x_i' = \frac{dx_i}{dx_1} \qquad i = 2, \ldots, m$$

and write the Lie derivative as an operator acting on the $2m - 1$ variables x_j, x_i' in

$$\pounds^{(1)}\psi = \left[\sum_{j=1}^{m} \xi_j \frac{\partial}{\partial x_j} + \sum_{i=2}^{m} \eta_i \frac{\partial}{\partial x_i'}\right]\psi \tag{7-12}$$

where, by Def. 7-3,

$$\eta_i = -\frac{d}{dt}\frac{dF_i(x_1,x_2(x_1), \ldots, x_m(x_1); a^{-1}(t))}{dF_1(x_1,x_2(x_1), \ldots, x_m(x_1); a^{-1}(t))}\bigg|_{t=0} \tag{7-13}$$

The whole setup makes sense only on a curve $x_i = x_i(x_1)(i = 2, \ldots, m)$ referred to x_1 as parameter.

Equation (7-13) is not a practical formula for the computation of η_j. $\pounds^{(1)}$ is the directional derivative along the trajectories of $\mathbf{F}$ not in R^m but in a space R^{2m-1} of coordinates $x_1, \ldots, x_m, x_2', \ldots, x_m'$. The trajectories are all in a certain subset of R^{2m-1} for which

$$dx_j - x_j'\, dx_1 = 0 \tag{7-14}$$

If d denotes differentiation along the curve (in R^{2m-1}), two directional derivatives commute by the symmetry property of second partial derivatives. Hence

$$d\pounds^{(1)} = \pounds^{(1)}d \tag{7-15}$$

and the functions η_j may be computed from

$$\begin{aligned}\pounds^{(1)}(dx_j - x_j'\, dx_1) &= d\pounds^{(1)}x_j - (\pounds^{(1)}x_j')\, dx_1 - x_j'\, d\pounds^{(1)}x_1 \\ &= d\xi_j - \eta_j\, dx_1 - x_j'\, d\xi_1 = 0\end{aligned}$$

$\pounds^{(1)}$ is the *first prolongation* of the Lie derivative (7-11). The second prolongation $\pounds^{(2)}$ will be the derivative, according to Def. 7-3, of a function $\phi(x_1, \ldots, x_m, x_2', \ldots, x_m', x_2'', \ldots, x_m'')$ in a space of $3m - 2$ dimensions. The explicit form

$$\pounds^{(2)} = \sum_{i=1}^{m} \xi_i \frac{\partial}{\partial x_i} + \sum_{j=2}^{m} \eta_j \frac{\partial}{\partial x_j'} + \sum_{j=2}^{m} \zeta_j \frac{\partial}{\partial x_j''}$$

may again be computed from (7-15) and

$$dx_j' - x_j''\, dx_1 = 0$$

The prolongation of a differential operator depends on the functions and on the types of derivatives that are considered. A surface in R^3 may be given locally as a function of two coordinates, $x_3 = x_3(x_1,x_2)$. Of interest here are the partial derivatives $p = \partial x_3/\partial x_1$, $q = \partial x_3/\partial x_2$, $r = \partial^2 x_3/\partial x_1^2$, $s = \partial^2 x_3/\partial x_1\,\partial x_2$, $t = \partial^2 x_3/\partial x_2^2$, ... A transformation group with Lie derivative $\pounds = \Sigma \xi_i(\partial/\partial x_i)$ will act on $\phi(x_1,x_2,x_3,p,q,r,s,t)$ (in some eight-dimensional space) by

$$\pounds^{(2)} = \xi_1 \frac{\partial}{\partial x_1} + \xi_2 \frac{\partial}{\partial x_2} + \xi_3 \frac{\partial}{\partial x_3} + \pi \frac{\partial}{\partial p} + \varkappa \frac{\partial}{\partial q} + \rho \frac{\partial}{\partial r} + \sigma \frac{\partial}{\partial s} + \tau \frac{\partial}{\partial t}$$

The functions π, $\varkappa$, ρ, σ, τ are obtained from (7-15) and

$$\begin{aligned} dx_3 &= p\,dx_1 + q\,dx_2 \\ dp &= r\,dx_1 + s\,dx_2 \\ dq &= s\,dx_1 + t\,dx_2 \end{aligned} \tag{7-16}$$

***Example* 7-5.** The transformation group of euclidean motions in space R^3 is

$$\mathbf{y} = \mathbf{x}A + \mathbf{b} \qquad A \in O_3,\ \mathbf{b} \in R_3$$

The Lie derivatives of the action of O_3 have been found in example 7-2. Those of the action of R_3 as a group of parallel translations in R^3 are

$$\pounds_4 = \frac{\partial}{\partial x_1} \qquad \pounds_5 = \frac{\partial}{\partial x_2} \qquad \pounds_6 = \frac{\partial}{\partial x_3}$$

(Compare example 7-1 for the two-dimensional case.)

In the theory of space curves, based on Eq. (7-14), the three operators of the action of O_3 prolong to

$$\begin{aligned} \pounds_1^{(1)} &= -x_3 \frac{\partial}{\partial x_2} + x_2 \frac{\partial}{\partial x_3} - x_3' \frac{\partial}{\partial x_2'} + x_2' \frac{\partial}{\partial x_3'} \\ \pounds_2^{(1)} &= -x_3 \frac{\partial}{\partial x_1} + x_1 \frac{\partial}{\partial x_3} + (1 + x_3'^2) \frac{\partial}{\partial x_3'} + x_2' x_3' \frac{\partial}{\partial x_2'} \\ \pounds_3^{(1)} &= -x_2 \frac{\partial}{\partial x_1} + x_1 \frac{\partial}{\partial x_2} + (1 + x_2'^2) \frac{\partial}{\partial x_2'} + x_2' x_3' \frac{\partial}{\partial x_3'} \end{aligned}$$

The translation operators have only trivial prolongations $\pounds_i^{(1)} = \pounds_i$, $i = 4, 5, 6$.

In the theory of surfaces, the action of the prolongation of $\pounds_1$ on the first equation (7-16) gives

$$\begin{aligned} dx_2 &= \pi\,dx_1 + \varkappa\,dx_2 - q\,dx_3 \\ &= (\pi - pq)\,dx_1 + (\varkappa - q^2)\,dx_2 \end{aligned}$$

This relation must be an identity in dx_1, dx_2; hence $\pi = pq$, $\varkappa = 1 + q^2$. The prolongations are

$$£_1^{*(1)} = -x_3 \frac{\partial}{\partial x_2} + x_2 \frac{\partial}{\partial x_3} + pq \frac{\partial}{\partial p} + (1 + q^2) \frac{\partial}{\partial q}$$

$$£_2^{*(2)} = -x_3 \frac{\partial}{\partial x_1} + x_1 \frac{\partial}{\partial x_3} + (1 + p^2) \frac{\partial}{\partial p} + pq \frac{\partial}{\partial q}$$

$$£_3^{*(3)} = -x_2 \frac{\partial}{\partial x_1} + x_1 \frac{\partial}{\partial x_2} - q \frac{\partial}{\partial p} + p \frac{\partial}{\partial q}$$

This example shows clearly that prolongation is not an intrinsic notion; nevertheless it is a useful one. The prolonged operations act in a higher-dimensional space of coordinates and derivatives. In general, a transformation in an $(m + s)$-dimensional space with coordinates $x_1, \ldots, x_s, p_{ki}$ which leaves invariant a complete set of relations

$$dx_k - \sum_i p_{ki}\, dx_i = 0$$

$i = 1, \ldots, n_0$; $k = n_0 + 1, \ldots, m$, is called a *contact transformation.* We have seen that a transformation group may be prolonged to a group of contact transformations in many ways. In any case, the matrix $V(\mathbf{x})$ will be prolonged into a matrix $V^{(k)}$ depending on the kth derivatives. The rank of $V^{(k)}(\mathbf{x})$ is that r' of $V(\mathbf{x})$; the dimension of the space in which it acts increases to infinity with k.

Theorem 7-20. *Every transformation group has differential invariants.*

The theorem is false for geometric invariants (exercise 7-2, Prob. 1).

***Example* 7-6.** Plane euclidean differential geometry is the theory of differential invariants of curves in R^2, with respect to the group of euclidean motions. Here $m = 2$, $r = r' = 3$, and each prolongation adds one dimension to the space in which the group is acting. We may expect a first differential invariant for $k = 2$, another for $k = 3$. The prolonged matrix is

$$V^{(3)} = \begin{pmatrix} -x_2 & x_1 & 1 + x_2'^2 & 3x_2'x_2'' & 3x_2''^2 + 4x_2'x_2''' \\ 1 & 0 & 0 & 0 & 0 \\ 0 & 1 & 0 & 0 & 0 \end{pmatrix}$$

The matrix acts on the third extension of the gradient

$$\text{grad}^{(3)} = \{\partial/\partial x_1, \partial/\partial x_2, \partial/\partial x_2', \partial/\partial x_2'', \partial/\partial x_2'''\}$$

The second-order differential invariant $f^{(2)}(x_1,x_2,x_2',x_2'')$ is a solution of

$$-x_2 \frac{\partial f}{\partial x_1} + x_1 \frac{\partial f}{\partial x_2} + (1 + x_2'^2) \frac{\partial f}{\partial x_2'} + 3x_2'x_2'' \frac{\partial f}{\partial x_2''} = 0$$

$$\frac{\partial f}{\partial x_1} = 0$$

$$\frac{\partial f}{\partial x_2} = 0$$

The last two equations show that no differential invariant of plane euclidean geometry depends directly on the coordinates x_1, x_2. The first equation thus reduces to

$$(1 + x_2'^2)\frac{\partial f}{\partial x_2'} + 3x_2'x_2''\frac{\partial f}{\partial x_2''} = 0$$

for which a first integral (see example 7-4) is found from an identification of this equation with $f_{x_2'}\,dx_2' + f_{x_2''}\,dx_2'' = 0$, that is,

$$\frac{dx_2'}{1 + x_2'^2} = \frac{dx_2''}{3x_2'x_2''}$$

Some care has to be taken in the integration. If $x_2'' = 0$, $dx_2' = 0$ appears as a first integral. This condition characterizes straight lines.

Theorem 7-21. *The property of being a straight line is an invariant of euclidean geometry.*

While one cannot say that this theorem presents a startling new discovery, it is nice to see how group theory automatically yields the basic features of a geometry, even in this nonintrinsic approach which gives different meanings to the coordinates.

If $x_2'' \neq 0$, the last equation may be written

$$\frac{3(1 + x_2'^2)^{\frac{1}{2}}x_2'}{x_2''}\,dx_2' - \frac{(1 + x_2'^2)^{\frac{3}{2}}}{x_2''^2}\,dx_2'' = 0$$

or

$$f^{(2)} = \frac{x_2''}{(1 + x_2'^2)^{\frac{3}{2}}} = \text{const}$$

This gives a new proof of theorem 2-10: *The curvature is a differential invariant of the group of euclidean motions.*

The third-order invariant may also be found by the method of the first integral. The corresponding equation is

$$\frac{dx_2'}{1 + x_2'^2} = \frac{dx_2''}{3x_2'x_2''} = \frac{dx_2'''}{3x_2''^2 + 4x_2'x_2'''}$$

The first equation has just been integrated. The new one,

$$\frac{dx_2'}{1 + x_2'^2} = \frac{dx_2'''}{3x_2''^2 + 4x_2'x_2'''}$$

may be written as a first-order linear differential equation

$$\frac{dx_2'''}{dx_2'} - 4\frac{x_2'}{1 + x_2'^2}x_2''' - \frac{3x_2''^2}{1 + x_2'^2} = 0$$

or

$$f^{(3)} = \frac{(1 + x_2'^2)x_2''' - 3x_2'x_2''^2}{(1 + x_2'^2)^3} = \text{const}$$

Each subsequent prolongation gives rise to one new differential invariant. But $df^{(3)}/df^{(2)}, \ldots, d^jf^{(3)}/df^{(2)j}$, $\ldots$, $j = 1, 2, 3, \ldots$, are invariant functions of order $3 + j$; hence all differential invariants of plane euclidean geometry are functions of $f^{(2)}$, $f^{(3)}$ and their derivatives. The same argument holds for any transitive group in the plane.

Theorem 7-22. *All differential invariants of a transitive transformation group in a plane are functions of the two invariants of lowest orders and their mutual derivatives.*

Euclidean geometry has a two-point invariant $|\mathbf{x}_1 - \mathbf{x}_0|$. It follows immediately from Def. 2-2 that the arc length s also is an invariant. The curvature $k = f^{(2)}$ is invariant; hence dk/ds must be an invariant of order three. In fact, $f^{(3)} = dk/ds$. The arc length $s = \int df^{(2)}/f^{(3)}$ is an *integral invariant*. If such an integral invariant is known, it may serve as an invariant parameter, and the theory is greatly simplified. In the euclidean case, theorem 7-22 leads to the following comment on theorem 2-13:

Corollary 7-23. *In plane euclidean geometry, every differential invariant may be computed from the natural equation $k = k(s)$.*

Theorem 7-22 resembles theorem 7-16 in that it states that there exists a finite set of invariants from which all other invariants may be deduced and which therefore contains all possible information on the geometry in question. A simple count of dimensions shows that the same statement holds for the differential invariants of any Klein geometry. If the number of independent variables is n, the number of independent invariants is $< n + m$. If invariants $J_1, \ldots, J_N$ are given, the number of invariant partial derivatives $\partial J_k/\partial J_j$ ($j = 1, \ldots, n; k = n + 1, \ldots, N$) increases faster than the number of dimensions added in prolongation, for $N \geq n + m$.

Definition 7-24. *A complete system of invariants of a geometric object in a Klein geometry is a set of differential invariants on which all other differential invariants are functionally dependent.*

Theorem 7-25. *Every Klein geometry has a finite complete set of invariants.*

Theorem 7-25 leads to a rational method of finding natural conditions of differentiability for geometric theorems. Let k_0 be the minimum (taken over all possible complete sets) of the maximal order of the invariants in a complete set. Then all possible information furnished by the invariants may be computed for C^{k_0} functions, whereas such information must be incomplete for C^k maps, $k < k_0$. No new information will be won by admitting C^l maps, $l > k_0$. In this sense, C^{k_0} is the natural hypothesis of differentiability in the particular question of differential geometry. For example, for plane euclidean geometry we see that the theorems of differential geometry should hold for C^2 curves referred to their arc length as an invariant parameter, or for C^3 curves without reference to s. Our experiences in Chaps. 2 and 3 confirm this expectation. The fact that s is an integral invariant is reflected in the shift from differentiation to integration techniques in the proof of theorems (Chap. 3) if the differentiability assumption is reduced from C^3 to C^2. In general, the existence of an invariant parameter allows a reduction of k_0. Unfortunately, such an invariant parameter need not always exist (as in the case of real projective geometry of surfaces of elliptic curvature).

For curves referred to the arc length s, derivatives of orders ≥ 3 do not have geometric significance in euclidean geometry. But this does not mean that higher derivatives have no geometric meaning at all. The unimodular affine group (exercise 7-1, Prob. 6) is a five-parameter group acting in the plane. It will be shown in the next section that it admits an invariant parameter. The properties of a plane curve that are not changed in a unimodular affinity therefore depend on a unique fourth-order invariant $(2 + 4 - 5 = 1)$. This "affine curvature" gives a geometric interpretation of fourth derivatives of functions of a real variable. One cannot expect a satisfactory theory of affine properties of plane curves defined by maps that are not piecewise C^4. On the other hand, one should be able to prove all relevant theorems for C^5 curves, even without using the invariant parameter.

Although arguments of this kind are very important, they should be used with care because the methods of this section do not permit discrimination between geometric and nongeometric invariants. Also, the count of constants may be misleading. There may exist differential invariants of orders smaller than expected. Their derivatives with respect to an invariant parameter do not yield new independent invariants but will take up some of the dimensions of the prolonged spaces.

Exercise 7-2

1. The transformation group described in exercise 7-1, Prob. 8, has the invariant x_1. Is it a geometric invariant? Find all differential invariants of the group.

2. Discuss the analogue of theorem 7-17 for solid euclidean geometry. Is the bound in theorem 7-16 the best possible?

3. Four points in a plane define six mutual distances. Show that five of them suffice to fix the relative position of the points in the plane.

4. $n \geq 4$ points in a plane define $\binom{n}{2}$ two-point invariants. How many of these form a complete set of n-point invariants?

5. Show that congruence of angles in euclidean geometry may be defined using only lengths of segments.

6. The unimodular affine transformation group is defined in exercise 7-1, Prob. 6, and example 5-6. Show that it has a three-point invariant, the area of the triangle defined by three points. Formulate a theorem of plane affine geometry similar to theorem 7-17.

7. Show that the projective group on the line (example 7-3) has a four-point invariant. Check on the differential equations that this invariant is the cross ratio

$$\frac{x_1 - x_3}{x_2 - x_3} : \frac{x_1 - x_4}{x_2 - x_4}$$

Formulate and prove a theorem analogous to theorem 7-17.

8. Use Prob. 7 to prove the *fundamental theorem of projective geometry:* A projective map of a line onto itself is determined if three pairs of corresponding points are given.

* **9.** The projective plane (example 5-7) may be given by homogeneous coordinates $\mathbf{x} = x_1 : x_2 : x_3$. A plane projective transformation is the action in the projective plane of an element A of GL_3 in $\mathbf{y} = \mathbf{x}A$. Two proportional matrices A and cA give the same projective transformation. Since $r' = 8$, $m = 2$, plane projective geometry has two five-point invariants. One of them is the cross ratio of four points on a line (Prob. 7). Five points determine a conic $\mathbf{x}Q\mathbf{x}^t = 0$. Two conics are projective images of one another if the ranks of their matrices Q are the same. This rank, then, is the second invariant of plane projective geometry. Investigate the dependence of the other k-point invariants on these two.

10. The intransitive group in R^2

$$x_1^* = x_1 \qquad x_2^* = \frac{ax_2 + b}{cx_2 + d} \qquad ad - bc = 1$$

has the invariant x_1. Example 7-3 shows that the Lie derivatives of the transformation group are

$$\pounds_1 = \frac{\partial}{\partial x_2} \qquad \pounds_2 = x_2 \frac{\partial}{\partial x_2} \qquad \pounds_3 = x_2{}^2 \frac{\partial}{\partial x_2}$$

The group has a third-order differential invariant. Compute the prolongations $\pounds_i{}^{(3)}$ and check that the invariant is

$$\{x_2\} = \frac{x_2'''}{x_2'} - \frac{3}{2}\left(\frac{x_2''}{x_2'}\right)^2$$

(the *Schwarzian derivative*). Show also that

$$\{x_2\} = \left\{\frac{ax_2 + b}{cx_2 + d}\right\}$$

The last proof should not need computations.

11. How many differential invariants of order two has a surface in R^3 in euclidean geometry? Write the differential equations of these invariants.

***12.** In the complex plane, the Moebius group or group of circular transformations

$$z^* = \frac{az + b}{cz + d} \qquad ad - bc = 1$$

is an action of the complex unimodular affine group in R^2. Show that the cross ratio of four complex numbers is invariant in a Moebius map.

13. Show that, for any algebra of Lie derivatives, the algebra of the kth prolongations is isomorphic to the original algebra.

14. s vector fields $\mathbf{v}_1(\mathbf{x}), \ldots, \mathbf{v}_s(\mathbf{x})$ are *connected* if there exist differentiable functions $g_1(\mathbf{x}), \ldots, g_s(\mathbf{x})$, not all zero, such that

$$g_1(\mathbf{x})\mathbf{v}_1(\mathbf{x}) + \cdots + g_s(\mathbf{x})\mathbf{v}_s(\mathbf{x}) = 0$$

Linearly dependent vector fields are connected. If s vector functions are connected, then at every point $\mathbf{x}_0$ the s vectors $\mathbf{v}_1(\mathbf{x}_0), \ldots, \mathbf{v}_s(\mathbf{x}_0)$ are linearly dependent. Prove that *a transformation group is locally transitive if and only if the number r of nonconnected row vectors of $V(\mathbf{x})$ is $\geq m$.*

◆15. Use the theorem of Prob. 14 to check whether the transformation groups of exercise 7-1, Probs. 10 to 12, are locally transitive.

7-3. Affine Differential Geometry

This section is devoted to the study of the properties of plane curves invariant under the group of area-preserving affine transformations $\mathbf{y} = \mathbf{x}A + \mathbf{b}$, $\|A\| = 1$. The group is generated by the action of the unimodular linear group SL_2 and the translation group R_2. The number of essential parameters (= dimension of the Lie algebra) is five; we have a three-point invariant, the area of the triangle defined by three points; and we may expect to characterize the unimodular affine properties of plane curves by an invariant parameter and a fourth-order invariant. The natural assumption of differentiability is C^4.

Elie Cartan's method of finding the differential invariants of a Klein geometry is a generalization of the technique employed in Chap. 2. The theory is, in some ways, dual to that of S. Lie. In the preceding chapters we always kept fixed a system of coordinates and transformed the points within that given system. Therefore the natural tools were coordinate (row) vectors. From now on, we shall keep a point fixed in a moving plane and move that plane over a fixed one. The natural tool here is frames, column vectors of vectors. The changeover in the type of vectors is not a whim of the author; it expresses the contrast between an analytical theory (Lie) and a geometrical one (Cartan).

As in Chap. 2, a fixed cartesian system of coordinates in the plane is determined by two orthogonal unit vectors $\mathbf{e}_1,\mathbf{e}_2$. The square with sides $\mathbf{e}_1,\mathbf{e}_2$ has area $\|\mathbf{e}_1,\mathbf{e}_2\| = \mathbf{e}_1 \times \mathbf{e}_2 = 1$. We try to define a frame $\{\mathbf{a}_1,\mathbf{a}_2\}$ at every point of a C^4 curve

$$\mathbf{x}(t) = x_1(t)\mathbf{e}_1 + x_2(t)\mathbf{e}_2 \tag{7-17}$$

such that $\mathbf{a}_1$ would be on the tangent to $\mathbf{x}(t)$ at that point and that the frame matrix $A(t)$,

$$\{\mathbf{a}_1(t),\mathbf{a}_2(t)\} = A(t)\{\mathbf{e}_1,\mathbf{e}_2\} \tag{7-18}$$

would be in SL_2, $\|A(t)\| = 1$ (Fig. 7-1). The curve (7-17) determines at

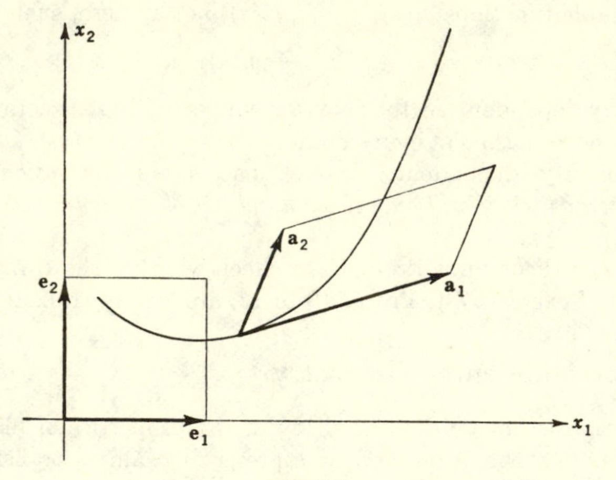

Fig. 7-1

every point an oblique system of coordinates (with unequal unit lengths on the coordinate axes) by the frame $\{\mathbf{x}^{\cdot}(t),\mathbf{x}^{\cdot\cdot}(t)\}$, if the two vectors are linearly independent; that is, $\|\mathbf{x}^{\cdot}(t),\mathbf{x}^{\cdot\cdot}(t)\| \neq 0$ for all t. In euclidean geometry this condition means that $k(t) \neq 0$ (exercise 2-2, Prob. 1). Therefore we restrict our attention to arcs with curvature of constant sign, for the time being.

The frame matrix $B(t)$ in

$$\{\mathbf{x}^{\cdot}(t),\mathbf{x}^{\cdot\cdot}(t)\} = B(t)\{\mathbf{e}_1,\mathbf{e}_2\}$$

will not, in general, be in SL_2; it will not characterize unimodular affine properties of $\mathbf{x}(t)$. (We need not be concerned with the translation group, as any frame matrix is always invariant under parallel translations; see theorem 2-6.) Without loss of generality, it may be assumed that

$\|\mathbf{x}^{\cdot}(t),\mathbf{x}^{\cdot\cdot}(t)\| > 0$. [If the determinant is <0, a change of parameter $t = -u$ makes $\|\mathbf{x}^{\cdot}(u),\mathbf{x}^{\cdot\cdot}(u)\| > 0$.] The vectors

$$\mathbf{a}_1 = \|\mathbf{x}^{\cdot}(t),\mathbf{x}^{\cdot\cdot}(t)\|^{-\frac{1}{3}}\mathbf{x}^{\cdot}(t)$$
$$\mathbf{a}_2 = \|\mathbf{x}^{\cdot}(t),\mathbf{x}^{\cdot\cdot}(t)\|^{-\frac{2}{3}}\mathbf{x}^{\cdot\cdot}(t)$$

determine a parallelogram of area $\|\mathbf{a}_1(t),\mathbf{a}_2(t)\| = 1$; the frame matrix

$$A(t) = \frac{1}{(x_1^{\cdot}x_2^{\cdot\cdot} - x_1^{\cdot\cdot}x_2^{\cdot})^{\frac{1}{3}}}\begin{pmatrix} x_1^{\cdot} & x_2^{\cdot} \\ x_1^{\cdot\cdot} & x_2^{\cdot\cdot} \end{pmatrix} \tag{7-19}$$

appearing in $\{\mathbf{a}_1,\mathbf{a}_2\} = A\{\mathbf{e}_1,\mathbf{e}_2\}$ is affine unimodular. The main idea now is to look for invariants among the elements of the Cartan matrix $C(A) = A^{\cdot}A^{-1}$. The reason is that the Cartan map $A(t) \to C(A)(t)$ of a group G into $\mathfrak{L}(G)$ is itself invariant under operations of G:

Theorem 7-26. *Let $A(t)$ be a differentiable curve in a Lie group G and M a constant element of G. Then $C(A) = C(AM)$.*

PROOF: Eq. (1-5).

It follows that *the nonzero terms of the Cartan matrix will be invariants if we succeed in fixing parameters and frames by unique and invariant conditions.* The Cartan matrix of $A(t)$ [Eq. (7-19)] is

$$C(A) = \begin{pmatrix} -\dfrac{1}{2}\dfrac{\|\mathbf{x}^{\cdot},\mathbf{x}^{\cdot\cdot\cdot}\|}{\|\mathbf{x}^{\cdot},\mathbf{x}^{\cdot\cdot}\|} & 1 \\ -\dfrac{\|\mathbf{x}^{\cdot\cdot},\mathbf{x}^{\cdot\cdot\cdot}\|}{\|\mathbf{x}^{\cdot},\mathbf{x}^{\cdot\cdot}\|} & +\dfrac{1}{2}\dfrac{\|\mathbf{x}^{\cdot},\mathbf{x}^{\cdot\cdot\cdot}\|}{\|\mathbf{x}^{\cdot},\mathbf{x}^{\cdot\cdot}\|} \end{pmatrix}$$

The trace of the matrix is zero, as it must be for elements of $\mathfrak{L}(SL_2)$. The essence of Cartan's method now is to try to find (by invariant conditions) a family of frames for which as many coefficients of $C(A)$ as possible are zero. In our case, the diagonal elements of $C(A)$ are zero if

$$\|\mathbf{x}^{\cdot},\mathbf{x}^{\cdot\cdot\cdot}\| = \|\mathbf{x}^{\cdot},\mathbf{x}^{\cdot\cdot}\|^{\cdot} = 0$$

Therefore we look for a parameter σ for which the area

$$\left\|\frac{d\mathbf{x}}{d\sigma},\frac{d^2\mathbf{x}}{d\sigma^2}\right\| = 1$$

This invariant parameter, the *affine arc length,* is obtained from

$$\left\|\frac{d\mathbf{x}}{dt},\frac{d^2\mathbf{x}}{dt^2}\right\| = \left\|\frac{d\mathbf{x}}{d\sigma},\frac{d^2\mathbf{x}}{d\sigma^2}\right\|\left(\frac{d\sigma}{dt}\right)^3 = \left(\frac{d\sigma}{dt}\right)^3$$

that is,
$$\sigma = \int_{t_0}^{t} \|\mathbf{x}^{\cdot}(t),\mathbf{x}^{\cdot\cdot}(t)\|^{\frac{1}{3}}\,dt \tag{7-21}$$

Differentiation with respect to the invariant parameter σ will again be denoted by primes. Then $\|\mathbf{x}',\mathbf{x}''\| = 1$, $\mathbf{a}_1 = \mathbf{x}'$, $\mathbf{a}_2 = \mathbf{x}''$, the moving-frame matrix is

$$A(\sigma) = \begin{pmatrix} x_1' & x_2' \\ x_1'' & x_2'' \end{pmatrix}$$

and its Cartan matrix, appearing in $\{\mathbf{a}_1,\mathbf{a}_2\}' = C(A)\{\mathbf{a}_1,\mathbf{a}_2\}$, is

$$C(A) = \begin{pmatrix} 0 & 1 \\ -\|\mathbf{x}'',\mathbf{x}'''\| & 0 \end{pmatrix} \tag{7-22}$$

This matrix is completely determined by the *affine curvature*

$$\varkappa(\sigma) = \|\mathbf{x}'',\mathbf{x}'''\| = x_1''x_2''' - x_1'''x_2'' \tag{7-23}$$

Theorem 7-26 implies that the affine curvature is a differential invariant of unimodular affine geometry. Lemmas 2-11 and 2-12 permit the construction of $A(\sigma)$ and $\mathbf{x}(\sigma)$ for given $\varkappa(\sigma)$.

Theorem 7-27. *A plane curve is determined up to a unimodular affine transformation by its natural equation $\varkappa = \varkappa(\sigma)$ of the affine curvature as a function of the affine arc length.*

As in the euclidean case, the frame for which all coefficients of the Cartan matrix are either constant or invariant will be called the *Frenet frame* of the curve $\mathbf{x}(t)$. In the defining equation (7-23), only third derivatives appear. Nevertheless, $\varkappa$ is a fourth-order invariant, as a function of a general parameter t. By Eq. (7-21),

$$\begin{aligned}
\mathbf{x}' &= \dot{\mathbf{x}}\|\dot{\mathbf{x}},\ddot{\mathbf{x}}\|^{-\frac{1}{3}} \\
\mathbf{x}'' &= \ddot{\mathbf{x}}\|\dot{\mathbf{x}},\ddot{\mathbf{x}}\|^{-\frac{2}{3}} - \tfrac{1}{3}\dot{\mathbf{x}}\|\dot{\mathbf{x}},\ddot{\mathbf{x}}\|^{-\frac{5}{3}}\|\dot{\mathbf{x}},\dddot{\mathbf{x}}\| \\
\mathbf{x}''' &= \dddot{\mathbf{x}}\|\dot{\mathbf{x}},\ddot{\mathbf{x}}\|^{-1} - \ddot{\mathbf{x}}\|\dot{\mathbf{x}},\ddot{\mathbf{x}}\|^{-2}\|\dot{\mathbf{x}},\dddot{\mathbf{x}}\| + \tfrac{5}{9}\dot{\mathbf{x}}\|\dot{\mathbf{x}},\ddot{\mathbf{x}}\|^{-3}\|\dot{\mathbf{x}},\dddot{\mathbf{x}}\|^2 \\
&\quad - \tfrac{1}{3}\dot{\mathbf{x}}\|\dot{\mathbf{x}},\ddot{\mathbf{x}}\|^{-2}\|\ddot{\mathbf{x}},\dddot{\mathbf{x}}\| - \tfrac{1}{3}\dot{\mathbf{x}}\|\dot{\mathbf{x}},\ddot{\mathbf{x}}\|^{-2}\|\dot{\mathbf{x}},\ddddot{\mathbf{x}}\|
\end{aligned}$$

Hence

$$\varkappa(\sigma) = -\tfrac{5}{9}\|\dot{\mathbf{x}},\ddot{\mathbf{x}}\|^{-\frac{8}{3}}\|\dot{\mathbf{x}},\dddot{\mathbf{x}}\|^2 + \tfrac{4}{3}\|\dot{\mathbf{x}},\ddot{\mathbf{x}}\|^{-\frac{5}{3}}\|\ddot{\mathbf{x}},\dddot{\mathbf{x}}\| + \tfrac{1}{3}\|\dot{\mathbf{x}},\ddot{\mathbf{x}}\|^{-\frac{5}{3}}\|\dot{\mathbf{x}},\ddddot{\mathbf{x}}\| \tag{7-24}$$

For constant $\varkappa$, Eq. (2-8) is easily integrated by the methods of Sec. 2-2:

$$A(\sigma) = \begin{pmatrix} \cos \varkappa^{\frac{1}{2}}\sigma & \varkappa^{-\frac{1}{2}} \sin \varkappa^{\frac{1}{2}}\sigma \\ -\varkappa^{\frac{1}{2}} \sin \varkappa^{\frac{1}{2}}\sigma & \cos \varkappa^{\frac{1}{2}}\sigma \end{pmatrix}$$

for $\varkappa \neq 0$ and

$$A(\sigma) = \begin{pmatrix} 1 & \sigma \\ 0 & 1 \end{pmatrix}$$

for $\varkappa = 0$. The first row of these matrices is the coordinate vector of $\mathbf{a}_1 = d\mathbf{x}/d\sigma$. Hence for $\varkappa = 0$

$$\mathbf{x}(\sigma) - \mathbf{x}(0) = \sigma\mathbf{e}_1 + \tfrac{1}{2}\sigma^2\mathbf{e}_2$$

The curve is a *parabola*. For $\varkappa > 0$, the integration gives

$$x_1(\sigma) - x_1(0) = \varkappa^{-\frac{3}{4}} \sin \varkappa^{\frac{1}{2}}\sigma$$
$$x_2(\sigma) - x_2(0) = -\varkappa^{-1} \cos \varkappa^{\frac{1}{2}}\sigma$$

an *ellipse*. Finally, for $\varkappa < 0$,

$$x_1(\sigma) - x_1(0) = |\varkappa|^{-\frac{3}{4}} \operatorname{Cosh} |\varkappa|^{\frac{1}{2}}\sigma$$
$$x_2(\sigma) - x_2(0) = |\varkappa|^{-1} \operatorname{Sinh} |\varkappa|^{\frac{1}{2}}\sigma$$

is a *hyperbola*.

Theorem 7-28. *The curves with constant affine curvature are the conics.*

Area is a unimodular affine invariant. Therefore we may expect a connection between area and affine curvature, at least in the case of the ellipse. The preceding equations show that the area of an ellipse of affine curvature $\varkappa$ is $\pi\varkappa^{-\frac{3}{2}}$.

The moving frame at a point $x(\sigma_0)$ of the curve may be taken to define the fixed system of oblique cartesian coordinates (with unequal unit lengths on the axes). In this case

$$\mathbf{e}_1 = \mathbf{a}_1(\sigma_0) \qquad \mathbf{e}_2 = \mathbf{a}_2(\sigma_0)$$

and

$$\mathbf{x}(\sigma_0) = 0 \qquad \mathbf{x}'(\sigma_0) = \mathbf{e}_1 \qquad \mathbf{x}''(\sigma_0) = \mathbf{e}_2 \qquad \mathbf{x}'''(\sigma_0) = -\varkappa_0\mathbf{e}_1$$

Therefore, the Taylor expansion of the curve in $x(\sigma_0)$ is

$$\mathbf{x}(\sigma) = (\sigma - \sigma_0)\mathbf{e}_1 + \frac{1}{2}(\sigma - \sigma_0)^2\mathbf{e}_2 - \frac{\varkappa_0}{6}(\sigma - \sigma_0)^3\mathbf{e}_1 + R_4 \quad (7\text{-}25)$$

Two curves $\mathbf{x}(\sigma)$ and $\mathbf{y}(\sigma)$ are *in contact of order* n *if* $\mathbf{x}(\sigma_0) = \mathbf{y}(\tau_0)$, $\mathbf{x}'(\sigma_0) = \mathbf{y}'(\tau_0)$, . . . , $\mathbf{x}^{(n-1)}(\sigma_0) = \mathbf{y}^{(n-1)}(\tau_0)$, $\mathbf{x}^{(n)}(\sigma_0) \neq \mathbf{y}^{(n)}(\tau_0)$. Equation (7-25) shows that the parabola

$$\mathbf{p}(\sigma) = (\sigma - \sigma_0)\mathbf{a}_1 + \tfrac{1}{2}(\sigma - \sigma_0)^2\mathbf{a}_2$$

is in contact of order ≥ 3 with the curve $\mathbf{x}(\sigma)$ at $\mathbf{x}(\sigma_0)$. $\mathbf{p}(\sigma)$ is the *osculating parabola* of $\mathbf{x}$ in $\mathbf{x}(\sigma_0)$. Its equation gives a simple geometric interpretation for the direction of the "*affine normal*" $\mathbf{a}_2$.

Theorem 7-29. *The direction of the axis of an osculating parabola is the direction of the "affine normal"* $\mathbf{a}_2$.

The conic $\mathbf{c}(\sigma)$ of affine curvature $\varkappa_c = \varkappa(\sigma_0)$ and in contact with $\mathbf{x}(\sigma)$ at $\mathbf{x}(\sigma_0)$,

$$\mathbf{c}(0) = \mathbf{x}(\sigma_0) \qquad \mathbf{c}'(0) = \mathbf{x}'(\sigma_0)$$

by Eq. (7-25) has contact of order ≥ 4 with $\mathbf{x}$. $\mathbf{c}$ is the *hyperosculating* conic of $\mathbf{x}$ in $\mathbf{x}(\sigma_0)$. $\mathbf{x}$ is said to be of elliptic, parabolic, or hyperbolic curvature at a point, if the hyperosculating conic is an ellipse, a parabola, or a hyperbola ($\varkappa \gtreqless 0$).

Convex curves are defined by a property invariant in an affinity (Def. 1-5). It makes sense to speak of closed convex curves in affine geometry.

Definition 7-30. *A sextactic point on a C^4 curve is one in which the affine curvature has a relative extremum.*

This notion appears in an analogon of the four-vertices theorem:

Theorem 7-31. *A closed convex C^4 curve has at least six sextactic points.*

The proof is parallel to that of the four-vertices theorem given on page 39.

Lemma 7-32. *For any quadratic polynomial $Q(x_1,x_2)$ with constant coefficients and any closed C^4 curve* $\mathbf{x}$, $\oint Q(x_1,x_2)\, d\varkappa = 0$.

We have to show that

$$\oint d\varkappa = \oint x_1\, d\varkappa = \oint x_2\, d\varkappa = \oint x_1^2\, d\varkappa = \oint x_1 x_2\, d\varkappa = \oint x_2^2\, d\varkappa = 0$$

The first integral vanishes trivially. The other integrals are converted into integrals over differentials by integration by parts, e.g.,

$$\oint x_1\, d\varkappa = -\oint \varkappa x_1'\, d\sigma = \oint x_1'''\, d\sigma = \oint dx_1'' = 0$$
$$\oint x_1^2\, d\varkappa = -2\oint \varkappa x_1 x_1'\, d\sigma = 2\oint x_1 x_1'''\, d\sigma = -2\oint x_1' x_1''\, d\sigma = \oint d(x_1')^2 = 0$$

By the proof of the four-vertices theorem on page 39, the lemma implies the existence of at least four sextactic points. If there are only four, we denote them by P_1, P_2, P_3, P_4 in cyclic order. We may assume that $\varkappa$ is maximal in P_1 and P_3 and minimal in P_2, P_4. Let $L_1(x_1,x_2) = 0$ be the equation of the line P_1P_3 and assume that $L_1(P_2) > 0$. Also, let $L_2(x_1,x_2) = 0$ be the equation of the line P_2P_4, $L_2(P_1) < 0$. On the curve,

$$\operatorname{sign} L_1L_2 = \operatorname{sign} d\varkappa$$

Hence $\oint L_1L_2\, d\varkappa > 0$. This contradicts lemma 7-32.

As a last problem of affine geometry we find the trajectories of one-parameter groups of unimodular affine transformations

$$\mathbf{x}(t) = \mathbf{x}_0 A(t_0) \qquad \|A(t)\| = 1$$

Such a trajectory is transformed in itself by the group. The affine curvature is conserved in the process. It must be constant along any trajectory, if it is defined at all:

Theorem 7-33. *The trajectories of a one-parameter group of unimodular affine transformations are conics or straight lines.*

Arcs with inflection points cannot be treated directly in affine geometry; they must be divided into locally convex arcs. Even for analytic curves there is, in general, no continuity of the moving frame at an inflection point.

Exercise 7-3

1. Use formula (7-24) to compute the affine curvature for a curve given by $x_2 = x_2(x_1)$.

2. For a C^5 curve, find the fourth-order Taylor approximation in the neighborhood of a point [see (7-25)].

3. A diameter of a conic is the locus of the midpoints of the segments cut off by the conic on the lines in the conjugate direction. (For the ellipse, see exercise 5-2, Prob. 15.) For the parabola, the midpoints of the chords parallel to a tangent are on the parallel to the axis through the point of contact. For a C^4 curve, the "midcurve" is the locus of the midpoints of the chords parallel to the tangent at a given point of the curve. Use (7-25) to show that the tangent to the midcurve at the point of the curve is the affine normal.

4. Show that a curve may be obtained from its affine curvature by integration of $\mathbf{t} = d\mathbf{x}/d\sigma$, $\mathbf{t}'' = -\varkappa\mathbf{t}$. Find the curve of the natural equation $\varkappa = \sigma$. (The differential equation may be integrated in terms of Bessel functions of orders $\pm\frac{1}{3}$.)

5. The *affine evolute* $\mathbf{A}_x$ of $\mathbf{x}$ is the envelope of the affine normals. $\mathbf{A}_x(\sigma) = \mathbf{x}(\sigma) + \lambda\mathbf{x}''(\sigma)$. Show that $\lambda(\sigma) = \varkappa(\sigma)^{-1}$. Why does the parabola not have an affine evolute?

6. Show that a C^4 curve is a conic if and only if all its affine normals are either parallel or concurrent. (Use either Prob. 4 or 5.)

7. Show that any two parabolas are unimodular affine images of one another.

8. State and prove the theorem of euclidean geometry which corresponds to theorem 7-33.

9. Show that the sign of the euclidean curvature is a unimodular affine invariant.

◆**10.** A *general affine transformation*

$$\mathbf{x}^* = \mathbf{x}M + \mathbf{b} \qquad \|M\| > 0$$

can be viewed as a unimodular affine transformation followed by a homothety of ratio $\|M\|$. Show that general affine transformations have four-point invariants and a fifth-order differential invariant. Find the four-point invariants. (HINT: How are areas transformed in a homothety?)

11. In a homothety (Sec. 3-4) the length of a vector is multiplied by the ratio $\|M\|$ of the homothety. Show that, in a general affinity, σ, $\varkappa$, $\varkappa'$ are multiplied by $\|M\|^{\frac{2}{3}}$, $\|M\|^{-\frac{4}{3}}$, $\|M\|^{-2}$, respectively, and that therefore $\mu = \varkappa'\varkappa^{-\frac{3}{2}}$ is the fifth-order differential invariant of general affine transformations.

12. Show that a curve remains unimodular affine to itself in any general affine transformation if and only if $\varkappa$ is a homogeneous function of degree -2 of σ, $\varkappa = c\sigma^{-2}$. Find the curves. (See Sec. 3-4.) NOTE: Differential equations like $d^2y/dx^2 = cyx^{-2}$ may be integrated by a substitution $y = xz$, $x = e^\theta$. The resulting equation in z, θ is linear.

13. Find the trajectories of one-parameter groups of general affine transformations.

14. Let $\mathbf{q}(\sigma)$ be a convex arc having the same hyperosculating conic $\mathbf{c}$ at its two endpoints A,B. Show that $\mathbf{q}$ has at least five sextactic points (Fabricius-Bjerre).

15. Let $\mathbf{q}(\sigma)$ be a closed convex curve which has a conic $\mathbf{c}$ as a hyperosculating conic at two distinct points. Show that $\mathbf{q}$ has at least 10 sextactic points (Fabricius-Bjerre). (Use Prob. 14.)

16. A curve is *parabolically convex* if in the neighborhood of any of its points it is contained in the interior of the osculating parabola at that point. Show that a closed parabolically convex C^4 curve meets any of its osculating conics *only* at the point of osculation (Carleman).

17. Show that an osculating parabola which is not hyperosculating intersects the curve at the point of contact. [Discuss the sign of $x_{2\,\text{curve}} - x_{2\,\text{parabola}}$ in (7-25).]

18. Show that a hyperosculating ellipse or hyperbola never intersects the curve at the point of contact unless that point is sextactic. (Use Prob. 2 for C^5 curves.)

References

SEC. 7-1

Aczél, J., and S. Gołab: "Funktionalgleichungen der Theorie der geometrischen Objekte," Monografie matematyczne, vol. 39, Panstwowe Wydawnictwo Naukowe, Warsaw, 1960.

Bianchi, L.: "Lezioni sulla teoria dei gruppi continui finiti di trasformazioni," Nichola Zanichelli, Bologna, 1919.

Ehresmann, C.: Sur la théorie des espaces fibrés, "Colloque de topologie algébrique," Centre National de la Recherche Scientifique, Paris, 1947.

Eisenhart, L. P.: "Continuous Groups of Transformations," Princeton University Press, Princeton, N.J., 1933.

Lie, S., and F. Engel: "Theorie der Transformationsgruppen," 3 vols., B. G. Teubner, Verlagsgesellschaft, mbH, Leipzig, 1888–1893.

SEC. 7-2

Cartan, E.: La théorie des groupes continus et la géométrie, "Oeuvres complètes," vol. III/2, 1727–1861, Gauthier-Villars, Paris, 1955.

Klein, F.: Vergleichende Betrachtungen über neuere geometrische Forschungen, *Math. Ann.*, **43:** 63 (1893); English translation, *Bull. N. Y. Math. Soc.*, **2:** 215 (1892).

For a geometric theory of prolongations, see the following:

Ehresmann, C.: Les prolongements d'une variété différentiable, *Compt. Rend.*, **233**: 598, 777, 1081 (1951); **234**: 1028, 1424 (1952).

SEC. 7-3

Blaschke, W.: Vorlesungen über Differentialgeometrie und geometrische Grundlagen von Einsteins Relativitätstheorie, II, "Affine Differentialgeometrie," Springer-Verlag OHG, Berlin, 1923.

Carleman, T.: Sur les courbes paraboliquement convexes, Festschrift R. Fueter, *Vierteljschr. Naturf. Ges. Zürich*, 1939, pp. 61–63.

Fabricius-Bjerre, F.: Note on a Theorem of G. Bol, *Arch. Math.*, **3**: 31–33 (1952).

Favard, J.: "Cours de géométrie différentielle locale," Gauthier-Villars, Paris, 1957.

Mohrmann, H.: Ueber beständig elliptisch, parabolisch oder hyperbolisch gekrümmte Kurven, *Math. Ann.*, **72**: 285–291; 593–595 (1912).

8
SPACE CURVES

8-1. Space Curves in Euclidean Geometry

The group of euclidean motions in R^3 has six parameters (three each for rotation and translation). In the theory of curves, each prolongation adds two dimensions to the space in which the group is acting. Given an invariant parameter, we expect the theory to depend on one invariant of second and one of third order.

The invariant parameter s of a C^3 space curve $\mathbf{x}(u) = (x_1(u),x_2(u),x_3(u))$ is the *arc length* $s = \int_{u_0}^{u} |\mathbf{x}^{\cdot}(u)|\,du$. The tangent vector $\mathbf{t} = \mathbf{x}'(s)$ is a unit vector (compare Sec. 2-1). The fixed cartesian system of coordinates is given by a frame of orthonormal vectors $\mathbf{e}_1,\mathbf{e}_2,\mathbf{e}_3$, $\mathbf{e}_i \cdot \mathbf{e}_i = 1$, $\mathbf{e}_i \cdot \mathbf{e}_j = 0$ $(i \neq j)$. To the point $\mathbf{x}(s)$ we attach another orthonormal frame $\mathbf{a}_1(s) = \mathbf{t}(s)$, $\mathbf{a}_2(s)$, $\mathbf{a}_3(s)$, derived from the fixed frame by a rotation

$$\{\mathbf{t},\mathbf{a}_2,\mathbf{a}_3\} = A(s)\{\mathbf{e}_1,\mathbf{e}_2,\mathbf{e}_3\} \qquad A(s) \in O_3^*$$

Then
$$\{\mathbf{t},\mathbf{a}_2,\mathbf{a}_3\}' = C(A)\{\mathbf{t},\mathbf{a}_2,\mathbf{a}_3\} \tag{8-1}$$

The Cartan matrix is skew-symmetric:

$$C(A) = \begin{pmatrix} 0 & p_{12} & p_{13} \\ -p_{12} & 0 & p_{23} \\ -p_{13} & -p_{23} & 0 \end{pmatrix}$$

If $p_{12}(s) = p_{13}(s) = 0$, $\mathbf{t}' = \mathbf{x}'' = 0$, $\mathbf{x}(s) = \mathbf{a}s + \mathbf{b}$, the curve is a straight line. In this case one cannot assign an intrinsic meaning to p_{23}. $p_{12} = p_{13} = 0$, $p_{23} \neq 0$ means that the frame $\{\mathbf{t},\mathbf{a}_2,\mathbf{a}_3\}$ moves in the direction of the straight line while the axes $\mathbf{a}_2,\mathbf{a}_3$ rotate in the normal plane to the line.

If $\mathbf{t}'(s) \neq 0$, we choose as a second vector $\mathbf{a}_2 = \mathbf{t}'(s)/|\mathbf{t}'(s)| = \mathbf{n}(s)$.

This is the *principal normal* to the curve. The third vector $\mathbf{a}_3 = \mathbf{b}(s)$, the *binormal*, then is the unique vector normal to both $\mathbf{t}$ and $\mathbf{n}$ such that the volume $\|\mathbf{t},\mathbf{n},\mathbf{b}\|$ of the cube spanned by the three vectors is $+1$. By construction, $p_{13}(s) = 0$; $C(A)$ is reduced to

$$C(A) = \begin{pmatrix} 0 & k(s) & 0 \\ -k(s) & 0 & t(s) \\ 0 & -t(s) & 0 \end{pmatrix} \tag{8-2}$$

Since the frame is now uniquely determined, $k(s)$ and $t(s)$ are invariants.

Definition 8-1. *$k(s)$ is the curvature and $t(s)$ the torsion of a space curve.*

Natural equations $k = k(s)$, $t = t(s)$ determine a space curve up to a congruence, by the integration lemma 2-12. This is a local theorem: If $\mathbf{x}''(s) = \mathbf{x}'''(s) = 0$ at some point, no continuous family of frames $\{\mathbf{t},\mathbf{n},\mathbf{b}\}$ needs to exist.

If $t(s) = 0$, A is easily computed to be an orthogonal matrix of the form (up to multiplication with a constant matrix)

$$\begin{pmatrix} a_{11} & a_{12} & 0 \\ a_{21} & a_{22} & 0 \\ 0 & 0 & 1 \end{pmatrix}$$

The vector $\mathbf{b}$ is constant; hence the plane defined by $\mathbf{t}(s)$ and $\mathbf{n}(s)$ also is constant:

Theorem 8-2. *A C^3 curve (other than a straight line) is a plane curve if and only if its torsion vanishes.*

It follows from the *Frenet equations* (8-1) and (8-2) that

$$\begin{aligned} \mathbf{x}' &= \mathbf{t} \\ \mathbf{x}'' &= k\mathbf{n} \\ \mathbf{x}''' &= -k^2\mathbf{t} + k'\mathbf{n} + kt\mathbf{b} \end{aligned}$$

The curvature

$$k = |\mathbf{x}''| \tag{8-3}$$

is a second-order invariant, and the torsion

$$t = \frac{1}{k^2} \|\mathbf{x}',\mathbf{x}'',\mathbf{x}'''\| \tag{8-4}$$

is of order three, as expected. By definition, the curvature of a space curve (even of a plane curve imbedded in three space) is always positive; the definition is not completely equivalent to that of Chap. 2. The torsion t has a sign which in itself is an invariant; by (8-4) it indicates whether the three vectors $\mathbf{x}',\mathbf{x}'',\mathbf{x}'''$ define a right-hand or a left-hand system of oblique coordinates.

***Example* 8-1.** A *circular helix* is the curve

$$\mathbf{x} = (a \cos u, a \sin u, bu)$$

Its arc length is $s = u(a^2 + b^2)^{\frac{1}{2}}$. The invariants are

$$k = \frac{a}{a^2 + b^2} \qquad t = \frac{b}{a^2 + b^2}$$

The circular helices therefore are the space curves with constant curvature and torsion. This includes as plane curves circles ($b = 0$) and lines ($a = 0$). The proof of theorem 7-33 shows that *the circular helices are the trajectories of one-parameter groups of euclidean motions.*

The vector function $\mathbf{t}(s)$ determines the curve, $\mathbf{x}(s) = \mathbf{x}_0 + \int_{s_0}^{s} \mathbf{t}(\sigma)\, d\sigma$. We may ask whether the function $\mathbf{n}(s)$ also determines the curve. Here a preliminary question is: May two distinct curves $\mathbf{x}(s_x)$, $\mathbf{y}(s_y)$ have the same principal normals? In this case, $\mathbf{y}(s_y)$ must be on the principal normal through $\mathbf{x}(s_x)$,

$$\mathbf{y} = \mathbf{x}(s_x) + \lambda(s_x)\mathbf{n}_x(s_x)$$

λ is a differentiable function, and $\mathbf{n}_x$ and $\mathbf{n}_y$ are the principal normals to the respective curves. By differentiation,

$$\mathbf{t}_y = [(1 - k_x\lambda)\mathbf{t}_x + \lambda'\mathbf{n}_x + \lambda t_x\mathbf{b}_x]\frac{ds_x}{ds_y}$$

$\mathbf{n}_x = \mathbf{n}_y$ is orthogonal to $\mathbf{t}_y$; hence

$$\lambda' = 0$$

λ is a constant, and $\mathbf{t}_y$ is in the plane spanned by $\mathbf{t}_x$ and $\mathbf{b}_x$,

$$\mathbf{t}_y = \mathbf{t}_x \cos \alpha(s_x) + \mathbf{b}_x \sin \alpha(s_x)$$

Hence

$$\mathbf{n}_y \frac{ds_y}{ds_x} = (k_x \cos \alpha - t_x \sin \alpha)\mathbf{n}_x + \mathbf{t}_x(\cos \alpha)' + \mathbf{b}_x(\sin \alpha)'$$

As before, it follows that $\alpha' = 0$, and the angle between $\mathbf{t}_x$ and $\mathbf{t}_y$ is constant. The two equations for $\mathbf{t}_y$,

$$\mathbf{t}_y = [(1 - k_x\lambda)\mathbf{t}_x + \lambda t_x\mathbf{b}_x]\frac{ds_x}{ds_y}$$
$$\mathbf{t}_y = \cos \alpha\, \mathbf{t}_x + \sin \alpha\, \mathbf{b}_x$$

are consistent if and only if the determinant vanishes:

$$\sin \alpha - \lambda(k_x \sin \alpha + t_x \cos \alpha) = 0$$

Theorem 8-3 (Bertrand). *A curve in R^3 is characterized (up to a translation) by its principal normals if and only if no linear relation with constant coefficients holds between its curvature and torsion.*

The theory of curves in R^m is parallel to that in R^3. A euclidean motion $\mathbf{x}^* = \mathbf{x}A + \mathbf{b}$, $A \in O_m$, depends on $m + m(m-1)/2$ parameters. Each prolongation adds $m - 1$ dimensions to the space of contact elements; given an invariant parameter, there are $m - 1$ independent invariants. The natural order of differentiability is $\geq (m + 1)/2 + 1$. The invariant parameter is the arc length $s = \int_{u_0}^{u} |\mathbf{x}^{\cdot}(\nu)|\, d\nu$; $\mathbf{x}'(s)$ is a unit vector.

The fixed system of cartesian coordinates is given by m orthonormal vectors $\mathbf{e}_i$, $i = 1, \ldots, m$. To the running point on the curve we attach a frame of orthonormal vectors $\mathbf{a}_1(s) = \mathbf{x}'(s)$, $\mathbf{a}_2(s), \ldots, \mathbf{a}_m(s)$. The frame matrix $A(s)$,

$$\{\mathbf{a}_1(s), \ldots, \mathbf{a}_m(s)\} = A(s)\{\mathbf{e}_1, \ldots, \mathbf{e}_m\}$$

is orthogonal; $C(A) = (p_{ij}(s))$ is skew-symmetric. If $p_{1j}(s) = 0$, $j = 1, \ldots, m$, $\mathbf{x}(s) = \mathbf{a}s + \mathbf{b}$ is a straight line. In this case $A(s)$ may be taken as constant. If there exists a $p_{1j}(s) \neq 0$, one may choose $\mathbf{a}_2 = \mathbf{a}_1'(s)/|\mathbf{a}_1'(s)|$. Then

$$C(A) = \begin{pmatrix} 0 & p_{12} & 0 & \cdots & 0 \\ -p_{12} & 0 & p_{23} & \cdots & p_{2m} \\ 0 & -p_{23} & & & \\ \cdot & \cdot & & & \\ \cdot & \cdot & & p_{ij} & \\ 0 & -p_{2m} & & & \end{pmatrix} \tag{8-5}$$

If $p_{2j}(s) = 0, j > 2$, for all s, the tangent to the curve remains in the plane of $\mathbf{a}_1(s_0)$, $\mathbf{a}_2(s_0)$; $\mathbf{x}(s)$ is a plane curve. If some $p_{2j}(s) \neq 0, j > 2$, we take $\mathbf{a}_3$ as the unit vector in the direction of $\mathbf{a}_2'(s) + p_{12}(s)\mathbf{a}_1(s)$. In this new system,

$$C(A) = \begin{pmatrix} 0 & p_{12} & 0 & 0 & \cdots & 0 \\ -p_{12} & 0 & p_{23} & 0 & \cdots & 0 \\ 0 & -p_{23} & 0 & & & \\ 0 & 0 & & & p_{ij} & \\ \cdot & \cdot & & & & \\ \cdot & \cdot & & & & \\ 0 & 0 & & & & \end{pmatrix}$$

Proceeding in this manner, either one obtains a proof that $\mathbf{x}(s)$ is situated wholly in some R^n, $n < m$, or else one finds a moving frame $\{\mathbf{a}_1, \ldots, \mathbf{a}_m\}$ for which the Cartan matrix will take the form

$$\begin{pmatrix} 0 & k_1(s) & 0 & 0 & 0 & \cdot & \cdots & 0 & 0 \\ -k_1(s) & 0 & k_2(s) & 0 & 0 & \cdot & \cdots & 0 & 0 \\ 0 & -k_2(s) & 0 & k_3(s) & 0 & \cdot & \cdots & 0 & 0 \\ \cdots & \cdots & \cdots & \cdots & \cdots & \cdots & \cdots & \cdots & \cdots \\ 0 & 0 & \cdot & \cdots & & & -k_{n-2}(s) & 0 & k_{n-1}(s) \\ 0 & 0 & \cdot & \cdots & & & 0 & -k_{n-1}(s) & 0 \end{pmatrix} \tag{8-6}$$

$k_j(s)$ is the jth *curvature* of $\mathbf{x}(s)$. By construction, the curvatures are invariants and characterize $\mathbf{x}(s)$ up to a euclidean motion. The determinant formed by the row vectors $\mathbf{x}'(s), \ldots, \mathbf{x}^{(m)}(s)$ is easily seen to be

$$\|\mathbf{x}', \ldots, \mathbf{x}^{(m)}\| = k_1^{m-1}k_2^{m-2} \cdots k_{m-2}^2 k_{m-1}$$

This shows that k_{m-1} is an mth-order invariant. The natural order of differentiability is C^m. This is more than the expected $C^{(m+1)/2+1}$ since the derivatives of the invariants of lower order k_j $(j<m-1)$ take up the remaining dimensions of the prolonged spaces (see the remark at the end of Sec. 7-2).

Exercise 8-1

1. Given values $k(s_0)$, $k'(s_0)$, $t(s_0)$, find the first terms of the Taylor expansion of $\mathbf{x}(s)$ in the moving system at $\mathbf{x}(s_0)$. Draw the projections of the curve onto the coordinate planes.
2. A *helix* is a curve whose tangent makes a constant angle with a fixed direction $\mathbf{d}$, $\mathbf{t}(s) \cdot \mathbf{d} = \cos \alpha$. Show that, for all s, $\mathbf{d}$ is in the plane determined by $\mathbf{t}(s)$, $\mathbf{b}(s)$, and that $\tan \alpha = k/t$.
3. If $k/t = \tan \alpha$ is constant, show that the vector $\mathbf{d} = \mathbf{t} \cos \alpha + \mathbf{b} \sin \alpha$ is constant. Deduce from this and Prob. 2 that $k/t =$ const characterizes helices.
4. $k/t =$ const is a linear relation, but the helices (Prob. 2) do not come under Bertrand's theorem. Explain.
5. Given two curves $\mathbf{f}(s)$, $\mathbf{g}(s)$ such that $\mathbf{n}_f(s) = \mathbf{n}_g(s)$, then

$$k_f(s) = ak_g(s) + bt_g(s)$$

   a,b constant. If $\mathbf{f}$ is of constant curvature, show that $\mathbf{g}$ is a *Bertrand* curve (theorem of Bianchi).
- ◆ 6. Find the curvature and torsion of the following curves:
   (*a*) (u,u^2,u^3)
   (*b*) $(\int\phi(\sigma) \sin \sigma \, d\sigma, \int\phi(\sigma) \cos \sigma \, d\sigma, \int\phi(\sigma)\psi(\sigma) \, d\sigma)$ if $\phi^2(\sigma) + \psi^2(\sigma) = 1$
- ◆ 7. Find the curve given by $k = t = 1/s$.

8. The definition of the contact of curves given in Sec. 7-3 carries over to space problems without change. There exists a sphere such that $\mathbf{x}(s)$ has a contact of order >3 with a curve on that sphere. If $\mathbf{c}$ is the radius vector of the center of the sphere and r its radius, the condition is that $g(s) = |\mathbf{x}(s) - \mathbf{c}|^2 - r^2$ and its first three derivatives vanish for $s = s_0$. Show that

$$\mathbf{c}(s) = \mathbf{x}(s) + \frac{1}{k}\mathbf{n}(s) - \frac{k'}{k^2t}\mathbf{b}(s)$$

   and that $r(s)^2 = k^{-2} + k'^2k^{-4}t^{-2}$.

9. Use Prob. 8 to show that a space curve is situated completely on a sphere if and only if either $k' \neq 0$, $k^{-2} + k'^2k^{-4}t^{-2} = \text{const}$, or $k' = t = 0$.

10. Show that $\mathbf{x}(u) = (a \sin^2 u,\ a \sin u \cos u,\ a \cos u)$ lies completely on a sphere (Prob. 9).

11. Give a condition for a helix (Prob. 2) to be a spherical curve (Prob. 9).

♦**12.** Four vectors in R^3 are always linearly dependent. Find the coefficients in

$$\mathbf{x}^{\text{IV}} = a\mathbf{x}' + b\mathbf{x}'' + c\mathbf{x}'''$$

for a space curve $\mathbf{x}(s)$.

13. Use Prob. 12 to show that a space curve is a helix if and only if $\|\mathbf{x}'',\mathbf{x}''',\mathbf{x}^{\text{IV}}\| = 0$.

14. Show that the binormals of a curve are the binormals of another curve [that is, $\mathbf{b}_f(s_f) = \mathbf{b}_g(s_g)$] if and only if both curves are plane.

15. If a family of straight lines is the family of principal normal lines to an infinity of curves, show that those curves are circular helices.

♦**16.** Find the curvature and torsion of the curve intersection of the two surfaces $x_2^2 = x_3x_1$, $x_1^2 + x_2x_3 = 3px_2$. (HINT: Take one of the coordinates as parameter.)

17. Given a curve $\mathbf{f}(s)$ of curvature and torsion $k(s)$ and $t(s)$, find the invariants of the curve $\mathbf{g} = \mathbf{f} + \lambda\mathbf{n}$, λ constant.

♦**18.** Find the curvature of

$$\mathbf{x} = (c\textstyle\int u(\sigma)\, d\sigma,\ c\int v(\sigma)\, d\sigma,\ c\int w(\sigma)\, d\sigma)$$

subject to $u^2 + v^2 + w^2 = u'^2 + v'^2 + w'^2 = 1$, c constant.

♦**19.** Find the torsion of

$$\mathbf{x} = (c\textstyle\int (vw' - wv')\, d\sigma,\ c\int (wu' - uw')\, d\sigma,\ c\int (uv' - vu')\, d\sigma)$$

under the conditions of Prob. 18.

20. The vector $\mathbf{n}(s)$ determines a spherical curve, the *normal indicatrix* of $\mathbf{x}(s)$. If a spherical curve is given referred to its arc length σ and if $w(\sigma) = d\sigma/ds$ is prescribed, show that there exists a unique $\mathbf{x}(s)$ having the given curve as its normal indicatrix. [HINT: Use (8-1) to compute $k(s)$, $t(s)$ from $\mathbf{n},\mathbf{n}',\mathbf{n}''$.]

***21.** Show that there exists a vector $\mathbf{d}$ such that the Frenet equations (8-1) and (8-2) may be written $\mathbf{a}_i' = \mathbf{d} \times \mathbf{a}_i$ (Darboux).

22. In R^n, the k_i's $(1 \leq i \leq n - 2)$ are lengths, and they are all nonnegative; k_{n-1} has a sign which is itself an invariant. Explain the difference in the meaning of the curvature in the two- and three-dimensional cases.

23. If $\mathbf{x}(s)$ is only C^2, the frame to (8-2) may not be defined. If $\mathbf{a}_1(s) = \mathbf{t}(s)$, show that $k = (p_{12}^2 + p_{13}^2)^{\frac{1}{2}}$ is invariant in any rotation of the frame that leaves $\mathbf{a}_1$ unchanged, i.e., under

$$\begin{pmatrix} 1 & 0 & 0 \\ 0 & \cos\theta & \sin\theta \\ 0 & -\sin\theta & \cos\theta \end{pmatrix}$$

24. A spherical curve (Prob. 9) may be referred to a frame $\mathbf{b}_1 = \mathbf{x}/r$, $\mathbf{b}_2 = \mathbf{t}$, $\mathbf{b}_3$ orthogonal to $\mathbf{b}_1$ and $\mathbf{b}_2$. Show that the corresponding Frenet formula is (for $s^* = s/r$)

$$\frac{d}{ds^*}\begin{pmatrix} \mathbf{b}_1 \\ \mathbf{b}_2 \\ \mathbf{b}_3 \end{pmatrix} = \begin{pmatrix} 0 & 1 & 0 \\ -1 & 0 & \sigma \\ 0 & -\sigma & 0 \end{pmatrix}\begin{pmatrix} \mathbf{b}_1 \\ \mathbf{b}_2 \\ \mathbf{b}_3 \end{pmatrix}$$

σ is the *spherical curvature* (Saban).

25. Prove that curvature, torsion, and spherical curvature (Prob. 24) are connected by $k = r^{-1}(1 + \sigma^2)^{\frac{1}{2}}$, $t = r^{-1}(1 + \sigma^2)^{-1}(d\sigma/ds^*)$.

26. Use the formulas of Prob. 25 to show that, for all closed C^3 curves on a sphere, $\oint k^n t\, ds = 0$ for all positive and negative integers n (Saban). [HINT: Express the integral in the form $\oint df(\sigma)$; it vanishes because σ appears with the same value in both limits of the integral.]

27. Investigate the natural differentiability conditions for the theory of space curves on a fixed sphere (Prob. 24).

28. The properties of space curves that are invariants of the group of similitudes in R^3 depend on one invariant of order three and one of order four. Show that these are $t^{-2}t'$ and either $k^{-2}k'$ or k/t.

29. Use Prob. 28 to find all curves that are trajectories of one-parameter groups of similitudes in R^3.

◆**30.** Find an explicit formula for k_2 of a space curve in R^4.

8-2. Ruled Surfaces

A Klein geometry is the theory of invariants of a transitive transformation group. The structure of the underlying Lie group is an essential element in the geometry. This fact is illustrated nicely in the theory of ruled surfaces. The presentation here is based on work by E. Study and W. Blaschke. The fundamental idea is to replace points by lines as fundamental building blocks of geometric beings. Points are then defined by the totality of (straight) lines passing through them.

We start the discussion with some algebraic preliminaries. The algebra $\mathcal{L}(O_3)$ is of dimension three. The map

$$M(x_1,x_2,x_3) = \begin{pmatrix} 0 & x_1 & -x_2 \\ -x_1 & 0 & x_3 \\ x_2 & -x_3 & 0 \end{pmatrix} \tag{8-7}$$

is an isomorphism of the vector group R_3 onto the additive group of $\mathfrak{L}(O_3)$. This isomorphism allows transporting the structure of the algebra $\mathfrak{L}(O_3)$ into the space R^3. The multiplication is the vector product

$$\mathbf{x} \times \mathbf{y} = M^{-1}(M\mathbf{x}M\mathbf{y} - M\mathbf{y}M\mathbf{x}) \tag{8-8a}$$

in coordinates

$$(x_1,x_2,x_3) \times (y_1,y_2,y_3) = (x_2y_3 - x_3y_2,\ x_3y_1 - x_1y_3,\ x_1y_2 - x_2y_1) \tag{8-8b}$$

We check that

$$\mathbf{x} \cdot (\mathbf{y} \times \mathbf{z}) = \|\mathbf{x},\mathbf{y},\mathbf{z}\| \tag{8-9}$$

Next, we introduce an indeterminate τ subject to the relation

$$\tau^2 = 0 \tag{8-10}$$

If a and b are real numbers, the combination $A = a + \tau b$ is called a *dual number*. Dual numbers are considered as polynomials in τ, subject to the defining relation (8-10). This means that two dual numbers A and $A^* = a^* + \tau b^*$ are added componentwise,

$$A + A^* = (a + a^*) + \tau(b + b^*)$$

and that they are multiplied by $AA^* = aa^* + \tau(ab^* + a^*b)$. Dual numbers form an algebra, not a field. The "pure dual" numbers τb are zero divisors, $(\tau b)(\tau b^*) = 0$. No number τb has an inverse in the algebra.

An oriented line in R^3 may be given by two points on it, $\mathbf{x}$ and $\mathbf{y}$. If ρ is any nonzero constant, we put

$$\mathbf{a} = \rho(\mathbf{y} - \mathbf{x}) \qquad \bar{\mathbf{a}} = \rho\mathbf{x} \times \mathbf{y} \tag{8-11}$$

The six components a_i, $\bar{a}_i (i = 1,2,3)$ of $\mathbf{a}$ and $\bar{\mathbf{a}}$ are *Plücker's homogeneous line coordinates*. The two vectors $\mathbf{a}$ and $\bar{\mathbf{a}}$ are not independent of one another, by Eq. (8-9):

$$\mathbf{a} \cdot \bar{\mathbf{a}} = 0 \tag{8-12}$$

The constant ρ will now be fixed by the normalization

$$|\mathbf{a}| = 1 \tag{8-13}$$

The unit vector $\mathbf{a}$ is the direction of the line; it is independent of the choice of the points $\mathbf{x}$ and $\mathbf{y}$. For any two points on the line,

$$\mathbf{y} = \mathbf{x} + |\mathbf{y} - \mathbf{x}|\mathbf{a}$$

It follows from (8-9) that two vectors $\mathbf{m},\mathbf{n}$ are linearly dependent (i.e., parallel) if and only if $\mathbf{m} \times \mathbf{n} = 0$. Hence

$$\mathbf{y} \times \mathbf{a} = (\mathbf{x} + |\mathbf{y} - \mathbf{x}|\mathbf{a}) \times \mathbf{a} = \mathbf{x} \times \mathbf{a} = \bar{\mathbf{a}} \tag{8-14}$$

for any point on the line; and vice versa, if $\mathbf{z} \times \mathbf{a} = \bar{\mathbf{a}}$, then $\mathbf{z} - \mathbf{x}$ is parallel to the direction of $\mathbf{a}$. $\mathbf{z} = \mathbf{x} \pm |\mathbf{z} - \mathbf{x}|\mathbf{a}$ is on the line. *A point* $\mathbf{z}$ *is on the line of vectors* $\mathbf{a},\bar{\mathbf{a}}$ *if and only if* $\mathbf{z} \times \mathbf{a} = \bar{\mathbf{a}}$. The set of oriented lines in R^3 is in one-to-one correspondence with pairs of vectors in R^3 subject to two conditions, and so we may expect to represent it as a certain four-dimensional set in R^6. Instead of the space R^6 of sextuples of real numbers, Study now takes the space D^3 of triples of dual numbers with coordinates

$$X_1 = x_1 + \tau\bar{x}_1 \qquad X_2 = x_2 + \tau\bar{x}_2 \qquad X_3 = x_3 + \tau\bar{x}_3$$

Each line in R^3 is represented by the *dual vector* in D^3

$$\mathbf{A} = \mathbf{a} + \tau\bar{\mathbf{a}} \tag{8-15}$$

$\mathbf{A}$ is a dual unit vector: $\mathbf{A} \cdot \mathbf{A} = \mathbf{a} \cdot \mathbf{a} + 2\tau\mathbf{a} \cdot \bar{\mathbf{a}} = 1$, if we carry over the formal definition of the products of vectors to dual space.

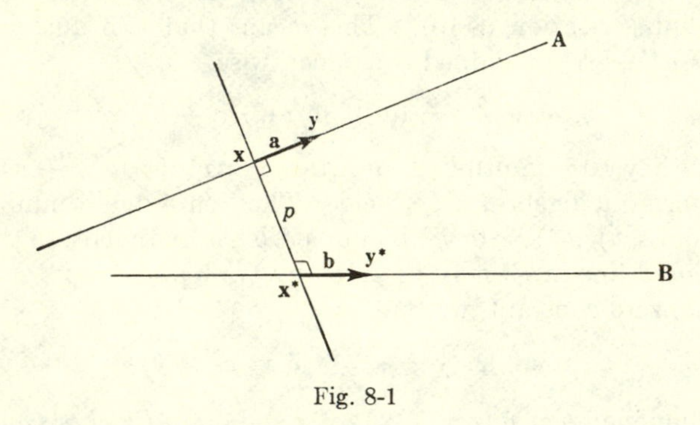

Fig. 8-1

Theorem 8-4 (Study). *The oriented lines in* R^3 *are in one-to-one correspondence with the points of the dual unit sphere* $\mathbf{A} \cdot \mathbf{A} = 1$ *in* D^3.

The scalar product of two lines $\mathbf{A} = \mathbf{a} + \tau\bar{\mathbf{a}}$, $\mathbf{B} = \mathbf{b} + \tau\bar{\mathbf{b}}$ is $\mathbf{A} \cdot \mathbf{B} = \mathbf{a} \cdot \mathbf{b} + \tau(\mathbf{a} \cdot \bar{\mathbf{b}} + \bar{\mathbf{a}} \cdot \mathbf{b})$. Both $\mathbf{a}$ and $\mathbf{b}$ are unit vectors; $\mathbf{a} \cdot \mathbf{b}$ is the cosine of the angle ϕ of the two lines. Two skew lines in space have a unique common perpendicular (Fig. 8-1). Let it meet $\mathbf{A}$ in $\mathbf{x}$, $\mathbf{B}$ in $\mathbf{x}^*$. If the lines $\mathbf{A}$ and $\mathbf{B}$ intersect, we choose $\mathbf{x} = \mathbf{x}^* =$ point of intersection. If the two lines are parallel, $\mathbf{x}$ and $\mathbf{x}^*$ may be any pair of points on a common perpendicular. In any case, the distance p between the two lines is defined as the distance *of* $\mathbf{x}$ *from* $\mathbf{x}^*$. In order to evaluate the scalar product by Eq. (8-11), we choose points $\mathbf{y} = \mathbf{x} + \mathbf{a}$ on $\mathbf{A}$, $\mathbf{y}^* = \mathbf{x}^* + \mathbf{b}$ on $\mathbf{B}$. Then

$$\begin{aligned}\mathbf{a} \cdot \bar{\mathbf{b}} + \bar{\mathbf{a}} \cdot \mathbf{b} &= \|\mathbf{a},\mathbf{x}^*,\mathbf{y}^*\| + \|\mathbf{x},\mathbf{y},\mathbf{b}\| = -\|\mathbf{a},\mathbf{b},\mathbf{x}^*\| + \|\mathbf{a},\mathbf{b},\mathbf{x}\| \\ &= \|\mathbf{a},\mathbf{b},\mathbf{x} - \mathbf{x}^*\| = -p \sin\phi\end{aligned}$$

The last equation follows from the interpretation of the determinant as the (oriented) volume of the parallelepipedon spanned by its row vectors. The formula for dual vectors now is

$$\mathbf{A} \cdot \mathbf{B} = \cos \phi - \tau p \sin \phi \tag{8-16}$$

It is possible to write this equation in a more concise way. A *dual angle* is defined by a pair of angle and length,

$$\Phi = \phi + \tau p \tag{8-17}$$

The formal definition of a dual trigonometric function

$$\cos \Phi = 1 - \frac{1}{2!}\Phi^2 + \frac{1}{4!}\Phi^4 - \cdots$$

gives

$$\begin{aligned}\cos \Phi &= 1 - \frac{1}{2!}\phi^2 + \frac{1}{4!}\phi^4 - \cdots - \tau p\left(\phi - \frac{1}{3!}\phi^3 + \frac{1}{5!}\phi^5 - \cdots\right)\\ &= \cos \phi - \tau p \sin \phi\end{aligned}$$

Hence

$$\mathbf{A} \cdot \mathbf{B} = \cos \Phi \tag{8-16a}$$

A euclidean motion in R^3 will leave unchanged the angle and the distance (hence also the dual angle) between two lines. The corresponding transformation in D^3 therefore will leave the scalar product $\mathbf{A} \cdot \mathbf{B} = \mathbf{A}^t\mathbf{B}$ invariant. It is the action of an orthogonal matrix with dual coefficients. The transformation group in D^3, the image of the euclidean motions in R^3, does not contain any translations, since the center of the dual unit sphere must remain fixed.

Theorem 8-5. *The euclidean motions in R^3 are represented in D^3 by the dual orthogonal matrices $\mathfrak{X} = (X_{ij})$, $\mathfrak{X}^t\mathfrak{X} = U$, X_{ij} dual numbers.*

The Lie algebra $\mathfrak{L}(O_{D^3})$ of the group of 3×3 orthogonal dual matrices is the algebra of skew-symmetric 3×3 dual matrices, as is seen by differentiation of $\mathfrak{X}^t\mathfrak{X} = U$.

A differentiable curve $\mathbf{A}(u)$ on the dual unit sphere, depending on a *real* parameter u, represents a differentiable family of straight lines in R^3: a *ruled surface.* The lines $\mathbf{A}(u)$ are the *generators* or *rulings* of the surface. We are not worried by the fact that as yet we have not defined surfaces, since our primary geometric objects now are lines, not points. In some sense, this treatment of ruled surfaces is more general than that based on points, since quite regular curves on the dual sphere may represent ruled surfaces with complicated point singularities in R^3.

In the theory of ruled surfaces, a six-parameter group acts on a four-dimensional space with one independent real variable. Thus we may expect one invariant of first and two of second orders.

An orthonormal frame for a C^2 curve $\mathbf{A}(u)$ is found immediately:

$$\mathbf{A}_1 = \mathbf{A}(u) \qquad \mathbf{A}_2 = \frac{\dot{\mathbf{A}}_1(u)}{\sqrt{\dot{\mathbf{A}}_1^2}} \qquad \mathbf{A}_3 = \mathbf{A}_1 \times \mathbf{A}_2 \tag{8-18}$$

It is necessary to suppose that $\mathbf{a}(u)$ in $\mathbf{A}(u) = \mathbf{a}(u) + \tau\bar{\mathbf{a}}(u)$ is not a constant vector, i.e., that the ruled surface does not contain a cylinder. In this case, $\dot{\mathbf{A}}_1^2 = s + \tau t$ has a positive real part s; it is possible to extract the root: $s + \tau t = (x + \tau y)^2$ if $x = \sqrt{s}$, $y = t/2x$. By hypothesis, $x \neq 0$; $\mathbf{A}_2$ is well defined. Differentiation of $\mathbf{A}_1^2 = 1$ shows that $\mathbf{A}_1 \cdot \mathbf{A}_2 = 0$. The remaining orthogonality relations $\mathbf{A}_3 \cdot \mathbf{A}_1 = \mathbf{A}_3 \cdot \mathbf{A}_2 = 0$ follow from Eq. (8-9). It may be checked by direct computation that the vector product of two dual unit vectors is again a unit vector, $\mathbf{A}_3^2 = \|\mathbf{A}_3,\mathbf{A}_1,\mathbf{A}_2\| = 1$.

The frame (8-18) has an intuitive interpretation in R^3. $\mathbf{A} \cdot \mathbf{B} = 0$ means, by (8-16), that the two lines $\mathbf{A}$, $\mathbf{B}$ meet at a right angle. $\mathbf{A}_1$, $\mathbf{A}_2$, $\mathbf{A}_3$ therefore are three concurrent mutually orthogonal lines in R^3. Their point of intersection is the *point of striction* $\mathbf{s}(u)$ on the ruling $\mathbf{A}_1(u)$. The locus of $\mathbf{s}(u)$ is the *curve of striction* on the ruled surface. $\mathbf{A}_3(u)$ is the limit position of the common perpendicular to $\mathbf{A}_1(u)$ and

$$\begin{aligned}\mathbf{A}_1(u + \Delta u) &= \mathbf{A}_1(u) + \Delta u\, \dot{\mathbf{A}}_1(u + \theta\, \Delta u) \\ &= \mathbf{A}_1(u) + h(u, \Delta u)\mathbf{A}_2(u + \theta\, \Delta u)\end{aligned}$$

$0 \leq \theta \leq 1$, since it is perpendicular to both $\mathbf{A}_1$ and $\mathbf{A}_2$. This shows that the point of striction is the point on $\mathbf{A}(u)$ "with minimal distance from neighboring generators," $\mathbf{A}_3(u)$ is the tangent to the surface at the point of striction, and $\mathbf{A}_2(u)$ is the normal to the surface at $\mathbf{s}(u)$. By construction, the Frenet equation is

$$\frac{d}{du}\begin{pmatrix}\mathbf{A}_1 \\ \mathbf{A}_2 \\ \mathbf{A}_3\end{pmatrix} = \begin{pmatrix}0 & K & 0 \\ -K & 0 & T \\ 0 & -T & 0\end{pmatrix}\begin{pmatrix}\mathbf{A}_1 \\ \mathbf{A}_2 \\ \mathbf{A}_3\end{pmatrix} \tag{8-19}$$

where $$K = \dot{\mathbf{A}}_1^2 \qquad K^2T = \|\mathbf{A},\dot{\mathbf{A}},\ddot{\mathbf{A}}\| \tag{8-20}$$

The reduced form of the Cartan matrix ($p_{13} = 0$) is obtained without the introduction of an invariant parameter. $K(u) = k_1(u) + \tau k_2(u)$ and $T(u) = t_1(u) + \tau t_2(u)$ are not invariants. The integrals $\int K\, du$ and $\int T\, du$ are dual arc lengths, in the sense of Def. 2-1, of the curves A_1 and A_3. Hence

$$\int k_1\, du \qquad \int k_2\, du \qquad \int t_1\, du \qquad \int t_2\, du$$

are *integral invariants*.

If $k_1(u) = 0$, all rulings are parallel, and the surface is a cylinder. The frame (8-18) is not defined, since there is no distinguished point of striction. If $k_1(u) \neq 0$, it may be used for the definition of an invariant parameter σ by

$$k_1(\sigma) = 1$$

In terms of this invariant parameter, the invariant of order one is

$$k_2 = \mathbf{a}' \cdot \bar{\mathbf{a}}' \tag{8-21a}$$

and those of order two are

$$t_1 = \|\mathbf{a},\mathbf{a}',\mathbf{a}''\| \tag{8-21b}$$
$$t_2 = \|\bar{\mathbf{a}},\mathbf{a}',\mathbf{a}''\| + \|\mathbf{a},\bar{\mathbf{a}}',\mathbf{a}''\| + \|\mathbf{a},\mathbf{a}',\bar{\mathbf{a}}''\| - 2k_2t_1 \tag{8-21c}$$

The Frenet formulas (8-19) may be split into real and dual parts:

$$\frac{d}{du}\begin{pmatrix}\mathbf{a}_1\\ \mathbf{a}_2\\ \mathbf{a}_3\end{pmatrix} = \begin{pmatrix}0 & k_1 & 0\\ -k_1 & 0 & t_1\\ 0 & -t_1 & 0\end{pmatrix}\begin{pmatrix}\mathbf{a}_1\\ \mathbf{a}_2\\ \mathbf{a}_3\end{pmatrix} \tag{8-19a}$$

$$\frac{d}{du}\begin{pmatrix}\bar{\mathbf{a}}_1\\ \bar{\mathbf{a}}_2\\ \bar{\mathbf{a}}_3\end{pmatrix} = \begin{pmatrix}0 & k_1 & 0\\ -k_1 & 0 & t_1\\ 0 & -t_1 & 0\end{pmatrix}\begin{pmatrix}\bar{\mathbf{a}}_1\\ \bar{\mathbf{a}}_2\\ \bar{\mathbf{a}}_3\end{pmatrix} + \begin{pmatrix}0 & k_2 & 0\\ -k_2 & 0 & t_2\\ 0 & -t_2 & 0\end{pmatrix}\begin{pmatrix}\mathbf{a}_1\\ \mathbf{a}_2\\ \mathbf{a}_3\end{pmatrix} \tag{8-19b}$$

The point of striction satisfies

$$\mathbf{s} \times \mathbf{a}_1 = \bar{\mathbf{a}}_1 \qquad \mathbf{s} \times \mathbf{a}_2 = \bar{\mathbf{a}}_2 \qquad \mathbf{s} \times \mathbf{a}_3 = \bar{\mathbf{a}}_3$$

by Eq. (8-14). If the tangent to the curve of striction is written as $d\mathbf{s}/du = \alpha\mathbf{a}_1 + \beta\mathbf{a}_2 + \gamma\mathbf{a}_3$, the coefficients may be computed by differentiation of the three equations for $\mathbf{s}$ (using Prob. 3, exercise 8-2). The result of the computation is

$$\frac{d\mathbf{s}}{du} = t_2\mathbf{a}_1 + k_2\mathbf{a}_3 \tag{8-22}$$

If $k_2 = 0$, the direction of $\mathbf{s}'$ is $\mathbf{a}_1$, the generator is tangent to the curve of striction, and the ruled surface is the surface of tangents to a space curve. In this case the curve of striction is called the *edge of regression*, and the surface itself is *developable*. $\mathbf{a}_2$ is the normal and $\mathbf{a}_3$ the binormal of $\mathbf{s}$.

If $t_1 \equiv 0$, $\mathbf{a}_3 = \text{const.}$ The vector $\mathbf{a}_1$ then is always parallel to a fixed plane, normal to $\mathbf{a}^3$.

If $t_2 \equiv k_2 \equiv 0$, $\mathbf{s}' = 0$, the point of striction is constant. All generators of the surface pass through one fixed point $\mathbf{s}$, and the surface is a *cone*.

If $k_2 \neq 0$, $t_2 \equiv 0$, the curve of striction is tangent to $\mathbf{a}_3$; it is normal to the ruling through $\mathbf{s}(u)$. The surface $\mathbf{A}_3(u)$ is developable; from the discussion of the case $k_2 = 0$ it is seen that here $\mathbf{a}_1$ is the binormal of $\mathbf{s}(u)$, and the surface is generated by the binormals of a space curve.

***Example* 8-2.** The equilateral hyperbolic paraboloid

$$x_3 = x_1^2 - x_2^2$$

is a ruled surface, generated, for example, by the family of lines

$$x_1 + x_2 = ux_3 \qquad x_1 - x_2 = \frac{1}{u}$$

The ruling of parameter u passes through the two points

$$\left(\frac{1}{2u}, -\frac{1}{2u}, 0\right) \qquad \left(\frac{u^2+1}{2u}, \frac{u^2-1}{2u}, 1\right)$$

Hence it is the point on the dual sphere

$$\mathbf{A} = (4 + 2u^2)^{-\frac{1}{2}}[(u,u,2) + \tau u^{-1}(-1,-1,u)]$$

From

$$\mathbf{A}^{\cdot} = 4(4 + 2u^2)^{-\frac{3}{2}}\left[(1,1,-u) + \tau\left(\frac{u^2+1}{u^2}, \frac{u^2+1}{u^2}, -\frac{u}{2}\right)\right]$$

one has $$\mathbf{A}^{\cdot 2} = \frac{2}{(2+u^2)^2}\left(1 + \tau\frac{u^2+2}{u^2}\right)$$

Hence $$|\mathbf{A}^{\cdot}| = \frac{\sqrt{2}}{2+u^2} + \tau\frac{1}{\sqrt{2}\,u^2}$$

and $$\mathbf{A}_2 = (2+u^2)^{\frac{1}{2}}\left[(1,1,-u) + \tau\left(\frac{1}{2}, \frac{1}{2}, \frac{1}{u}\right)\right]$$

$$\mathbf{A}_3 = 2^{-\frac{1}{2}}(-1,1,0)$$

$$k_1 = \frac{\sqrt{2}}{2+u^2} \qquad k_2 = \frac{1}{\sqrt{2}\,u^2} \qquad t_1 = t_2 = 0$$

The curve of striction is the straight line $x_1 = x_2$, $x_3 = 0$. (The paraboloid is a ruled surface also by a second family of lines $x_1 + x_2 = u$, $x_1 - x_2 = x_3/u$ for which the line of striction is $x_1 = -x_2$, $x_3 = 0$.)

Exercise 8-2

1. In R^3, let $\mathbf{e}_1,\mathbf{e}_2,\mathbf{e}_3$ be an orthonormal frame. Find the dual vectors
   (*a*) Of the coordinate axes
   (*b*) Of the diagonal of the unit cube
   (*c*) Of the parallel to the x_1 axis through (0,1,2)
2. Find the inverse of a dual number $a + \tau b$, $a \neq 0$.
3. With the notation of Prob. 1, show that $\mathbf{e}_1 \times \mathbf{e}_2 = \mathbf{e}_3$, $\mathbf{e}_2 \times \mathbf{e}_3 = \mathbf{e}_1$, $\mathbf{e}_3 \times \mathbf{e}_1 = \mathbf{e}_2$.
4. Show that the dual algebra $D = R + \tau R$ has the ideal τR (exercise 6-2, Prob. 2).
5. ♦ Compute the distance and angle of the lines
   (*a*) l_1 through (1,0,0) and (0,5,0) and l_2 through (1,1,1) and (0,0,0)
   (*b*) l_1 through (1,2,3) in direction (0,1,1) and l_2 through (3,4,5) in direction (1,1,0)

6. Find the mutual perpendicular to the lines l_1 and l_2 of Prob. 5*a*.

7. Find the mutual perpendicular $\mathbf{P} = \mathbf{p} + \tau\bar{\mathbf{p}}$ to two lines

$$\mathbf{A} = \mathbf{a} + \tau\bar{\mathbf{a}} \qquad \mathbf{B} = \mathbf{b} + \tau\bar{\mathbf{b}}$$

8. Characterize the dual vectors of lines through the origin.

9. Show that $\sin \Phi = \sin \phi + \tau p \cos \phi$.

◆**10.** Find the dual vector $\mathbf{A}(x_1)$ of the line through $(x_1,0,0)$ parallel to the (x_2,x_3) plane and making an angle $\arctan k/x_1$ with the x_2 axis.

11. Find the invariants of the surface generated by the tangents to the space curve $(a \cos u, a \sin u, bu)$.

12. Find all dual orthogonal matrices that correspond to euclidean motions in which the x_1 axis is invariant.

13. Characterize the ruled surfaces that are transformed into themselves by a one-parameter group of euclidean motions.

◆**14.** Find all ruled surfaces that are transformed into themselves by a one-parameter group of euclidean motions and for which at least one invariant vanishes.

15. What is the image of a plane on the dual sphere?

16. Find the dual representation of the right circular cone through the three coordinate axes.

17. Compute the frame, the invariants, and the curve of striction for the hyperboloid $x_1^2/a^2 - x_3^2/c^2 = 1 - x_2^2/b^2$.

18. Characterize the surfaces

(*a*) With a plane curve of striction

(*b*) Whose curve of striction is a circle

(*c*) Whose curve of striction is a straight line

19. Compute the curvature and torsion of the edge of regression as functions of the invariants $t_1(\sigma)$, $t_2(\sigma)$ of a developable surface.

20. Show that $k_2(\sigma)$, $t_1(\sigma)$, $t_2(\sigma)$ determine a ruled surface up to a euclidean motion.

◆**21.** For a ruled surface,

$$\mathbf{A}(u) \cdot \mathbf{A}(u + \Delta u) = \cos \phi(u,\Delta u) - p(u,\Delta u) \sin \phi(u,\Delta u)$$

The *distribution parameter* of the surface is

$$d(u) = \lim_{\Delta u \to 0} \frac{p(u,\Delta u)}{\phi(u,\Delta u)}$$

Compute $d(u)$ as a function of the invariants of the surface.

22. The *tangent cone* of a ruled surface is the cone through the origin generated by parallels to the tangents of the curve of striction. If the tangents to the curve of striction are all tangent to a fixed cylinder, show that either the surface is developable or the tangent cone is a circular cone (Dini).

8-3. Space Curves in Affine Geometry

The group of unimodular affine transformations in R^m

$$\mathbf{x}^* = \mathbf{x}A + \mathbf{b} \qquad \|A\| = 1$$

has $m^2 + m - 1$ parameters. In the theory of curves each prolongation adds $m - 1$ dimensions to the contact space. The natural condition of differentiability is C^{m+2}, and so one may expect $m - 1$ invariants of orders $\leq m + 2$.

For the reasons discussed in Sec. 7-3, we treat in this section only curves $\mathbf{x}(u)$ in R^m for which

$$\|\mathbf{x}^{\cdot}(u), \ldots, \mathbf{x}^{(m)}(u)\| \neq 0$$

at all points. As a consequence, the arc in question is not contained wholly in some subspace R^n of R^m, $n < m$. The invariant parameter, the *affine arc length* σ, is defined by

$$\left\| \frac{d\mathbf{x}}{d\sigma}, \frac{d^2\mathbf{x}}{d\sigma^2}, \ldots, \frac{d^m\mathbf{x}}{d\sigma^m} \right\| = \pm 1 \tag{8-23}$$

The derivative $du/d\sigma = j$ is computed from

$$\|\mathbf{x}^{\cdot}, \mathbf{x}^{\cdot\cdot}, \ldots, \mathbf{x}^{(m)}\| j^{\frac{1}{2}m(m+1)} = \pm 1$$

For $m = 4k$ or $m = 4k + 3$ the sign of the determinant (8-23) is an affine invariant. The curve is *dextrorse* for $+1$ and *sinistrorse* for -1. j is a differential expression of order m.

A complete set of unimodular affine invariants may be obtained by following the procedure of Sec. 7-3. The fixed cartesian system of coordinates is given by an orthonormal frame $\mathbf{e}_1, \ldots, \mathbf{e}_m$. The moving frame along the curve $\mathbf{x}(u)$ is formed by the vectors $\mathbf{y}_i = d^i\mathbf{x}/d\sigma^i$, $i = 1, \ldots, m$. The $\mathbf{y}_i$ are invariant, they are linearly independent, and by Eq. (8-23) they span an m-dimensional volume 1. The frame matrix $A(\sigma)$

$$\{\mathbf{y}_1, \ldots, \mathbf{y}_m\} = A(\sigma)\{\mathbf{e}_1, \ldots, \mathbf{e}_m\}$$

is unimodular. $C(A)$ is of vanishing trace. By definition,

$$\mathbf{y}_i' = \mathbf{y}_{i+1} \qquad i < m$$

hence

$$C(A) = \begin{pmatrix} 0 & 1 & 0 & \cdot & \cdot & 0 \\ 0 & 0 & 1 & 0 & \cdot & 0 \\ \cdot & \cdot & \cdot & \cdot & \cdot & \cdot \\ 0 & \cdot & \cdot & \cdot & 0 & 1 \\ k_1 & k_2 & \cdot & \cdot & k_{m-1} & 0 \end{pmatrix} \tag{8-24}$$

Therefore the k_i $(1 \leq i \leq m - 1)$ are $m - 1$ invariants which characterize $\mathbf{x}(\sigma)$ up to a constant unimodular affine transformation. The computation of

$$k_i = \left\| \frac{d\mathbf{x}}{d\sigma}, \ . \ . \ . \ , \frac{d^{i-1}\mathbf{x}}{d\sigma^{i-1}}, \frac{d^{m+1}\mathbf{x}}{d\sigma^{m+1}}, \frac{d^{i+1}\mathbf{x}}{d\sigma^{i+1}}, \ . \ . \ . \ , \frac{d^m\mathbf{x}}{d\sigma^m} \right\|$$

involves that of $d^{m+1}\mathbf{x}/d\sigma^{m+1} = d^m(\mathbf{x}^{\cdot}j)/d\sigma^m$. This is an expression containing the mth derivative of j, hence the $2m$th derivatives of $\mathbf{x}$. The invariants k_i are of order $2m$; they are *not* natural invariants (see exercise 8-3, Prob. 1). The computations needed for the construction of a frame of minimal order are complicated. We carry them out only in the case $m = 3$.

The invariants k_i are obtained from the frame vectors $\mathbf{y}_i$,

$$\begin{aligned}
\mathbf{y}_1 &= \mathbf{x}' = \mathbf{x}^{\cdot}j \\
\mathbf{y}_2 &= \mathbf{x}'' = \mathbf{x}^{\cdot\cdot}j^2 + \mathbf{x}^{\cdot}j^{\cdot}j \\
\mathbf{y}_3 &= \mathbf{x}''' = \mathbf{x}^{\cdot\cdot\cdot}j^3 + 3\mathbf{x}^{\cdot\cdot}j^{\cdot}j^2 + \mathbf{x}^{\cdot}(j^{\cdot\cdot}j^2 + j^{\cdot 2}j) \\
\mathbf{x}^{\mathrm{IV}} &= \mathbf{x}^{(4)}j^4 + 6\mathbf{x}^{\cdot\cdot\cdot}j^{\cdot}j^3 + \mathbf{x}^{\cdot\cdot}(4j^{\cdot\cdot}j^3 + 7j^{\cdot 2}j^2) + \mathbf{x}^{\cdot}(j^{\cdot\cdot\cdot}j^3 + 4j^{\cdot\cdot}j^{\cdot}j^2 + j^{\cdot 3}j)
\end{aligned}$$

Roman numerals indicate prime (′) derivations; Arabic numbers in parentheses indicate dot (·) derivations. The highest derivative of j appears always in the factor multiplying $\mathbf{x}^{\cdot}$. This is the reason why

$$k_2 = \|\mathbf{y}_1,\mathbf{x}^{\mathrm{IV}},\mathbf{y}_3\| = 4j^{\cdot\cdot}j - 7j^{\cdot 2} - \|\mathbf{x}^{\cdot},\mathbf{x}^{\cdot\cdot\cdot},\mathbf{x}^{(4)}\|j^8$$

is only of order five while

$$k_1 = \|\mathbf{x}^{\mathrm{IV}},\mathbf{y}_2,\mathbf{y}_3\| = j^{\cdot\cdot\cdot}j^2 + j^{\cdot 3} + \|\mathbf{x}^{\cdot},\mathbf{x}^{\cdot\cdot\cdot},\mathbf{x}^{(4)}\|j^{\cdot}j^8 + \|\mathbf{x}^{\cdot\cdot},\mathbf{x}^{\cdot\cdot\cdot},\mathbf{x}^{(4)}\|j^9$$

is of order six. The term of highest order (five) in $\mathbf{y}_3$ is $\mathbf{x}^{\cdot}j^{\cdot\cdot}j^2 = \mathbf{y}_1 j^{\cdot\cdot}j$; hence the vector

$$\mathbf{z}_3 = -\frac{k_2}{4}\mathbf{y}_1 + \mathbf{y}_3 = \mathbf{x}^{\cdot\cdot\cdot}j^3 + 3\mathbf{x}^{\cdot\cdot}j^{\cdot}j^2 + \mathbf{x}^{\cdot}\left(\frac{11}{4}j^{\cdot 2}j + \frac{1}{4}\|\mathbf{x}^{\cdot},\mathbf{x}^{\cdot\cdot\cdot},\mathbf{x}^{(4)}\|j^9\right)$$

is only of order four. Its derivative is a vector of order five,

$$\mathbf{z}_3' = (k_1 - \tfrac{1}{4}k_2')\mathbf{y}_1 + \frac{3k_2}{4}\mathbf{y}_2$$

The coefficient of $\mathbf{y}_1$, the difference of two functions containing sixth derivatives, is only of order five,

$$\begin{aligned}
\varkappa_1 = k_1 - \tfrac{1}{4}k_2' = \tfrac{5}{2}j^{\cdot\cdot}j^{\cdot}j + j^{\cdot 3} + 9\|\mathbf{x}^{\cdot},\mathbf{x}^{\cdot\cdot},\mathbf{x}^{(4)}\|j^{\cdot}j^8 + \tfrac{5}{4}\|\mathbf{x}^{\cdot\cdot},\mathbf{x}^{\cdot\cdot\cdot},\mathbf{x}^{(4)}\|j^9 \\
+ \tfrac{1}{4}\|\mathbf{x}^{\cdot},\mathbf{x}^{\cdot\cdot\cdot},\mathbf{x}^{(5)}\|j^9
\end{aligned}$$

With the definitions

$$\mathbf{z}_1 = \mathbf{y}_1 \qquad \mathbf{z}_2 = \mathbf{y}_2 \qquad \varkappa_2 = \tfrac{1}{4}k_2$$

the invariants of minimal order in the unimodular affine geometry of space curves appear in

$$\begin{pmatrix} \mathbf{z}_1 \\ \mathbf{z}_2 \\ \mathbf{z}_3 \end{pmatrix}' = \begin{pmatrix} 0 & 1 & 0 \\ -\kappa_2 & 0 & 1 \\ \kappa_1 & 3\kappa_2 & 0 \end{pmatrix} \begin{pmatrix} \mathbf{z}_1 \\ \mathbf{z}_2 \\ \mathbf{z}_3 \end{pmatrix} \tag{8-25}$$

The construction makes use of the sixth derivatives; it is valid for C^6 curves. No proof for C^5 seems to be known. The same situation holds in R^m. The methods at our disposal today need C^{2m} curves to construct differential invariants of order $m + 2$.

As in euclidean geometry, it is possible to find a frame of order five for C^5 curves with fifth derivatives of bounded variation if one looks only for Cartan matrices whose elements are fixed up to an additive constant (integral invariants). The frame itself then is not unique but is defined up to a certain constant unimodular affinity. The construction leading to (8-25) suggests the search for a convenient frame $\{\mathbf{a}_1,\mathbf{a}_2,\mathbf{a}_3\}$ derived from $\{\mathbf{y}_1,\mathbf{y}_2,\mathbf{y}_3\}$ by

$$\begin{pmatrix} \mathbf{a}_1 \\ \mathbf{a}_2 \\ \mathbf{a}_3 \end{pmatrix} = \begin{pmatrix} 1 & 0 & 0 \\ 0 & 1 & 0 \\ a & 0 & 1 \end{pmatrix} \begin{pmatrix} \mathbf{y}_1 \\ \mathbf{y}_2 \\ \mathbf{y}_3 \end{pmatrix}$$

By Eq. (1-5) the Cartan matrix of the new frame becomes

$$\begin{pmatrix} 0 & 1 & 0 \\ -a & 0 & 1 \\ a' + k_1 & a + k_2 & 0 \end{pmatrix}$$

which suggests the determination of a by

$$a = -\int k_1\, d\sigma = -\int \|d\mathbf{x}''',\mathbf{x}'',\mathbf{x}'''\|$$

for a frame with the Cartan matrix

$$\begin{pmatrix} 0 & 1 & 0 \\ q_1 & 0 & 0 \\ 0 & q_2 & 0 \end{pmatrix} \qquad q_1 + q_2 = k_2 \qquad q_1' = -k_1 \tag{8-26}$$

In R^3, the curves with vanishing *affine curvatures* k_1, k_2 are the integral curves of $\mathbf{y}_3' = \mathbf{x}^{\mathrm{IV}} = 0$. They are twisted cubics $\mathbf{x} = \mathbf{a}\sigma^3 + \mathbf{b}\sigma^2 + \mathbf{c}\sigma + \mathbf{d}$ (all quadric curves are plane and are excluded from the three-dimensional theory).

The trajectories of one-parameter groups of unimodular affine transformations are the curves with constant affine curvatures k_i. In R^3, they are integral curves of

$$\mathbf{x}^{\mathrm{IV}} - k_2\mathbf{x}'' - k_1\mathbf{x}' = 0 \tag{8-27}$$

The solution of this vector differential equation with constant coefficients depends on its characteristic equation

$$r^3 - k_2 r - k_1 = 0 \tag{8-28}$$

I. $k_1 \neq 0$.

A. If (8-28) has three distinct real roots r_1, r_2, r_3, a particular solution of (8-27) is

$$\mathbf{x}' = (ar_1 e^{r_1\sigma}, br_2 e^{r_2\sigma}, cr_3 e^{r_3\sigma}) \qquad r_1 + r_2 + r_3 = 0$$

or

$$\mathbf{x} = (ae^{r_1\sigma}, be^{r_2\sigma}, ce^{r_3\sigma})$$

The condition $\|\mathbf{x}',\mathbf{x}'',\mathbf{x}'''\| = 1$ is $abck_1(4k_2^3 - 27k_1^3) = 1$. The curve is wholly on the cubic surface

$$x_1 x_2 x_3 = \frac{1}{k_1(4k_2^3 - 27k_1^2)}$$

B. If (8-28) has a double root, $\mathbf{x}'$, $\mathbf{x}''$, $\mathbf{x}'''$ cannot be linearly independent. This case $4k_2^3 - 27k_1^2 = 0$ is to be excluded.

C. If (8-28) has only one real root r_1, the other two roots are

$$r_2 = m_1 + im_2 \qquad r_3 = m_1 - im_2$$

A particular solution of (8-27) is

$$\mathbf{x} = (ae^{r_1\sigma}, be^{m_1\sigma}\cos m_2\sigma, ce^{m_1\sigma}\sin m_2\sigma)$$

where

$$r_1 + 2m_1 = 0$$
$$abcr_1 m_1[(m_1^2 + m_2^2)^2 + r_1(m_1 - m_2)^2 - 2r_1 m_1^3] = 1$$

II. $k_1 = 0$.

A. $k_2 > 0$. A solution of (8-27) is

$$\mathbf{x} = \frac{1}{k_2}(\operatorname{Sinh} k_2^{\frac{1}{2}}\sigma, \operatorname{Cosh} k_2^{\frac{1}{2}}\sigma, k_2^{\frac{1}{2}}\sigma)$$

B. $k_2 < 0$.

$$\mathbf{x} = -\frac{1}{k_2}(\sin(-k_2)^{\frac{1}{2}}\sigma, \cos(-k_2)^{\frac{1}{2}}\sigma, (-k_2)^{\frac{1}{2}}\sigma)$$

Since the invariants characterize the curves up to a unimodular affinity, a particular solution of (8-27) does the same service as the general one.

Exercise 8-3

1. Show that k_i is of order $2m + 1 - i$.

2. In the moving frame at $\mathbf{x}_0$, find the first terms of the Taylor expansion of $x_2(x_1)$, $x_3(x_1)$ for a space curve referred to x_1 as parameter. Show that a space curve in R^3 has an osculating twisted cubic in each point.

3. Prove that the affine curvatures k_i are constant if and only if there exist constants α, β, γ and a constant vector $\mathbf{q}$ such that

$$\mathbf{x}(u) + \alpha\mathbf{y}_1(u) + \beta\mathbf{y}_2(u) + \gamma\mathbf{y}_3(u) = \mathbf{q}$$

◆ 4. Find the curves for which $\beta = 0$ (Prob. 3).

5. Characterize the position in space R^3 of $\mathbf{x}'''$ for curves with $k_1 = 0$.

6. Find the one-parameter groups that have as trajectories the curves with constant affine curvatures (page 173). Show that the one-parameter group is completely determined by the invariants if at least one invariant $\neq 0$.

7. Use Prob. 6 to find all space curves invariant under a two-parameter group of unimodular affinities in R^3.

8. Find all curves for which $\mathbf{z}_3$ has constant direction.

9. Construct the frame with invariants of minimal order in R^4.

◆10. Find the curves of vanishing affine curvatures in R^4.

◆11. Find invariants for the group of *general* affine transformations in R^3.

12. Find the trajectories of one-parameter groups of general affine transformations in R^3.

References

SEC. 8-1

Blaschke, W.: "Einführung in die Differentialgeometrie," Springer-Verlag OHG, Berlin, 1950.

Eisenhart, L. P.: "An Introduction to Differential Geometry with Use of Tensor Calculus," Princeton University Press, Princeton, N.J., 1947.

Saban, G.: Nuove caratterizzazioni della sfera, *Atti Accad. Naz. Lincei, Rend., Classe Sci. Fis. Mat. Nat.*, Ser. 8, **25**: 457–464 (1958).

SEC. 8-2

Blaschke, W.: "Vorlesungen über Differentialgeometrie und geometrische Grundlagen von Einsteins Relativitätstheorie," 3d ed., vol. 1, Springer-Verlag OHG, Berlin, 1930.

Study, E.: "Geometrie der Dynamen," B. G. Teubner Verlagsgesellschaft, mbH, Leipzig, 1903.

SEC. 8-3

The texts of Blaschke and Favard quoted in the references for Sec. 7-3, and

Berwald, L.: Differentialinvarianten in der Geometrie; Riemannsche Mannigfaltigkeiten und ihre Verallgemeinerungen, "Enzyklopädie der math. Wissenschaften," vol. III/3, pp. 73–181, B. G. Teubner Verlagsgesellschaft, mbH, Leipzig, 1927.

9
TENSORS

9-1. Dual Spaces

This chapter is devoted to a generalization of the notions of the vector and the matrix of linear algebra. This technical preliminary to the study of surfaces and of higher-dimensional spaces will involve also some change in our notations for vectors.

We recall some fundamental definitions. A *vector space* V (over the real numbers) is an *abelian* group for whose elements multiplication by a real number is defined:

$$\text{For all } \mathbf{x} \in V,\ \alpha \in R,\ \alpha\mathbf{x} \in V \text{ is defined}$$

This "outer" multiplication by a real number is *free*, i.e.,

$$\alpha\mathbf{x} = 0,\ \alpha \neq 0 \text{ implies } \mathbf{x} = 0$$

and it is distributive,

$$\alpha(\mathbf{x} + \mathbf{y}) = \alpha\mathbf{x} + \alpha\mathbf{y}$$

A set of *vectors* $\mathbf{e}_1, \ldots, \mathbf{e}_n \in V$ is *linearly independent* if

$$\alpha_1\mathbf{e}_1 + \alpha_2\mathbf{e}_2 + \cdots + \alpha_n\mathbf{e}_n = 0 \text{ implies } \alpha_1 = \alpha_2 = \cdots = \alpha_n = 0$$

V is of *dimension* n if there exists a set of n linearly independent vectors $\mathbf{e}_1, \ldots, \mathbf{e}_n$ (a *basis*) which is maximal, i.e., any vector $\mathbf{x} \in V$ is a linear combination with real coefficients of the basis vectors,

$$\mathbf{x} = x^1\mathbf{e}_1 + \cdots + x^n\mathbf{e}_n$$

$x^1, \ldots, x^n$ are the *coordinates* of $\mathbf{x}$ in the basis $\{\mathbf{e}_i\}$. The coordinates are uniquely determined; $\mathbf{x} = y^1\mathbf{e}_1 + \cdots + y^n\mathbf{e}_n$ implies

$$\mathbf{x} - \mathbf{x} = (x^1 - y^1)\mathbf{e}_1 + \cdots + (x^n - y^n)\mathbf{e}_n = 0$$

Hence $x^i - y^i = 0$, $i = 1, 2, \ldots, n$. A basis $\{\mathbf{e}_i\}$ defines a map

$$\mathbf{x} \rightarrow (x^1, \ldots, x^n) \tag{9-1}$$

of the vector space V onto the space of n-tuples of real numbers R^n. R^n is a vector space

$$(x^1, \ldots, x^n) + (y^1, \ldots, y^n) = (x^1 + y^1, \ldots, x^n + y^n)$$
$$\alpha(x^1, \ldots, x^n) = (\alpha x^1, \ldots, \alpha x^n)$$

The map (9-1) is an *isomorphism* of V and R^n. This shows that any two real vector spaces of the same (finite) dimension n are isomorphic. Two vector spaces of different dimensions are never isomorphic (see exercise 9-1, Prob. 2).

The set of all real-valued linear functions defined on V is denoted by by V^*. $\phi \in V^*$ if $\phi(\mathbf{x})$ is a real number and

$$\phi(\alpha\mathbf{x} + \beta\mathbf{y}) = \alpha\phi(\mathbf{x}) + \beta\phi(\mathbf{y})$$

A vector-space structure is defined in V^* by

$$(\phi + \psi)(\mathbf{x}) = \phi(\mathbf{x}) + \psi(\mathbf{x})$$
$$(\alpha\phi)(\mathbf{x}) = \alpha\phi(\mathbf{x})$$

The zero element of V^* is the function which is identically zero for all $\mathbf{x} \in V$. This defines linear independence in V^*.

Given a basis $\{\mathbf{e}_i\}$ of V, *the functions* $\boldsymbol{\varepsilon}^i$

$$\boldsymbol{\varepsilon}^i(\mathbf{e}_j) = \begin{cases} 0 & i \neq j \\ 1 & i = j \end{cases} \qquad i = 1, \ldots, n \tag{9-2}$$

are a basis of V^*. To prove this statement one shows first that the functions $\boldsymbol{\varepsilon}^i$ are well defined, i.e., that the value $\boldsymbol{\varepsilon}^i(\mathbf{x})$ may be computed in a unique way:

$$\begin{aligned}\boldsymbol{\varepsilon}^i(\mathbf{x}) &= \boldsymbol{\varepsilon}^i(x^1\mathbf{e}_1 + \cdots + x^n\mathbf{e}_n) \\ &= x^1\boldsymbol{\varepsilon}^i(\mathbf{e}_1) + \cdots + x^n\boldsymbol{\varepsilon}^i(\mathbf{e}_n) = x^i\end{aligned}$$

Next, one verifies that the $\boldsymbol{\varepsilon}^i$'s are linearly independent. If

$$\lambda_1\boldsymbol{\varepsilon}^1 + \cdots + \lambda_n\boldsymbol{\varepsilon}^n = 0$$

then for all j, $1 \leq j \leq n$,

$$(\lambda_1\boldsymbol{\varepsilon}^1 + \cdots + \lambda_n\boldsymbol{\varepsilon}^n)(\mathbf{e}_j) = \lambda_j = 0$$

Finally, one shows that all $\phi \in V^*$ are linear combinations of the $\boldsymbol{\varepsilon}^i$'s. For any $\phi \in V^*$, put

$$w_i = \phi(\mathbf{e}_i)$$

Then

$$\phi = w_1\boldsymbol{\varepsilon}^1 + \cdots + w_n\boldsymbol{\varepsilon}^n \tag{9-3}$$

because

$$\begin{aligned}\boldsymbol{\phi}(\mathbf{x}) &= \boldsymbol{\phi}(x^1\mathbf{e}_1 + \cdots + x^n\mathbf{e}_n) = x^1\boldsymbol{\phi}(\mathbf{e}_1) + \cdots x^n\boldsymbol{\phi}(\mathbf{e}_n) \\ &= x^1 w_1 + \cdots + x^n w_n\end{aligned}$$

and

$$\begin{aligned}(w_1\boldsymbol{\varepsilon}^1 + \cdots + w_n\boldsymbol{\varepsilon}^n)(\mathbf{x}) &= (w_1\boldsymbol{\varepsilon}^1 + \cdots + w_n\boldsymbol{\varepsilon}^n)(x^1\mathbf{e}_1 + \cdots + x^n\mathbf{e}_n) \\ &= w_1x^1 + \cdots + w_nx^n\end{aligned}$$

Both sides of Eq. (9-3) represent the same function. The dimensions of V and V^* are identical, and the two spaces are isomorphic (the statement is not true for vector spaces that are not of finite dimension).

In tensor algebra, V is called the space of *contravariant* vectors. In a fixed basis $\{\mathbf{e}_i\}$, contravariant vectors are given by row vectors (x^i) of *contravariant* coordinates. For computations we use not only the matrix notation $\mathbf{x} = (x^i)\{\mathbf{e}_i\}$ but also *Einstein's summation convention:* A repeated index which appears once as a *superscript* and once as a *subscript* implies summation over the whole range of the index:

$$\mathbf{x} = x^i\mathbf{e}_i = x^j\mathbf{e}_j = \sum_{\sigma=1}^{n} x^\sigma\mathbf{e}_\sigma$$

The summation convention for vectors is just matrix notation without parentheses and braces. V^* is the *dual space* to V, $(\boldsymbol{\varepsilon}^i)$ is the dual base to $\{\mathbf{e}_i\}$, and the linear functions

$$\boldsymbol{\phi} = (\boldsymbol{\varepsilon}^i)\{w_i\} = \boldsymbol{\varepsilon}^i w_i$$

are the *covariant vectors.* The value of $\boldsymbol{\phi}(\mathbf{x}) = x^i w_i$ is also written $\mathbf{x} \cdot \boldsymbol{\phi}$ or $\langle \mathbf{x}, \boldsymbol{\phi} \rangle$; it is the *scalar product* defined in $V \times V^*$. In euclidean geometry based on a system of orthonormal vectors $\mathbf{e}_i$ it is possible to identify covariant and contravariant vectors. Any vector $\mathbf{x}$ is also a linear function $\cdot\, \mathbf{x}$ whose value on $\mathbf{y}$ is $\mathbf{y} \cdot \mathbf{x}$, and any linear function on V may be represented as a scalar product by a fixed vector. $\mathbf{y} \cdot \mathbf{x}$ is invariant only with respect to orthogonal transformations. For a scalar product invariant under the action of GL_n the distinction between V and V^* and between superscripts and subscripts becomes essential.

Given a basis $\{\mathbf{e}_i\}$ of V, any other basis $\{\mathbf{e}_{i'}\}$ may be obtained from $\{\mathbf{e}_i\}$ by a linear transformation $A \in GL_n$:

$$\{\mathbf{e}_{i'}\} = A\{\mathbf{e}_i\} \tag{9-4}$$

Here we follow the *kernel-index notation* of J. A. Schouten. The letter $\mathbf{e}$ is chosen once and for all to represent a basis vector in V; it is the kernel of the symbol. The different bases are distinguished by primes or other signs attached to the indices. In our case, $i = 1, 2, \ldots, n$, $i' = 1', 2', \ldots, n'$. The elements of the matrix A are $A_{i'}{}^{i}$. In the Einstein convention, Eq. (9-4) is written

$$\mathbf{e}_{i'} = A_{i'}{}^{j}\mathbf{e}_j \tag{9-4a}$$

The kernel-index notation is the easiest way to avoid an excess of symbols and at the same time to eliminate all ambiguities. We shall use it in general, with two exceptions. If a matrix A is considered not as an element of some vector space of matrices but as a map, the inverse of $A_{i'}{}^{j}$ will be denoted by $(A^{-1})_j{}^{i'}$ and not by $A_j{}^{i'}$. If a matrix or any other quantity is referred to an *invariant* choice of parameters, frames, etc., special symbols will be chosen for their components. The kernel-index method is best suited for the derivation of formulas that are valid in *any* system of coordinates and referred to *any* admissible set of parameters; special symbols will then be used to single out functions, frames, etc., referred to *special* situations and *special* parameters.

A vector $\mathbf{x}$ is an element of V; as such it is given without reference to a basis. But the coordinate vector (x^i) which represents $\mathbf{x}$ in the basis $\{\mathbf{e}_i\}$ clearly depends on the choice of the basis vectors. A change of basis (9-4) induces a change of coordinates by

$$\mathbf{x} = (x^i)\{\mathbf{e}_i\} = (x^{i'})\{\mathbf{e}_{i'}\} = (x^{i'})A\{\mathbf{e}_i\}$$

Hence
$$(x^i) = (x^{i'})A \tag{9-5a}$$
or
$$(x^{i'}) = (x^i)A^{-1} \tag{9-5b}$$
in components
$$x^{i'} = x^i(A^{-1})_i{}^{i'} \tag{9-5c}$$

The dual basis $(\boldsymbol{\varepsilon}^{i'})$ is defined by

$$\boldsymbol{\varepsilon}^{i'}(\mathbf{e}_{j'}) = \delta_{j'}{}^{i'} \qquad (\delta_{j'}{}^{i'}) = U$$

For $(\boldsymbol{\varepsilon}^{i'}) = (\boldsymbol{\varepsilon}^i)B$ we obtain

$$\boldsymbol{\varepsilon}^{i'}(\mathbf{e}_{j'}) = \boldsymbol{\varepsilon}^i(A_{j'}{}^{k}\mathbf{e}_k)B_i{}^{i'} = A_{j'}{}^{k}\boldsymbol{\varepsilon}^i(\mathbf{e}_k)B_i{}^{i'} = A_{j'}{}^{k}B_k{}^{i'} = \delta_{j'}{}^{i'}$$
or
$$AB = U$$
Hence
$$\boldsymbol{\varepsilon}^{i'} = \boldsymbol{\varepsilon}^i(A^{-1})_i{}^{i'} \tag{9-6}$$

Finally, the covariant coordinates of a $\boldsymbol{\phi} \in V^*$ are given by

$$\boldsymbol{\phi} = (\boldsymbol{\varepsilon}^i)\{w_i\} = (\boldsymbol{\varepsilon}^{i'})\{w_{i'}\} = (\boldsymbol{\varepsilon}^i)A^{-1}\{w_{i'}\}$$
hence
$$w_{i'} = A_{i'}{}^{i}w_i \tag{9-7}$$

To sum up: Given a transformation of the basis of V, the covariant coordinate vectors transform by the matrix A of the change of basis (this explains the word *co*variant), and the *contra*variant coordinate vectors and the covariant basis vectors transform by the inverse matrix A^{-1}. The scalar product is invariant,

$$x^{i'}w_{i'} = x^i(A^{-1})_i{}^{i'}A_{i'}{}^{j}w_j = x^iw_i$$

A vector $\mathbf{x} \in V$ may serve as a function $\mathbf{x}^{**}$ on V^* by $\mathbf{x}^{**}(\phi) = \phi(\mathbf{x})$. This shows that V may be identified with $(V^*)^*$ in a natural way.† "Natural" means that the isomorphism $\mathbf{x} \rightarrow \mathbf{x}^{**}$ depends only on the spaces V and V^* but not on the choice of bases in these spaces. On the other hand, we know from a count of dimensions that V is isomorphic to V^*. A possible realization of this isomorphism is $x^i\mathbf{e}_i \rightarrow \boldsymbol{\varepsilon}^i w_i$, $x^i = w_i$. But no such map can be natural if we admit changes of coordinates that are not orthogonal transformations, since the scalar product $\Sigma x^i y^i$ cannot then remain invariant. This fact again emphasizes the unique position of euclidean geometry.

In tensor calculus, the function $\boldsymbol{\varepsilon}^i$ is often denoted by dx^i. A covariant vector $\omega = w_i\, dx^i$ is then called a *differential form* (or *pfaffian*). The notation is very adequate. The total differential of a given function $f(x^1, \ldots ,x^n)$ is the differential form $df = (\partial f/\partial x^j)\, dx^j$ (for the summation convention, differentiation as regards a contravariant coordinate makes the corresponding index a subscript). If $\mathbf{n} = n^i\mathbf{e}_i$ is a unit vector, the directional derivative of f in the direction $\mathbf{n}$ is $D_{\mathbf{n}}f = df(\mathbf{n}) = n^j\, \partial f/\partial x^j$. In general, the value of df on a vector $\mathbf{a}$ is $df(\mathbf{a}) = \mathbf{a}$ grad f. Integrals also fit in. Let $\omega = w_i\, dx^i$ be a differential form whose coefficients $w_i = w_i(\mathbf{x})$ are C^0 functions of a point $\mathbf{x}$ in some region of R^n. Let $\mathbf{f}\colon I \rightarrow R^n$ be a C^0 curve in that region whose coordinate functions are of bounded variation. The integral $\int_{\mathbf{f}} \omega = \lim \Sigma\ \omega(\mathbf{f}(t_{k+1}) - \mathbf{f}(t_k))$ is a Riemann-Stieltjès integral as defined in Sec. 3-1. The differential appears clearly as a function on vectors.

Exercise 9-1

1. Let $\mathbf{e}_1, \mathbf{e}_2, \mathbf{e}_3$ be an orthonormal basis in R^3. Show that the vectors

$$\begin{aligned} \mathbf{e}_{1'} &= 5\mathbf{e}_1 + 3\mathbf{e}_2 \\ \mathbf{e}_{2'} &= \mathbf{e}_1 \qquad\quad - 7\mathbf{e}_3 \\ \mathbf{e}_{3'} &= \mathbf{e}_1 + \mathbf{e}_2 + \mathbf{e}_3 \end{aligned}$$

also form a basis. Find the matrix A and the functions $\boldsymbol{\varepsilon}^{i'}$.

2. Let $\mathbf{e}_1, \mathbf{e}_2, \mathbf{e}_3, \mathbf{e}_4$ be a basis of R^4. Given the vectors

$$\begin{aligned} \mathbf{e}_{1'} &= 3\mathbf{e}_1 + 2\mathbf{e}_2 + \mathbf{e}_3 + 6\mathbf{e}_4 \\ \mathbf{e}_{2'} &= \qquad\quad \mathbf{e}_2 \qquad\quad - \mathbf{e}_4 \\ \mathbf{e}_{3'} &= \mathbf{e}_1 \qquad\qquad\qquad\quad + \mathbf{e}_4 \end{aligned}$$

find some vector $\mathbf{e}_{4'}$ such that the $\mathbf{e}_{i'}$ become a basis in R^4.

† Again this is true only for finite-dimensional vector spaces. For infinite-dimensional spaces it follows only that there exists a natural isomorphism of V into V^{**}.

3. If a vector space has a basis formed by k vectors, show that any $k + 1$ vectors of the space are linearly dependent. [The coordinates of the $k + 1$ vectors form a $(k + 1) \times k$ matrix.]

4. Use Prob. 3 to show that all bases of a finite-dimensional vector space contain the same number of vectors.

5. Show that the real-valued continuous functions defined in $0 \leq s \leq 1$ form a vector space. Is this space finite-dimensional? Show also that on this space the map

$$f \to \int_0^1 f(s)\, ds$$

is a linear function.

◆ 6. What is the coordinate vector of $\mathbf{e}_i$ in the basis $\{\mathbf{e}_j\}$?

7. In the change of basis introduced in Prob. 1, give the $\varepsilon^{i'}$ as functions of the ε^i.

8. In R^3, referred to an orthonormal frame $\{\mathbf{e}_i\}$, is given the differential form $\omega = w_i\, dx^i = x^3\, dx^1 + (x^1)^2\, dx^2 + x^2\, dx^3$.
(*a*) Find the covariant coordinates $w_{i'}$ for the coordinates defined by the frame $\{\mathbf{e}_{i'}\}$ of Prob. 1.
(*b*) Compute the integral of ω along the twisted cubic $\mathbf{f} = (t,t^2,t^3)$ for $0 \leq t \leq 1$.

◆ 9. In R^2 let $\alpha = 2\, dx^1 + 5\, dx^2$ be given. Find all (contravariant) vectors $\mathbf{x}$ such that $\alpha(\mathbf{x}) = 0$.

10. A *subspace* $T^* \subset V^*$ is said to be *orthogonal* to a subspace $S \subset V$ if $\phi(\mathbf{x}) = 0$ for all $\phi \in T^*$ and all $\mathbf{x} \in S$. The biggest subspace of V^* orthogonal to S is called *the* space S' orthogonal to S.
(*a*) Show that S' exists for any $S \subset V$. (One has to show that the set of all functions orthogonal to a set S forms a vector space.)
(*b*) Show that dimension S + dimension S' = dimension V. (This is equivalent to one of the basic results in the theory of systems of linear equations.)

11. (See definitions in Prob. 10.) Prove that if $T \subset S$, then $S' \subset T'$.

12. Let $f(x^1, \ldots ,x^n)$ be a differentiable function of n variables such that df is not identically zero at $x_0^1, \ldots , x_0^n$. What is the R^{n-1} defined by $[df(\mathbf{x}_0)](\mathbf{x}) = 0$?

13. The *transpose* M^t of a map of vector spaces $M: V \to W$ is the map M^t: $W^* \to V^*$ defined by $M^t\phi(\mathbf{x}) = \phi(M\mathbf{x})$ for $\mathbf{x} \in V$, $\phi \in W^*$. Show that for linear maps of finite-dimensional vector spaces given by matrices the transpose map is defined by the transpose matrix.

14. Prove that $(MN)^t = N^tM^t$.

15. Show that a linear map M of a vector space onto itself is orthogonal (i.e., its matrix in any basis is an orthogonal matrix) if and only if $M\phi(M\mathbf{x}) = \phi(\mathbf{x})$ for all $\mathbf{x} \in V$ and all $\phi \in V^*$.

16. Show that $A_j^{i'}(A^{-1})_{i'}^k = \delta_j^k$.

9-2. The Tensor Product

The preceding section was devoted to the study of dual spaces V^* of linear functions on vector spaces V. In this section we study *bilinear* functions on the product $V \times W$ of two vector spaces V, W of dimensions v and w. The product of two sets V, W is the set of ordered couples $(\mathbf{x},\mathbf{y})$, $\mathbf{x} \in V$, $\mathbf{y} \in W$.

Definition 9-1. *A function $F\colon V \times W \to R$ is bilinear if*

$$F(a\mathbf{x}_1 + b\mathbf{x}_2, \mathbf{y}) = aF(\mathbf{x}_1,\mathbf{y}) + bF(\mathbf{x}_2,\mathbf{y})$$
$$F(\mathbf{x}, a\mathbf{y}_1 + b\mathbf{y}_2) = aF(\mathbf{x},\mathbf{y}_1) + bF(\mathbf{x},\mathbf{y}_2)$$

for all $a,b \in R$; $\mathbf{x},\mathbf{x}_1,\mathbf{x}_2 \in V$; $\mathbf{y},\mathbf{y}_1,\mathbf{y}_2 \in W$.

The set of bilinear functions F on $V \times W$ may be given the structure of a vector space by the same definitions that were used for the set of linear functions in Sec. 9-1. The functions in V^* and W^* give rise to bilinear functions on $V \times W$ by

$$\boldsymbol{\phi} \otimes \boldsymbol{\psi}(\mathbf{x},\mathbf{y}) = \boldsymbol{\phi}(\mathbf{x})\boldsymbol{\psi}(\mathbf{y}) \qquad \boldsymbol{\phi} \in V^*, \boldsymbol{\psi} \in W^* \tag{9-8}$$

The symbol $\otimes$ is read "tensor." Any bilinear function on $V \times W$ is a linear combination of tensor products $\boldsymbol{\phi} \otimes \boldsymbol{\psi}$. To make the statement more precise, we choose a basis $\{\mathbf{e}_i\}$ of V and one $\{\mathbf{c}_\sigma\}$ of W. The dual bases are $(\boldsymbol{\varepsilon}^i)$ in V^* and $(\boldsymbol{\gamma}^\sigma)$ in W^*. For any bilinear function F, one may find the real numbers

$$F_{i\sigma} = F(\mathbf{e}_i,\mathbf{c}_\sigma)$$

Then

$$F = F_{i\sigma}\boldsymbol{\varepsilon}^i \otimes \boldsymbol{\gamma}^\sigma \tag{9-9}$$

In fact,

$$F(\mathbf{x},\mathbf{y}) = F(x^i\mathbf{e}_i,y^\sigma\mathbf{c}_\sigma) = x^iy^\sigma F(\mathbf{e}_i,\mathbf{c}_\sigma) = x^iy^\sigma F_{i\sigma}$$
$$F_{i\sigma}\boldsymbol{\varepsilon}^i \otimes \boldsymbol{\gamma}^\sigma(\mathbf{x},\mathbf{y}) = F_{i\sigma}\boldsymbol{\varepsilon}^i(\mathbf{x})\boldsymbol{\gamma}^\sigma(\mathbf{y}) = F_{i\sigma}x^iy^\sigma$$

The set of functions $\boldsymbol{\varepsilon}^i \otimes \boldsymbol{\gamma}^\sigma$ is linearly independent: From $\lambda_{i\sigma}\boldsymbol{\varepsilon}^i \otimes \boldsymbol{\gamma}^\sigma = 0$ it follows that $\lambda_{k\tau} = \lambda_{i\sigma}\boldsymbol{\varepsilon}^i \otimes \boldsymbol{\gamma}^\sigma(\mathbf{e}_k,\mathbf{c}_\tau) = 0$ for all couples of indices.

Definition 9-2. *The vector space of the bilinear functions on $V \times W$ is the tensor product $V^* \otimes W^*$ of V^* and W^*.*

Theorem 9-3. *The dimension of $V^* \otimes W^*$ is the product of the dimensions of V^* and W^*; for any bases $(\boldsymbol{\varepsilon}^i)$ of V^* and $(\boldsymbol{\gamma}^\sigma)$ of W^* the set of functions $\boldsymbol{\varepsilon}^i \otimes \boldsymbol{\gamma}^\sigma$ is a basis of $V^* \otimes W^*$.*

Since V may be identified with V^{**}, the same construction and results will hold if one or more of the given spaces are spaces of covariant vectors. If we retain the previous notation, $V^* \otimes W^*$ is the space of "twice-covariant" tensors $F_{i\sigma}\boldsymbol{\varepsilon}^i \otimes \boldsymbol{\gamma}^\sigma$, and $V \otimes W$ that of "twice-contravariant" tensors $F^{i\sigma}\mathbf{e}_i \otimes \mathbf{c}_\sigma$. Here the vectors $\mathbf{e}_i$, $\mathbf{c}_\sigma$ act as functions on $\boldsymbol{\varepsilon}_j$, $\boldsymbol{\gamma}^\sigma$. There are two spaces of once-covariant, once-contravariant tensors, viz., the space $V^* \otimes W$ of tensors $a_i{}^\sigma\boldsymbol{\varepsilon}^i \otimes \mathbf{c}_\sigma$ and the space $V \otimes W^*$ of tensors

$b_\sigma{}^i \mathbf{e}_i \otimes \boldsymbol{\gamma}^\sigma$. The names characterize the transformation behavior of the *coordinate tensors* $F_{i\sigma}$, $F^{i\sigma}$, $a_i{}^\sigma$, $b_\sigma{}^i$ under changes of bases $\{\mathbf{e}_{i'}\} = A\{\mathbf{e}_i\}$, $\{\mathbf{c}_{\sigma i}\} = B\{\mathbf{c}_\sigma\}$. By formulas (9-4) to (9-7),

$$F_{i\sigma}\boldsymbol{\varepsilon}^i \otimes \boldsymbol{\gamma}^\sigma = F_{i'\sigma'}\boldsymbol{\varepsilon}^{i'} \otimes \boldsymbol{\gamma}^{\sigma'} = \boldsymbol{\varepsilon}^i(A^{-1})_i{}^{i'}F_{i'\sigma'}\boldsymbol{\gamma}^\sigma(B^{-1})_\sigma{}^{\sigma'}$$

Hence

$$(F_{i\sigma}) = A^{-1}(F_{i'\sigma'})B^{-1t}$$

or

$$F_{i'\sigma'} = A_{i'}{}^iB_{\sigma'}{}^\sigma F_{i\sigma} \tag{9-10}$$

In the same way, one obtains

$$F^{i'\sigma'} = (A^{-1})_i{}^{i'}(B^{-1})_\sigma{}^{\sigma'}F^{i\sigma} \tag{9-10a}$$

$$a_{i'}{}^{\sigma'} = A_{i'}{}^i(B^{-1})_\sigma{}^{\sigma'}a_i{}^\sigma \tag{9-10b}$$

$$b_{\sigma'}{}^{i'} = (A^{-1})_i{}^{i'}B_{\sigma'}{}^\sigma b_\sigma{}^i \tag{9-10c}$$

The tensor product of vector spaces is associative, since for three vector spaces U, V, W one has, by defining equation (9-8),

$$\begin{aligned}[\boldsymbol{\phi} \otimes \boldsymbol{\psi}(\mathbf{x},\mathbf{y})] \otimes \boldsymbol{\chi}(\mathbf{z}) &= \boldsymbol{\phi}(\mathbf{x}) \otimes [\boldsymbol{\psi} \otimes \boldsymbol{\chi}(\mathbf{y},\mathbf{z})] \\ &= \boldsymbol{\phi}(\mathbf{x})\boldsymbol{\psi}(\mathbf{y})\boldsymbol{\chi}(\mathbf{z}) \\ &= \boldsymbol{\phi} \otimes \boldsymbol{\psi} \otimes \boldsymbol{\chi}(\mathbf{x},\mathbf{y},\mathbf{z})\end{aligned}$$

where $\mathbf{x} \in U$, $\mathbf{y} \in V$, $\mathbf{z} \in W$, $\boldsymbol{\phi} \in U^*$, $\boldsymbol{\psi} \in V^*$, $\boldsymbol{\chi} \in W^*$. From this it will be clear how one defines p-times covariant, q-times contravariant tensors as vectors in a space of $(p + q)$-linear functions on ordered $(p + q)$-tuples of vectors.

If in Eq. (9-8) one of the arguments is fixed, say $\mathbf{x}$, the result is an element of W^*: With $\boldsymbol{\phi} \otimes \boldsymbol{\psi} \in V^* \otimes W^*$ is associated the function $\boldsymbol{\phi}(\mathbf{x})\boldsymbol{\psi}(\ \ ) \in W^*$. This operation is called the *contraction* of $\boldsymbol{\phi} \otimes \boldsymbol{\psi}$ by $\mathbf{x}$. Matrices of linear maps $V \to W$ are "mixed" tensors of $V^* \otimes W$. The map $M\mathbf{x}$ is the value of the tensor $M = M_i{}^\sigma\boldsymbol{\varepsilon}^i \otimes \mathbf{c}_\sigma$ on the vector $\mathbf{x} \in V$:

$$M\mathbf{x} = M_i{}^\sigma\boldsymbol{\varepsilon}^i \otimes \mathbf{c}_\sigma(\mathbf{x},\ \ ) = M_i{}^\sigma\boldsymbol{\varepsilon}^i(\mathbf{x})\mathbf{c}_\sigma = M_i{}^\sigma x^i\mathbf{c}_\sigma$$

Matrix algebra is a special case of tensor algebra.

In the sequel, we shall deal mostly with tensor products of a space with itself or with its dual. The repeated tensor product of a space with itself will be written as a tensor power, $\overset{p}{\otimes} V = V \otimes V \otimes \cdots \otimes V$, of p spaces V. A tensor *of type* (p,q) is an element of $(\overset{p}{\otimes} V) \otimes (\overset{q}{\otimes} V^*)$. It is a p-times-contravariant, q-times-covariant tensor

$$T = T^{i_1\cdots i_p}{}_{j_1\cdots j_q}\mathbf{e}_{i_1} \otimes \cdots \otimes \mathbf{e}_{i_p} \otimes \boldsymbol{\varepsilon}^{j_1} \otimes \cdots \otimes \boldsymbol{\varepsilon}^{j_q}$$

Such a tensor may be considered as a function on $(\overset{p}{\otimes} V^*) \otimes (\overset{q}{\otimes} V)$. The *contraction* of T by a tensor a of type (r,s), $r \leq q$, $s \leq p$, is a tensor of type $(p - s, q - r)$. It is the following function, for $\boldsymbol{\phi}_j \in V^*$ and $x_j \in V$, on $\boldsymbol{\phi}_1 \otimes \cdots \otimes \boldsymbol{\phi}_{p-s} \otimes \mathbf{x}_1 \otimes \cdots \otimes \mathbf{x}_{q-r} \quad (\boldsymbol{\phi}_j \in V^*, \mathbf{x}_j \in V)$:

$$\begin{aligned} &T^{i_1 \cdots i_p}{}_{j_1 \ldots j_q} \mathbf{e}_{i_1} \otimes \cdots \otimes \mathbf{e}_{i_p} \otimes \boldsymbol{\varepsilon}^{j_1} \otimes \cdots \otimes \boldsymbol{\varepsilon}^{j_q} \\ &\quad (a^{k_1 \cdots k_r}{}_{l_1 \ldots l_s} \mathbf{e}_{k_1} \otimes \cdots \otimes \mathbf{e}_{k_r} \otimes \boldsymbol{\varepsilon}^{l_1} \otimes \cdots \otimes \boldsymbol{\varepsilon}^{l_s} \otimes \boldsymbol{\phi}_1 \otimes \cdots \otimes \boldsymbol{\phi}_{p-s} \otimes \mathbf{x}_1 \otimes \cdots \otimes \mathbf{x}_{q-r}) \\ &\quad = T^{i_1 \cdots i_s i_{s+1} \cdots i_p}{}_{j_1 \ldots j_r j_{r+1} \ldots j_q} a^{j_1 \cdots j_r}{}_{i_1 \ldots i_s} \mathbf{e}_{i_{s+1}} \otimes \cdots \otimes \mathbf{e}_{i_p} \otimes \boldsymbol{\varepsilon}^{j_{r+1}} \otimes \cdots \otimes \boldsymbol{\varepsilon}^{j_q} \\ &\qquad\qquad (\boldsymbol{\phi}_1 \otimes \cdots \otimes \boldsymbol{\phi}_{p-s} \otimes \mathbf{x}_1 \otimes \cdots \otimes \mathbf{x}_{q-r}) \end{aligned}$$

The identity map of V into itself defined by

$$U\mathbf{x} = \mathbf{x}$$

has as its coordinate tensor the Kronecker tensor $\delta_i{}^j$ ($\delta_i{}^i = 1$, $\delta_i{}^j = 0$ for $i \neq j$) in *any* basis. This follows also from Eq. (9-10*b*) in the case $A = B$ (that is, $V = W$):

$$\delta_{i'}{}^{j'} = A_{i'}{}^i (A^{-1})_j{}^{j'} \delta_i{}^j = A_{i'}{}^i (A^{-1})_i{}^{j'} = \begin{cases} 0 & i' \neq j' \\ 1 & i' = j' \end{cases}$$

The contraction of a tensor T of type (p,q) with U is a tensor of type $(p-1, q-1)$, usually called *the* contraction of T,

$$T^{ii_1 \cdots i_{p-1}}{}_{ij_1 \ldots j_{q-1}} \mathbf{e}_{i_1} \otimes \cdots \otimes \mathbf{e}_{i_{p-1}} \otimes \boldsymbol{\varepsilon}^{j_1} \otimes \cdots \otimes \boldsymbol{\varepsilon}^{j_{q-1}}$$

Naturally, it would be possible to prescribe that the contracting tensor should act on any given set of indices, not just on the first ones. Then it is possible to contract any pair of upper and lower indices by the Kronecker tensor.

The contraction of a matrix $M_i{}^j$ is its trace $M_i{}^i$. It is a real number which, by the definition of a general contraction, does not depend on the basis chosen in V. Equation (9-10*b*) shows that the following holds true:

Theorem 9-4. *The traces of a square matrix M and of AMA^{-1} are equal for any nonsingular matrix A.*

If the basis in V is fixed, a tensor may be given by its coordinate tensor, just as a vector is given by its coordinate (row) vector.

Exercise 9-2

1. In a finite-dimensional vector space we associate with any basis $\{\mathbf{e}_i\}$ a system of numbers $T^{i_1 \cdots i_k}{}_{j_1 \ldots i_s}$. Show that these numbers are the coordinates of a tensor of type (k,s) if and only if for any s contravariant vectors $\underset{a}{\mathbf{v}}$ and k covariant vectors $\underset{b}{\boldsymbol{\alpha}}$ the number

$$T^{i_1 \cdots i_k}{}_{j_1 \ldots j_s} \underset{1}{v}^{j_1} \cdots \underset{s}{v}^{j_s} \underset{1}{\alpha}_{i_1} \cdots \underset{k}{\alpha}_{i_k}$$

is independent of the choice of the basis. (The "only if" part is trivial; for the "if" part show that the numbers T obey the right transformation law under changes of basis.)

◆ **2.** Compute $\phi \otimes \psi$ and $\psi \otimes \phi$ for

$$\phi = dx^1 + 5\,dx^2 + 7\,dx^3 \qquad \psi = dx^1 - dx^3$$

in R^{3*}. Write your result (*a*) as a matrix and (*b*) as a quadratic differential form. What are the values of the functions on the basis vectors $\mathbf{e}_i \times \mathbf{e}_j$?

◆ **3.** Find the value of $(7\mathbf{e}_1 + 3\mathbf{e}_2 + \mathbf{e}_3) \otimes (2\mathbf{e}_1 + 5\mathbf{e}_1 - 21\mathbf{e}_3)$ on $(4\boldsymbol{\varepsilon}^1 - 6\boldsymbol{\varepsilon}^2 - 9\boldsymbol{\varepsilon}^3) \times (\boldsymbol{\varepsilon}^1 + 17\boldsymbol{\varepsilon}^2 - 6\boldsymbol{\varepsilon}^3)$.

4. Show that *the* contraction of the tensor product $\mathbf{x} \otimes \boldsymbol{\alpha}$ of a contravariant vector $\mathbf{x}$ and a covariant vector $\boldsymbol{\alpha}$ is the value of $\boldsymbol{\alpha}$ on $\mathbf{x}$.

5. Let M and N be two linear maps of a space V onto itself. The matrix of MN is the matrix product of the two matrices. Show that trace (MN) = trace (NM).

6. Find an example of three matrices for which trace $(ABC) \neq$ trace (ACB).

7. Show that the result of Prob. 5 implies theorem 9-4.

8. Prove that $V \otimes W$ is isomorphic to $W \otimes V$ and that for $V = W$ of dimension >1, $\mathbf{x} \otimes \mathbf{y} \neq \mathbf{y} \otimes \mathbf{x}$ for $\mathbf{x} \neq \mathbf{y}$. [Note that the condition implies that there exists $\boldsymbol{\alpha}$ such that $\boldsymbol{\alpha}(\mathbf{x}) \neq \boldsymbol{\alpha}(\mathbf{y})$.]

9. The *tensor algebra* $\mathfrak{J}(V)$ is the set of all finite linear combinations of elements of R, V, $\overset{2}{\otimes} V, \ldots, \overset{n}{\otimes} V, \ldots$ under vector addition and tensor product as operations. Use Prob. 8 to show that $\mathfrak{J}(V)$ is not commutative if the dimension of V is >1.

10. The *tensor algebra* (Prob. 9) has no *zero divisors*. This means that $\mathbf{x} \otimes \mathbf{y} = 0$ implies that either $\mathbf{x} = 0$ or $\mathbf{y} = 0$. Prove this statement by referral to the definitions.

11. Find the coordinate tensors of the basis vectors $\boldsymbol{\varepsilon}^i \otimes \mathbf{e}_j$ of $R^{3*} \otimes R^3$. Write the tensors in matrix form.

12. The duality of V and V^* may be realized by a tensor

$$\delta = \delta_{ij}\boldsymbol{\varepsilon}^i \otimes \boldsymbol{\varepsilon}^j$$

$\delta_{ii} = 1$, $\delta_{ij} = 0$ for $i \neq j$. In fact, $\delta(\mathbf{e}_k) = \delta_{ij}\boldsymbol{\varepsilon}^i(\mathbf{e}_k)\boldsymbol{\varepsilon}^j = \boldsymbol{\varepsilon}_k$.

(*a*) Compute the coordinates of δ in a basis $\{\mathbf{e}_{i'}\} = A\{\mathbf{e}_i\}$. Find $\delta(\mathbf{e}_{k'})$.

(*b*) Show that the coordinate transformations A which leave invariant the coordinates of δ form a group. Find this group as a subgroup of GL_n.

13. Any matrix g_{ij} may be used to define a tensor $g = g_{ij}\boldsymbol{\varepsilon}^i \otimes \boldsymbol{\varepsilon}^j$. Prove that $g(\mathbf{x}) = g_{ij}\boldsymbol{\varepsilon}^i(\mathbf{x})\boldsymbol{\varepsilon}^j$ is an isomorphism $V \to V^*$ if and only if $\|g_{ij}\| \neq 0$.

14. (Sequel to Prob. 13.) If $\|g_{ij}\| \neq 0$ and $g_{ij} = g_{ji}$, find a new basis $\{\mathbf{e}_{i'}\}$ such that the coordinate tensor of g becomes $\delta_{i'j'}$.

15. Define the matrix product of two matrices M_i^j, N_i^j in terms of tensor operations on $M = M_i^j\varepsilon^i \otimes \mathbf{e}_j$ and $N = N_i^j\varepsilon^i \otimes \mathbf{e}_j$.

16. Let T be a tensor of type (q,q). Show that the contraction of T by $\overset{q}{\otimes} U$ is a number independent of the basis in which the coordinates of T are expressed. Write this number in terms of the coordinates of T.

17. For the two tensors of Prob. 15, find the coordinate tensor of $M \otimes N$.

18. The tensor $M \otimes N$ constructed in Prob. 17 can be viewed, in the case of $n \times n$ matrices M_i^j and N_i^j, as representing a linear map of $V \otimes V$ into itself. Show that here

$$\text{trace } (M \otimes N) = \text{trace } M \cdot \text{trace } N$$

(Introduce a new, linearly ordered numbering of the basis vectors of $V \otimes V$.)

19. A tensor $T \in \overset{2}{\otimes} V^*$ is *symmetric* if $T(\mathbf{x},\mathbf{y}) = T(\mathbf{y},\mathbf{x})$ for all $\mathbf{x},\mathbf{y} \in V$. Prove that

(*a*) T is symmetric if and only if $T_{ij} = T_{ji}$ for all pairs of indices.

(*b*) If $T_{ij} = T_{ji}$ for all pairs of indices, then $T_{i'j'} = T_{j'i'}$ holds in any basis.

20. If both T and S are symmetric twice-covariant tensors such that

$$T_{ij}S_{kl} - T_{il}S_{jk} + T_{jk}S_{il} - T_{kl}S_{ij} = 0$$

prove that there exists a real number c such that $T_{ij} = cS_{ij}$ (Schouten's theorem).

21. Show that a mixed tensor of coordinates P_i^j is of the form

$$P_i^j = c\delta_i^j + p_iq^j$$

if and only if $P_i^jv^i = cv^j$ for all vectors (v^j) orthogonal (exercise 9-1, Prob. 10) to $\{p_j\}$. (The tensor $P_i^j - c\delta_i^j$ is orthogonal to *the* orthogonal space of $\{p_j\}$.)

22. If $aT_{ij} + bT_{ji} = 0$ holds for all coordinates of a twice-covariant tensor with coefficients independent of the basis, prove that either $ab = 0$ or $(a/b)^2 = 1$.

23. Let $D^1(\mathbf{x}_0)$ be the space of differentials $df = (\partial f/\partial x^i)\, dx^i$ of C^∞ functions $f(x^1, \ldots ,x^n)$, the coefficients being evaluated at a point $\mathbf{x}_0 \in R^n$. In the same way, $D^2(\mathbf{x}_0)$ is the space of second differentials

$$d^2f = \frac{\partial^2 f}{\partial x^i\, \partial x^j}\, dx^i \otimes dx^j$$

evaluated at $\mathbf{x}_0$. Show that $D^1(\mathbf{x}_0) \otimes D^1(\mathbf{x}_0) = D^2(\mathbf{x}_0)$ but that $D^1(\mathbf{x}) \otimes D^1(\mathbf{x}) \subset D^2(\mathbf{x})$ if $\mathbf{x}$ is allowed to vary in an open domain of

R^n. (In the first case the coefficients of the tensors are numbers; in the second, they are functions.)

24. The product $V \times W$ of two vector spaces may be given the structure of a vector space by the definitions

$$(\mathbf{x}_1 + \mathbf{x}_2, \mathbf{y}) = (\mathbf{x}_1,\mathbf{y}) + (\mathbf{x}_2,\mathbf{y})$$
$$(\mathbf{x}, \mathbf{y}_1 + \mathbf{y}_2) = (\mathbf{x},\mathbf{y}_1) + (\mathbf{x},\mathbf{y}_2)$$
$$a(\mathbf{x},\mathbf{y}) = (a\mathbf{x},\mathbf{y}) = (\mathbf{x},a\mathbf{y})$$

Prove that the vector space defined in this way is isomorphic to $V \otimes W$.

25. An *isomer* of a tensor is a quantity obtained from a tensor by the permutation of basis vectors. For instance, from

$$T = T_{ijk}{}^l \boldsymbol{\varepsilon}^i \otimes \boldsymbol{\varepsilon}^j \otimes \boldsymbol{\varepsilon}^k \otimes \mathbf{e}_l$$

one may obtain $S = S^{\;l}_{i\cdot kj} \boldsymbol{\varepsilon}^i \otimes \mathbf{e}_l \otimes \boldsymbol{\varepsilon}^k \otimes \boldsymbol{\varepsilon}^j$ by $T_{ijk}{}^l = S^{\;l}_{i\cdot kj}$. Prove that any isomer of a tensor is a tensor of the same type (in another tensor space; see Prob. 8).

9-3. Exterior Calculus

From now on, all tensor products are on a fixed vector space V and its dual V^*.

The tensor product is not commutative (see exercise 9-2, Prob. 9). We study the deviation from commutativity. For two vectors, we put

$$2\mathbf{x} \wedge \mathbf{y} = \mathbf{x} \otimes \mathbf{y} - \mathbf{y} \otimes \mathbf{x} \tag{9-11}$$

(to be read "x exterior y" or "x wedge y"). The definition may be generalized to products of k vectors. The *generalized Kronecker symbol* is

$$\delta^{i_1 \cdots i_k}_{j_1 \cdots j_k} = \begin{cases} 0 & \text{if } (i_1, \ldots, i_k) \text{ is not a permutation of } (j_1, \ldots, j_k) \\ +1 & \text{if } (i_1, \ldots, i_k) \text{ is an } \textit{even} \text{ permutation of } (j_1, \ldots, j_k) \\ -1 & \text{if } (i_1, \ldots, i_k) \text{ is an } \textit{odd} \text{ permutation of } (j_1, \ldots, j_k) \end{cases}$$

The notions of even and odd permutations are the same as those used for the definition of determinants. Defining equation (9-11) may be written

$$2\mathbf{x}_1 \wedge \mathbf{x}_2 = \delta^{i_1 i_2}_{1\ 2} \mathbf{x}_{i_1} \otimes \mathbf{x}_{i_2}$$

The definition of the exterior product of k vectors is

$$k!\mathbf{x}_1 \wedge \cdots \wedge \mathbf{x}_k = \delta^{i_1 \cdots i_k}_{1 \cdots k} \mathbf{x}_{i_1} \otimes \cdots \otimes \mathbf{x}_{i_k} \tag{9-12}$$

This is an average of the function $\mathbf{x}_1 \otimes \cdots \otimes \mathbf{x}_k$ over all $k!$ permutations of the indices $1, \ldots, k$.

The exterior product is distributive. The finite sums of exterior products of k vectors form a vector space $\overset{k}{\wedge} V$ which is a subspace of $\overset{k}{\otimes} V$. If V is a space of contravariant vectors, the elements of $\overset{k}{\wedge} V$ are *k-vectors;* for V as a space of covariant vectors they are *k-forms.* A k-vector (or k-form) which is the exterior product of k vectors (or k pfaffians) is said to *split.* The exterior product will be associative if we define the exterior product of split vectors $X = \mathbf{x}_1 \wedge \cdots \wedge \mathbf{x}_j$ and $Y = \mathbf{y}_1 \wedge \cdots \wedge \mathbf{y}_k$ as

$$X \wedge Y = \mathbf{x}_1 \wedge \cdots \wedge \mathbf{x}_j \wedge \mathbf{y}_1 \wedge \cdots \wedge \mathbf{y}_k \tag{9-13}$$

The product of multivectors

$$f = f^{i_1 \cdots i_k} \mathbf{x}_{i_1} \wedge \cdots \wedge \mathbf{x}_{i_k} \qquad \text{and} \qquad g = g^{j_1 \cdots j_l} \mathbf{y}_{j_1} \wedge \cdots \wedge \mathbf{y}_{j_l}$$

is

$$f \wedge g = f^{i_1 \cdots i_k} g^{j_1 \cdots j_l} \mathbf{x}_{i_1} \wedge \cdots \wedge \mathbf{x}_{i_k} \wedge \mathbf{y}_{j_1} \wedge \cdots \wedge \mathbf{y}_{j_l}$$

An important property of the exterior product is that a split vector changes sign if two of its vectors are permuted:

$$\begin{aligned} &\mathbf{x}_1 \wedge \cdots \wedge \mathbf{x}_{s-1} \wedge \mathbf{x}_s \wedge \mathbf{x}_{s+1} \wedge \mathbf{x}_{s+2} \wedge \cdots \wedge \mathbf{x}_k \\ &\qquad = -\mathbf{x}_1 \wedge \cdots \wedge \mathbf{x}_{s-1} \wedge \mathbf{x}_{s+1} \wedge \mathbf{x}_s \wedge \mathbf{x}_{s+2} \wedge \cdots \wedge \mathbf{x}_k \end{aligned} \tag{9-14a}$$

or, what amounts to the same,

$$\mathbf{x} \wedge \mathbf{x} = 0 \tag{9-14b}$$

for vectors. This is an immediate consequence of the definition of the generalized Kronecker symbol. For multivectors the commutation rule is

$$f \wedge g = (-1)^{kl} g \wedge f \tag{9-15}$$

A basis $\{\mathbf{e}_i\}$ $(i = 1, \ldots, v)$ of V induces a basis $\mathbf{e}_{i_1} \otimes \cdots \otimes \mathbf{e}_{i_k}$ of $\overset{k}{\otimes} V$; by (9-12) there corresponds to it a basis of $\overset{k}{\wedge} V$

$$e_{i_1 \ldots i_k} = \mathbf{e}_{i_1} \wedge \cdots \wedge \mathbf{e}_{i_k} \qquad i_1 < i_2 < \cdots < i_k \tag{9-16}$$

The ordering of the indices assures that each basis element appears once and only once. As in the preceding sections, one shows that the tensors $e_{i_1 \ldots i_k}$ are linearly independent for different (ordered) sets of indices. The number of basis k vectors of $\overset{k}{\wedge} V$ is the number $\binom{v}{k}$ of sets of k elements taken out of v. Since $\binom{v}{k} = \binom{v}{v-k}$, $\overset{k}{\wedge} V$ and $\overset{v-k}{\wedge} V$ are isomorphic vector spaces, and the isomorphism may be realized by

$$*e_{i_1 \ldots i_k} = \delta^{1 \cdots \cdots v}_{i_1 \cdots i_k j_1 \cdots j_{v-k}} e_{j_1 \cdots j_{v-k}} \tag{9-17}$$

We define $\overset{0}{\wedge} V = R$, $\overset{1}{\wedge} V = V$. If $k > v$, $\overset{k}{\wedge} V = 0$ since it cannot have nonvanishing basis vectors, by (9-16). The union of all spaces of k-vectors over V,

$$\wedge V = \overset{0}{\wedge} V + \overset{1}{\wedge} V + \cdots + \overset{v}{\wedge} V$$

is an algebra under vector addition and the exterior product (9-13); it is the *Grassmann algebra* of V.

The exterior product of two vectors $\mathbf{x} = x^1\mathbf{e}_1 + x^2\mathbf{e}_2$, $\mathbf{y} = y^1\mathbf{e}_1 + y^2\mathbf{e}_2$ in R^2 is

$$\begin{aligned}\mathbf{x} \wedge \mathbf{y} &= x^1y^2\mathbf{e}_1 \wedge \mathbf{e}_2 + x^2y^1\mathbf{e}_2 \wedge \mathbf{e}_1 \\ &= (x^1y^2 - x^2y^1)e_{12} = \|\mathbf{x},\mathbf{y}\|e_{12}\end{aligned}$$

This gives a new definition for the determinant of the coordinate vectors,

$$\|\mathbf{x},\mathbf{y}\| = *(\mathbf{x} \wedge \mathbf{y})$$

if the defining equation (9-17) is extended to include $*e_{1\ldots v} = 1$. The same relation holds good for n vectors in R^n:

$$\mathbf{x}_1 \wedge \cdots \wedge \mathbf{x}_n = \|\mathbf{x}_1, \ldots, \mathbf{x}_n\|e_{1\ldots n} \tag{9-18}$$

In fact, the determinant of n coordinate vectors in R^n is the *unique* function of n vectors which satisfies the following (see exercise 9-3, Prob. 22):

I. The determinant is a multilinear function of the n vectors.

II. $\|\mathbf{x}_1, \ldots, \mathbf{x}_n\| = 0$ if two vectors are equal.

III. $\|\mathbf{e}_1, \ldots, \mathbf{e}_n\| = 1$.

All three properties are easily verified for Eq. (9-18).

For the remainder of this section we deal with differential forms. (The first two theorems are equally valid for vectors.)

Theorem 9-5. *r pfaffians $\omega^1, \ldots, \omega^r$ are linearly dependent if and only if $\omega^1 \wedge \omega^2 \wedge \cdots \wedge \omega^r = 0$.*

If the $\omega^i (i = 1, \ldots, r)$ are linearly independent, there exist covariant vectors $\omega^{r+1}, \ldots, \omega^v$ such that $\omega^1, \ldots, \omega^v$ form a basis of V^*. Then we know that $\|\omega^1, \ldots, \omega^v\| \neq 0$; hence

$$\begin{aligned}\omega^1 \wedge \cdots \wedge \omega^v &= \|\omega^1, \ldots, \omega^v\|\epsilon_{1\ldots v} \\ &= \|\omega^1, \ldots, \omega^v\|\, dx^1 \wedge dx^2 \wedge \cdots \wedge dx^v \neq 0\end{aligned}$$

and, a fortiori, $\omega^1 \wedge \cdots \wedge \omega^r \neq 0$. On the other hand, if the ω^i are linearly dependent, one may assume without loss of generality that $\omega^1 = \sum_2^r c_j\omega^j$; hence $\omega^1 \wedge \cdots \wedge \omega^r = 0$, by (9-14).

Theorem 9-6 (Elie Cartan). *If $\omega^1, \ldots, \omega^r$ are r linearly independent pfaffians in V^* and if there exist r pfaffians $\pi^1, \ldots, \pi^r \in V^*$ such that*

$\sum_{s=1}^{r} \omega^s \wedge \pi^s = 0$, *then there exists an* $r \times r$ *symmetric matrix* $(c_t{}^s)$ *such that*

$\pi^s = \sum_{t=1}^{r} c_t{}^s \omega^t$.

As before, a basis $\omega^1, \ldots, \omega^r, \omega^{r+1}, \ldots, \omega^v$ may be found for V^*. By theorem 9-5, $\omega^i \wedge \omega^j \neq 0$ for $i \neq j$. There exist numbers $c_t{}^s$ such that $\pi^s = c_t{}^s \omega^t (s = 1, \ldots, r; t = 1, \ldots, v)$. By hypothesis,

$$\sum_{s=1}^{r} c_t{}^s \omega^s \wedge \omega^t = 0$$

hence $c_t{}^s = 0$ for $t > r$, and

$$\sum_{s=1}^{r} c_t{}^s \omega^s \wedge \omega^t = \sum_{s=1}^{r} \sum_{\substack{t=1 \\ s<t}}^{r} (c_t{}^s - c_s{}^t) \omega^s \wedge \omega^t = 0$$

or $c_t{}^s = c_s{}^t$.

In previous chapters we studied matrix functions, i.e., maps of a parameter interval into a matrix group G. Here we discuss *differential k-forms*. These are differentiable maps of open domains $\mathfrak{U} \subset R^n$ into the space $\overset{k}{\wedge} R^{n*}$ of k-forms in n differentials

$$\omega^{(k)}: \mathfrak{U} \to \overset{k}{\wedge} R^{n*}$$

Explicitly

$$\omega^{(k)} = \omega_{i_1 \ldots i_k}(x^1, \ldots, x^n)\, dx^{i_1} \wedge \cdots \wedge dx^{i_k} \tag{9-19}$$

the $\omega_{i_1 \ldots i_k}$ being differentiable functions of the point $(x^1, \ldots, x^n) \in \mathfrak{U} \subset R^n$. Any such differential k-form gives rise to a differential $(k + 1)$-form

$$d\omega^{(k)}: \mathfrak{U} \to \overset{k+1}{\wedge} R^{n*}$$

by

$$\begin{aligned} d\omega^{(k)} &= (d\omega_{i_1 \ldots i_k}) \wedge dx^{i_1} \wedge \cdots \wedge dx^{i_k} \\ &= \frac{\partial \omega_{i_1 \ldots i_k}}{\partial x^j} dx^j \wedge dx^{i_1} \wedge \cdots \wedge dx^{i_k} \end{aligned} \tag{9-20}$$

The *exterior differential* $d\omega^{(k)}$ is the sum of the exterior products of the total differentials of the coordinate functions $\omega_{i_1 \ldots i_k}$ with the basis vectors $dx^{i_1} \wedge \cdots \wedge dx^{i_k}$. We have emphasized before that the coordinate tensor has no intrinsic meaning, only the k-form built on it. Furthermore, the definition of the exterior differential uses the coordinate tensor in a very special way, and it is possible to mix the classical total differential with the k-forms only by the trick of a formal substitution $\varepsilon^i \to dx^i$. Therefore it is necessary to show that a calculus based on (9-20) makes sense. This we do by proving that the operation d defined in Eq. (9-20) is invariant in a differentiable change of coordinates; it has geometric meaning.

If two systems of coordinates are given, each in an open domain of R^n, a change of coordinates is a differentiable homeomorphism $\mathbf{f}: \mathcal{V} \to \mathcal{U}$ of a domain $\mathcal{V} \subset R^n$ onto a domain $\mathcal{U} \subset R^n$. Both $\mathbf{f}(\mathbf{y})$, $\mathbf{y} \in \mathcal{V}$ and $\mathbf{f}^{-1}(\mathbf{x})$, $\mathbf{x} \in \mathcal{U}$ are one-to-one differentiable functions. By the inverse-function theorem, the jacobian matrix $J_y\mathbf{f} = (\partial f^i/\partial y^j)$ has a determinant $\neq 0$ everywhere in $\mathcal{V}$. The differential form which results from $\omega^{(k)}$ by the substitution $\mathbf{x} = \mathbf{f}(\mathbf{y})$ is a map

$$f^*\omega^{(k)}: \mathcal{V} \to \overset{k}{\wedge} R^{n*}$$

defined by

$$\mathbf{f}^*\omega^{(k)}(\mathbf{y},d\mathbf{y}) = \omega^{(k)}(\mathbf{f}(\mathbf{y}),d\mathbf{f}(\mathbf{y})) \tag{9-21}$$

or, explicitly,

$$\begin{aligned} f^*\omega^{(k)}(\mathbf{y},d\mathbf{y}) &= \omega_{i_1 \cdots i_k}(f^1(\mathbf{y}), \ . \ . \ . \ ,f^n(\mathbf{y}))\, df^{i_1}(\mathbf{y}) \wedge \cdots \wedge df^{i_n}(\mathbf{y}) \\ &= \omega_{i_1 \cdots i_k} \frac{\partial f^{i_1}}{\partial y^{j_1}} \cdots \frac{\partial f^{i_k}}{\partial y^{j_k}}\, dy^{j_1} \wedge \cdots \wedge dy^{j_k} \end{aligned}$$

Here again we have used the total differential of the mapping functions f^i. The next theorem states that the exterior differential of the transformed form is equal to the transformed form of the exterior differential.

Theorem 9-7. $df^*\omega = f^*\, d\omega$.

The proof depends on two important formulas. By the defining equation (9-20),

$$\begin{aligned} dd\omega^{(k)} &= \frac{\partial^2 \omega_{i_1 \cdots i_k}}{\partial x^l\, \partial x^j}\, dx^l \wedge dx^j \wedge dx^{i_1} \wedge \cdots \wedge dx^{i_k} \\ &= -\frac{\partial^2 \omega_{i_1 \cdots i_k}}{\partial x^j\, \partial x^l}\, dx^j \wedge dx^l \wedge dx^{i_1} \wedge \cdots \wedge dx^{i_k} \\ &= -dd\omega^{(k)} \end{aligned}$$

Hence

$$dd\omega^{(k)} = 0 \tag{9-22}$$

This result is known as *Poincaré's formula.* Another consequence of Eq. (9-20) is

$$d(\omega^{(k)} \wedge \pi^{(l)}) = d\omega^{(k)} \wedge \pi^{(l)} + (-1)^k \omega^{(k)} \wedge d\pi^{(l)} \tag{9-23}$$

To understand the sign $(-1)^k$ in this formula one has only to point out that, by definition, the differential resulting from the total differentiation must be *first*, whereas in $\omega^{(k)} \wedge d\pi^{(l)}$ it is written only after the k differentials of $\omega^{(k)}$.

If in (9-20) the variable $\mathbf{x} = (x^1, \ . \ . \ . \ ,x^n)$ is a function of $\mathbf{y}$,

$$dx^i = \frac{\partial x^i}{\partial y^j}\, dy^j$$

then, by Eqs. (9-22) and (9-23),

$$\begin{aligned} df^*\omega &= d(\omega_{i_1\ldots i_k}\, dx^{i_1} \wedge \cdots \wedge dx^{i_k}) \\ &= d\omega_{i_1\ldots i_k} \wedge dx^{i_1} \wedge \cdots \wedge dx^{i_k} + \omega_{i_1\ldots i_k}\, d(dx^{i_1} \wedge \cdots \wedge dx^{i_k}) \\ &= d\omega_{i_1\ldots i_k} \wedge dx^{i_1} \wedge \cdots \wedge dx^{i_k} \\ &= f^*\, d\omega \end{aligned}$$

The invariance theorem 9-7 is important in the theory of multiple integrals. For a double integral in the plane $\iint F(x^1,x^2)\, dx^1\, dx^2$ we study the two-form $\omega^{(2)} = F(x^1,x^2)\, dx^1 \wedge dx^2$. Changing coordinates by

$$dx^1 = \frac{\partial x^1}{\partial y^1}\, dy^1 + \frac{\partial x^1}{\partial y^2}\, dy^2$$
$$dx^2 = \frac{\partial x^2}{\partial y^1}\, dy^1 + \frac{\partial x^2}{\partial y^2}\, dy^2$$

one has, by (9-18),

$$f^*\omega^{(2)} = F(y^1,y^2)\,\frac{\partial(x^1,x^2)}{\partial(y^1,y^2)}\, dy^1 \wedge dy^2 = F(y^1,y^2)\|J_y\mathbf{x}\|\, dy^1 \wedge dy^2$$

This is just the substitution rule for the change of variables in double integrals! One sees that in reality integration is not on functions but on differential forms.

The integral of an n-form $\omega^{(n)} = \omega_{1\ldots n}\, dx^1 \wedge \cdots \wedge dx^n$ over the unit cube I^n $(0 \leq x^j \leq 1, j = 1, \ldots, n)$ in R^n is best defined as a kind of Stieltjès integral. Each unit interval is subdivided,

$$0 = x_0{}^j < x_1{}^j < \cdots < x_{N_j}{}^j = 1$$

This subdivision creates small cubes of edges $\Delta x_k{}^j = x_{k+1}{}^j - x_k{}^j$. The value of $dx^1 \wedge \cdots \wedge dx^n$ on $\Delta x_{k_1}{}^1 \wedge \cdots \wedge \Delta x_{k_n}{}^n$ by (9-18) is the volume of the cube of sides $\Delta x_{k_i}{}^i$, $i = 1, \ldots, n$. For a lattice of intermediate points of coordinates $\xi_{k_i}{}^i$, $x_{k_i}{}^i \leq \xi_{k_i}{}^i \leq x^i_{k_{i+1}}$, one has a Riemann sum

$$\Sigma\omega_{1\ldots n}(\xi_{k_1}{}^1, \ldots, \xi_{k_n}{}^n)\, dx^1 \wedge \cdots \wedge dx^n\, (\Delta x_{k_1}{}^1, \ldots, \Delta x_{k_n}{}^n)$$

The integral

$$\int_{I^n} \omega^{(n)} = \int_{0\cdots}^{1\cdots} \int_0^1 \omega_{1\ldots n}\, dx^1 \wedge \cdots \wedge dx^n$$

is the limit of the Riemann sums if all the mesh sizes tend to zero. Its existence may be shown by methods analogous to those of Sec. 3-1. Theorem 9-7 implies that the integral is also defined for all piecewise differentiable homeomorphic images of the unit cube and that its value is unchanged if the total differentials of a coordinate transformation are treated as exterior differentials.

If $\omega^{(n)} = d\alpha^{(n-1)}$, one defines the integral of

$$\alpha^{(n-1)} = \alpha_{1\ldots\hat{\imath}\ldots n}\, dx^1 \wedge \cdots d\hat{x}^i \cdots \wedge dx^n\dagger$$

over the boundary ∂I^n of the unit cube as

$$\int_{\partial I^n} \alpha^{(n-1)} = \sum_{i=1}^{n} \int_{0\cdots}^{1\cdots} \int_0^1 \alpha_{1\ldots\hat{\imath}\ldots n}(x^1, \ldots, x^{i-1}, 1, x^{i+1}, \ldots, x^n)\, dx^1 \wedge \cdots d\hat{x}^i \cdots \wedge dx^n$$

$$- \sum_{i=1}^{n} \int_{0\cdots}^{1\cdots} \int_0^1 \alpha_{1\ldots\hat{\imath}\ldots n}(x^1, \ldots, x^{i-1}, 0, x^{i+1}, \ldots, x^n)\, dx^1 \wedge \cdots d\hat{x}^i \cdots \wedge dx^n$$

$\omega^{(n)} = d\alpha^{(n-1)}$ means

$$\omega_{1\ldots n} = \sum_{i=1}^{n} (-1)^{i-1} \frac{\partial \alpha_{1\ldots\hat{\imath}\ldots n}}{\partial x^i}$$

Hence, by integration by parts,

$$\int_{I^n} d\alpha^{(n-1)} = \int_{\partial I^n} \alpha^{(n-1)} \tag{9-24}$$

This formula, known as *Stokes's formula*, is the basis of the theory of closed surfaces and manifolds. By theorem 9-7, it is valid also for all differentiable homeomorphic images of the interior of the unit cube.

The use of exterior calculus is explained in some examples.

***Example* 9-1.** Let ω_1, ω_2 be two linear differential forms (pfaffians) in two variables x^1, x^2; they are maps $R^2 \to R^{2*}$. With a third variable θ we put

$$\pi_1 = \cos\theta\, \omega_1 + \sin\theta\, \omega_2 \qquad \pi_2 = -\sin\theta\, \omega_1 + \cos\theta\, \omega_2$$

By exterior differentiation,

$$d\pi_1 = d\theta \wedge (-\sin\theta\, \omega_1 + \cos\theta\, \omega_2) + \cos\theta\, d\omega_1 + \sin\theta\, d\omega_2$$
$$d\pi_2 = -d\theta \wedge (\cos\theta\, \omega_1 + \sin\theta\, \omega_2) - \sin\theta\, d\omega_1 + \cos\theta\, d\omega_2$$

Now assume ω_1, ω_2 to be linearly independent in every point of their domain of definition. Then $\pi_1 \wedge \pi_2 = \omega_1 \wedge \omega_2 \neq 0$ may serve as the basis for $\overset{2}{\wedge} R^{2*}$, and one may write

$$d\omega_1 = a(x^1,x^2)\omega_1 \wedge \omega_2 \qquad d\omega_2 = b(x^1,x^2)\omega_1 \wedge \omega_2$$

There exists a linear form π_{12} such that

$$d\pi_1 = \pi_{12} \wedge \pi_2 \qquad d\pi_2 = -\pi_{12} \wedge \pi_1$$

viz., $\pi_{12} = (a\cos\theta + b\sin\theta)\pi_1 - (a\sin\theta - b\cos\theta)\pi_2 + d\theta$

† The symbol ^ indicates that the quantity under it *must be left out.*

A new differentiation gives, by (9-14),

$$d\pi_{12} \wedge \pi_2 = d\pi_{12} \wedge \pi_1 = 0$$

By theorem 9-5, $d\pi_{12}$ is linearly dependent on π_1 and π_2 (hence also on ω_1 and ω_2). This means that $d\pi_{12}$ cannot contain any term in $d\theta$; $d\pi_{12} = -K(x^1,x^2,\theta)\pi_1 \wedge \pi_2$. By (9-22), $0 = -dK \wedge \pi_1 \wedge \pi_2$. Again dK cannot contain any term in $d\theta$; $\partial K/\partial\theta = 0$, $K = K(x^1,x^2)$ *does not depend on* θ. This result will be important in the theory of surfaces.

***Example* 9-2.** Theorem 9-7 indicates that differentials in k-forms may be treated as total differentials of functions. For a one-parameter group in R^n with vector Ξ tangent to its trajectories we want to compute the Lie derivative of a pfaffian $\omega = \omega_i\, dx^i$. According to the principles of tensor algebra, the vector Ξ has contravariant coordinates Ξ^i. It is convenient to indicate by $i(\Xi)$ the contraction of a tensor by a vector Ξ. In this notation, the Lie derivative of a function as defined in Sec. 7-1 may be written

$$£_\Xi f = i(\Xi)\, df = \Xi^i \frac{\partial f}{\partial x^i}$$

For the Lie derivative of the linear form one has accordingly

$$\begin{aligned} £_\Xi\omega &= (£_\Xi\omega_i)\, dx^i + \omega_i\, d£_\Xi x^i \\ &= \left(\Xi^j \frac{\partial\omega_i}{\partial x^j} + \omega_j \frac{\partial\Xi^j}{\partial x^i}\right) dx^i \end{aligned}$$

Since

$$i(\Xi)\, d\omega = \left(\Xi^j \frac{\partial\omega_i}{\partial x^j} - \Xi^j \frac{\partial\omega_j}{\partial x^i}\right) dx^i$$

one sees that

$$£_\Xi\omega = i(\Xi)\, d\omega + di(\Xi)\omega \tag{9-25}$$

Exercise 9-3

◆ **1.** Compute $\delta^{k\ \cdot\ n\ 1\ \cdot\ k-1}_{1\ \cdot\ \cdot\ \cdot\ \cdot\ n}$.

◆ **2.** Compute $\delta^{1\ \cdot\ \cdot\ \cdot\ \cdot\ \cdot\ 2n}_{1\ n+1\ 2\ n+2\ \cdot\ n\ 2n}$.

3. Show that $**\alpha^{(k)} = (-1)^{k(n-k)}\alpha^{(k)}$. (Use Prob. 1.)

◆ **4.** Compute the following exterior products:

(*a*) $(x^1\, dx^1 + 7(x^3)^2\, dx^2) \wedge (x^2\, dx^1 - \sin 3x^1\, dx^2 + dx^3)$

(*b*) $(5\, dx^1 + 3\, dx^2) \wedge (3\, dx^1 + 2\, dx^2)$

(*c*) $(6\, dx^1 \wedge dx^2 + 27\, dx^1 \wedge dx^3) \wedge (dx^1 + dx^2 + dx^3)$

◆ **5.** Compute the following exterior differentials:

(*a*) $d(\cos x^2\, dx^1 - \sin x^1\, dx^2)$

(*b*) $d(2x^1x^2\, dx^1 + (x^1)^2\, dx^2)$

(*c*) $d(6x^3\, dx^1 \wedge dx^2 - x^1x^2\, dx^1 \wedge dx^3)$

6. Use theorem 9-5 to check the linear dependence or independence of the vectors

$$\mathbf{a}_1 = \mathbf{e}_1 + 3\mathbf{e}_3 + 5\mathbf{e}_4 + 7\mathbf{e}_5 \qquad \mathbf{a}_2 = 2\mathbf{e}_1 + \mathbf{e}_2 + 6\mathbf{e}_3 + 3\mathbf{e}_4 + 3\mathbf{e}_5$$
$$\mathbf{a}_3 = 8\mathbf{e}_1 + 4\mathbf{e}_2 - 2\mathbf{e}_3 + 6\mathbf{e}_5 \qquad \mathbf{a}_4 = 5\mathbf{e}_1 + 3\mathbf{e}_2 - 11\mathbf{e}_3 - 8\mathbf{e}_4 - 4\mathbf{e}_5$$

7. Use theorem 9-5 and Eq. (9-18) to prove that the number of linearly independent row vectors of a matrix M is equal to the maximal order of a minor with a nonzero determinant extracted from M.

8. Verify the converse of theorem 9-6.

9. Formula (6-29) defines a vector of invariant differential forms on a Lie group: $(\omega^i) = d\mathbf{x}\,Q(\mathbf{x})$, $\omega^i(\mathbf{x},d\mathbf{x}) = \omega^i(\mathbf{xa},d\mathbf{xa})$.

(*a*) Show that $d\omega^i = c_{jk}{}^i(\mathbf{x})\omega^j \wedge \omega^k$.

(*b*) Show that $c_{jk}{}^i(\mathbf{x}) = c_{jk}{}^i(\mathbf{xa})$; hence the $c_{jk}{}^i$ are constants.

(*c*) Show that the Jacobi identity (6-24) is a consequence of Poincaré's formula for the ω^i.

NOTE: The equations $d\omega^i = c_{jk}{}^i\omega^j \wedge \omega^k$ are the *Maurer-Cartan* equations of the Lie group.

10. Poincaré's formula has a converse: *If* $d\omega^{(k)} = 0$ *in* $\mathfrak{U}$, *then for all* $\mathbf{x} \subset \mathfrak{U}$ *there exists a neighborhood* $\mathfrak{V}(\mathbf{x}) \subset \mathfrak{U}$ *such that* $\omega^{(k)} = d\alpha^{(k-1)}$ *in* $\mathfrak{V}(\mathbf{x})$. Prove this statement in the following steps:

(*a*) For a monomial $\pi = a(\mathbf{x})\,dx^1 \wedge \cdots \wedge dx^k$, define

$$\tilde{\pi} = b(\mathbf{x})\Sigma\, dx^1 \wedge \cdots x^i \cdots \wedge dx^k,\ b(\mathbf{x}) = \int_0^1 a(\mathbf{x}t)t^{k-1}\,dt.$$

Show that $\pi = d\tilde{\pi} + \widetilde{d\pi}$.

(*b*) For any $\omega^{(k)}$, define $\tilde{\omega}$ by linearity from monomials. If $d\omega = 0$, $\omega = d\tilde{\omega}$. Is the map $\omega \rightarrow \tilde{\omega}$ natural?

11. In R^3, show that $\mathbf{x}_1 \times \mathbf{x}_2 = *(\mathbf{x}_1 \wedge \mathbf{x}_2)$.

12. In R^3, show that $(\mathbf{x} \times \mathbf{y}) \times \mathbf{z} = -\mathbf{x}(\mathbf{y} \cdot \mathbf{z}) + \mathbf{y}(\mathbf{z} \cdot \mathbf{x})$. (Use Prob. 11.)

13. Define $(t_{ij})^2 = \sum_{i,j} t_{ij}t_{ij}$ and prove that for any two vectors in R^n, $\mathbf{x}^2\mathbf{y}^2 = (\mathbf{x} \cdot \mathbf{y})^2 + \frac{1}{2}(\mathbf{x} \wedge \mathbf{y})^2$.

14. Show that $\omega \wedge *\omega = 0$ if and only if $\omega = 0$.

15. In R^3, one writes grad $f = df$, curl $\alpha = *d\alpha$, div $\alpha = *d*\alpha$ for any differentiable function f and any linear differential form α. Compute explicitly curl α, div α, curl grad α, div grad f, and (grad div $-$ curl curl) α.

16. Show that a pfaffian is a gradient, $\alpha =$ grad f, if and only if curl $\alpha = 0$ (see Probs. 15 and 10). Make explicit your differentiability assumptions.

17. Show that for a pfaffian α,

$$\int_{I^3} * \operatorname{div} \alpha = \int_{\partial I^3} * \alpha$$
$$\int_{I^2} * \operatorname{curl} \alpha = \int_{\partial I^2} \alpha$$

18. Given a C^2 function $u(x^1,x^2)$, let $\mathbf{n} = \{n^i\}$ be the unit normal to a level curve $u = \text{const}$ in R^2. Use exercise 2-2, Prob. 8, and the preceding Prob. 15 to show that the curvature of the level curve is $k = \operatorname{div}\,(n^1\,dx^1 + n^2\,dx^2)$. Why is the mixing of covariant and contravariant coordinates possible in this problem?

19. Given a two-form $\tau = \tau_{ij}\epsilon^{ij}$, $\tau_{12} \neq 0$:

(*a*) Show that $\tau = \alpha \wedge \alpha' + \phi$,

where
$$\alpha = \boldsymbol{\varepsilon}^1 - \sum_3^n \frac{\tau_{2i}}{\tau_{12}} \boldsymbol{\varepsilon}^i \qquad \alpha' = \sum_2^n \tau_{1i}\,\boldsymbol{\varepsilon}^i$$
$$\phi = \sum_{i,j \geq 3} \phi_{ij}\epsilon^{ij}$$

(*b*) Use the result of part (*a*) to show that any exterior two form may be decomposed into a sum of products of linearly independent pfaffians, $\tau = \alpha_1 \wedge \alpha_1' + \cdots + \alpha_r \wedge \alpha_r'$.

(*c*) The number r appearing in part (*b*) is characterized by $\overset{r}{\wedge}\tau \neq 0$, $\overset{r+1}{\wedge}\tau = 0$. Hence it is independent of the special choice of the forms α_i, α_i'.

20. Prove Eq. (9-25) for k-forms.

21. Given two vector functions $\mathbf{x}$, $\mathbf{y}$, show that for any linear differential form

$$d\alpha(\mathbf{x},\mathbf{y}) = \pounds_x\alpha(\mathbf{y}) - \pounds_y\alpha(\mathbf{x}) - \alpha([\mathbf{x},\mathbf{y}])$$

where
$$[\mathbf{x},\mathbf{y}] = \pounds_x\mathbf{y} - \pounds_y\mathbf{x}$$

22. Prove that conditions I to III (page 188) characterize determinants. (HINT: It follows from I and II that the determinant is not changed if to one row is added a linear combination of other rows; this may be used to bring the matrix in diagonal form. By I and III, the determinant of a diagonal matrix is the product of the diagonal elements; it is unique.)

23. Show that $\pounds(\Xi)[\omega^{(p)} \wedge \omega^{(q)}] = [\pounds(\Xi)\omega^{(p)}] \wedge \omega^{(q)} + \omega^{(p)} \wedge [\pounds(\Xi)\omega^{(q)}]$.

24. The operation $i(\Xi)$ defined in example 9-2 can be extended to p-forms by the definition

$$i(\Xi)[\alpha \wedge \beta] = [i(\Xi)\alpha] \wedge \beta + (-1)^{\text{dimension}\,\alpha}\alpha \wedge i(\Xi)\beta$$

(*a*) Show that (9-25) holds for p-forms. (Use Prob. 23.)
(*b*) Show that $i(\Xi)i(\mathrm{H}) + i(\mathrm{H})i(\Xi) = 0$.

25. Show that a twice-covariant (twice-contravariant) tensor is the sum of a two vector and a symmetric tensor.

26. Use the defining equation (9-18) of the determinant of a square matrix (composed of row vectors) to show that $\|AB\| = \|A\|\,\|B\|$.

9-4. Manifolds and Tensor Fields

This section contains the basic topological notions which underlie both local and global differential geometry. A good grasp of the subject of this section is essential to the understanding of modern geometry.

In Chap. 1 a curve was defined as a map $\mathbf{f}\colon I \to R^n$ of a parameter interval into a cartesian space R^n, subject to eventual differentiability conditions. By analogy, a map

$$\mathbf{p}\colon \overset{0}{I^k} \to R^n \tag{9-26}$$

of the interior of a parameter cube

$$\overset{0}{I^k}\colon\quad 0 < u^1 < 1 \cdot\cdot\cdot 0 < u^k < 1$$

into R^n could be called a Peano variety in R^n. We know from the theory of plane curves that this notion is too general to be useful in differential geometry. Therefore we restrict our attention to maps (9-26) that are C^r ($r \geq 1$, to be fixed) and for which the jacobian matrix $J_u\mathbf{p} = (\partial p^i/\partial u^j)$ is of maximum rank k for all points $\mathbf{u}$ of $\overset{0}{I^k}$. This last condition generalizes $\mathbf{f}^{\boldsymbol{\cdot}}(t) \neq 0$ in the theory of curves. By the inverse-function theorem, for any $\mathbf{u} \in \overset{0}{I^k}$ there exists a neighborhood $U(\mathbf{u})$ such that $\mathbf{p}(\mathbf{u})$ is one-to-one in $U(\mathbf{u})$ and that $\mathbf{p}^{-1}$ is a C^r map on the subset $\mathbf{p}(U(\mathbf{u}))$ of R^n. The geometry of $\mathbf{p}(\overset{0}{I^k})$ does not depend on the parametrization of $\overset{0}{I^k}$. Accordingly, the *local subspace* of R^n defined by the mapping $\mathbf{p}$ is the equivalence class of all maps $\mathbf{p}H$, where H is any C^r homeomorphism of $\overset{0}{I^k}$ onto itself. For $k = n - 1$ the local subspace is called a *local hypersurface;* the local hypersurfaces of R^3 are the *local surfaces.*

Local differential geometry is concerned mostly with local subspaces and with neighborhoods in R^n. But for all questions connected with topology some more general definitions are needed.

A C^r *manifold* of dimension k is a set M which is covered by a family of subsets U_α such that for each of the subsets there exists a one-to-one mapping

$$q_\alpha\colon\quad U_\alpha \to V_\alpha \subset R^k$$

and that at the intersection $U_\alpha \cap U_\beta$ of two subsets the two functions

$$q_\alpha q_\beta^{-1} \qquad \text{and} \qquad q_\beta q_\alpha^{-1} \tag{9-27}$$

define C^r *homeomorphisms* of the *open* sets $q_\alpha(U_\alpha \cap U_\beta)$ and $q_\beta(U_\alpha \cap U_\beta)$ in R^k.

For $\alpha = \beta$ it follows that the V_α must be open sets in R^k. A couple (U_α, q_α) is a *chart* of the manifold. The set of charts is the *atlas.* A manifold may have many different atlases. The manifold is *compact* if its point set M is compact, i.e., if every atlas has a finite subatlas whose neighborhoods U_α cover M. In many global investigations it is necessary to assume at least the existence of an atlas of countably many charts. A *closed surface* is a compact 2-manifold each of whose U_α is a local surface. A local subspace is not, in general, a manifold; the set $\mathbf{p}(U(\mathbf{u}))$ as previously defined always is one. Local subspaces are *varieties.* These are sets with charts defined like manifolds, except that the functions (9-27) define homeomorphisms *only if* at least one of the sets $q_\alpha(U_\alpha \cap U_\beta)$ and $q_\beta(U_\alpha \cap U_\beta)$ is open. Varieties may have self-intersections and other singularities; manifolds may not. If two atlases on a set M are both subatlases of a third atlas defined on M, they are said to define *the same C^r structure.*†

***Example* 9-3.** The cylinder of revolution

$$\mathbf{p} = (r \cos 4\pi u^1, r \sin 4\pi u^1, \tan \pi(u^2 - \tfrac{1}{2}))$$

is a local surface. Its point set $(x^1)^2 + (x^2)^2 = r^2$ may also be given the structure of a manifold. An atlas is obtained, for example, by the sets U_i mapped by $\mathbf{p}^{-1}$ into

$$\begin{array}{lll} 0 < u^1 < \frac{1}{5} & 0 < u^2 < 1 & \text{for } U_1 \\ \frac{1}{6} < u^1 < \frac{2}{5} & 0 < u^2 < 1 & \text{for } U_2 \\ \frac{1}{3} < u^1 < \frac{3}{5} & 0 < u^2 < 1 & \text{for } U_3 \end{array}$$

It is clear that in the definition of a local surface the cube $\overset{0}{I^k}$ may be replaced by any domain C^3 homeomorphic to it.

Although the cylinder may be covered by an atlas of a finite number (three) of charts, it is not a compact surface. An infinite atlas is obtained, for example, by replacing the U_i with sets $U_{i,j}$ mapped by $\mathbf{p}^{-1}$ onto the same strip parallel to the u^2 axis as the U_i but then restricted to

$$j - \epsilon < \tan \pi(u^2 - \tfrac{1}{2}) < j + \epsilon$$

for all integer j. No finite subatlas of this atlas covers the cylinder.

† Milnor has given examples of manifolds on which distinct differentiable structures exist. Kervaire has shown that there exists a C^0 manifold on which no C^1 structure is possible. These problems on structures need very advanced tools of algebraic topology.

***Example* 9-4.** If a plane curve has a double point (Fig. 9-1), the right cylinder constructed over it is a variety but not a manifold. On this local surface two neighborhoods of the ruling through the double point may intersect in that ruling only (which is not an open set).

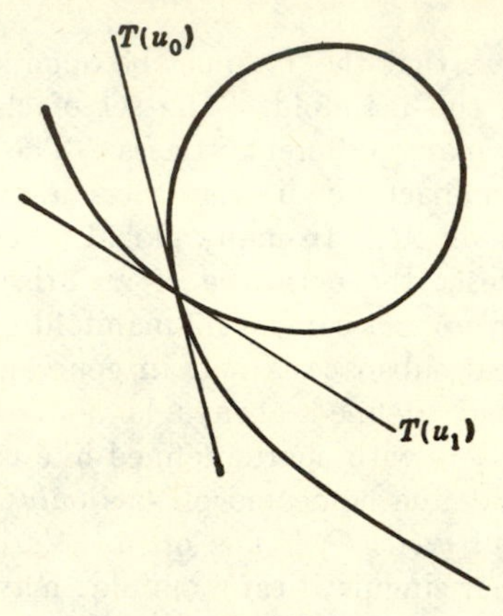

Fig. 9-1

The unit circle $(x^1)^2 + (x^2)^2 < 1$ is a manifold. A possible atlas consists of one chart, viz., itself as a set and the identity mapping as mapping. It is not a closed surface since it is not compact. The unit disk $(x^1)^2 + (x^2)^2 \leq 1$ is a compact set but not a manifold. No sets which include boundary points can be varieties.

***Example* 9-5.** The sphere S^2

$$(\sin \phi \sin \theta, \sin \phi \cos \theta, \cos \phi)$$

defined on

$$-\epsilon < \phi < \pi + \epsilon \qquad -\epsilon < \theta < 2\pi + \epsilon$$

is a closed surface but not a local surface. For $\phi = 0$ and $\phi = \pi$ the jacobian of the mapping is

$$\begin{pmatrix} \pm \sin \theta & 0 \\ \pm \cos \theta & 0 \\ 0 & 0 \end{pmatrix}$$

It is only of rank one. To save the situation, we first introduce other parameters by

$$p^1 = \frac{u^1}{[(u^1)^2 + (u^2)^2]^{\frac{1}{2}}} \sin \pi[(u^1)^2 + (u^2)^2]^{\frac{1}{2}} \qquad p^1(0,0) = 0$$

$$p^2 = \frac{u^2}{[(u^1)^2 + (u^2)^2]^{\frac{1}{2}}} \sin \pi[(u^1)^2 + (u^2)^2]^{\frac{1}{2}} \qquad p^2(0,0) = 0$$

$$p^3 = \cos \pi[(u^1)^2 + (u^2)^2]^{\frac{1}{2}}$$

The image of the unit disk $(u^1)^2 + (u^2)^2 < 1$ is the sphere minus its south pole $(0,0,-1)$. This is a local surface. The whole unit circumference $(u^1)^2 + (u^2)^2 = 1$ is mapped onto the south pole; there the jacobian has rank one. A second map

$$q^1(u^1,u^2) = p^1(u^1,u^2) \qquad q^2(u^1,u^2) = p^2(u^1,u^2)$$
$$q^3(u^1,u^2) = -p^3(u^1,u^2)$$

covers S^2 minus the north pole $(0,0,1)$. At any point different from the north and south poles, the map $(u^1,u^2) \rightarrow (t^1,-t^2)$ defined by

$$\mathbf{p}(\mathbf{t}) = \mathbf{q}(\mathbf{u})$$

is

$$u^1 = \frac{1-r}{r}t^1 \qquad u^2 = \frac{1-r}{r}(-t^2) \qquad r^2 = (t^1)^2 + (t^2)^2$$

This is a homeomorphism of the circular disk $0 < r < 1$ onto itself. Its jacobian $J_t\mathbf{u}$ has the determinant $(1 - r)/r > 0$. We have found an atlas for S^2 consisting of two charts, viz., (S^2 − south pole, $\mathbf{p}$) and (S^2 − north pole, $\mathbf{q}$). The maps (9-27) are real analytic functions.

A variety which has an atlas such that the jacobians of all maps (9-27) have positive determinants is called *orientable*. A sphere is an orientable manifold. Orientable manifolds are important because on them integrals may be defined globally without difficulties. On nonorientable manifolds the use of different charts at the intersection of two neighborhoods may result in different signs for an integral evaluated there since the Jacobian determinant appears in the change of coordinates.

***Example* 9-6.** A projective plane (example 5-7) is a nonorientable manifold. $(x^1{:}x^2{:}x^3)$ is the same point of P^2 as $(-x^1{:}-x^2{:}-x^3)$. Since the jacobian determinant $\|J_{(x^i)}(-x^i)\| = -1$, the sign of such a determinant on P^2 cannot have any geometric sense.

Let $\{\mathbf{e}_i\}$ be the orthonormal frame of the system of cartesian coordinates in R^n. The jacobian $J_u\mathbf{p}$ defines the differential $d\mathbf{p}$ of the local subspace $\mathbf{p}$,

$$d\mathbf{p} = d\mathbf{u}\, J_u\mathbf{p}\, \{\mathbf{e}_i\} = \frac{\partial p^i}{\partial u^j}du^j\mathbf{e}_i \tag{9-28}$$

$d\mathbf{p}$ is a *vector-valued differential form* or *differential vector*. Its value on a contravariant vector $\mathbf{t} \in \overset{0}{R^k}$ is a contravariant vector $d\mathbf{p}(\mathbf{t}) \in R^n$. For any fixed $\mathbf{u}_0 \in I^k$ the vectors $d\mathbf{p}(\mathbf{t})_{\mathbf{u}=\mathbf{u}_0}$ form a vector space T_0 of dimension k, the *tangent space* of $\mathbf{p}$ in $\mathbf{p}_0 = \mathbf{p}(\mathbf{u}_0)$. In a differentiable change of parameters $\mathbf{u} \rightarrow \mathbf{u}^*$ the differentials change by

$$d\mathbf{u}^* = d\mathbf{u}\, J_u\mathbf{u}^*$$

The tangent space T_0 does not depend on the system of parameters to which the local subspace is referred; a change of coordinates induces in T_0 a linear transformation of the coordinate vectors whose matrix is the jacobian of the change of parameters evaluated at $\mathbf{p}_0$. A space curve $\mathbf{c}$ in $\mathbf{p}$ passing through $\mathbf{p}_0$ is the image of a space curve $\mathbf{u}(\sigma)$ in $\overset{0}{I}{}^k$. The tangent vector in $\mathbf{p}_0$ to $\mathbf{c} = \mathbf{p}[\mathbf{u}(\sigma)]$ is

$$\frac{\partial p^i}{\partial u^j}\frac{du^j}{d\sigma}\mathbf{e}_i \in T_0$$

This shows that *the tangent space in* $\mathbf{p}_0$ *may be identified with the set of tangent vectors to curves in* $\mathbf{p}$ *through* $\mathbf{p}_0$, *or*, in a more abstract manner, *with the set of equivalence classes of curves in* $\mathbf{p}$ *through* $\mathbf{p}_0$, two curves being in the same class if their Taylor expansions in $\mathbf{p}_0$ have identical linear terms. If the local subspace is not a manifold, a point on it may have different tangent spaces belonging to distinct parameter values $\mathbf{u}_0$ (Fig. 9-1).

The definition of the tangent space may be carried over to manifolds that are not subspaces of a cartesian space. For a given chart (U_α,q_α) the set $V_\alpha = q_\alpha(U_\alpha)$ is open in R^k; its points carry the cartesian coordinates in R^k. The couple (V_α,q_α^{-1}) is a *coordinate system* in U_α. With each point $\mathbf{x} \in U_\alpha$ it associates the coordinates of $q_\alpha(\mathbf{x}) \in V_\alpha$. Differentiability is defined with respect to these coordinates. If a function on M is differentiable in one coordinate system of an atlas, it is differentiable in any coordinate system of the same differentiable structure defined on U_α. In a fixed coordinate system, the differentiable curves through a fixed point $\mathbf{x}_0 \in U$ have Taylor expansions

$$x^i = x_0{}^i + a^i(\sigma - \sigma_0) + R_2 \tag{9-29}$$

The space T_0 of vectors (a^i) gives a vector-space structure on the set of equivalence classes of differentiable curves through $\mathbf{x}_0$, as in the case of local subspaces. The coordinates in T_0 are uniquely defined by the coordinate system (V_α,q_α^{-1}). A change of coordinate system $(V_\alpha,q_\alpha^{-1}) \rightarrow (V_\beta,q_\beta^{-1})$ induces a change of the coordinate vectors by a linear transformation of matrix

$$J_{q_\alpha q_\beta^{-1}} q_\alpha^{-1} = J(\alpha,\beta) \tag{9-30}$$

The *tangent space* $T(M)$ of the manifold M is the set of all pairs $(\mathbf{x},\mathbf{r})$, where $\mathbf{x} \in M$ and $\mathbf{r} \in T_x$.

***Example* 9-7.** The Lie algebra $\mathfrak{L}(G)$ of a Lie group G is defined in the tangent space T_e of the unit of G. Since any curve $b(t)$, $b(0) = a$, starting from a point $a \in G$, may be written as

$$b(t) = [b(t)a^{-1}]a \qquad [b(t)a^{-1}]_{t=0} = e$$

it follows that the Cartan map

$$b'(0) \to b'(0)b(0)^{-1}$$

is an isomorphism of T_a onto T_e.

Theorem 9-8. *The tangent space of a C^r manifold is a C^{r-1} manifold.*

The sets U_α determine sets $\tilde{U}_\alpha = \{(\mathbf{x},\mathbf{r}) | \mathbf{x} \in U_\alpha, \mathbf{r} \in T_x\}$ which cover $T(M)$. The maps q_α determine maps $\tilde{q}_\alpha$: $(\mathbf{x},\mathbf{r}) \to (q_\alpha\mathbf{x},\mathbf{r}) \in R^k \times R^k$. The functions (9-27) are $(q_\alpha q_\beta^{-1}, J(\alpha,\beta))$; they are C^{r-1} homeomorphisms of open sets in $R^{2k} = R^k \times R^k$.

It is possible to construct tensor spaces like $\overset{a}{\otimes} T_0 \otimes \overset{b}{\otimes} T_0^*$ or $\overset{s}{\wedge} T_0^*$ starting from the vector space T_0. A *tangent tensor space* is a set of couples consisting of a point $\mathbf{x} \in M$ and an element of some fixed tensor space constructed on T_x. It will follow from Eq. (9-31) that all tangent tensor spaces of a C^r manifold are C^{r-1} manifolds.

Definition 9-9. *A tensor field is a map*

$$G: \mathbf{x} \to (\overset{a}{\otimes} T_0 \otimes \overset{b}{\otimes} T_0^*)_x$$

which associates with each point $\mathbf{x} = (x^1, \ldots, x^k) \in M$ a coordinate tensor $G_{i_1 \ldots i_b}{}^{j_1 \cdots j_b}$ in a tangent tensor space in $\mathbf{x}$.

The next theorem follows immediately from Eqs. (9-30) and (9-10).

Theorem 9-10. *In a change of coordinates on M $(x^1, \ldots, x^k) \to (x^{1'}, \ldots, x^{k'})$ the coordinate tensor of a tensor field G is transformed according to*

$$G_{i_1' \ldots i_b'}{}^{j_1' \cdots j_a'} = (J_x\mathbf{x}')^{j_1'}_{j_1} \cdots (J_x\mathbf{x}')^{j_a'}_{j_a} (J_{x'}\mathbf{x})^{i_1}_{i_1'} \cdots (J_{x'}\mathbf{x})^{i_b}_{i_b'} G_{i_1 \ldots i_b}{}^{j_1 \cdots j_a}$$

If the change of coordinates $q_\alpha q_\beta^{-1}$ is written as $\mathbf{x} \to \mathbf{x}'$, the last equation means

$$G_{i_1' \ldots i_b'}{}^{j_1' \cdots j_a'} = \frac{\partial x^{j_1'}}{\partial x^{j_1}} \cdots \frac{\partial x^{j_a'}}{\partial x^{j_a}} \frac{\partial x^{i_1}}{\partial x^{i_1'}} \cdots \frac{\partial x^{i_b}}{\partial x^{i_b'}} G_{i_1 \ldots i_b}{}^{j_1 \cdots j_b} \quad (9\text{-}31)$$

In general, let $\mathfrak{J}$ be any tensor space constructed over T, and let $\mathfrak{J}(M)$ be the corresponding tangent tensor space of the ordered couples $(\mathbf{x},t)$, $\mathbf{x} \in M$, $t \in \mathfrak{J}_x$. The *projection map* of $\mathfrak{J}(M)$ onto M is π: $(\mathbf{x},t) \to \mathbf{x}$. A map s: $\mathbf{x} \to (\mathbf{x},s(\mathbf{x}))$ defined over a U_α is a *section* in $\tilde{U}_\alpha$ if $\pi s(\mathbf{x}) = \mathbf{x}$. A tensor field defines a section in any $\tilde{U}_\alpha$. M itself may be identified with the section $(\mathbf{x},0)$ in any one of its tangent tensor spaces. There are many important sections that are not tensor fields. For example, Cartan matrices are elements of tensor spaces but they transform by Eq. (1-5) and not by (9-31). They have no sense independent of the system of coordinates and of the frame to which they belong. What really matters is that the transformation behavior of a section under coordinate transformations of M must be known. If this is the case, the section is called a *geometric object.* In this text, no geometric objects more general than Cartan matrices will be treated. Tangent tensor spaces will be discussed further in Chaps. 11, 13, and 14.

A note on terminology: In older books, tensor fields are referred to as tensors. What we have called coordinate tensors sometimes appear as affine tensors or affinors.

The usual notations for vector algebra mix very well with those of tensor algebra but some special signs are needed if differentials are to be multiplied exteriorally. In this case we shall put quotation marks on the symbol of the algebraic operation to be performed. For example, if we have differential vectors $d\mathbf{p}_k = \omega_k{}^i\mathbf{e}_i$, where the $\omega_k{}^i$ are pfaffians, we shall write

$$d\mathbf{p}_1 \times d\mathbf{p}_2 = \omega_1{}^i \otimes \omega_2{}^j\mathbf{e}_i \times \mathbf{e}_j$$

but

$$d\mathbf{p}_1 \text{“}\times\text{”}\, d\mathbf{p}_2 = \omega_1{}^i \wedge \omega_2{}^j\mathbf{e}_i \times \mathbf{e}_j$$

In the same way the determinant “$\|$” $d\mathbf{p}_1, \ldots, d\mathbf{p}_n$ “$\|$” is to be computed as “$\|$” $\omega_i{}^k$ “$\|$” $\|\mathbf{e}_1, \ldots, \mathbf{e}_n\|$, where in the determinant

$$\text{“}\|\text{”}\ \omega_i{}^k\ \text{“}\|\text{”} = \delta^{k_1\cdots k_n}_{1\cdots n}\omega_1{}^{k_1} \wedge \cdots \wedge \omega_n{}^{k_n}$$

great care has to be taken not to disturb the prescribed order of multiplication without due change in sign.

The euclidean arc length of a curve $\mathbf{c}(\sigma)$ in R^n is the integral over $ds = |(d\mathbf{c}/d\sigma)^2|^{\frac{1}{2}}\, d\sigma$. The *element of arc length* on a local subspace is

$$ds^2 = d\mathbf{p} \cdot d\mathbf{p} = \sum_{i=1}^{n} \frac{\partial p^i}{\partial u^j}\frac{\partial p^i}{\partial u^k}\, du^j\, du^k$$

$ds^2 = g_{jk}\, du^j\, du^k$ gives the euclidean arc length for any curve $\mathbf{u} = \mathbf{u}(\sigma)$ on the local subspace. Since $g_{jk} = g_{kj}$ is symmetric, no harm can come from our writing $du^j\, du^k$ instead of $du^j \otimes du^k$ [see defining equation (9-8)]. For surfaces (two-subspaces in R^3)

$$d\mathbf{p} = du^1\mathbf{p}_{u^1} + du^2\mathbf{p}_{u^2}$$

$$\mathbf{p}_{u^i} = \frac{\partial p^1}{\partial u^i}\,\mathbf{e}_1 + \frac{\partial p^2}{\partial u^i}\,\mathbf{e}_2 + \frac{\partial p^3}{\partial u^i}\,\mathbf{e}_3$$

one usually writes

$$ds^2 = E(du^1)^2 + 2F\, du^1\, du^2 + G(du^2)^2$$

where

$$E = \mathbf{p}_{u^1} \cdot \mathbf{p}_{u^1} \qquad F = \mathbf{p}_{u^1} \cdot \mathbf{p}_{u^2} \qquad G = \mathbf{p}_{u^2} \cdot \mathbf{p}_{u^2}$$

ds^2 is also called the *first fundamental form* of the surface. $d\mathbf{p}(1,0) = \mathbf{p}_{u^1}$ is the tangent vector to the curve $u^1 = s$, $u^2 =$ const, the u^1 *parameter line* (it is an unfortunate tradition to call special *curves* on surfaces by the name of *lines*). $F = 0$ if and only if the two systems of parameter lines are orthogonal trajectories of one another. The two-form α defined by

$$d\mathbf{p} \text{ ``}\times\text{'' } d\mathbf{p} = 2\alpha\mathbf{n} \qquad \mathbf{n}\cdot\mathbf{n} = 1 \tag{9-32}$$

is the *area element* of the surface. Since

$$\begin{aligned} d\mathbf{p} \text{ ``}\times\text{'' } d\mathbf{p} &= (\mathbf{p}_{u^1}\, du^1 + \mathbf{p}_{u^2}\, du^2) \text{ ``}\times\text{'' } (\mathbf{p}_{u^1}\, du^1 + \mathbf{p}_{u^2}\, du^2) \\ &= 2\mathbf{p}_{u^1} \times \mathbf{p}_{u^2}\, du^1 \wedge du^2 \\ &= 2\left[\left(\frac{\partial p^2}{\partial u^1}\frac{\partial p^3}{\partial u^2} - \frac{\partial p^2}{\partial u^2}\frac{\partial p^3}{\partial u^1}\right)\mathbf{e}_1 + \left(\frac{\partial p^3}{\partial u^1}\frac{\partial p^1}{\partial u^2} - \frac{\partial p^3}{\partial u^2}\frac{\partial p^1}{\partial u^1}\right)\mathbf{e}_2 \right. \\ &\qquad \left. + \left(\frac{\partial p^1}{\partial u^1}\frac{\partial p^2}{\partial u^2} - \frac{\partial p^1}{\partial u^2}\frac{\partial p^2}{\partial u^1}\right)\mathbf{e}_3\right] du^1 \wedge du^2 \end{aligned}$$

a simple computation shows that the length of the vector in brackets is $(EG - F^2)^{\frac{1}{2}}$; hence

$$\alpha = (EG - F^2)^{\frac{1}{2}}\, du^1 \wedge du^2 \tag{9-32a}$$

$\iint\alpha$ gives the value of the surface area. The subject of surface area is a difficult one; therefore we shall avoid using in proofs a geometric meaning of this term.

***Example* 9-8.** For the unit sphere (example 9-5) given by the angles ϕ and θ one has

$$\begin{aligned} d\mathbf{p} = (\cos\theta\sin\phi,\ -\sin\theta\sin\phi,\ 0)\, d\theta \\ + (\sin\theta\cos\phi,\ \cos\theta\cos\phi,\ -\sin\phi)\, d\phi \end{aligned}$$

Hence

$$ds^2 = d\mathbf{p}\cdot d\mathbf{p} = \sin^2\phi\, d\theta^2 + d\phi^2$$

and
$$\begin{aligned} d\mathbf{p} \text{ ``}\times\text{'' } d\mathbf{p} &= 2(\sin\theta\sin^2\phi,\ \cos\theta\sin^2\phi,\ \cos\phi\sin\phi)\, d\theta \wedge d\phi \\ &= 2\sin\phi\, d\theta \wedge d\phi\, \mathbf{p} \end{aligned}$$

Since $\mathbf{p}$ is a unit vector, $\alpha = \sin\phi\, d\theta \wedge d\phi$.

Exercise 9-4

1. Find the parameters ϕ,θ on the sphere S^2 as functions of the parameters u^1,u^2 (example 9-5).

2. Find an atlas for the unit three-sphere S^3 in R^4 [locus of the points $(x^1)^2 + (x^2)^2 + (x^3)^2 + (x^4)^2 = 1$].

3. The right circular cone $\mathbf{c}(u^1,u^2) = (u^1\cos u^2,\ u^1\sin u^2,\ u^1)$, $u^1 \geq 0$, $0 \leq u^2 \leq 2\pi$, is a C^0 surface in R^3. Show that the tangent vectors to $\mathbf{c}$ in (0,0,0) do not generate a plane (but a cone).
The normal projection $(x^1,x^2,x^3) \to (x^1,x^2,0)$ is a homeomorphism of a cone onto a plane (a C^∞ manifold). Use the previous result to show that *on the cone* the normal projection is a C^0 homeomorphism but no C^r homeomorphism for $r \geq 1$ (in the neighborhood of the origin).

* **4.** Use Zorn's lemma to prove that each differentiable structure has a unique *maximal atlas* of which every atlas of the structure is a subatlas.

5. Show that the projective line P^1 is an orientable manifold.

6. Show that the surface $(u^1 \sin 2\pi u^2, u^1 \cos 2\pi u^2, \sin \pi u^2)$, $(u^1,u^2) \varepsilon I^2$, is a nonorientable variety.

7. Discuss the right cylinder erected on the lemniscate $r = a \cos 2\theta$ [formed by all parallels to the x^3 axis through points of the lemniscate in the (x^1,x^2) plane]. Find an atlas for it and show that it is not a manifold.

8. A *surface of revolution* (axis of revolution x^3) is one that has a parametric representation

$$\mathbf{p} = (f(u^1) \cos u^2, f(u^1) \sin u^2, g(u^1))$$

Find its *meridians* (intersections of the surface with planes through the x^3 axis), parallel circles (p^3 = const), and its ds^2. Why is $F = 0$?

9. Apply the formula for the ds^2 found in Prob. 8 to a right circular cylinder and to a sphere (examples 9-3 and 9-5).

10. The one-parameter group of motions in R^3

$$\mathbf{x}^* = \mathbf{x}\begin{pmatrix} \cos t & \sin t & 0 \\ -\sin t & \cos t & 0 \\ 0 & 0 & 1 \end{pmatrix} + (a^1,a^2,a^3)t$$

is one of *helicoidal movements.* Describe the surface generated in R^3 by the movement of a space curve $\mathbf{c}(u^1)$ in helicoidal movements.

11. (Sequel to Prob. 6.) A *helicoid* is the surface obtained by the movement of a line $\mathbf{c} = \mathbf{d} + \mathbf{b}u^1$ in the helicoidal group, e.g., for

$$a^1 = a^2 = 0$$

Give its parameter representation in u^1, $u^2 = t$, compute its ds^2, and find the conditions to be imposed on the vectors $\mathbf{d},\mathbf{b}$ for the surface to be a manifold.

12. Represent the equilateral paraboloid $x^3 = (x^1)^2 - (x^2)^2$ as a surface in our sense. Compute its ds^2.

13. Same question as Prob. 8 for any locus $x^3 = f(x^1,x^2)$. When is the locus a surface? When a manifold?

14. Discuss the surface $\mathbf{p} = \mathbf{a}u^1 + \mathbf{b}u^2 + \mathbf{c}$, $\mathbf{a} \times \mathbf{b} \neq 0$. Compute its ds^2.

◆**15.** The *tangent surface* of a space curve $\mathbf{c}(\sigma)$ is

$$\mathbf{p}(u,\sigma) = \mathbf{c}(\sigma) + u\mathbf{c}^{\cdot}(\sigma)$$

Describe the σ and the u parameter lines. The jacobian $J_{(u,\sigma)}\mathbf{p}$ has rank one for $u = 0$. $\mathbf{p}$ is a ruled surface in the terminology of Sec. 8-2; it is not a surface unless we exclude the edge of regression. The jacobian may also vanish along certain tangents. Find the condition for this in terms of the curvature of the space curve.

16. Show that the *Moebius strip*

$$\left(\cos\theta + u\sin\frac{\theta}{2}, \sin\theta - u\cos\frac{\theta}{2}, u\right)$$
$$0 \le \theta \le 2\pi \qquad -\tfrac{1}{2} < u < \tfrac{1}{2}$$

is a nonorientable manifold.

17. A *catenoid* is the surface of revolution (Prob. 8) generated by the rotation of $x^3 = \operatorname{Cosh} x^1$ about the x^3 axis. Show that its ds^2 is equal to that of a *right helicoid* (Prob. 11: $\mathbf{d} = 0$, $\mathbf{b} = \mathbf{e}_1$, $a^3 = 1$). A right helicoid is not a surface of revolution.

18. Show that $d\mathbf{p}$ "×" $d\mathbf{q} = d\mathbf{q}$ "×" $d\mathbf{p}$.

19. Let V and W be two local subspaces of R^n such that V is an image of $\overset{0}{I^v}$, W of $\overset{0}{I^w}$, $v < w$, and that every point "on" V is "on" W. Show that, for all points on V, the tangent space of V is a subspace of that of W.

20. Show that the contraction of a tensor field by another tensor field is again a tensor field.

21. Show that the cartesian plane is a noncompact C^∞ manifold.

References

SECS. 9-1, 9-2

Bourbaki, N.: "Eléments de mathématique," pt. 1, Book II, Algèbre, chap. 2, Algèbre linéaire, 3d ed.; chap. 3, Algèbre multilinéaire, 2d ed., Hermann & Cie, Paris, 1948, 1954.

Halmos, P.: "Finite Dimensional Vector Spaces," Princeton University Press, Princeton, N.J., 1953.

Willmore, T. J.: "An Introduction to Differential Geometry," Clarendon Press, Oxford, 1959.

SEC. 9-3

Cartan, E.: "Les systèmes différentiels extérieurs et leurs applications géométriques," Hermann & Cie, Paris, 1945.

Favard, J.: see references for Sec. 7-3.

SEC. 9-4

Auslander, Louis, and Robert E. MacKenzie: "Introduction to Differentiable Manifolds," McGraw-Hill Book Company, Inc., New York, 1963.

Ehresmann, C.: Introduction à la théorie des structures infinitésimales et des pseudogroupes de Lie, *Colloq. Intern. Centre Natl. Rech. Sci.* (*Paris*), LII, Géométrie différentielle, 1953, pp. 97–110.

Kervaire, M.: A Manifold Which Does Not Admit Any Differentiable Structure, *Comment. Math. Helv.*, **34**: 257–270 (1960).

Milnor, J.: On Manifolds Homeomorphic to the 7-Sphere, *Ann. of Math.*, Ser. 2, **64**: 399–405 (1956).

10

SURFACES

10-1. Curvatures

This chapter is devoted to the study of the euclidean geometry of local surfaces in R^3. The group of euclidean motions has six parameters; two prolongations of R^3 (one dependent, two independent variables) yield eight dimensions. The theory will depend on two invariants of the second order. Nevertheless we shall assume first that the surfaces are C^3 since we shall study surfaces through the curves drawn on them and we know that the natural assumption of differentiability in the euclidean geometry of space curves is C^3.

If we succeed in finding an invariant set of frames on the surface, the elements of the corresponding Cartan matrix will be invariants. We assume a given fixed system of cartesian coordinates in R^3. Through a point $\mathbf{p}_0$ on the surface $\mathbf{p} = \mathbf{p}(u^1,u^2)$ we take a space curve $\mathbf{x}$, given by $u^1 = u^1(s)$, $u^2 = u^2(s)$. We may restrict our attention to a neighborhood of coordinates $(u_0{}^1,u_0{}^2)$ of $\mathbf{p}_0$ such that the corresponding piece of surface is a manifold. Along $\mathbf{x}$ we have the *Frenet frame* $\{\mathbf{t},\mathbf{n},\mathbf{b}\}$. Its Cartan matrix is

$$C_1 = \begin{pmatrix} 0 & k & 0 \\ -k & 0 & t \\ 0 & -t & 0 \end{pmatrix}$$

The vectors of the frame have no connection with the surface except $\mathbf{t} \in T_x$. This is not appropriate for the study of surface properties. By hypothesis, the two vectors tangent to the parameter lines at any point of $\mathbf{x}$ are linearly independent, and the vector

$$\mathbf{N} = \frac{\mathbf{p}_{u^1} \times \mathbf{p}_{u^2}}{|\mathbf{p}_{u^1} \times \mathbf{p}_{u^2}|} = *\left(\frac{\mathbf{p}_{u^1}}{|\mathbf{p}_{u^1}|} \wedge \frac{\mathbf{p}_{u^2}}{|\mathbf{p}_{u^2}|}\right) \tag{10-1}$$

is the *normal* to T_x (it is usually called the *normal to the surface* in **x**). A rotation of angle $\alpha = \angle(\mathbf{b},\mathbf{N})$ about **t** will bring **b** onto **N** and **n** onto a vector **T** normal to the curve ($\mathbf{n} \cdot \mathbf{t} = \mathbf{T} \cdot \mathbf{t} = 0$) and in T_x ($\mathbf{n} \cdot \mathbf{b}$

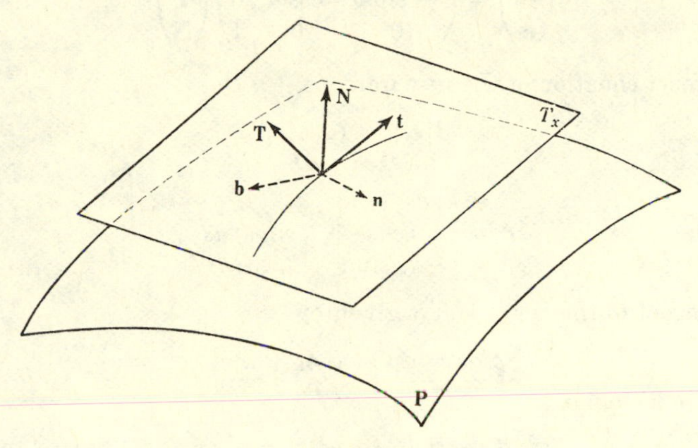

Fig. 10-1

$= \mathbf{T} \cdot \mathbf{N} = 0$). **T** is the *tangent normal* of the curve $\mathbf{x}(s)$ on the surface $\mathbf{p}(u^1,u^2)$. The rotation (Fig. 10-1)

$$\begin{pmatrix} \mathbf{t} \\ \mathbf{T} \\ \mathbf{N} \end{pmatrix} = \begin{pmatrix} 1 & 0 & 0 \\ 0 & \cos\alpha & \sin\alpha \\ 0 & -\sin\alpha & \cos\alpha \end{pmatrix} \begin{pmatrix} \mathbf{t} \\ \mathbf{n} \\ \mathbf{b} \end{pmatrix}$$

induces a change of the Cartan matrix, by Eq. (1-5). The corresponding Frenet equation is

$$d\begin{pmatrix} \mathbf{t} \\ \mathbf{T} \\ \mathbf{N} \end{pmatrix} = \begin{pmatrix} 0 & k\cos\alpha\, ds & -k\sin\alpha\, ds \\ -k\cos\alpha\, ds & 0 & t\, ds + d\alpha \\ k\sin\alpha\, ds & -t\, ds - d\alpha & 0 \end{pmatrix} \begin{pmatrix} \mathbf{t} \\ \mathbf{T} \\ \mathbf{N} \end{pmatrix}$$

$$= \begin{pmatrix} 0 & k_g\, ds & k_n\, ds \\ -k_g\, ds & 0 & t_r\, ds \\ -k_n\, ds & -t_r\, ds & 0 \end{pmatrix} \begin{pmatrix} \mathbf{t} \\ \mathbf{T} \\ \mathbf{N} \end{pmatrix} \tag{10-2}$$

$k_g = k\cos\alpha$ is the *geodesic curvature*, $k_n = -k\sin\alpha$ the *normal curvature*, and $t_r = t + \alpha'$ the *relative torsion*† of the curve on the surface. It is clear that in any other frame attached to the surface we should like to retain the unique normal **N** as a third vector $\mathbf{e}_3$. Any frame $\{\mathbf{e}_1,\mathbf{e}_2,\mathbf{e}_3\}$ such that $\mathbf{e}_1$ and $\mathbf{e}_2$ are tangent to the surface and $\mathbf{e}_3 = \mathbf{N}$ may be obtained

† In the literature, t_r is often denoted t_g and called the *geodesic* torsion. However, t_r is not a "geodesic" quantity in the sense of Chap. 11 which today is standard.

from the *tangent normal frame* $\{\mathbf{t},\mathbf{T},\mathbf{N}\}$ by a rotation about $\mathbf{N}$ through an angle $\theta = \theta(s)$,

$$\begin{pmatrix}\mathbf{e}_1\\ \mathbf{e}_2\\ \mathbf{e}_3\end{pmatrix} = \begin{pmatrix}\cos\theta & \sin\theta & 0\\ -\sin\theta & \cos\theta & 0\\ 0 & 0 & 1\end{pmatrix}\begin{pmatrix}\mathbf{t}\\ \mathbf{T}\\ \mathbf{N}\end{pmatrix}$$

The Frenet equation of the new frame is

$$d\{\mathbf{e}_i\} = (\omega_i{}^j)\{\mathbf{e}_j\} \tag{10-3}$$

where

$$\omega_i{}^j = -\omega_j{}^i \tag{10-4}$$

and

$$\begin{aligned}\omega_1{}^2 &= k_g\, ds + d\theta\\ \omega_1{}^3 &= (k_n \cos\theta + t_r \sin\theta)\, ds\\ \omega_2{}^3 &= (-k_n \sin\theta + t_r \cos\theta)\, ds\end{aligned} \tag{10-5}$$

The tangent to the curve $\mathbf{x}(s)$ is given by

$$\mathbf{t}\, ds = d\mathbf{p} = \omega^1\mathbf{e}_1 + \omega^2\mathbf{e}_2 \tag{10-6}$$

where we have put

$$\omega^1 = \cos\theta\, ds \qquad \omega^2 = -\sin\theta\, ds \tag{10-7}$$

By definition,

$$ds^2 = (\omega^1)^2 + (\omega^2)^2 \tag{10-8}$$

Theorem 10-1. *For any differentiable frame on a local surface such that* $\mathbf{e}_3 = \mathbf{N}$ *there exist linear differential forms* ω^1, ω^2 *that diagonalize the first fundamental form*

$$ds^2 = (du^1,du^2)\begin{pmatrix}E & F\\ F & G\end{pmatrix}\begin{pmatrix}du^1\\ du^2\end{pmatrix} = (\omega^1,\omega^2)\begin{pmatrix}1 & 0\\ 0 & 1\end{pmatrix}\begin{pmatrix}\omega^1\\ \omega^2\end{pmatrix}$$

and for which the vector-valued differential form $d\mathbf{p}$ *is*

$$d\mathbf{p} = \omega^1\mathbf{e}_1 + \omega^2\mathbf{e}_2$$

The five linear differential forms

$$\omega^1,\ \omega^2,\ \omega_1{}^2,\ \omega_1{}^3,\ \omega_2{}^3$$

are not independent of one another. They must satisfy several Poincaré relations (9-22), for example,

$$\begin{aligned}dd\mathbf{p} &= d\omega^1\mathbf{e}_1 - \omega^1 \wedge d\mathbf{e}_1 + d\omega^2\mathbf{e}_2 - \omega^2 \wedge d\mathbf{e}_2\\ &= (d\omega^1 - \omega^2 \wedge \omega_2{}^1)\mathbf{e}_1 + (d\omega^2 - \omega^1 \wedge \omega_1{}^2)\mathbf{e}_2\\ &\qquad - (\omega^1 \wedge \omega_1{}^3 + \omega^2 \wedge \omega_2{}^3)\mathbf{e}_3\\ &= 0\end{aligned}$$

Hence

$$\begin{aligned}d\omega^1 &= -\omega^2 \wedge \omega_1{}^2\\ d\omega^2 &= \omega^1 \wedge \omega_1{}^2\\ 0 &= \omega^1 \wedge \omega_1{}^3 + \omega^2 \wedge \omega_2{}^3\end{aligned} \tag{10-9}$$

The Poincaré relation from (10-3) is

$$dd\mathbf{e}_i = d\omega_i{}^j\mathbf{e}_j - \omega_i{}^j \wedge d\mathbf{e}_j = (d\omega_i{}^j - \omega_i{}^k \wedge \omega_k{}^j)\mathbf{e}_j = 0$$

that is,

$$\begin{aligned} d\omega_1{}^2 &= \omega_1{}^3 \wedge \omega_3{}^2 = -\omega_1{}^3 \wedge \omega_2{}^3 \\ d\omega_1{}^3 &= \omega_1{}^2 \wedge \omega_2{}^3 \\ d\omega_2{}^3 &= \omega_2{}^1 \wedge \omega_1{}^3 = -\omega_1{}^2 \wedge \omega_1{}^3 \end{aligned} \tag{10-10}$$

$dd\omega^i = 0$ and $dd\omega_i{}^j = 0$ are consequences of Eqs. (10-9) and (10-10), the *Gauss-Codazzi-Mainardi* equations of surface theory.

The main problem of surface theory is to find invariant parameters and invariant frames. Both may be found at the same time from Eqs. (10-3) and (10-8) to (10-10). It follows from the last equation (10-9) and Cartan's theorem 9-6 that one may write

$$(\omega_1{}^3,\omega_2{}^3) = (\omega^1,\omega^2)\begin{pmatrix} l_{11} & l_{12} \\ l_{12} & l_{22} \end{pmatrix} \tag{10-11}$$

and one checks in the formulas (10-5) that the l_{ij} are continuous. (Theorem 9-6 is purely algebraic; it cannot give information on continuity questions.) We ask whether it is possible to find differentiable frames for which the matrix (l_{ij}) becomes diagonal. This condition has intrinsic geometric meaning; it says that the *symmetric* (not exterior!) quadratic differential form

$$\mathrm{II} = -d\mathbf{N}\cdot d\mathbf{p} = \omega_1{}^3\omega^1 + \omega_2{}^3\omega^2 = (\omega^1,\omega^2)\begin{pmatrix} l_{11} & l_{12} \\ l_{12} & l_{22} \end{pmatrix}\begin{pmatrix} \omega^1 \\ \omega^2 \end{pmatrix} \tag{10-12}$$

the *second fundamental form* of $\mathbf{p}$, is diagonal. The admissible rotations of the frame $\{\mathbf{e}_i\}$ are

$$\begin{pmatrix} \mathbf{e}_{1'} \\ \mathbf{e}_{2'} \\ \mathbf{e}_{3'} \end{pmatrix} = \begin{pmatrix} \cos\phi & \sin\phi & 0 \\ -\sin\phi & \cos\phi & 0 \\ 0 & 0 & 1 \end{pmatrix}\begin{pmatrix} \mathbf{e}_1 \\ \mathbf{e}_2 \\ \mathbf{e}_3 \end{pmatrix}$$

The new forms $\omega^{i'}$ are obtained from (10-6):

$$d\mathbf{p} = \omega^1\mathbf{e}_1 + \omega^2\mathbf{e}_2 = \omega^{1'}\mathbf{e}_{1'} + \omega^{2'}\mathbf{e}_{2'}$$

as

$$(\omega^{1'},\omega^{2'}) = (\omega^1,\omega^2)\begin{pmatrix} \cos\phi & -\sin\phi \\ \sin\phi & \cos\phi \end{pmatrix} \tag{10-13}$$

whereas the forms $\omega_{i'}{}^{j'}$, by Eq. (10-5), become

$$\begin{aligned} \omega_{1'}{}^{3'} &= \omega_1{}^3\cos\phi + \omega_2{}^3\sin\phi \\ &= [l_{11}\cos^2\phi + l_{22}\sin^2\phi + 2l_{12}\cos\phi\sin\phi]\omega^{1'} \\ &\qquad + [(l_{22} - l_{11})\cos\phi\sin\phi + l_{12}(\cos^2\phi - \sin^2\phi)]\omega^{2'} \end{aligned} \tag{10-14a}$$

$$\begin{aligned} \omega_{2'}{}^{3'} &= -\omega_1{}^3\sin\phi + \omega_2{}^3\cos\phi \\ &= [(l_{22} - l_{11})\cos\phi\sin\phi + l_{12}(\cos^2\phi - \sin^2\phi)]\omega^{1'} \\ &\qquad + [l_{11}\sin^2\phi + l_{22}\cos^2\phi - 2l_{12}\cos\phi\sin\phi]\omega^{2'} \end{aligned} \tag{10-14b}$$

The second fundamental form will be diagonal if

$$\tan 2\phi = \frac{2l_{12}}{l_{11} - l_{22}}$$

2ϕ is defined up to a multiple of π unless $l_{12} = 0, l_{11} = l_{22}$. In the general case, ϕ is determined up to a multiple of $\pi/2$; this means that the frame $\{\mathbf{e}_{i'}\}$ is defined up to either a change of terms $\mathbf{e}_1 \to \mathbf{e}_2$, $\mathbf{e}_2 \to -\mathbf{e}_1$, or a change in sign $\mathbf{e}_i \to -\mathbf{e}_i$ $(i = 1,2)$.

Definition 10-2. *A line in T_x for which the second fundamental form becomes diagonal is a principal direction in* $\mathbf{x}$. *A curve all of whose tangents are in principal directions is a curvature line. The normal curvature of a curvature line is a principal curvature of the surface. A point at which two principal curvatures coincide is an umbilic.*

Theorem 10-3. *If a point $\mathbf{p}_0$ is not an umbilic, there are exactly two mutually orthogonal principal directions in T_{p_0}. Every direction in the tangent plane of an umbilic is principal.*

The last assertion in the theorem remains to be proved, but first we introduce some special notations for the invariant frame we have found. $\{\mathbf{a}_i\}$ will denote an orthonormal frame such that $\mathbf{a}_1$, $\mathbf{a}_2$ are *principal vectors* (unit vectors in principal directions) and $\mathbf{a}_3 = \mathbf{N}$ is the normal to the surface. Such a frame is called a *Darboux* frame. The pfaffians belonging to this frame will be π^i, $\pi_i{}^j$. The principal curvatures k_1, k_2 are given by

$$\pi_i{}^3 = k_i \pi^i \qquad \textit{no summation, } i = 1, 2$$

The Frenet equations of the Darboux frame are

$$d\mathbf{p} = \pi^1 \mathbf{a}_1 + \pi^2 \mathbf{a}_2 \tag{10-15}$$

$$d\begin{pmatrix} \mathbf{a}_1 \\ \mathbf{a}_2 \\ \mathbf{a}_3 \end{pmatrix} = \begin{pmatrix} 0 & \pi_1{}^2 & k_1\pi^1 \\ -\pi_1{}^2 & 0 & k_2\pi^2 \\ -k_1\pi^1 & -k_2\pi^2 & 0 \end{pmatrix} \begin{pmatrix} \mathbf{a}_1 \\ \mathbf{a}_2 \\ \mathbf{a}_3 \end{pmatrix} \tag{10-16}$$

These formulas hold also in an umbilic since, by (10-14), it is always possible to get a frame such that $l_{12} = 0$. If $k_1 = k_2$, formulas (10-14) show that the Cartan matrix is the same for *all* frames in the umbilic. This proves theorem 10-3. If a local surface has an umbilic, it is not possible to define the Darboux frames in a unique and differentiable way on the whole local surface.

It is always possible to choose the terms for the frame vectors so that a given curvature line $\mathbf{x}(s)$ is tangent to $\mathbf{a}_1$, that is, $\pi^1(d\mathbf{x}) = ds$, $\pi^2(d\mathbf{x}) = 0$. In this case the Darboux frame becomes the tangent normal frame to the curvature line; a comparison with (10-2) shows that the following holds:

Theorem 10-4 (Rodrigues). *A curve on a surface is a curvature line if and only if $t_r = 0$ or if and only if $d\mathbf{N} + k_1\mathbf{t}\, ds = 0$.*

The second condition holds in a slightly more general form: *If for some curve on* $\mathbf{p}$,

$$d\mathbf{N} + \lambda \mathbf{t}\, ds = 0$$

then the curve is a curvature line and λ *is a principal curvature.* In this case $\mathrm{II} = -d\mathbf{N} \cdot d\mathbf{p} = \lambda\, ds^2 = \lambda[(\omega^1)^2 + (\omega^2)^2]$ is diagonal. Hence $\mathbf{t}$ is a principal vector.

Corollary 10-5. *The curvature lines are the solutions of the differential equation* $\|\mathbf{N}, d\mathbf{N}, d\mathbf{p}\| = 0$.

By (10-16), $d\mathbf{N} = d\mathbf{a}_3$ is linearly independent of $\mathbf{N}$ unless $k_1 = k_2 = 0$. In the latter case we are in the presence of an umbilic, the determinant vanishes identically, and every direction is principal. If $d\mathbf{N} \neq 0$, the determinant vanishes only if $d\mathbf{N} + \lambda\, d\mathbf{p} = 0$, since $d\mathbf{p}$ and $\mathbf{N}$ always are linearly independent.

The corollary gives the most practical method for the computation of the curvature lines and the Darboux frames. If the local surface is without an umbilic, the euclidean arc lengths of the two families of curvature lines may serve as invariant parameters. However, it is not worthwhile to give the surface in invariant parameters except for integration problems (Secs. 10-3 and 10-4).

Any frame $\{\mathbf{e}_i\}$ may be obtained from the Darboux frame by a rotation of angle $\phi = \angle(\mathbf{a}_1, \mathbf{e}_1)$ about $\mathbf{N}$. The reasoning which leads to Eqs. (10-14) may be applied to express the linear forms belonging to $\{\mathbf{e}_i\}$ by the principal curvatures, especially

$$\begin{aligned} \omega_1{}^3 &= (k_1 \cos^2\phi + k_2 \sin^2\phi)\omega^1 + (k_2 - k_1)\cos\phi \sin\phi\, \omega^2 \\ \omega_2{}^3 &= (k_2 - k_1)\cos\phi \sin\phi\, \omega^1 + (k_1 \sin^2\phi + k_2 \cos^2\phi)\omega^2 \end{aligned}$$

The frame $\{\mathbf{e}_i\}$ is the tangent normal frame for a curve $\mathbf{x}, d\mathbf{x} = \mathbf{t}\, ds$, for which the value of ω^2 on $d\mathbf{x}$ is zero. A comparison with (10-2) shows that

$$\begin{aligned} k_n &= k_1 \cos^2\phi + k_2 \sin^2\phi \\ t_r &= (k_2 - k_1)\cos\phi \sin\phi \end{aligned} \tag{10-17}$$

The first of these formulas, known as *Euler's* formula, explains the name "normal curvature." k_n depends only on the direction of $\mathbf{e}_1 = \mathbf{t}$; it is common to all curves tangent to $\mathbf{e}_1$ in $\mathbf{x}$. It may be computed, e.g., on the curve intersection of $\mathbf{p}$ with the normal plane spanned by $\mathbf{N}$ and $\mathbf{e}_1$, for this "normal section" $\alpha = \pi/2$ and $k_n = -k$. The normal curvature is the curvature of the plane normal section, up to a sign.

By construction, the principal curvatures are invariants. More useful than these are the invariants

$$\begin{aligned} K &= k_1 k_2 \\ H &= \tfrac{1}{2}(k_1 + k_2) \end{aligned} \tag{10-18}$$

K is the *Gauss curvature*, and H is the *mean curvature* of the surface. K is the determinant, and H is half the trace of the matrix of II in curvature-line parameters. The determinant and trace are the two invariants of a 2×2 matrix in any rotation; hence

$$\begin{aligned} K &= l_{11}l_{22} - l_{12}^2 \\ H &= \tfrac{1}{2}(l_{11} + l_{22}) \end{aligned} \tag{10-19}$$

By the first of Eqs. (10-10)

$$d\pi_1{}^2 = -K\pi^1 \wedge \pi^2$$

But we have shown in example 9-1 that K is independent of the angle $\phi = \angle(\mathbf{a}_1,\mathbf{e}_1)$. Hence also

$$d\omega_1{}^2 = -K\omega^1 \wedge \omega^2 \tag{10-20}$$

By Eqs. (10-9), $\omega_1{}^2$ may be computed from ω^1 and ω^2, that is, from $ds^2 = (\omega^1)^2 + (\omega^2)^2$. This leads to the following definition and theorem.

Definition 10-6. *Two local surfaces* $\mathbf{p}(u^1,u^2)$ *and* $\mathbf{q}(u^1,u^2)$ *are isometric if* $ds_p{}^2(u^1,u^2) = ds_q{}^2(u^1,u^2)$.

Problem 13, exercise 9-4, shows that isometric surfaces need not be congruent; a manifold may be isometric to a nonmanifold. Equation (10-20) and the remark on $\omega_1{}^2$ show the following:

Theorem 10-7 (Theorema Egregium; Gauss). *Isometric surfaces have identical Gauss curvatures.*

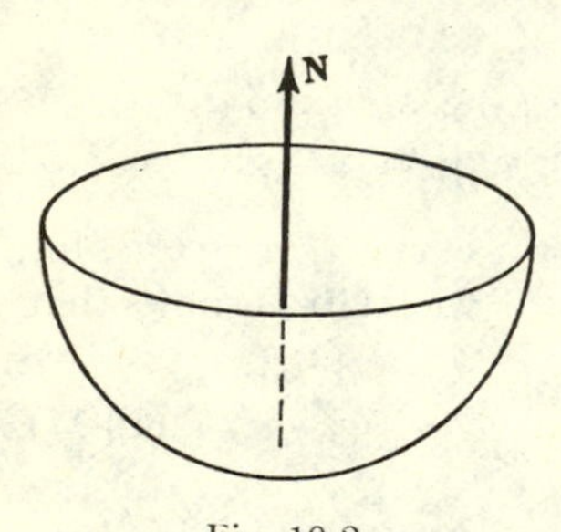

Fig. 10-2

The sign of $K(\mathbf{p}_0)$ gives important information about the shape of the surface in the neighborhood of $\mathbf{p}_0$.

If $K(\mathbf{p}_0) > 0$, $\mathbf{p}_0$ is an *elliptic* point of the surface. Here

$$\text{sign } k_1 = \text{sign } k_2$$

By (10-17) all normal sections through $\mathbf{N}$ have curvatures of the same sign. Together they form a kind of cup with the bottom (or top) at $\mathbf{p}_0$ (Fig. 10-2). It is possible to compare this cup with a spherical calotte. The element of surface area α has been defined in Sec. 9-4 by

$$d\mathbf{p} \text{ ``}\times\text{'' } d\mathbf{p} = 2\alpha\mathbf{N}$$

Equation (10-6) shows that

$$\alpha = \omega^1 \wedge \omega^2 \tag{10-21}$$

The mapping $\mathbf{p} \to \mathbf{N}(\mathbf{p})$ is the *spherical image* of the surface. $\mathbf{N}(\mathbf{p})$ is a point on the unit sphere. The corresponding surface area element is obtained from $d\mathbf{N}$ "$\times$" $d\mathbf{N} = 2\omega_1{}^3 \wedge \omega_2{}^3\mathbf{N}$; it is $K\omega^1 \wedge \omega^2$. Hence K *is the ratio of the area elements of* $\mathbf{p}$ *and its spherical image.* If $\mathbf{p}$ is a sphere of radius R, its area element is R^2 times that of the unit sphere, $K = R^{-2}$. The area of a small calotte about $\mathbf{p}_0$ is approximated by that of a calotte on a sphere of radius $K^{-\frac{1}{2}}$.

If $K(\mathbf{p}_0) = 0$, $\mathbf{p}_0$ is a *parabolic* point. At least one of the principal curvatures vanishes. If $k_1 = k_2 = 0$, $\mathbf{p}_0$ is a *planar umbilic* $d\mathbf{N} = 0$, and every normal section has a contact of order ≥ 2 with its tangent at $\mathbf{p}_0$. If $k_1 = 0$, $k_2 \neq 0$, sign k_n = sign k_2 for all directions except that of $\pm\mathbf{e}_1$ (Fig. 10-3).

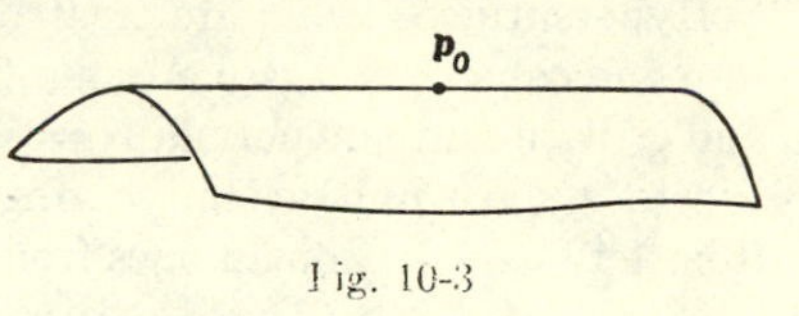

Fig. 10-3

If $K(\mathbf{p}_0) < 0$, $\mathbf{p}_0$ is a *hyperbolic* point. There are two distinct directions for which $k_n = 0$ (hence II = 0). They are called *asymptotic directions.* They divide the tangent plane into four sectors in each of which k_n is of constant sign. The sign of k_n changes from sector to sector. $\mathbf{p}_0$ is a saddle point (Fig. 10-4).

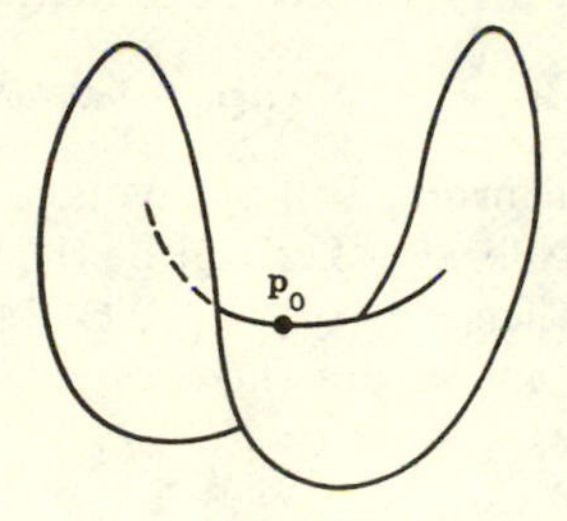

Fig. 10-4

For all practical computations it is important to know the normal $\mathbf{N}$ to the surface. It may be computed from the defining equation (10-1). The angle ϕ between the parameter lines may be obtained from

$$\mathbf{p}_{u^1} \cdot \mathbf{p}_{u^2} = |\mathbf{p}_{u^1}|\,|\mathbf{p}_{u^2}| \cos\phi = F$$

hence
$$|\mathbf{p}_{u^1} \times \mathbf{p}_{u^2}| = |\mathbf{p}_{u^1}|\,|\mathbf{p}_{u^2}| \sin\phi = (EG - F^2)^{\frac{1}{2}}$$

$$\mathbf{N} = (EG - F^2)^{-\frac{1}{2}}\mathbf{p}_{u^1} \times \mathbf{p}_{u^2} \tag{10-22}$$

If $\mathbf{N}$ is known, the Gauss and the mean curvatures are best computed by a set of formulas due to H. Hopf and K. Voss. They are easily verified from Eqs. (10-19) and (10-21), the Frenet equations, and the fact that $\|\mathbf{e}_1,\mathbf{e}_2,\mathbf{e}_3\| = 1$:

$$\text{“}\|\text{”}\ \mathbf{N},d\mathbf{p},d\mathbf{p}\ \text{“}\|\text{”} = 2\alpha \tag{10-23}$$
$$\text{“}\|\text{”}\ \mathbf{N},d\mathbf{p},d\mathbf{N}\ \text{“}\|\text{”} = -2H\alpha \tag{10-24}$$
$$\text{“}\|\text{”}\ \mathbf{N},d\mathbf{N},d\mathbf{N}\ \text{“}\|\text{”} = 2K\alpha \tag{10-25}$$

The linear forms ω^i are easily obtained only for orthogonal parameter lines, $F = 0$. In this case, $\omega^1 = \sqrt{E}\,du^1$, $\omega^2 = \sqrt{G}\,du^2$, and, by (10-9), $\omega_1{}^2 = \frac{1}{2}(EG)^{-\frac{1}{2}}(-E_{u^2}\,du^1 + G_{u^1}\,du^2)$.

Hypersurfaces in R^n are treated like surfaces in R^3. From any curve in $\mathbf{p}$ one constructs again a frame $\{\mathbf{e}_i\}$ such that $\mathbf{e}_\alpha \in T_x$ $(1 \leq \alpha \leq n-1)$ and $\mathbf{e}_n$ is the unique normal $\mathbf{N} = *(\mathbf{e}_1 \wedge \cdots \wedge \mathbf{e}_{n-1})$. $\mathbf{N}$ is the unique (up to a sign) unit vector solving $\mathbf{N} \cdot \mathbf{e}_\alpha = 0$. Here Greek indices run from 1 to $n - 1$, Roman ones from 1 to n. In this frame

$$d\mathbf{p} = \omega^\alpha \mathbf{e}_\alpha$$
$$d\{\mathbf{e}_i\} = (\omega_i{}^j)\{\mathbf{e}_j\} \qquad \omega_i{}^j = -\omega_j{}^i$$

The Gauss-Codazzi-Mainardi equations are

$$d\omega^\alpha = \omega^\beta \wedge \omega_\beta{}^\alpha \tag{10-26}$$
$$\omega^\beta \wedge \omega_\beta{}^n = 0 \tag{10-27}$$
$$d\omega_i{}^k = \omega_i{}^j \wedge \omega_j{}^k \tag{10-28}$$

By (10-27) and theorem 9-6, the second fundamental form

$$\mathrm{II} = -d\mathbf{N} \cdot d\mathbf{p} = \Sigma\omega_n{}^\alpha\omega^\alpha$$

is a symmetric quadratic form. It may be diagonalized by an orthogonal transformation leaving $\mathbf{N}$ fixed. Let $\{\mathbf{a}_i\}$ be the corresponding frame and π^α, $\pi_i{}^j$ its linear forms; then $\pi_n{}^\alpha = k^\alpha\pi^\alpha$. The k^α are the *principal curvatures.* Generalizing, H and K are the *elementary symmetric functions* of the principal curvatures

$$\binom{n-1}{1} H_1 = \Sigma k^\alpha$$
$$\binom{n-1}{2} H_2 = \Sigma k^{\alpha_1}k^{\alpha_2}$$
$$\cdots\cdots\cdots\cdots$$
$$\binom{n-1}{n-1} H_{n-1} = k^{\alpha_1}k^{\alpha_2} \cdots k^{\alpha_{n-1}}$$

The $\pi_\alpha{}^\beta$ may be computed from (10-26) and the principal curvatures from $d\pi_\alpha{}^\beta = \pi_\alpha{}^\gamma \wedge \pi_\gamma{}^\beta - k^\alpha k^\beta\pi^\alpha \wedge \pi^\beta$. The principal curvatures are determined by the ds^2 alone (no corresponding result holds for surfaces). The remaining equations (10-28)

$$d\pi_n{}^\alpha = d(k^\alpha\pi^\alpha) = k^\alpha k^\beta \pi^\alpha \wedge \pi^\beta$$

are a set of partial differential equations for the k^α, hence also for the coefficients $g_{\alpha\beta}$ of the first fundamental form $ds^2 = d\mathbf{p} \cdot d\mathbf{p} = g_{\alpha\beta}\,du^\alpha\,du^\beta$.

Exercise 10-1

1. Prove that the distance from the origin of the tangent plane T_0 is $\mathbf{p}_0 \cdot \mathbf{N}(\mathbf{p}_0)$.

2. Find the tangent plane and **N** for a quadric in R^3:

$$a_{ij}x^ix^j + b_ix^i + c = 0$$

◆ 3. A local surface is given by an equation $x^3 = f(x^1,x^2)$. $d\mathbf{p}$ is the vector $[dx^1, dx^2, (\partial f/\partial x^1)\,dx^1 + (\partial f/\partial x^2)\,dx^2]$. Compute **N**, α, H, and K.

◆ 4. A point is an umbilic if $\|\mathbf{N},d\mathbf{N},d\mathbf{p}\| = 0$. Check whether

$$(x^2 + y^2 + z^2)^2 = 4a^2(x^2 + y^2)$$

has umbilics.

5. Compute H and K for the surface of Prob. 4. (Use the approach of Prob. 3.)

6. Same as Prob. 5 for an ellipsoid.

◆ 7. Show that the parameter lines on a *right conoid* $(u^1 \cos u^2, u^1 \sin u^2, \phi(u^2))$ are orthogonal. Compute the Cartan matrix for the frames whose vectors $\mathbf{e}_1$, $\mathbf{e}_2$ are tangent to the parameter lines. Find H and K.

8. Diagonalize the matrix $\begin{pmatrix} E & F \\ F & G \end{pmatrix}$ of the first fundamental form and compute ω^1, ω^2, and $\omega_1{}^2$ for the frame so obtained.

9. *Dupin's indicatrix* is the curve $\mathbf{i}(\phi) = k_n(\phi)^{-\frac{1}{2}}(\mathbf{e}_1 \cos \phi + \mathbf{e}_2 \sin \phi)$ in T_p. Show that sign $k_n = l_{mn}i^mi^n$ and that the indicatrix is an ellipse at an elliptic point, a pair of conjugate hyperbolas at a hyperbolic point, and a pair of parallel lines at a parabolic point.

10. The principal curvatures are the two numbers k for which the determinant of the matrix of the quadratic form $k\,ds^2 - \text{II}$ vanishes. Prove this statement [using (10-8)] and derive from it a formula for K and H as functions of E, F, G and the l_{ij}.

11. In this problem, a surface is referred to the arc lengths s^1, s^2 of its curvature lines as parameters. The normals to the surface along a curvature line $s^2 = s^2{}_0$ form a ruled surface.
(*a*) Compute the dual vector of the line $\mathbf{p}(s^1,s^2{}_0) + \lambda\mathbf{N}(s^1,s^2{}_0)$.
(*b*) Show that the surface of the normals is developable and that its edge of regression is $\mathbf{x} = \mathbf{p} + k_1{}^{-1}\,\mathbf{N}$. $\mathbf{x}$ is the *line of centers* of the curvature line $s^2 = \text{const}$. The locus of the line of centers for varying s^2 is a *center surface* of $\mathbf{p}$. A surface has two center surfaces.

12. Find the center surfaces (Prob. 10) of a sphere and of a right circular cylinder.

13. If $\mathbf{p}_0$ is not an umbilic, show that the tangent planes to the two center surfaces at the points $\mathbf{p}_0 + k_1^{-1}\mathbf{N}_0$, $\mathbf{p}_0 + k_2^{-1}\mathbf{N}_0$ are perpendicular (see Prob. 10).

14. If a center surface (Prob. 10) is reduced to a curve, show that the corresponding curvature lines must be circles.

15. Show that in a parabolic point which is not a planar umbilic both l_{11} and l_{22} may not vanish together.

16. Prove that $H^2 - K \geq 0$ for any surface.

17. The *parallel surface* in distance c to a surface $\mathbf{p}$ is

$$\mathbf{p}_c(u^1,u^2) = \mathbf{p}(u^1,u^2) + c\mathbf{N}(u^1,u^2)$$

(*a*) Prove that at corresponding points a surface and its parallel surfaces have parallel tangent planes.

(*b*) Prove that the surface element of $\mathbf{p}_c$ is $\alpha_c = (1 - 2cH + c^2K)\alpha$.

(*c*) If $\mathbf{p}$ is a closed convex surface which bounds a volume V_0, prove that the volume bounded by $\mathbf{p}_c$ is $V_0 + A_0c - c^2\int H\alpha + \frac{1}{3}c^3\int K\alpha$, where A_0 is the surface area of $\mathbf{p}$ (Steiner).

18. Show that the sum of the normal curvatures in two perpendicular directions is always $2H$.

19. Let $\mathbf{c}$ be a plane section of a surface $\mathbf{p}$ through a point $\mathbf{p}_0$, and let ϕ be the angle between $\mathbf{a}_1$ and the intersection of T_0 with the plane of $\mathbf{c}$, and θ the angle between $\mathbf{N}$ and the normal to the plane of $\mathbf{c}$.

(*a*) Compute the curvature k_c of $\mathbf{c}$ as a function of k_1, k_2, ϕ, θ.

(*b*) Show that the line connecting the center of curvature of $\mathbf{c}(\theta)$ with the center of curvature of the corresponding normal section $\mathbf{c}(0)$ is orthogonal to $\mathbf{N}$ (Meusnier, 1776).

20. The *third fundamental form* of a surface is $\mathrm{III} = d\mathbf{N} \cdot d\mathbf{N}$. Prove that $K\,ds^2 - 2H\,\mathrm{II} + \mathrm{III} = 0$.

21. Prove that T_p is constant along the rulings of a *developable* surface. Use this fact and Rodrigues's theorem to show that the rulings are curvature lines on developable surfaces. What is the value of K for developable surfaces?

22. Two families of curves on a surface are *conjugate* if their tangents at any point of intersection of two curves from different families are symmetric with respect to the principal directions. Show that the only conjugate families whose curves are orthogonal trajectories of one another are the curvature lines (Dupin). (HINT: Take the curves as parameter lines and show that $l_{12} = 0$.)

23. If $\mathbf{x}(s)$ is a curve on a sphere, put $D = \mathbf{x}''' \cdot \mathbf{N}$. Compute D as a function of k_n, k_g, and t_r and show that for a closed curve $\oint D\,ds = 0$ (Saban). (See exercise 8-1, Probs. 25 to 27.)

24. A *canal surface* is the envelope of a (one-parameter) family of spheres of constant radius. Show that each sphere has a circle in common with the envelope and that the circles are curvature lines (Enneper). (Use Rodrigues's theorem.)

25. Let $\mathbf{p}(u^1,u^2)$ be a local subspace (two-dimensional) in R^4. To a family of frames $\{\mathbf{e}_i\}$, $\mathbf{e}_1,\mathbf{e}_2 \in T_x$, $\mathbf{e}_3,\mathbf{e}_4$ normal to the subspace, there belongs a Frenet equation $d\mathbf{e}_i = \omega_i{}^j\mathbf{e}_j$, $\omega_i{}^j = -\omega_j{}^i$. Study the action of a transformation of the frames by

$$\begin{pmatrix} \cos\phi & \sin\phi & 0 & 0 \\ -\sin\phi & \cos\phi & 0 & 0 \\ 0 & 0 & \cos\theta & \sin\theta \\ 0 & 0 & -\sin\theta & \cos\theta \end{pmatrix}$$

on the Cartan matrix. Deduct the equations corresponding to (10-9) and (10-10). Show that

$$(\omega_1{}^3,\omega_2{}^3,\omega_1{}^4,\omega_2{}^4) = (\omega^1,\omega^2)\begin{pmatrix} a & b & f & g \\ b & c & g & h \end{pmatrix}$$

Show that, in general, an invariant frame will be obtained by the condition $a + c = 0$, $b = 0$. How many invariants does one obtain? State the conditions on a,b,c under which no rotation can be found to satisfy $a + c = b = 0$.

10-2. Examples

I. *Constant curvatures.* What are the surfaces that are transformed into themselves by a group of euclidean motions transitive on the surface? Such a group in R^3 which may bring a given point $\mathbf{p}_0$ onto any other point $\mathbf{p}$ of the surface must have at least two parameters and the curvatures of the surface must be constant. In this case

$$\begin{aligned} d\pi_1{}^3 &= k_1\, d\pi^1 = k_1\pi_1{}^2 \wedge \pi^2 \\ &= \pi_1{}^2 \wedge \pi_2{}^3 = k_2\pi_1{}^2 \wedge \pi^2 \end{aligned}$$

Hence either

$$k_1 = k_2$$

and all points of the surface are umbilics, or

$$\pi_1{}^2 \equiv 0$$

π^2 cannot be the zero form if ds^2 is positive-definite.

A. If all points of the surface are umbilics, Rodrigues's equation $d\mathbf{N} + \lambda\, d\mathbf{p} = 0$ must be an identity in $d\mathbf{p}$.

1. If $\lambda = 0$, then $d\mathbf{N} = 0$, $\mathbf{N}$ is constant, and the local surface $\mathbf{p}$ is an open set in the plane $(\mathbf{p} - \mathbf{p}_0) \cdot \mathbf{N} = 0$.

2. If $\lambda \neq 0$, then $d(d\mathbf{N} + \lambda d\mathbf{p}) = d\lambda \wedge d\mathbf{p} = 0$ must be an identity; hence $d\lambda = 0$, $\lambda = \text{const}$. We may take the fixed system of coordinates such that $\mathbf{N}_0 = \mathbf{p}_0$. Then for all points of the surface, $\mathbf{N}(\mathbf{p}) = \lambda\mathbf{p}$. The surface is the sphere $\mathbf{p} \cdot \mathbf{p} = |\lambda|^{-1}$, the orthogonal surface to the rays through the origin (or an open set on the sphere). It follows from the proof that a surface with only umbilics must be of constant curvature. For the plane, $d\mathbf{N} = 0$ implies $k_1 = k_2 = 0$. For the sphere it was shown in the previous section that $K = k_1k_2 = R^{-2}$; hence

$$k_1 = k_2 = R^{-1} = \text{const}$$

Theorem 10-8. *A local surface all of whose points are umbilics is part either of a plane or of a sphere.*

If the surface is transformed into itself by a group of euclidean motions, it must be a complete plane or a complete sphere.

B. If $\pi_1{}^2 = 0$, then $d\pi^1 = d\pi^2 = 0$, by Eqs. (10-9). If one writes $\pi^i = a_1\,du^1 + a_2\,du^2$, $d\pi^i = 0$ means that $\partial a_1/\partial u^2 = \partial a_2/\partial u^1$; this is the necessary and sufficient condition for the existence of a function q^i such that $dq^i = \pi^i$ (see also exercise 9-3, Prob. 10, and theorem 10-25). By a change of parameters the ds^2 becomes $ds^2 = (dq^1)^2 + (dq^2)^2$, and the surface is *isometric with a plane.* The q^i are curvature-line parameters since $q^i = \text{const}$ implies $\pi^i = 0$. If the surface is not a plane (case *A*1) one may assume $k_1 \neq 0$. The curvature lines $q^2 = \text{const}$ are obtained from the Frenet equation

$$\frac{d}{dq^1}\begin{pmatrix}\mathbf{a}_1\\ \mathbf{a}_2\\ \mathbf{a}_3\end{pmatrix} = \begin{pmatrix}0 & 0 & k_1\\ 0 & 0 & 0\\ -k_1 & 0 & 0\end{pmatrix}\begin{pmatrix}\mathbf{a}_1\\ \mathbf{a}_2\\ \mathbf{a}_3\end{pmatrix}$$

They are circles of radius $k_1{}^{-1}$ in the $(\mathbf{a}_1,\mathbf{a}_3)$ plane. By (10-20),

$$K = k_1k_2 = 0$$

hence $k_2 = 0$. The Frenet equation of the curvature lines $q^1 = \text{const}$ is $d\{\mathbf{a}_i\} = 0$. These are straight lines perpendicular to the circles $q^2 = \text{const}$. $\mathbf{p}$ is a right circular cylinder.

Once the surfaces are known, the groups are easily determined.

Theorem 10-9. *The only groups of euclidean motions that act transitively on a surface in R^3 are* (up to isomorphisms):

a. The group of plane euclidean motions

$$\mathbf{x}^* = \mathbf{x}\begin{pmatrix}\cos\theta & \sin\theta & 0\\ -\sin\theta & \cos\theta & 0\\ 0 & 0 & 1\end{pmatrix} + (a,b,0)$$

acting transitively on the plane $(\mathbf{p} - \mathbf{p}_0) \cdot \mathbf{e}_3 = 0$

b. The group O_3, $\mathbf{x}^* = \mathbf{x}A$, *acting transitively on the sphere* $\mathbf{p} \cdot \mathbf{p} = \mathbf{R}^2$

c. The two-parameter group

$$\mathbf{x}^* = \mathbf{x}\begin{pmatrix} \cos\theta & 0 & \sin\theta \\ 0 & 1 & 0 \\ -\sin\theta & 0 & \cos\theta \end{pmatrix} + (0,c,0)$$

acting transitively on the right circular cylinder $(x^1)^2 + (x^3)^2 = R^2$

The surfaces described here are the only ones with constant principal curvatures.

An interesting consequence of the theorem is that no surface with hyperbolic points may have constant principal curvatures.

II. $K = 0$. Surfaces of constant Gauss curvature will be treated in detail in Chap. 11. Here we want to find only the surfaces with vanishing Gauss curvature. Excluding the trivial case of the plane, we may assume that $k_1 \neq 0$, $k_2 = 0$. By the last equation (10-10),

$$\pi_1{}^2 \wedge \pi_1{}^3 = k_1\pi_1{}^2 \wedge \pi^1 = 0$$

hence $\pi_1{}^2 = a\pi^1$. The curvature lines $s^1 = \text{const}$ are the integral curves of $\pi^1 = 0$. Their Frenet equation is $d\{\mathbf{a}_i\} = 0$; they are straight lines in the direction $d\mathbf{p} = \pi^2\mathbf{a}_2$. The surface is either a developable surface with an edge of regression $\mathbf{s}(s^2) = \mathbf{s}_0 + \int \mathbf{a}_2(s^2)\,ds^2$, or a cone $\mathbf{s}(s^2) = \mathbf{s}_0$, or a cylinder (no $\mathbf{s}_0$ exists). Since the edge of regression (or the vertex of the cone) is not part of the local surface, by the definitions of Sec. 9-4, the methods of Sec. 8-2 are much better suited for the study of these surfaces. The curvature lines $s^2 = \text{const}$ are generalizations of involutes; they are orthogonal trajectories to the tangents of a space curve. On the other hand, $K = 0$ for any developable surface, cone, or cylinder (see Exercise 10-1, Prob. 21) since the rulings are curvature lines.

Theorem 10-10. $K = 0$ *is a necessary and sufficient condition for a local surface to be part of a developable surface, a cone, or a cylinder.*

III. $H = 0$. The surfaces with vanishing mean curvature are called *minimal surfaces.* We give only very few details on the theory of minimal surfaces. Almost all deeper results of that theory are based on the properties of complex analytic functions.

Minimal surfaces are solutions of the Euler equations of the following problem in the calculus of variations. Given a closed space curve $\mathbf{c}$, let $\mathbf{p}$ be a surface whose boundary is $\mathbf{c}$. [$\mathbf{c}$ is the set of those cluster points on R^3 of the point set $\mathbf{p}(u^1,u^2)$ which are not points of $\mathbf{p}$.] Let $\eta(\mathbf{p})$ be any differentiable function on the bounded surface which vanishes on $\mathbf{c}$. $\mathbf{p}^* = \mathbf{p} + \epsilon\eta\mathbf{N}$ is a family of surfaces spanned into $\mathbf{c}$ and near to $\mathbf{p}$ for small ϵ. We look for a condition for $\mathbf{p}$ to be a surface of *minimal surface area* spanned into $\mathbf{c}$, compared with the nearby surfaces $\mathbf{p}^*$. (It may be intuitively clear that there are no surfaces of maximal area.) The line element of $\mathbf{p}^*$ is (neglecting terms in ϵ^2)

$$d\mathbf{p}^* \cdot d\mathbf{p}^* = E^*(du^1)^2 + 2F^*\, du^1\, du^2 + G^*(du^2)^2$$
$$= d\mathbf{p} \cdot d\mathbf{p} + 2\epsilon\eta\, d\mathbf{N} \cdot d\mathbf{p} + \cdots$$

Hence

$$E^* = E - 2\epsilon\eta l_{11} + \cdots$$
$$F^* = F - 2\epsilon\eta l_{12} + \cdots$$
$$G^* = G - 2\epsilon\eta l_{22} + \cdots$$

The element of surface area of $\mathbf{p}^*$ results in

$$\alpha^* = (E^*G^* - F^{*2})^{\frac{1}{2}}\, du^1 \wedge du^2$$
$$= (EG - F^2 - 2\epsilon\eta H + \cdots)^{\frac{1}{2}}\, du^1 \wedge du^2$$

The Euler equation for the variational problem is the condition

$$\left.\frac{d\alpha^*}{d\epsilon}\right|_{\epsilon=0} = 0 \qquad \text{for all } \eta$$

that is,

$$H = 0$$

IV. *A study of the ellipsoid.* The ellipsoid of half axes $a^1 > a^2 > a^3$ is the set of points satisfying

$$\frac{(x^1)^2}{(a^1)^2} + \frac{(x^2)^2}{(a^2)^2} + \frac{(x^3)^2}{(a^3)^2} = 1 \tag{10-29}$$

Its tangential directions $d\mathbf{p} = (dx^i)$ are solutions of

$$\frac{x^1\, dx^1}{(a^1)^2} + \frac{x^2\, dx^2}{(a^2)^2} + \frac{x^3\, dx^3}{(a^3)^2} = 0 \tag{10-30}$$

It follows that $\mathbf{N}$ is the unit vector in the direction $(x^i/(a^i)^2)$; therefore $d\mathbf{N}$ is the sum of a vector in the direction of $(dx^i/(a^i)^2)$ and a vector in the direction of $\mathbf{N}$. By corollary 10-5, the curvature lines are integral curves of

$$\left\| \frac{x^i}{(a^i)^2}, \frac{dx^i}{(a^i)^2}, dx^i \right\| = 0$$

We multiply the ith column by $(a^i)^2$ and expand:

$$((a^1)^2 - (a^3)^2)x^2\, dx^1\, dx^3 + ((a^2)^2 - (a^1)^2)x^3\, dx^2\, dx^1$$
$$+ ((a^3)^2 - (a^2)^2)x^1\, dx^3\, dx^2 = 0$$

First we assume $x^3 \neq 0$. We multiply by $x^3/(a^3)^2$ and eliminate x^3 and dx^3 with the help of Eqs. (10-29) and (10-30):

$$dx^1\, dx^2 \left\{ (x^1)^2 - \frac{(a^1)^2[(a^2)^2 - (a^3)^2]}{(a^2)^2[(a^1)^2 - (a^3)^2]}(x^2)^2 - \frac{(a^1)^2[(a^1)^2 - (a^2)^2]}{(a^1)^2 - (a^3)^2} \right\}$$
$$+ x^1x^2 \left\{ \frac{(a^1)^2[(a^2)^2 - (a^3)^2]}{(a^2)^2[(a^1)^2 - (a^3)^2]}(dx^2)^2 - (dx^1)^2 \right\} = 0 \tag{10-31}$$

In an umbilic, (10-31) holds for arbitrary dx^1 and dx^2. An umbilic satisfies

$$(x^1)^2 - \frac{(a^1)^2[(a^2)^2 - (a^3)^2]}{(a^2)^2[(a^1)^2 - (a^3)^2]}(x^2)^2 - \frac{(a^1)^2[(a^1)^2 - (a^2)^2]}{(a^1)^2 - (a^3)^2} = 0$$

$$x^1x^2 = 0$$

There are only four real umbilics:

$$(x^1)^2 = (a^1)^2\frac{(a^1)^2 - (a^2)^2}{(a^1)^2 - (a^3)^2} \qquad (x^2)^2 = 0 \qquad (x^3)^2 = (a^3)^2\frac{(a^2)^2 - (a^3)^2}{(a^1)^2 - (a^3)^2} \tag{10-32}$$

They are in the plane of the greatest and the smallest half axes. For $x^3 = 0$ one may eliminate x^2, dx^2, instead of x^3 and dx^3. No real umbilics are found.

The tangent plane (10-30) may also be written as

$$\frac{x^1X^1}{(a^1)^2} + \frac{x^2X^2}{(a^2)^2} + \frac{x^3X^3}{(a^3)^2} = 1$$

where (x^i) is the point of contact with the ellipsoid and (X^i) the running point in the plane. The distance of a plane $A_iX^i = B$ from the origin is $D = B/(\Sigma A_i^2)^{-\frac{1}{2}}$. For the ellipsoid one has

$$\frac{1}{D^2} = \sum \frac{(x^i)^2}{(a^i)^4} \tag{10-33}$$

The plane $\Sigma x^iX^i/(a^i)^2 = 0$ is the *central plane* to (x^i); it is the plane through the center of the ellipsoid parallel to the tangent plane in (x^i). Its intersection with the surface of the ellipsoid is the *central section* to (x^i). All intersections of the ellipsoid with planes parallel to the central plane are images in a normal affinity of circular sections of a sphere with a family of parallel planes; hence they are all homothetic to one another. We now take a new system of cartesian coordinates: The origin is at $\mathbf{x} = (x^i)$, the x and y axes are in T_x, and the z axis is the normal line to T_x pointing to the interior. The equation of the ellipsoid remains quadratic; it will be of the form

$$z = rx^2 + 2sxy + ty^2$$

The equation of a plane section parallel to the central section to $\mathbf{x}$ is $rx^2 + 2sxy + ty^2 = \text{const.}$ From $\mathbf{N} \cdot d\mathbf{p} = 0$ it follows by differentiation with symmetric differentials that $d\mathbf{N} \cdot d\mathbf{p} + \mathbf{N} \cdot d^2\mathbf{p} = 0$. This gives a formula which is often convenient for the computation of the second fundamental form:

$$\begin{aligned} \text{II} &= -\mathbf{N} \cdot d^2\mathbf{p} \\ &= (EG - F^2)^{-\frac{1}{2}}\|\mathbf{p}_{u^1}, \mathbf{p}_{u^2}, \mathbf{p}_{u^1u^1}(du^1)^2 + 2\mathbf{p}_{u^1u^2}\,du^1\,du^2 + \mathbf{p}_{u^2u^2}(du^2)^2\| \end{aligned} \tag{10-34}$$

For $u^1 = x$, $u^2 = y$, one obtains $\mathrm{II}(x) = rx^2 + 2sxy + ty^2$. The equation of an ellipse is diagonal ($s = 0$) if the curve is referred to its axes as coordinate axes.

Theorem 10-11. *The principal directions at a point* $\mathbf{x}$ *of an ellipsoid are the directions of the axes of the central section to* $\mathbf{x}$.

Corollary 10-12. *The central sections to the umbilics are circles.*

The corollary admits a converse. Let a circular central section of radius r be given. The sphere of radius r concentric with the ellipsoid is tangent to the ellipsoid along the given circle; hence one of the axes must be the diameter of the sphere. Under our hypotheses, either $r = a^2$ or $r = 0$. Along the circle $\Sigma(x^i)^2/(a^i)^2 = r^{-2}\Sigma(x^i)^2$. The equation of the two central planes whose sections are circular is

$$(x^1)^2\,\frac{(a^1)^2 - (a^2)^2}{(a^1)^2(a^2)^2} = (x^3)^2\,\frac{(a^2)^2 - (a^3)^2}{(a^2)^2(a^3)^2}$$

One checks easily that the normals of the planes are the normals to the ellipsoid at the umbilics (10-32).

V. *Theorems of Joachimsthal, Dupin, and Liouville.* The angle of two surfaces is the angle of their normals at a point of intersection.

Theorem 10-13 (Joachimsthal). *If the curve of intersection of two surfaces is a line of curvature on both, then their angle is constant along the curve.*

If the angle of two surfaces is constant along their curve of intersection and if that curve is a curvature line on one surface, it is a curvature line also on the other one.

Proof: If a curve $\mathbf{c}$ is a curvature line on two surfaces $\mathbf{p}$ and $\mathbf{q}$, then by Rodrigues's formula $d\mathbf{N}_p + \lambda\, d\mathbf{c} = d\mathbf{N}_q + \lambda^*\, d\mathbf{c} = 0$. It follows for the angle ϕ of the two surfaces that

$$d\cos\phi = d(\mathbf{N}_p\cdot\mathbf{N}_q) = -\lambda\, d\mathbf{c}\cdot\mathbf{N}_q - \lambda^*\mathbf{N}_p\cdot d\mathbf{c} = 0$$

Conversely, if $d(\mathbf{N}_p\cdot\mathbf{N}_q) = 0$ and $\mathbf{c}$ is the curvature line of $\mathbf{p}$, then by Rodrigues's formula $d\mathbf{N}_p\cdot\mathbf{N}_q = -\lambda\, d\mathbf{c}\cdot\mathbf{N}_q = 0$ and by hypothesis $\mathbf{N}_p\cdot d\mathbf{N}_q = 0$. This means that, at any point of $\mathbf{c}$, $d\mathbf{N}_q \in T_p(\mathbf{c})$. We also know that $d\mathbf{N}_q \in T_q(\mathbf{c})$. The intersection of the tangent planes T_p and T_q is the line in the direction $d\mathbf{c}$ of the tangent to $\mathbf{c}$ unless $\phi = 0$. If $\phi \neq 0$, the argument shows that $d\mathbf{N}_q + \lambda^*\, d\mathbf{c} = 0$; $\mathbf{c}$ is the curvature line on $\mathbf{q}$. If $\phi = 0$, $\mathbf{N}_p = \mathbf{N}_q$ along $\mathbf{c}$, $d\mathbf{N}_q + \lambda\, d\mathbf{c} = 0$.

Definition 10-14. *A* C^r *map* $\mathbf{F}(u^1,u^2;\sigma)\colon R^3 \to R^3$ *is a family of surfaces if* $\mathbf{F}(u^1,u^2,\sigma_0)$ *is a surface for all values* σ_0. *Three families of surfaces* $\mathbf{F}_i(u^1,u^2;\sigma^i)$ *are triply orthogonal if there is a region in* R^3 *in which*

$$\mathbf{F}_1(u^1,u^2;\sigma^1) = \mathbf{F}_2(v^1,v^2;\sigma^2) = \mathbf{F}_3(w^1,w^2;\sigma^3) = \mathbf{x}$$

has a unique solution for each $\mathbf{x}$ *and if* $\mathbf{N}_i(\mathbf{x}) \cdot \mathbf{N}_j(\mathbf{x}) = \delta_{ij}$, *where* $\mathbf{N}_i$ *is the normal to the surface* $\mathbf{F}_i$ *through* $\mathbf{x}$.

Each one of the families of a triply orthogonal system contains a unique surface through any given point in the region considered. The values $(\sigma^1,\sigma^2,\sigma^3)$ associated with $\mathbf{x}$ may serve as coordinates in that region.

***Example* 10-1.** Assume $a^1 > a^2 > a^3$. Bolzano's theorem shows that the equation

$$\frac{(x^1)^2}{\sigma - a^1} + \frac{(x^2)^2}{\sigma - a^2} + \frac{(x^3)^2}{\sigma - a^3} = 1 \tag{10-35}$$

has three real roots $\sigma^1 > a^1 > \sigma^2 > a^2 > \sigma^3 > a^3$ for any given point $(x^i) \in R^3$ not in one of the coordinate planes. The three families of quadrics so defined are triply orthogonal. $\mathbf{N}_i$ is the unit vector in the direction of $\mathbf{v}_i = (x^j/(\sigma^i - a^j))$ and

$$\begin{aligned}\mathbf{v}_i \cdot \mathbf{v}_k &= \sum_{j=1}^{3} \frac{(x^j)^2}{(\sigma^i - a^j)(\sigma^k - a^j)} = \frac{1}{\sigma^k - \sigma^i} \sum_j \left[\frac{(x^j)^2}{\sigma^i - a^j} - \frac{(x^j)^2}{\sigma^k - a^j}\right] \\ &= 0\end{aligned}$$

$\mathbf{F}_1$ is a family of ellipsoids, $\mathbf{F}_2$ of one-sheeted hyperboloids, and $\mathbf{F}_3$ of two-sheeted hyperboloids.

Theorem 10-15 (Dupin). *The surfaces of a triply orthogonal system intersect in curvature lines.*

A curve is a curvature line if and only if its relative torsion vanishes. By Eq. (10-17), mutually orthogonal curves have opposite relative torsions, $t_r(\phi) + t_r(\phi + \pi/2) = 0$. Let t_i be the relative torsion on the curve $\sigma^j = \text{const}$, $j \neq i$. These are the curves of intersection of the families of the triply orthogonal system; they are mutually orthogonal and lie by pairs on one surface. Hence $t_i + t_j = 0$ $(i \neq j;\ i,j = 1,2,3)$ or $t_i = 0$ $(i = 1,2,3)$.

Dupin's theorem admits a converse in the sense that every curvature line on a surface may be obtained as an intersection with a surface of a triply orthogonal system. If the given surface is referred to its curvature-line parameters s^1 and s^2, the family $\mathbf{F}(s^1,s^2;\sigma) = \mathbf{p}(s^1,s^2) + \sigma\mathbf{N}(s^1,s^2)$ contains two families of mutually orthogonal developables of normals ($s^1 = \text{const}$ and $s^2 = \text{const}$) (see exercise 10-1, Prob. 10) and the parallel surfaces $\sigma = \text{const}$ $(|\sigma| < \epsilon)$. The normal vectors are $\mathbf{N}_i = \mathbf{a}_i$; they are mutually orthogonal.

As an application of Dupin's theorem, we determine all conformal transformations in R^3. It is well known that every analytic function of one complex variable $w = f(z)$ is a conformal transformation in some domain in the plane R^2. These functions do not form a transformation group in the Lie sense; no finite-dimensional algebra can be associated with these transformations. In higher dimensions, the situation is quite different.

Definition 10-16. *A C^6 transformation $F: R^3 \to R^3$ is conformal if non-oriented angles are invariants of F.*

F is conformal if for any two curves $\mathbf{x}(s)$, $\mathbf{y}(s)$, intersecting for $s = s_0$,

$$\cos \phi = \mathbf{x}'(s_0) \cdot \mathbf{y}'(s_0) = \frac{F'(\mathbf{x}(s_0)) \cdot F'(\mathbf{y}(s_0))}{|F'(\mathbf{x}(s_0))|\,|F'(\mathbf{y}(s_0))|}$$

Trivial examples of conformal mappings are the euclidean motions and the homotheties—in short, the similitudes (Sec. 3-4).

Transformation groups define their own spaces of action. The construction of these spaces for a given group is a problem of global geometry or differential topology. But for conformal mappings the problem admits a simple solution by elementary geometry.

First we convince ourselves that it is not convenient to discuss conformal mappings in R^3. A simple conformal transformation which is not a similitude is the *inversion*

$$\mathbf{y} = \frac{\mathbf{x}}{|\mathbf{x}|^2} \tag{10-36}$$

$\mathbf{y}$ is the vector of length $|\mathbf{x}|^{-1}$ on the ray of $\mathbf{x}$. The unit sphere is invariant in an inversion. An equation $a\mathbf{x}^2 + \mathbf{b} \cdot \mathbf{x} + c = 0$ represents a sphere for $a \neq 0$ and a plane for $a = 0$, $\mathbf{b} \neq 0$. Its image in an inversion (at least for the points $\mathbf{x} \neq 0$) is $a + \mathbf{b} \cdot \mathbf{y} + c\mathbf{y}^2 = 0$. This is a sphere for $c \neq 0$ and a plane for $c = 0$, $\mathbf{b} \neq 0$. Especially, spheres through the origin ($c = 0$) are transformed into planes and vice versa. Given two intersecting curves $\mathbf{x}_i(s)(i = 1,2)$, their transforms $\mathbf{x}_i(s)/|\mathbf{x}_i(s)^2|$ make angles [computed for $\mathbf{x} = \mathbf{x}_1(s_1) = \mathbf{x}_2(s_2)$]

$$\frac{\mathbf{x}^2\mathbf{x}_1' - 2(\mathbf{x} \cdot \mathbf{x}_1')\mathbf{x}}{\mathbf{x}^2} \cdot \frac{\mathbf{x}^2\mathbf{x}_2' - 2(\mathbf{x} \cdot \mathbf{x}_2')\mathbf{x}}{\mathbf{x}^2} = \mathbf{x}_1' \cdot \mathbf{x}_2' = \cos \phi$$

The inversion is conformal. It is an *involutive* transformation, which means that its square is the identity transform.

Equation (10-36) is not defined for $\mathbf{x} = 0$ nor is $\mathbf{y} = 0$ the image point of any $\mathbf{x} \in R^3$. But if $|\mathbf{x}| < \epsilon$, $|\mathbf{y}| > \epsilon^{-1}$; therefore it is natural to introduce a single "point at infinity" as the image of $\mathbf{x} = 0$ in the inversion. A "neighborhood of ∞" will be the exterior of any sphere in R^3. With this neighborhood relation, $R^3 + \infty$ may be given the coordinate structure of the three-dimensional sphere S^3 in R^4. S^3 is the surface

$$(\xi^1)^2 + (\xi^2)^2 + (\xi^3)^2 + (\xi^4)^2 = 1$$

in cartesian coordinates (ξ^i) in R^4. We project S^3 onto its equatorial R^3 by "stereographic projection" (see Fig. 5-2). The line through a point $\boldsymbol{\xi} = (\xi^i)$ on the sphere and the "south pole" $\boldsymbol{\sigma} = (0,0,0,-1)$ is the set of vectors $a\boldsymbol{\xi} + b\boldsymbol{\sigma}$, $a + b = 1$. Its intersection $\mathbf{x} = \Sigma(\boldsymbol{\xi})$ with the hyperplane $\xi^4 = 0$ is determined by $a\xi^4 - b = 0$, $a + b = 1$, or $a = (1 + \xi^4)^{-1}$. $\mathbf{x}$ is the stereographic projection of $\boldsymbol{\xi}$,

$$x^i = \frac{\xi^i}{1 + \xi^4} \qquad i = 1, 2, 3 \tag{10-37}$$

This mapping is one-to-one for all points in R^3 and all points in $S^3 - \boldsymbol{\sigma}$. The inverse mapping is

$$\xi^i = \frac{2x^i}{1 + \mathbf{x}^2} \quad i = 1, 2, 3 \qquad \xi^4 = \frac{1 - \mathbf{x}^2}{1 + \mathbf{x}^2} \tag{10-38}$$

If $\boldsymbol{\xi}$ approaches $\boldsymbol{\sigma}$, $|\mathbf{x}|$ tends to infinity. R^3 is represented as a local surface on S^3. The inversion $I: \mathbf{x} \to \mathbf{x}/|\mathbf{x}|^2$ is the image of the reflection

$$A = \begin{pmatrix} 1 & 0 & 0 & 0 \\ 0 & 1 & 0 & 0 \\ 0 & 0 & 1 & 0 \\ 0 & 0 & 0 & -1 \end{pmatrix}$$

in R^4,

$$I = \Sigma A \Sigma^{-1} \tag{10-39}$$

Introduction of the "point at infinity" reduces inversion geometry in R^3 to orthogonal geometry on S^3 in R^4 and makes I a one-to-one and differentiable transformation. Each point in R^3 may serve as the origin of the coordinates. There is an inversion defined for each point.

Theorem 10-17 (Liouville). *The conformal transformations in S^3 form a Lie transformation group generated by similitudes and inversions.*

We show that any conformal transformation of S^3 is either a similitude or an inversion or the composition of one inversion and one similitude.

A conformal map M transforms C^3 surfaces into C^3 surfaces and triply orthogonal systems into triply orthogonal systems. Curvature lines are mapped into curvature lines by Dupin's theorem and its converse; hence umbilics go over into umbilics, and, by theorem 10-8, planes and spheres are mapped into planes and spheres. M transforms the planes $x^i =$ const $(i = 1,2,3)$ either into a triply orthogonal system of planes or into a triply orthogonal system of spheres [all passing through $\mathbf{z} = M(\infty)$]. In the latter case, let I be the inversion of center $\mathbf{z}$. IM maps ∞ onto itself. Hence the planes $x^i =$ const are mapped onto some triply orthogonal system of planes. These new planes may be given by $y^i =$ const in a new cartesian system of coordinates. A coordinate axis $x^i = x^j = 0$

$(i \neq j)$ is mapped onto the corresponding axis $y^i = y^j = 0$. The map, M in the first case, IM in the second case, therefore has an analytic representation $y^i = y^i(x^i)$. The map transforms spheres $\mathbf{x} \cdot \mathbf{x} = \text{const}$ into spheres $\mathbf{y} \cdot \mathbf{y} = \text{const}$; the equations $\Sigma x^i \, dx^i = 0$ and $\Sigma y^i y^{i\cdot} \, dx^i = 0$ must hold simultaneously. Since $d\mathbf{x}$ may be any vector orthogonal to $\mathbf{x}$, the vector $(y^i y^{i\cdot})$ must be linearly dependent on (x^i). The differential equations

$$y^i \frac{dy^i}{dx^i} = c^2 x^i \qquad y^i(0) = 0$$

imply $\mathbf{y} = c\mathbf{x}$. Therefore either M or IM is a translation bringing the origin of the x^i coordinates on that of the y^i coordinates followed by a rotation of the parallels to the x^i axes onto the y^i axes followed by a homothety of ratio c; either M or IM is a similitude.

VI. *Asymptotic lines.* The curvature lines are not the only invariant system of curves that may serve to define an invariant family of frames on a surface, at least not on hyperbolic surfaces $K < 0$. Through every point of a hyperbolic surface there pass two *asymptotic lines* whose tangents are asymptotic directions (page 213). From Eq. (10-2) it is seen that *a curve is asymptotic* ($k_n = 0$) *if and only if its tangent normal frame is its Frenet frame as a space curve.* For given principal curvatures one may compute the asymptotic directions from the first equation (10-17) and insert the result into the second equation; one obtains a formula due to Enneper:

$$t_r = \sqrt{-K} \tag{10-40}$$

Exercise 10-2

1. Use the results of I to show that O_3 has no two-dimensional subgroup.
2. In cylindrical coordinates (r,θ,z) the group of motions which acts transitively on a right circular cylinder has the equations of a translation group. Explain.
3. If the rulings of a ruled surface are curvature lines, show that the surface is developable or is a cylinder or a cone.
4. Prove that the Gauss curvature of a ruled surface is

$$K = -\left(\frac{k_1 k_2}{k_1^2 r^2 + k_2^2}\right)^2$$

where k_1 and k_2 are the functions introduced in Sec. 8-2 and r is the distance of the point from the line of striction measured on the ruling.

5. Show that the tangents to a curve on a surface and to its spherical image are parallel at corresponding points if and only if the curve is a curvature line.

6. What is the spherical image of a plane line of curvature?

♦ **7.** Parallel surfaces have been defined in exercise 10-1, Prob. 17.

(*a*) Show that every frame of $\mathbf{p}_c$ is parallel to a frame of $\mathbf{p}$.

(*b*) Compute the forms $\omega_c{}^i$ in $d\mathbf{p}_c = \omega_c{}^i\mathbf{e}_i$.

(*c*) Using part (*a*), show that $\pi_i{}^j{}_{(c)} = \pi_i{}^j$. Use this relation to compute the curvatures of $\mathbf{p}_c$, given those of $\mathbf{p}$ ($\omega_c{}^i \neq \omega^i$!).

(*d*) Prove Bonnet's theorem: Among all parallel surfaces to a surface of constant Gauss curvature ($K \neq 0$) there are two of constant mean curvature.

(*e*) From exercise 10-1, Prob. 17*c*, show that a minimal surface has *greater* surface area than its nearby parallel surfaces (Steiner).

8. A *surface of revolution* may be given by

$$x^1 = r(u)\cos v \qquad x^2 = r(u)\sin v \qquad x^3 = \phi(u)$$

u becomes the arc length of the meridian if we normalize $r'^2 + \phi'^2 = 1$. $u = \text{const}$ are parallel circles; $v = \text{const}$ are meridians. The fixed frame is denoted by $\{\mathbf{i}_i\}$.

(*a*) Show that the frame adapted to this orthogonal system of parameter lines is

$$\begin{pmatrix}\mathbf{e}_1\\ \mathbf{e}_2\\ \mathbf{e}_3\end{pmatrix} = \begin{pmatrix} r'\cos v & r'\sin v & \phi' \\ -\sin v & \cos v & 0 \\ -\phi'\cos v & -\phi'\sin v & r'\end{pmatrix}\begin{pmatrix}\mathbf{i}_1\\ \mathbf{i}_2\\ \mathbf{i}_3\end{pmatrix}$$

(*b*) Compute the Cartan matrix to the frame $\{\mathbf{e}_i\}$ and show that parallel circles and meridians are curvature lines.

(*c*) Show that a surface of revolution is a minimal surface only if $r'' = 1$ [use the result of part (*b*)]. Find the corresponding minimal surface.

(*d*) When are there umbilics on a surface of revolution?

(*e*) Show that the only surfaces of revolution with vanishing Gauss curvature are right circular cones ($r'' = 0$) and right circular cylinders ($r' = 0$).

(*f*) The first fundamental form of the surface is $ds^2 = du^2 + r^2\,dv^2$. The shape of the surface depends also on ϕ. Verify by direct computation that $K = f(r)$.

9. Prove that

$(au^1 + \sin u^1 \operatorname{Cosh} u^2,\ u^2 + a\cos u^1 \operatorname{Sinh} u^2,\ 1 - a^2\cos u^1 \operatorname{Cosh} u^2)$

is a minimal surface with plane curvature lines (Lie).

10. Prove that the surface $e^z = \cos y/\cos x$ is a minimal surface (Scherk).

11. The principal curvature radii of a surface are $R_i = k_i^{-1}$. Let $\mathbf{p}^* = \mathbf{p} + h\mathbf{N}$ be a parallel surface in distance h. The principal curvature radii of $\mathbf{p}^*$ are R_i^* (for the computation see Prob. 7). If on each normal to the surface the cross ratios (exercise 7-2, Prob. 7) $(R_1,R_2,0,h) = (R_1^*,R_2^*,0,h) = -1$, show that both $\mathbf{p}$ and $\mathbf{p}^*$ are of constant mean curvature $1/h$ (Darboux).

12. What are the curvature lines and the umbilics of an ellipsoid of revolution?

◆**13.** Find the umbilics of $x^1x^2x^3 = a^3$.

14. Prove that the following theorem of Joachimsthal is a consequence of theorem 10-13: If all curvature lines of a surface are plane, their spherical images are two mutually orthogonal systems of circles.

15. Show that the spherical images of the curvature lines of a surface form two orthogonal families of curves on the sphere.

16. If a curvature line is plane, show that the tangent plane to the surface at all points of the curve makes a constant angle with the plane of the curvature line (Joachimsthal).

17. Prove Dupin's theorem directly by showing that $l_{12} = 0$ in the second fundamental form of any one of the surfaces.

18. Does a triply orthogonal system always give *all* curvature lines on a surface of the system?

19. Use Dupin's theorem and example 10-1 to show that Eq. (10-31) must have as its solution a polynomial in x^1, x^2, of order ≤ 2. Find that polynomial by the method of indeterminate coefficients and show that the curvature lines of the ellipsoid are intersections of the surface with certain right cylinders.

20. An ellipsoid is given by $\sigma^1 =$ const in the coordinates of example 10-1. Show that, in these triply orthogonal coordinates, the first fundamental form of the ellipsoid is

$$ds^2 = \frac{1}{4}\frac{\sigma^2(\sigma^2 - \sigma^3)}{(a^1 - \sigma^2)(a^2 - \sigma^2)(a^3 - \sigma^2)}(d\sigma^2)^2 + \frac{1}{4}\frac{\sigma^3(\sigma^3 - \sigma^2)}{(a^1 - \sigma^3)(a^2 - \sigma^3)(a^3 - \sigma^3)}(d\sigma^3)^2$$

HINT: For easy computation, show first that from (10-35) there follows

$$2\frac{dx^i}{x^i} = \frac{d\sigma^1}{a^i - \sigma^1} + \frac{d\sigma^2}{a^i - \sigma^2} + \frac{d\sigma^3}{a^i - \sigma^3}$$

21. Show that the paraboloids

$$\frac{x^2}{p - t} + \frac{y^2}{q - t} = 2(z - t)$$

form two mutually orthogonal families for p and q constant.

22. Prove that a surface is invariant under a two-parameter group of similitudes if (in curvature-line parameters)

$$k_i^{-2}\frac{\partial k_i}{\partial s^j} = \text{const} \qquad i,j = 1,2$$

(see Sec. 3-4 and exercise 8-1, Prob. 28) (Lie).

◆**23.** Find the conditions on the functions $A(s)$, $B(s)$, $C(s)$ such that a surface with the first fundamental form

$$ds^2 = e^{2hs^1}[A(s^2)(ds^1)^2 + 2B(s^2)\,ds^1\,ds^2 + C(s^2)(ds^2)^2]$$

becomes invariant under a group of similitudes (Darboux).

24. Find the Lie algebra of the group of conformal mappings in R^3.

25. Find the group of conformal mappings in R^4.

26. What are the natural orders of differentiability for conformal geometry in R^3?

27. Prove that a ruling on a hyperbolic local surface is always an asymptotic line.

28. How many rulings may pass through a point of *hyperbolic* curvature? (Use Prob. 27.)

29. Show that asymptotic lines are transformed into asymptotic lines in an inversion.

10-3. Integration Theory

A basic problem of surface theory is to find a surface, given its invariants, and to show that the invariants determine a surface up to a euclidean motion. These and all similar questions of more dimensional differential geometry depend upon one fundamental integration theorem. It is important to have this theorem available in all dimensions and also for Cartan matrices that are not in the algebra of an orthogonal group. We shall see that no additional effort is needed to prove the theorem in the most general case.

In some convex neighborhood $\mathfrak{U} \subset R^k$ of variables $u^1, \ldots, u^k$ we assume to be given k linearly independent linear differential forms ω^α and n^2 linear differential forms $\omega_i{}^j$ in the same variables ($n \geq k$). Greek indices always will run from 1 to k, Roman ones from 1 to n. Given vectors $\overset{0}{\mathbf{p}}$, $\overset{0}{\mathbf{e}}_i \in R^n$, where the $\overset{0}{\mathbf{e}}_i$ form a basis in R^n ($\|\overset{0}{\mathbf{e}}_i\| \neq 0$), we want to find a local subspace $\mathbf{p}\colon \mathfrak{U} \to R^n$ and a family of frames $\{\mathbf{e}_i(\mathbf{u})\}$ such that

$$\begin{aligned} d\mathbf{p} &= \omega^\alpha \mathbf{e}_\alpha \\ d\mathbf{e}_i &= \omega_i{}^j \mathbf{e}_j \end{aligned} \tag{10-41}$$

$$\mathbf{p}(\mathbf{u}_0) = \overset{0}{\mathbf{p}} \qquad \mathbf{e}_i(\mathbf{u}_0) = \overset{0}{\mathbf{e}}_i$$

We have to show that a unique map $\mathbf{p}$ can be found under these conditions and that the jacobian of $\mathbf{p}$ is of rank k in a neighborhood of $\mathbf{u}_0$.

For the moment we assume that the coefficients of the ω^α are C^2 functions of $\mathbf{u} \in \mathfrak{U}$ and those of $\omega_i{}^j$ are C^1. The problem is impossible unless (1) the $\overset{0}{\mathbf{e}}_\alpha$ span T_{p_0} and (2) our data formally imply $dd\mathbf{p} = dd\mathbf{e}_i = 0$. Therefore we shall assume that the *integrability conditions*

$$\begin{aligned} d\omega^\alpha &= \omega^\beta \wedge \omega_\beta{}^\alpha \\ \omega^\alpha \wedge \omega_\alpha{}^j &= 0 \qquad j > k \\ d\omega_i{}^j &= \omega_i{}^s \wedge \omega_s{}^j \end{aligned} \tag{10-42}$$

hold in $\mathfrak{U}$.

We choose the origin of the coordinates in R^n at $\mathbf{p}_0$ to simplify the formulas. The fixed orthonormal frame in R^n is $\{\mathbf{i}_i\}$; then $\{\overset{0}{\mathbf{e}}_i\} = A_0\{\mathbf{i}_i\}$, $\|A_0\| \neq 0$. For any differentiable curve $\mathbf{c} = \mathbf{c}(s)$ in R^k passing through $\mathbf{u}_0$ we replace the differentials du^α by their values $(du^\alpha/ds)\,ds$ along $\mathbf{c}$. As a normalization, we assume that $\mathbf{c}(0) = \mathbf{u}_0$. The linear differential forms $\omega^\alpha = a_\beta{}^\alpha\, du^\beta$, $\omega_i{}^j = b_i{}^j{}_\alpha\, du^\alpha$ then may be written

$$\omega^\alpha = w^\alpha\, ds \qquad \omega_i{}^j = w_i{}^j\, ds$$

where

$$w^\alpha = a_\beta{}^\alpha \frac{dc^\beta}{ds} \qquad w_i{}^j = b_i{}^j{}_\beta \frac{dc^\beta}{ds}$$

for the purposes of integration along $\mathbf{c}$. Let $W(s) = (w_i{}^j(s))$. By lemma 2-12, the matrix differential equation $X'X^{-1} = W$ has a unique solution $X(s) = A_0Y(s)$ such that $Y(0) = U$. If we define

$$\{\mathbf{e}_i(s)\} = X(s)\{\mathbf{i}_i\}$$

the map

$$\mathbf{p}(s) = \int_0^s w^\alpha(\sigma)\mathbf{e}_\alpha(\sigma)\, d\sigma \tag{10-43}$$

is a differentiable function of s. We choose a point $\mathbf{p}_1 = \mathbf{p}(s_1)$ and join 0 and $u_1 = \mathbf{c}(s_1)$ by another curve $\mathbf{c}^*(s^*)$ in $\mathfrak{U}$. By the same process as before, we obtain another set of solutions $X^*(s^*)$, $\mathbf{p}^*(s^*)$. It is possible to normalize the parameters on $\mathbf{c}$ and $\mathbf{c}^*$ so that $\mathbf{u}_1 = \mathbf{c}(1) = \mathbf{c}^*(1)$ (s and s^* are *not* normally arc lengths). If we are able to show that $\mathbf{p}(1) = \mathbf{p}^*(1)$, it will follow that $\mathbf{p}$ as defined by Eq. (10-43) does not depend on the curve $\mathbf{c}$ chosen for the computation but only on the point $\mathbf{u} \in \mathfrak{U}$. For the proof we find X not only on the curves $\mathbf{c}(s)$ and $\mathbf{c}^*(s)$ but on the whole family

$$\mathbf{c}_t(s) = (1-t)\mathbf{c}(s) + t\mathbf{c}^*(s) \qquad 0 \le s, t \le 1 \tag{10-44}$$

$\mathfrak{U}$ being convex, all points $\mathbf{c}_t(s)$ are in $\mathfrak{U}$. Along a curve $\mathbf{c}_t$,

$$\frac{du^i}{ds} = (1 - t)\frac{dc^i}{ds} + t\frac{dc^{*i}}{ds} \tag{10-45}$$

This is a special case of

$$d\mathbf{c}_t = \left[(1 - t)\frac{d\mathbf{c}}{ds} + t\frac{d\mathbf{c}^*}{ds}\right] ds + (\mathbf{c}^* - \mathbf{c})\, dt \tag{10-46}$$

The matrix

$$W(\mathbf{c}_t(s)) = \left(b_i{}^j{}_\alpha(\mathbf{c}_t(s))\left[(1 - t)\frac{d\mathbf{c}^\alpha}{ds} + t\frac{d\mathbf{c}^{*\alpha}}{ds}\right]\right) \tag{10-47}$$

is a differentiable function of t, and so is the matrix $X(\mathbf{c}_t(s))$ obtained by integration from $W(\mathbf{c}_t(s))$ for constant t. The first step in the proof is to show that

$$\{\mathbf{e}_i(1,t)\} = X(\mathbf{c}_t(1))\{\mathbf{i}_i\} = \int_0^1 W(\mathbf{c}_t(\sigma))\, d\sigma\, \{\mathbf{e}_i(\sigma,t)\}$$

is independent of t; that is, the frame $\{\mathbf{e}_i(\mathbf{u})\}$ does not depend on the curve $\mathbf{c}$ used in its definition. It follows from (10-46) and (10-47) that

$$\omega_i{}^j = W_i{}^j\, ds + b_i{}^j{}_\alpha(c^{*\alpha} - c^\alpha)\, dt$$

From the definitions we have

$$\begin{aligned}\frac{\partial \mathbf{e}_i(1,t)}{\partial t} = \int_0^1 \frac{\partial b_i{}^j{}_\alpha}{\partial t}[(1 - t)\mathbf{c}' + t\mathbf{c}^{*\prime}]^\alpha\, d\sigma\, \mathbf{e}_j(\sigma,t) \\ + \int_0^1 b_i{}^j{}_\alpha(\mathbf{c}^{*\prime} - \mathbf{c}')^\alpha\, d\sigma\, \mathbf{e}_j(\sigma,t) \\ + \int_0^1 b_i{}^j{}_\alpha[(1 - t)\mathbf{c}' + t\mathbf{c}^{*\prime}]^\alpha\, d\sigma\, \frac{\partial \mathbf{e}_j}{\partial t}(\sigma,t)\end{aligned}$$

In $\mathbf{e}_i(1,1) - \mathbf{e}_i(1,0) = \int_0^1 (\partial \mathbf{e}_i/\partial t)\, dt$ the order of the integrations is inverted in the second integral and integration by parts is used. The constant term vanishes because of $\mathbf{c}(1) = \mathbf{c}^*(1)$, and it results that

$$\begin{aligned}&\mathbf{e}_i(1,1) - \mathbf{e}_i(1,0) \\ &\quad = \int_0^1\int_0^1 \left\{\frac{\partial b_i{}^j{}_\alpha}{\partial t}[(1 - \tau)\mathbf{c}' + \tau\mathbf{c}^{*\prime}]^\alpha - \frac{\partial b_i{}^j{}_\alpha}{\partial s}(\mathbf{c}^* - \mathbf{c})^\alpha\right\} \mathbf{e}_j\, d\tau \wedge d\sigma \\ &\qquad + \int_0^1\int_0^1 \left\{b_i{}^j{}_\alpha[(1 - \tau)\mathbf{c}' + \tau\mathbf{c}^{*\prime}]^\alpha \frac{\partial \mathbf{e}_j}{\partial t} - (\mathbf{c}^* - \mathbf{c})^\alpha \frac{\partial \mathbf{e}_j}{\partial s}\right\} d\tau \wedge d\sigma \\ &\quad = \int_{I^2} d(\omega_i{}^j \mathbf{e}_j) = 0\end{aligned}$$

by our hypotheses. In the same way, if we put

$$w^\alpha(s,t) = a_\beta{}^\alpha(\mathbf{c}_t(s))[(1 - t)\mathbf{c}' + t\mathbf{c}^{*\prime}]^\beta$$

the local variety is

$$\mathbf{p}(s,t) = \int_0^s w^\alpha(\sigma,t)\mathbf{e}_\alpha(\sigma,t)\,d\sigma$$

As before, we obtain

$$\frac{\partial \mathbf{p}(1,t)}{\partial t} = \int_0^1 \frac{\partial a_\beta{}^\alpha}{\partial t}\mathbf{e}_\alpha[(1-t)\mathbf{c}' + t\mathbf{c}^{*\prime}]^\beta\,d\sigma + \int_0^1 a_\beta{}^\alpha \mathbf{e}_\alpha(\mathbf{c}^{*\prime} - \mathbf{c}')^\beta\,d\sigma + \int_0^1 a_\beta{}^\alpha[(1-t)\mathbf{c}' + t\mathbf{c}^{*\prime}]^\beta \frac{\partial \mathbf{e}_\alpha}{\partial t}\,d\sigma$$

Integrating by parts, it follows that

$$\mathbf{p}^*(1) - \mathbf{p}(1) = \mathbf{p}(1,1) - \mathbf{p}(1,0) = \int_0^1 \frac{\partial \mathbf{p}}{\partial t}(1,\tau)\,d\tau = \int_{I^2} d(\omega^\alpha \mathbf{e}_\alpha) = 0$$

where $\qquad \omega^\alpha(s,t) = w^\alpha(s,t)\,ds + a_\beta{}^\alpha(\mathbf{c}^* - \mathbf{c})^\beta\,dt$

The forms ω^α are a basis in R^{*k}. The dual basis of R^k will be $\{\mathbf{x}_\alpha\}$. If $\mathbf{c}_\alpha$ is a curve whose tangent vector at 0 is $\mathbf{x}_\alpha$,

$$d\mathbf{p}\left(\frac{d\mathbf{c}_\alpha}{ds}\right)_{s=0} = \overset{0}{\mathbf{e}}_\alpha$$

This shows that the jacobian $J_u\mathbf{p}$ is of rank k at $\mathbf{u} = 0$. The jacobian is continuous; it is of rank k in a neighborhood of $\mathbf{u} = 0$. We have proved the following theorem:

Theorem 10-18. *k linearly independent linear differential C^1 forms ω^α and n^2 $(n \geq k)$ linear differential C^1 forms $\omega_i{}^j$ defined in an open domain $\mathcal{U} \subset R^k$ determine in a neighborhood of any $\mathbf{u}_0 \in \mathcal{U}$ a unique local subspace $\mathbf{p} \in R^n$ and a unique family of frames $\{\mathbf{e}_i\}$ such that*

$$\mathbf{p}(\mathbf{u}_0) = \mathbf{p}_0 \qquad \mathbf{e}_i(\mathbf{u}_0) = \overset{0}{\mathbf{e}}_i \qquad \|\overset{0}{\mathbf{e}}_i\| \neq 0$$
$$d\mathbf{p} = \omega^\alpha \mathbf{e}_\alpha \qquad d\mathbf{e}_i = \omega_i{}^j\mathbf{e}_j$$

if and only if

$$d\omega^\alpha = \omega^\beta \wedge \omega_\beta{}^\alpha \qquad d\omega_i{}^j = \omega_i{}^m \wedge \omega_m{}^j$$

We apply this theorem to a series of existence and unicity theorems on surfaces in R^3.

Theorem 10-19. *Differential forms in two variables ω^1, ω^2, $\omega_1{}^2$, $\omega_1{}^3$, $\omega_2{}^3$ determine a local surface up to a euclidean motion if and only if $\omega^1 \wedge \omega^2 \neq 0$ and Eqs.* (10-9) *and* (10-10) *hold.*

If we define $\omega_i{}^j = -\omega_j{}^i$, the matrix W belongs to $\mathfrak{L}(O_3)$. X will be an orthogonal matrix if A_0 is orthogonal; the frames $\{\mathbf{e}_i\}$ will be orthonormal. Two different sets of initial conditions determine congruent local surfaces. The geometric importance of theorem 10-18 lies in the fact that X will be in a Lie group G if A_0 is in G and $(\omega_i{}^j)$ is in $\mathfrak{L}(G)$. Theorem 10-18 implies a statement analogous to theorem 10-19 for any type of local subspace in any Klein geometry.

We may discuss here the question of the natural conditions of differentiability in surface theory. Curvatures, curvature lines, and Darboux frames may be computed from $d\mathbf{p}$ and $d\mathbf{N}$. Equations (10-17) show that even the relative torsion of a curve on a surface may be obtained from the principal curvatures, i.e., of second-order invariants. From the start, we could have taken C^2 space curves that do not have Cartan matrices (8-2) but only frames associated with $C(A) = (p_i{}^j)$, where

$$(p_1{}^2)^2 + (p_1{}^3)^2 = k^2$$

[compare page 156 and formula (10-2)]. The only place where derivatives of second-order quantities appear are the integrability conditions (10-10). The proof of theorem 10-18 shows that these are needed only in the integrated form $\iint d\omega_i{}^k = \iint \omega_i{}^m \wedge \omega_m{}^j$ for all smooth images of the unit square in R^k. This is a less stringent condition and does not require C^3 mapping functions. However, the determination of the exact conditions to be imposed on the mapping functions is a difficult problem.†

We ask next whether a surface is determined (up to a congruence) by its two fundamental forms. It is a classical theorem of linear algebra that two quadratic forms $ds^2 = g_{\alpha\beta}\, du^\alpha\, du^\beta$ and $\mathrm{II} = l_{\alpha\beta}\, du^\alpha\, du^\beta$, one of which is positive-definite, may be diagonalized simultaneously by a linear transformation $(\pi^\alpha) = (du^\beta)(A_\beta{}^\alpha)$ such that

$$ds^2 = \Sigma(\pi^\alpha)^2 \qquad \mathrm{II} = \Sigma k_\alpha(\pi^\alpha)^2$$

The principal curvatures are the roots of the characteristic equation $\|kg_{\alpha\beta} - l_{\alpha\beta}\| = 0$. If the two fundamental forms are given, the principal curvatures and the forms π^α may be obtained by an algebraic process.

The forms ω^α which diagonalize the first fundamental form are a basis of T_p^*. Therefore, the differential of a function f defined on the surface may be written

$$df = f_{,1}\omega^1 + f_{,2}\omega^2 \tag{10-48}$$

The functions $f_{,\alpha}$ are the *covariant derivatives* of f with respect to the frame $\{\mathbf{e}_i\}$ which belongs to the ω^α. If the frame is invariant, the $f_{,\alpha}$ are *invariant derivatives* on $\mathbf{p}$. In the sequel we shall use only invariant derivatives with respect to the Darboux frame on the surface.

If f is an invariant function on $\mathbf{p}$, the invariant derivatives are also invariant functions. For the principal curvatures one has the four new invariants $k_{\alpha,\beta}$ defined by

$$dk_\alpha = k_{\alpha,1}\pi^1 + k_{\alpha,2}\pi^2 \tag{10-49}$$

† See, for example, Hassler Whitney, "Geometric Integration Theory," Princeton University Press, Princeton, N.J., 1957.

Furthermore, the functions ρ_α in

$$\pi_1{}^2 = \rho_1\pi^1 + \rho_2\pi^2 \tag{10-50}$$

are invariants since all pfaffians in Eq. (10-50) are invariant forms. For a general frame one writes

$$\omega_1{}^2 = \Gamma_{11}{}^2\omega^1 + \Gamma_{12}{}^2\omega^2 \tag{10-50a}$$

The ρ_α may be computed from Eqs. (10-9):

$$d\pi^1 = \rho_1\pi^1 \wedge \pi^2 \qquad d\pi^2 = \rho_2\pi^1 \wedge \pi^2 \tag{10-51}$$

Equations (10-10) may be translated into relations between the functions just introduced:

$$\rho_{1,2} - \rho_{2,1} = k_1k_2 + \rho_1{}^2 + \rho_2{}^2 \tag{10-52}$$

$$\begin{aligned} k_{1,2} &= \rho_1(k_2 - k_1) \\ k_{2,1} &= \rho_2(k_1 - k_2) \end{aligned} \tag{10-53}$$

Equation (10-52) is the *theorema egregium;* it shows how the Gauss curvature is computed from the ρ_α which depend only on the first fundamental form. Theorem 10-19 may now be translated to apply to the problem at hand:

Theorem 10-20. *A positive-definite quadratic form ds^2 and a quadratic form* II *are the fundamental forms of a local surface (unique up to a congruence) if and only if Eqs.* (10-52) *and* (10-53) *hold.*

Many other existence and unicity theorems are possible. For example, one may give k_1, k_2, ρ_1, and ρ_2 as functions of u^1 and u^2. Equations (10-53) permit the computation of the π^α by a comparison of partial and invariant derivatives. Equations (10-51) and (10-52) remain as integrability conditions of the problem (but compare exercise 10-3, Prob. 7). Another important application is the following: If only the ds^2 is given, the ρ_α and the Gauss curvature may be computed by Eqs. (10-51) and (10-52) if some pfaffians ω^α $[ds^2 = (\omega^1)^2 + (\omega^2)^2]$ are arbitrarily designated as invariant forms π^α. If we eliminate k_2 from Eqs. (10-53) by $k_2 = K/k_1$, the second fundamental form may be obtained from $dk_1 = k_{1,1}\pi^1 + k_{1,2}\pi^2$. By construction, the integrability conditions are satisfied; the ds^2 is always the metric of some surface.

Theorem 10-21. *A positive-definite C^2 quadratic form in two variables is always the ds^2 of some local surface.*

This result illuminates the gap between the theory of surfaces and that of hypersurfaces in R^n $(n > 3)$. It was shown at the end of Sec. 10-1 that the $g_{\alpha\beta}$ of a hypersurface must satisfy a set of partial differential equations depending on the second fundamental form.

***Example* 10-2.** Find all local surfaces isometric with a plane. The $ds^2 = (du^1)^2 + (du^2)^2$ is first assumed to be in curvature-line parameters.

Here $\rho_1 = \rho_2 = K = 0$. At most, one principal curvature is $\neq 0$; let it be k_2. The only nontrivial integrability condition is $k_{2,1} = 0$; it means $k_2 = k_2(u^2)$. The matrix to be integrated is

$$W = \begin{pmatrix} 0 & 0 & 0 \\ 0 & 0 & ku^{2\prime} \\ 0 & -ku^{2\prime} & 0 \end{pmatrix}$$

With $\phi = \int k(u^2)\, du^2$ the Darboux frame becomes

$$\{\mathbf{a}_i\} = \begin{pmatrix} 1 & 0 & 0 \\ 0 & \cos\phi & \sin\phi \\ 0 & -\sin\phi & \cos\phi \end{pmatrix} \{\mathbf{i}_i\}$$

The surface is a cylinder.

By a coordinate transformation $u^\alpha = u^\alpha(v^1,v^2)$ the ds^2 of the plane becomes $ds^2 = E(dv^1)^2 + 2F\, dv^1\, dv^2 + G(dv^2)^2$. $F = 0$ if the v coordinate lines are mutually orthogonal. In order to simplify matters, we ask that $E = (u^1{}_{v^1})^2 + (u^2{}_{v^1})^2 = 1$. Then $K = \rho_1 = 0$, and Sec. 10-2, II, shows that all developable surfaces and all cones may be obtained by prescribing the different possible v^α coordinates as curvature-line parameters: *A surface is locally isometric to the plane if and only if* $K = 0$. (Compare also Sec. 10-4.)

***Example* 10-3.** A type of surfaces that has been studied extensively by G. Monge, one of the founders of differential geometry, is the "molding surfaces" (*surfaces moulures*) defined by

$$\rho_1\rho_2 = 0$$

The surfaces with $\rho_1 = \rho_2 = 0$ are the plane and the right circular cylinders (Sec. 10-2, IB). If $\rho_1 = 0$ and $\rho_2 \neq 0$, the curvature line $\pi^2 = 0$ is obtained by integration of the Frenet equation

$$d\{\mathbf{a}_i\} = \begin{pmatrix} 0 & 0 & k_1\pi^1 \\ 0 & 0 & 0 \\ -k_1\pi^1 & 0 & 0 \end{pmatrix} \{\mathbf{a}_i\}$$

It is a plane curve in a normal plane to the surface; the molding surfaces are orthogonal to a one-parameter family of planes.

***Example* 10-4.** For later use we want to find the ds^2 of a surface of constant negative curvature $K = -1$ in asymptotic-line parameters. Let the local surface be given first in curvature-line parameters,

$$ds^2 = E(ds^1)^2 + G(ds^2)^2$$

Since $\pi^1 = E^{\frac{1}{2}}\, ds^1$, $\pi^2 = G^{\frac{1}{2}}\, ds^2$, the invariant derivatives of a function f are

$$f_{,1} = E^{-\frac{1}{2}} \frac{\partial f}{\partial s^1} \qquad f_{,2} = G^{-\frac{1}{2}} \frac{\partial f}{\partial s^2}$$

By hypothesis, we may write $k_1 = \cot\sigma$ and $k_2 = -\tan\sigma$. From (10-51) we have

$$\rho_1 = -\frac{1}{2}E^{-1}G^{-\frac{1}{2}}\frac{\partial E}{\partial s^2} \qquad \rho_2 = -\frac{1}{2}G^{-1}E^{-\frac{1}{2}}\frac{\partial G}{\partial s^1}$$

Equations (10-53) become

$$E^{-1}\frac{\partial E}{\partial s^2} = 2\cot\sigma\frac{\partial\sigma}{\partial s^2} \qquad G^{-1}\frac{\partial G}{\partial s^1} = -2\tan\sigma\frac{\partial\sigma}{\partial s^1}$$

Both equations involve only derivatives with respect to one variable; they are easily integrated and give $\log E = 2\log|\sin\sigma| + \log f(s^1)$, $\log G = 2\log|\cos\sigma| + \log g(s^2)$. By a change of gauge on the curvature lines $du^1 = f(s^1)^{\frac{1}{2}}\,ds^1$, $du^2 = g(s^2)^{\frac{1}{2}}\,ds^2$, the first fundamental form is brought into

$$ds^2 = \sin^2\sigma(du^1)^2 + \cos^2\sigma(du^2)^2 \tag{10-54}$$

The second fundamental form is diagonal in all parameters defined on the curvature lines, $\mathrm{II} = l_{11}\,(du^1)^2 + l_{22}(du^2)^2 = k_1(\pi^1)^2 + k_2(\pi^2)^2$; hence $l_{11} = k_1E = \sin\sigma\cos\sigma$ and $l_{22} = k_2G = -\sin\sigma\cos\sigma$. The second fundamental form may now be written

$$\mathrm{II} = \sin\sigma\cos\sigma(du^1 + du^2)(du^1 - du^2) \tag{10-55}$$

New parameters q^α are defined by $u^1 = q^1 - q^2$, $u^2 = q^1 + q^2$. The curves $q^\alpha = \text{const}$ are asymptotic lines on the surface. In these new parameters the first fundamental form becomes

$$ds^2 = (dq^1)^2 + 2\cos 2\sigma\,dq^1\,dq^2 + (dq^2)^2 \tag{10-56}$$

A system of parameter lines for which it is possible to have $E = G = 1$ is a *Tchebychef net*. The q^α automatically are arc lengths on the curves of a Tchebychef net. 2σ is the angle of the two asymptotic lines at any point of the surface. It may be computed from the Gauss equation (10-52). This equation is in the curvature-line parameters u^α,

$$\frac{\partial^2\sigma}{\partial(u^2)^2} - \frac{\partial^2\sigma}{\partial(u^1)^2} = \sin\sigma\cos\sigma \tag{10-57}$$

and in asymptotic-line parameters q^α,

$$\frac{\partial^2(2\sigma)}{\partial q^1\,\partial q^2} = \sin 2\sigma \tag{10-58}$$

Exercise 10-3

◆ **1.** The line element of a sphere is $ds^2 = d\phi^2 + \cos^2\phi\,d\theta^2$. There exist surfaces (local) isometric to the sphere for which ϕ and θ are curvature-line parameters with principal curvatures

$$k_1 = (1 - c^2 \cos^2 \phi)^{-\frac{1}{2}} \qquad k_2 = (1 - c^2 \cos^2 \phi)^{\frac{1}{2}}$$

Prove this statement and find the coordinates of such a local surface in R^3 (but compare Sec. 10-5). (Use exercise 10-2, Prob. 8.)

2. Find a quadratic form that is admissible as II for the

$$ds^2 = \cos^2 \phi \, d\theta^2 + \cos^2 \theta \, d\phi^2$$

3. Find the surfaces with plane lines of centers (exercise 10-1, Prob. 11).

4. Investigate the validity of theorem 10-20 for surfaces with umbilics.

5. Formulate a theorem analogous to theorem 10-20 for hypersurfaces in R^n.

6. Find the condition for a positive-definite quadratic form in three variables to be the ds^2 of *more than one* hypersurface in R^4.

7. Show that k_1, k_2, ρ_1, ρ_2 do not define a unique surface if $k_1 = k_1(u^1)$, $k_2 = k_2(u^1)$.

8. Write the equations corresponding to (10-52) and (10-53) for parameters other than curvature-line parameters, in the case $F = 0$.

◆ **9.** If the two fundamental forms of two surfaces are identical, the surfaces are congruent. Does this notion of congruence include symmetry?

10. A *Weingarten surface*, or W surface, is one on which there exists a relation between the principal curvatures, $f(k_1,k_2) = 0$. In this case one may introduce an auxiliary variable h and express the principal curvatures as $k_1 = F(h)$, $k_2 = F(h)[F(h) - hF'(h)]$.

(*a*) Prove that $ds^2 = h^{-2}(ds^1)^2 + F'(h)^{-2}(ds^2)^2$ in curvature-line parameters.

(*b*) Find F for $K = \text{const}$ and for $H = \text{const}$ (Dini).

11. Show that a surface with constant mean curvature and plane curvature lines is a surface of revolution (Dini).

12. Show that a helicoid (exercise 9-4, Prob. 11) is a ruled W surface (Beltrami).

13. Prove that canal surfaces (exercise 10-1, Prob. 24) are W surfaces with $F = (h + \text{const})^{-1}$ (Weingarten).

14. Prove that the element of the surface area of a surface of constant curvature $K = -1$ is $\alpha = \sin 2\sigma \, dq^1 \wedge dq^2$.

15. Show that the area of a parallelogram formed by asymptotic lines on a surface of curvature $K = -1$ is equal to the excess of the sum of the interior angles of the parallelogram over 2π (Hazzidakis). [HINT: Use Prob. 14 and Eq. (10-58).]

◆**16.** Surfaces of revolution (see exercise 10-2, Prob. 8) of constant Gauss curvature -1 may be obtained from particular solutions of

(10-57) in which σ is a function only of u^1. In this case (10-57) implies $\sigma'^2 = a^2 - \cos^2 \sigma$. Find the surfaces of revolution in the three cases $a^2 \gtreqless 1$.

17. Let $\mathbf{p}$ be a local surface of Gauss curvature $K = -1$ in R^3. Show that another surface

$$\mathbf{q} = \mathbf{p}(u^1,u^2) + \cos \alpha[\cos \phi(u^1,u^2)\mathbf{a}_1 + \sin \phi(u^1,u^2)\mathbf{a}_2]$$

is of constant Gauss curvature -1 if and only if ϕ is a solution of Eq. (10-57). The constant α is the angle between the normals of $\mathbf{p}$ and $\mathbf{q}$ at corresponding points (Bianchi).

18. Prove that the curves on the center surfaces of surfaces of constant Gauss curvature -1 which correspond to asymptotic lines are again asymptotic lines.

19. Write the equivalent of Eqs. (10-51) to (10-53) for a general frame.

20. Write Eqs. (10-52) and (10-53) in terms of E, F, G, L, M, N.

10-4. Mappings and Deformations

The main problem in the theory of mappings can be formulated as follows: Given two local varieties

$$\mathbf{p}: U \to R^n \qquad \mathbf{q}: V \to R^n$$

a map of $\mathbf{p}$ onto $\mathbf{q}$ can be realized by a differentiable homeomorphism of the coordinate domains

$$F: U \to V$$

We shall write $\mathbf{q} = F\mathbf{p}$ as a shorthand for the map $\mathbf{x}_0 \to \mathbf{q}(F(\mathbf{p}^{-1}(\mathbf{x}_0)))$ of points of $\mathbf{p}(U)$ onto points of $\mathbf{q}(V)$. By theorem 10-18, the local varieties may be determined by the differential forms $\omega_p{}^\alpha$, $\omega_{pi}{}^j$; ω_q, $\omega_{qi}{}^j$ and some initial conditions (in the differentiable case). The matrix $(\omega_{qi}{}^j)$ is a Cartan matrix; it is known how it will change if the family of frames to which it refers undergoes a transformation. We now ask: Does there exist a map F and on $\mathbf{q}$ a family of frames with differential forms $\omega_q^{*\alpha}$ and ω_{qi}^{*j} such that *certain* of the equalities

$$\omega_p{}^\alpha(\mathbf{p},d\mathbf{p}) = \omega_q^{*\alpha}(F\mathbf{p},dF\mathbf{p}) \qquad \omega_{pi}{}^j(\mathbf{p},d\mathbf{p}) = \omega_{qi}^{*j}(F\mathbf{p},dF\mathbf{p})$$

hold? If we ask that *all* such equalities hold, the answer is given by theorem 10-18: F exists if and only if all invariants and all their invariant derivatives are equal at corresponding points of $\mathbf{p}$ and $\mathbf{q}$. In the general case the problem is much more difficult. The present state of the theory is not completely satisfactory. It seems difficult to compress the essence of E. Cartan's method into theorems. Therefore we shall treat only some examples in the text and in the exercises to present the right method of approach.

***Example* 10-5.** We want to find all C^2 isometric mappings of a plane Π_1 given in cartesian coordinates, x^1,x^2, onto a plane Π_2 referred to polar coordinates r and ϕ. The elements of arc length are $ds^2 = (dx^1)^2 + (dx^2)^2$ and $ds^2 = dr^2 + r^2\,d\phi^2$, respectively; the corresponding linear forms are $\omega_{(1)}{}^\alpha = dx^\alpha$, $\omega_{(2)}{}^1 = dr$, $\omega_{(2)}{}^2 = r\,d\phi$. The problem is to find a mapping $F(x^1,x^2) \to (r,\phi)$ and a family of frames in Π_2 (obtained by a rotation of angle θ from the frames whose vectors $\mathbf{e}_\alpha$ are tangent to the curves $r = \text{const}$, $\phi = \text{const}$) such that

$$
\begin{aligned}
dx^1 &= \omega^*_{(2)}{}^1 = \cos\theta(x^1,x^2)\,dr(x^1,x^2) - \sin\theta(x^1,x^2)\,r(x^1,x^2)\,d\phi(x^1,x^2)\\
dx^2 &= \omega^*_{(2)}{}^2 = \sin\theta(x^1,x^2)\,dr(x^1,x^2) + \cos\theta(x^1,x^2)\,r(x^1,x^2)\,d\phi(x^1,x^2)
\end{aligned}
$$

From $\omega_{(1)} = \omega^*_{(2)}$ it follows that $\omega_{(1)1}{}^2 = \omega^*_{(2)1}{}^2$. Clearly $\omega_{(1)1}{}^2 = 0$; hence $\omega^*_{(2)1}{}^2 = \omega_{(2)1}{}^2 - d\theta = d\phi - d\theta = 0$. The rotation of the frames is by an angle $\theta = \phi + \alpha$, α constant. The equations of the linear forms $\omega_{(i)}{}^\alpha$ become

$$
\begin{aligned}
dx^1 &= \cos\theta\,dr - \sin\theta\,r\,d\theta = d(r\cos\theta)\\
dx^2 &= \sin\theta\,dr + \cos\theta\,r\,d\theta = d(r\sin\theta)
\end{aligned}
$$

This means that all admissible maps $\Pi_1 \to \Pi_2$ are given by

$$x^1 = r\cos(\phi+\alpha) + c^1 \qquad x^2 = r\sin(\phi+\alpha) + c^2$$

If Π_1 is identical with Π_2, the result shows that the group of isometric automorphisms of the euclidean plane is the three-parameter group of euclidean motions.

***Example* 10-6.** We generalize the preceding example. Given two local surfaces $\mathbf{p}$ and $\mathbf{q}$ and a differentiable homeomorphism F of $\mathbf{p}$ onto $\mathbf{q}$, we want to recognize whether F conserves the element of length,

$$ds^2(\mathbf{p},d\mathbf{p}) = ds^2(F\mathbf{p},dF\mathbf{p})$$

Example 10-5 shows that we cannot simply compare the coefficients of the two quadratic forms since the coordinates on $\mathbf{p}$ and $\mathbf{q}$ need not be images of one another. Nevertheless, if the mapping is isometric, there exist forms $\omega_q{}^\alpha$ belonging to a certain frame on $\mathbf{q}$ such that

$$\omega_p{}^\alpha(\mathbf{p},d\mathbf{p}) = \omega_q{}^\alpha(F\mathbf{p},dF\mathbf{p})$$

Hence

$$
\begin{aligned}
d\omega_p{}^1 - d\omega_q{}^1 &= -\omega_p{}^2 \wedge \omega_{p1}{}^2 + \omega_q{}^2 \wedge \omega_{q1}{}^2 = -\omega_p{}^2 \wedge (\omega_{p1}{}^2 - \omega_{q1}{}^2) = 0\\
d\omega_p{}^2 - d\omega_q{}^2 &= \omega_p{}^1 \wedge \omega_{p1}{}^2 - \omega_q{}^1 \wedge \omega_{q1}{}^2 = \omega_p{}^1 \wedge (\omega_{p1}{}^2 - \omega_{q1}{}^2) = 0
\end{aligned}
$$

or, by theorem 9-5,

$$\omega_{p1}{}^2(\mathbf{p},d\mathbf{p}) = \omega_{q1}{}^2(F\mathbf{p},dF\mathbf{p})$$

and also

$$d\omega_{p1}{}^2 = -K_p\omega_p{}^1 \wedge \omega_p{}^2 = d\omega_{q1}{}^2 = -K_q\omega_q{}^1 \wedge \omega_q{}^2$$

that is,

$$K_p(\mathbf{p}) = K_q(F\mathbf{p})$$

In the tangent normal frame of a curve on $\mathbf{p}$, $\omega_{p1}{}^2 = k_g\,ds$. By hypothesis, ds is invariant in the mapping. Hence the geodesic curvatures of a curve and of its image in an isometric mapping must be equal at corresponding points.

Theorem 10-22 (**Gauss**). *The geodesic curvature of curves and the Gauss curvature of surfaces are invariant in isometric mappings of surfaces.*

For the Gauss curvature, this is a restatement of theorem 10-7. A curve for which $k_g = 0$ is a *geodesic.*

Corollary 10-23. *Geodesics are mapped into geodesics in an isometric mapping.*

The further discussion splits into that of a number of separate cases. This is a characteristic feature of most equivalence problems.

I. *K is not a constant.*

$K =$ const defines a family of curves on either surface. The curves are images of one another in the mapping F. At each point let $\mathbf{c}_2$ be the unit vector tangent to $K =$ const, $\mathbf{c}_1$ the tangent normal to that curve, and $\mathbf{c}_3$ the unit normal $-\mathbf{N}$. $\{\mathbf{c}_i\}$ is an invariant family of frames on either surface, and the frames correspond to one another in the mapping F. They are not defined in the relative extrema of K. The corresponding values of the differential forms ω^α will be γ^α, $d\mathbf{p} = \gamma_p{}^1\mathbf{c}_{p1} + \gamma_p{}^2\mathbf{c}_{p2}$. The previous result is translated into

$$\gamma_p{}^\alpha = \gamma_q{}^\alpha \qquad \gamma_{p1}{}^2 = \gamma_{q1}{}^2 \tag{10-59}$$

The invariant derivatives with respect to this new frame are denoted by a semicolon: $df = f_{;1}\gamma^1 + f_{;2}\gamma^2$. By construction, $K_{p;2} = K_{q;2} = 0$; it follows from $dK = K_{p;1}\gamma_p{}^1 = K_{q;1}\gamma_q{}^1$ that

$$K_{p;1} = K_{q;1} \tag{10-60}$$

Two cases are now possible.

A. $\|J_u(K_p,K_{p;1})\| \neq 0$.

K_p and $K_{p;1}$ may serve as coordinates instead of u^1 and u^2 on both $\mathbf{p}$ and $\mathbf{q}$. The index p may be dropped. $K =$ const are the parameter lines $\gamma^1 = 0$; $K_{;1} =$ const are the lines $\gamma^2 = 0$. The map F associates points with the same $(K,K_{;1})$ coordinates. The condition (10-60) is not sufficient to ensure that F is an isometry. Two additional conditions

$$K_{p;1;1} = K_{q;1;1} \qquad K_{p;1;2} = K_{q;1;2} \tag{10-61}$$

result from $dK_{;1} = K_{;1;1}\gamma^1 + K_{;1;1}\gamma^2$. Conditions (10-60) and (10-61) together with theorem 10-22, are sufficient for F to be an isometry.

γ^1 may be computed from $dK = K_{;1}\gamma^1$ and then γ^2 from

$$dK_{;1} = K_{;1;1}\gamma^1 + K_{;1;2}\gamma^2$$

By construction, $\gamma_p{}^\alpha = \gamma_q{}^\alpha$; that is, $ds^2(\mathbf{p},d\mathbf{p}) = ds^2(\mathbf{q},d\mathbf{q})$.

B. $\|J_u(K,K_{;1})\| = 0$.

The two functions K and $K_{;1}$ are dependent upon one another,

$$K_{;1} = f(K)$$

They cannot be used as coordinates. γ^1 may be computed from K and $K_{;1}$:

$$\gamma^1 = \frac{dK}{f(K)}$$

The function ρ_1 in $\gamma_1{}^2 = \rho_1\gamma^1 + \rho_2\gamma^2$ may also be computed. From $ddK = d(f(K)\gamma^1) = (K_{;1;2} - K_{;1}\rho_1)\gamma^2 \wedge \gamma^1 = 0$, it follows that

$$\gamma_1{}^2 = \frac{K_{;1;2}}{K_{;1}}\gamma^1 + \rho_2\gamma^2$$

ρ_2 is an invariant function which may be independent of K.

1. $\|J_u(K,\rho_2)\| \neq 0$.

K and ρ_2 may be used as coordinates on both $\mathbf{p}$ and $\mathbf{q}$. As in case A, they define a unique mapping F of points with identical coordinates. The mapping is isometric if and only if

$$\rho_{p2} = \rho_{q2} \qquad \rho_{p2;1} = \rho_{q2;1} \qquad \rho_{p2;2} = \rho_{q2;2} \tag{10-62}$$

2. $\|J_u(K,\rho_2)\| = 0$.

On the parameter lines $\gamma^2 = 0$ one may take K as a parameter; this amounts to a normalization $f(K) = K_{;1} = 1$. By hypothesis, $\rho_{p2} = g(K_p)$ in some neighborhood on $\mathbf{p}$; that is, $d\gamma_p{}^2 = g(K_p)\,dK_p \wedge \gamma_p{}^2$. If we define

$$\log G(K_p) = -\int_{K_0}^{K} g(K)\,dK$$

then $d(G(K_p)\gamma_p{}^2) = 0$, and there exists a function u such that

$$du = G(K_p)\gamma_p{}^2$$

(see exercise 9-3, Prob. 10, or exercise 10-4, Prob. 12). K and u may serve as coordinates on both surfaces. The metric becomes

$$ds^2 = dK^2 + \frac{1}{G(K)^2}\,du^2$$

The two surfaces are isometric if and only if

$$G(K_p) = G(K_q) \tag{10-63}$$

The surfaces are isometric with the surface of revolution $(G(K) \cos u, G(K) \sin u, \phi(K))$ if $G'(K)^2 + \phi'(K)^2 = 1$. Such a surface admits a one-parameter family of isometric mappings onto itself, the rotations $(K,u) \to (K,u + c)$.

II. *K is constant.*

Theorem 10-24. *Two local surfaces of constant Gauss curvature are isometric if and only if their Gauss curvatures are equal.*

The coordinates are (u^α) in $\mathbf{p}$ and (v^α) in $\mathbf{q}$. On each one of the surfaces some orthonormal frames are given. The pfaffians with respect to these frames are $\omega_p{}^\alpha(u^1,u^2)$ and $\omega_q{}^\alpha(v^1,v^2)$. We denote by $\omega_q{}^\alpha(v^1,v^2,\theta)$ the forms which belong to the frames obtained from those originally given on $\mathbf{q}$ by a rotation of angle $\theta(v^1,v^2)$. It is possible to find a map $v^\alpha = v^\alpha(u^1,u^2)$ such that a given point $\mathbf{p}_0$ is mapped onto a given point $\mathbf{q}_0$ (initial condition) and that

$$\begin{aligned} \omega_p{}^1(u^1,u^2,du^1,du^2) - \omega_q{}^1(v^1,v^2,\theta,dv^1,dv^2,d\theta) &= 0 \\ \omega_p{}^2(u^1,u^2,du^1,du^2) - \omega_q{}^2(v^1,v^2,\theta,dv^1,dv^2,d\theta) &= 0 \\ \omega_{p1}{}^2(u^1,u^2,du^1,du^2) - \omega_{q1}{}^2(v^1,v^2,\theta,dv^1,dv^2,d\theta) &= 0 \end{aligned} \qquad (10\text{-}64)$$

The result implies for $\mathbf{p} = \mathbf{q}$ that a transitive group of isometries acts on any surface of constant curvature. Since θ may be prescribed, the group also acts transitively on the vectors of the tangent plane. The number of parameters is at least three. We shall determine the groups in Sec. 11-2. All these results are purely local. Theorem 10-9 states that a right circular cylinder is invariant under a two-parameter group of motions. The first fundamental form of the cylinder over the unit circle is $ds^2 = d\phi^2 + dz^2$ in cylindrical coordinates (r,ϕ,z). The cylinder is locally isometric with the plane. A simply connected neighborhood on the cylinder admits a three-parameter pseudo group of euclidean motions. However, the group of plane rotations cannot be extended to a transformation group on a full cylinder. Only the two-parameter group of translations can be extended. A complete discussion needs advanced methods of algebraic topology.

The proof of our assertion on the system (10-64) is an immediate consequence of the following generalization of theorem 10-18: A system of linearly independent "linear" differential equations defined in $U \subset R^n$,

$$\omega^\alpha = a_i{}^\alpha(x^1, \ldots, x^n)\, dx^i = 0 \qquad (10\text{-}65)$$
$$\alpha = 1, \ldots, n - k \qquad i = 1, \ldots, n$$

is *totally integrable* if for any $\mathbf{y}_0 \in U$ there exists a local variety

$$\mathbf{x}: R^k \to R^{n-k}$$

such that $\mathbf{x}(\mathbf{u}_0) = \mathbf{y}_0$, R^{n-k} is the subspace of the first $n - k$, R^k that of the last k coordinates in R^n, and

$$a_i^\alpha(x^1(x^{n-k+1}, \ldots ,x^n), \ldots ,x^{n-k}(x^{n-k+1}, \ldots ,x^n),x^{n-k+1}, \ldots ,x^n)\, dx^i = 0$$

identically after replacement of dx^j $(j \leq n - k)$ by $(\partial x^j/\partial x^A)\, dx^A$ $(A > n - k)$.

Theorem 10-25 (Frobenius-Cartan). *The system* (10-65) *with* C^1 *functions* a_i^α *is completely integrable if and only if there exist pfaffians* $\omega_\beta{}^\alpha(x^1, \ldots ,x^n,dx^1, \ldots ,dx^n)$ *such that*

$$d\omega^\alpha = \omega_\beta{}^\alpha \wedge \omega^\beta \tag{10-66}$$

In the case of the system (10-64) we put

$$\sigma^\alpha = \omega_p{}^\alpha - \omega_q{}^\alpha, \qquad \sigma_1{}^2 = \omega_{p1}{}^2 - \omega_{q1}{}^2$$

By hypothesis,

$$\begin{aligned} d\sigma^1 &= \omega_{q1}{}^2 \wedge \sigma^2 - \omega_p{}^2 \wedge \sigma_1{}^2 \\ d\sigma^2 &= \omega_{q1}{}^2 \wedge \sigma^1 + \omega_p{}^1 \wedge \sigma_1{}^2 \\ d\sigma_1{}^2 &= (K\omega_p{}^1) \wedge \sigma^2 - (K\omega_q{}^2) \wedge \sigma^1 \end{aligned}$$

The system (10-64) of three equations in five variables, u^1, u^2, v^1, v^2, and θ, is completely integrable. Therefore it is possible to find (v^1,v^2,θ) as a function of (u^1,u^2) for given initial conditions.

The proof of the Frobenius theorem 10-25 is parallel to that of theorem 10-18 after some preparatory steps.

1. The condition (10-66) is necessary. It is possible to find a basis ω^i of R^{n*} which contains the ω^α as its first $n - k$ elements. Clearly there exist forms $\omega_i{}^j$ such that $d\omega^\alpha = \omega_i{}^\alpha \wedge \omega^i$. $\omega^\alpha = 0$ implies $d\omega^\alpha = 0$; hence $\sum_{i>n-k} \omega_i{}^\alpha \wedge \omega^i = 0$. The ω^i $(i > n - k)$ are to a large extent arbitrary; the condition $d\omega^\alpha = 0$ can hold identically only if $\omega_i{}^\alpha = 0$ $(i > n - k)$.

2. If (10-66) holds for the ω^α, then it holds also for $\theta^\alpha = A_\beta{}^\alpha\omega^\beta$, where $(A_\beta{}^\alpha)$ is any C^1 nonsingular matrix. The matrix $(\theta_\beta{}^\alpha)$ in $d\theta^\alpha = \theta_\beta{}^\alpha \wedge \theta^\beta$ is $\theta = dAA^{-1} + A\,(\omega_\beta{}^\alpha)A^{-1}$. It is possible to choose A so as to have

$$\theta^\alpha = dx^\alpha - \sum_{j>n-k} b_j{}^\alpha(x^1, \ldots ,x^n)\, dx^j$$

We change the term x^α $(\alpha \leq n - k)$ to y^α and x^A $(A > n - k)$ to u^{A-n+k}. The problem is to find $\mathbf{y}(\mathbf{u})$ from the initial condition $\mathbf{y}(\mathbf{u}_0) = \mathbf{y}_0$ and

$$dy^\alpha = b_j{}^\alpha(y^1, \ldots ,y^{n-k},u^1, \ldots ,u^k)\, du^j \qquad j = 1, \ldots , k \tag{10-67}$$

3. For any differentiable curve $\mathbf{c} = \mathbf{u}(s)$ starting from $\mathbf{u}_0$, one may solve the Volterra integral equation

$$y(s)^\alpha = y_0{}^\alpha + \int_0^s b_j{}^\alpha(\mathbf{y},\mathbf{c}(s))c'(s)^j\,ds \tag{10-68}$$

[In all standard books on differential equations it is shown that the successive approximations $\mathbf{y}_{n+1} = \mathbf{y}_0 + \int_0^s b_j{}^\alpha(\mathbf{y}_n,\mathbf{c})c'^j\,ds$ converge to a solution if the $b_j{}^\alpha$ satisfy a Lipschitz condition, a fortiori if they are differentiable.] If we choose a second function $\mathbf{c}^*(s)$ through $\mathbf{u}_0$ such that $\mathbf{c}(1) = \mathbf{c}^*(1)$, we may compute $\mathbf{y}(s,t)$ as the solution of Eq. (10-68) for the curve (10-44). The theorem is proved if $\mathbf{y}(1,t) = \mathbf{y}(1)$; the function $\mathbf{y}(\mathbf{u})$ is then defined in a neighborhood of $\mathbf{u}_0$ and satisfies all required conditions. Just as for theorem 10-18, one shows that $\mathbf{y}(1,1) - \mathbf{y}(1,0 = \iint db_j{}^\alpha \wedge du^j$. The hypothesis (10-66) implies that $d\theta^\alpha = d(dy^\alpha - b_j{}^\alpha\,du^j) = db_j{}^\alpha \wedge du^j = 0$ always if $\theta^\beta = 0$ $(\beta = 1, \ldots, n - k)$. Hence $\mathbf{y}(1,1) - \mathbf{y}(1,0) = 0$.

There is a serious gap in the proof because the standard theorems on differential equations do not imply that $\mathbf{y}(s,t)$ is a differentiable function of the parameter t. Theorem 10-18 is so much simpler than theorem 10-25 because there all differentiability questions may be settled by inspection. The existence of $\partial\mathbf{y}(s,t)/\partial t$ may be shown as follows: If $\partial\mathbf{y}(s,t)/\partial t$ exists and is continuous on the compact set $0 \leq s,t \leq 1$, the function $\mathbf{z}(s,t) = \partial\mathbf{y}(s,t)/\partial t$ is a solution of the integral equation

$$\begin{aligned}\mathbf{z}^\alpha(s,t) = \int_0^s \frac{\partial b_j{}^\alpha}{\partial y^\beta}\,[(1-t)\mathbf{c}' - t\mathbf{c}^{*\prime})]^j z^\beta(\sigma,t)\,d\sigma \\ + \int_0^s \frac{\partial b_j{}^\alpha}{\partial u^k}\,(\mathbf{c}^* - \mathbf{c})^k[(1-t)\mathbf{c}' - t\mathbf{c}^*]^j\,d\sigma \\ + \int_0^s b_j{}^\alpha(\mathbf{c}^{*\prime} - \mathbf{c}')^j\,d\sigma\end{aligned}$$

which is of the type

$$\mathbf{z}(s,t) = \mathbf{P}(s,t) + \int_0^s Q(\sigma,t)\mathbf{z}(\sigma,t)\,d\sigma$$

with continuous (hence bounded) $\mathbf{P}$ and Q in $0 \leq s,\ t \leq 1$. Therefore $\mathbf{z}(s,t)$ exists, and solutions may be obtained by successive approximations. The function $\mathbf{Z}(s,t) = \int_0^t \mathbf{z}(s,\tau)\,d\tau + \mathbf{y}(s,0) - \mathbf{y}(s,t)$ satisfies the integro-differential equation

$$\mathbf{Z}(s,t) = \int_0^s \int_0^t Q(\sigma,\tau)d_t\mathbf{Z}(\sigma,\tau) \wedge d\sigma \qquad \mathbf{Z}(s,0) = 0$$

Here d_t indicates that the integral is a Stieltjès integral with respect to the variable t, s being kept constant. The only possible value for $\mathbf{Z}$ is 0.

As for (2-7), it follows from the equation for $\mathbf{Z}$ that $|\mathbf{Z}| \leq k|Q|\,|\mathbf{Z}|s$. If this inequality is introduced under the integral sign, it follows that $|\mathbf{Z}| \leq k^2|Q|^2|\mathbf{Z}|s^2/2!, \ldots, |\mathbf{Z}| \leq k^r|Q|^r|\mathbf{Z}|s^r/r!$. The right-hand sides of the inequalities tend to zero as terms of the Taylor series of $e^{k|Q||Z|s}$. Since $\mathbf{Z}(s,t) = 0$, $\mathbf{y}$ is the integral of a continuous function of t; it is differentiable in t.†

A kind of mapping problem that admits a direct approach is that of infinitesimal deformations.

Definition 10-26. *A family of surfaces* $\mathbf{x}(u^1,u^2,\epsilon) = \mathbf{p}(u^1,u^2) + \epsilon\mathbf{z}(u^1,u^2)$ *is an infinitesimal deformation of the surface* $\mathbf{p}$ *if*

$$\lim_{\epsilon \to 0} \frac{ds^2(\mathbf{x},d\mathbf{x}) - ds^2(\mathbf{p},d\mathbf{p})}{\epsilon} = 0$$

An infinitesimal deformation is not strictly an isometric mapping; for small ϵ the lengths are disturbed only by an amount of order of magnitude ϵ^2. By definition,

$$ds^2(\mathbf{x},d\mathbf{x}) = ds^2(\mathbf{p},d\mathbf{p}) + 2\epsilon\, d\mathbf{p} \cdot d\mathbf{z} + \epsilon^2\, d\mathbf{z} \cdot d\mathbf{z}$$

The vector function $\mathbf{z}$ determines an infinitesimal deformation if

$$d\mathbf{p} \cdot d\mathbf{z} = 0 \tag{10-69}$$

From (10-69) it follows that there exists a vector $\mathbf{y}$ such that

$$d\mathbf{z} = \mathbf{y} \times d\mathbf{p} \tag{10-70}$$

If $d\mathbf{p} = \omega^1\mathbf{e}_1 + \omega^2\mathbf{e}_2$, then $d\mathbf{z}$ must be of the form

$$d\mathbf{z} = -r\omega^2\mathbf{e}_1 + r\omega^1\mathbf{e}_2 + \zeta^3\mathbf{e}_3$$

where r is a function of the u^i and ζ^3 is a linear combination of ω^1 and ω^2 which we may write $\zeta^3 = y^1\omega^2 - y^2\omega^1$. $\mathbf{y}$ is the vector field (y^1,y^2,r). Exterior differentiation of (10-70) gives $0 = d\mathbf{y}$ "$\times$" $d\mathbf{p}$; for $d\mathbf{y} = v^i\mathbf{e}_i$ this means that

$$v^3 = 0$$
$$v^1 \wedge \omega^2 - v^2 \wedge \omega^1 = 0$$

The second equation implies, by Cartan's theorem 9-6, that there exists a symmetric matrix (function) such that

$$(v^1, -v^2) = (\omega^2, \omega^1) \begin{pmatrix} a & b \\ b & c \end{pmatrix} \tag{10-71}$$

† For details of the proof see H. Guggenheimer, A Simple Proof of Frobenius's Integration Theorem, *Proc. Am. Math. Soc.*, **13**: 24–28 (1962). A proof under weaker differentiability conditions may be found in R. Nevanlinna and F. Nevanlinna, "Absolute Analysis," Springer-Verlag OHG, Berlin, 1959.

The relation $dd\mathbf{y} = 0$ is equivalent to

$$\begin{aligned} al_{11} - 2bl_{12} + cl_{22} &= 0 \\ a_{,1} - b_{,2} &= c\Gamma_{12}{}^2 - a\Gamma_{12}{}^2 \\ c_{,2} - b_{,1} &= c\Gamma_{11}{}^2 - a\Gamma_{11}{}^2 \end{aligned} \tag{10-72}$$

Conversely, any solution of the system (10-72) gives rise to an infinitesimal deformation. The deformed surface is congruent to the original one if and only if $\mathbf{y}$ is a constant vector, that is, $a = b = c = 0$. If the system (10-72) admits only this trivial solution, the surface is *infinitesimally rigid.* The theory is much simplified for surfaces that may be represented by an equation $x^3 = f(x^1,x^2)$. Here $\mathbf{y} = (z^3{}_{2}, - z^3{}_{1},\psi)$ where we have put $f_i = \partial f/\partial x^i$. With the usual notations of (7-16) for the partial derivatives of x^3,

$$d\mathbf{y} = (z^3{}_{21}\, dx^1 + z^3{}_{22}\, dx^2, -z^3{}_{11}\, dx^1 - z^3{}_{12}\, dx^2, \psi_1\, dx^1 + \psi_2\, dx^2)$$

and

$$d\mathbf{p} = (dx^1, dx^2, p\, dx^1 + q\, dx^2)$$

$d\mathbf{y}$ "$\times$" $d\mathbf{p} = 0$ means that

$$\begin{aligned} \psi_1 &= -qz^3{}_{11} + pz^3{}_{12} \\ \psi_2 &= pz^3{}_{22} - qz^3{}_{12} \end{aligned}$$

The integrability condition $\psi_{12} = \psi_{21}$ gives the unique equation for z^3:

$$rz^3{}_{22} - 2sz^3{}_{12} + tz^3{}_{11} = 0 \tag{10-73}$$

If z^3 is known as a solution of (10-73), ψ, $\mathbf{y}$, and $\mathbf{z}$ may be computed.

Exercise 10-4

1. Parameters v^1, v^2 are *isothermal* on a surface $\mathbf{p}$ if its first fundamental form is $ds^2 = g(v^1,v^2)[(dv^1)^2 + (dv^2)^2]$. Prove the theorem: A C^3 surface always has isothermal parameters. [Hint: The introduction of isothermal parameters corresponds to a mapping of the parameter domain onto itself so that $dv^1 = a(\cos\theta\, \pi^1 + \sin\theta\, \pi^2)$, $dv^2 = a(-\sin\theta\, \pi^1 + \cos\theta\, \pi^2)$.] If the forms on the right-hand sides of the equations are denoted by σ^i, one may obtain equations $d\sigma^i = \chi \wedge \sigma^i$ if one puts $da/a = \chi - (\theta_{,2} + \rho_2)\pi^1 + (\theta_{,1} + \rho_1)\pi^2$. A necessary and sufficient condition for the existence of isothermal parameters is $d\chi = 0$; that is, $\theta_{,2,2} + \theta_{,1,1} + \rho_{2,2} + \rho_{1,1} = 0$. (Note: *E. Cartan's method uses too many differentiations,* here and in general. It is possible to establish the existence of isothermal parameters for C^1 surfaces.)

2. Compute K for a surface in isothermal parameters (Prob. 1).

3. Prove that a mapping F is *conformal* if and only if

$$(\omega_p{}^1)^2 + (\omega_p{}^2)^2 = c(u^1,u^2)[(\omega_q{}^1)^2 + (\omega_q{}^2)^2]$$

(A mapping of two surfaces is called *conformal* if it induces homothetic maps of the tangent planes, i.e., preserves angles between tangents to curves on the surface.) Use Prob. 1 to show that any C^3 surface is conformal to the plane.

4. Solve the problem of O. Bonnet: To find the conditions for the existence of an isometric mapping $\mathbf{p} \to \mathbf{q}$ that conserves principal curvatures. (If the mapping also transforms curvature lines into curvature lines, it is a congruence, as shown in Sec. 10-3.) The Darboux frames on both surfaces must be connected by an equation

$$(\pi_q{}^1,\pi_q{}^2) = (\pi_p{}^1,\pi_p{}^2)\begin{pmatrix} \cos\theta & \sin\theta \\ -\sin\theta & \cos\theta \end{pmatrix}$$

Put $I = k_1 - k_2$, $dH = I(a\pi^1 + b\pi^2)$. Since $dI_p = dI_q$, it follows that $a_q \cos\theta = a_p - \theta_{,2}$; $b_q \cos\theta = b_p - \theta_{,1}$; $a_q \sin\theta = -\theta_{,1}$; $b_q \sin\theta = \theta_{,2}$. Put $t = \cot\theta$. A condition for the existence of the mapping is

$$dt = t(a_p\pi_p{}^1 - b_p\pi_p{}^2) - (b_p\pi_p{}^1 + a_p\pi_p{}^2) \qquad (*)$$

There exists a group of curvatures conserving isometric maps if and only if Eq. (*) is completely integrable. This is the case if and only if Eq. (*) may be written $dt = -t(dP/P) + dQ/Q$. This implies that $d(PI)^2\pi^1 = d(PI)^2\pi^2 = 0$; the ds^2 in curvature-line parameters is $ds^2 = (PI)^{-2}[(ds^1)^2 + (ds^2)^2]$. If the curvature lines may be provided with isothermal parameters, the surface is called *isothermal. A surface admits a group of isometries which preserve principal curvatures if and only if it is isothermal* (Bonnet; E. Cartan).

5. Prove that a surface admits a mapping into the plane so that geodesics ($\omega_1{}^2 = 0$) are mapped onto straight lines if and only if it is of constant Gauss curvature (Beltrami).

6. It is impossible to draw a map of any part of the surface of the earth which faithfully reproduces all lengths. Show that there exist geographic maps which faithfully reproduce all angles.

7. A map is *area-preserving* if

$$(\omega_q{}^1,\omega_q{}^2) = (\omega_p{}^1,\omega_p{}^2)\begin{pmatrix} a & b \\ c & d \end{pmatrix} \qquad ad - bc = 1$$

Justify this equation and show that any surface has area-preserving maps into the plane.

8. Prove that the surfaces of constant mean curvature are isothermal (Prob. 4).

9. Find the conditions on the function a_i such that

$$a_1\,dx^1 + a_2\,dx^3 = 0 \qquad a_2\,dx^1 - a_3\,dx^2 = 0$$

are completely integrable.

10. A partial differential equation

$$F\left(x^1, \ldots, x^n, z, \frac{\partial z}{\partial x^1}, \ldots, \frac{\partial z}{\partial x^n}, \frac{\partial^2 z}{(\partial x^1)^2}, \ldots, \frac{\partial^2 z}{(\partial x^n)^2}\right) = 0$$

is equivalent to the finite equation

$$F(x^1, \ldots, x^n, z, p_1, \ldots, p_n, r_{11}, \ldots, r_{nn}) = 0$$

and the $n + 1$ pfaffian equations

$$dz - p_i\,dx^i = 0 \qquad dp_i - r_{ik}dx^k = 0$$

State and prove a theorem on the existence of solutions of a second-order partial differential equation for prescribed initial values.

11. A function f is an invariant of a pseudo group of transformations if and only if it is a solution of

$$df - \omega^k \pounds_k f = 0$$

where the ω^k are the invariant forms introduced in exercise 9-3, Prob. 9. Show that the Frobenius-Cartan theorem implies the result stated on page 136.

12. Use theorem 10-25 to prove that a linear differential form ω is the differential of a function if and only if $d\omega = 0$. (HINT: Consider step 2 in the proof of theorem 10-25.)

13. A vector field $\mathbf{v}(x^1,x^2,x^3)$ is the field of normals to a surface $\mathbf{p}$ if $\omega = \mathbf{v} \cdot d\mathbf{p} = v^1\,dx^1 + v^2\,dx^2 + v^3\,dx^3 = 0$. Find a necessary condition for $\mathbf{v}$ to be the field of normals of some surface (by exterior differentiation of ω).

14. Show that, if an infinitesimal deformation is an isometric mapping of a surface on itself, the surface is isometric to a surface of revolution (Bianchi).

♦**15.** Find all the infinitesimal deformations of the paraboloid

$$2x^3 = (x^1)^2 + (x^2)^2$$

16. The interior of a plane simple closed curve is not infinitesimally rigid even if the boundary is kept fixed. Show that, if $f(x^1,x^2)$ is a function such that the given curve is one of its level lines 0, the surface $x = (x^1,x^2,\epsilon f(x^1,x^2))$ is an infinitesimal deformation of the plane $x^3 = 0$ which corresponds to the given boundary condition. The same example shows that an infinitesimal deformation is not, in general, an isometry.

17. Show that, for an infinitesimal deformation, $(\partial K/\partial\epsilon)_{\epsilon=0} = 0$.

10-5. Closed Surfaces

Definition 10-27. *A closed surface is a surface which is a compact manifold. A closed surface is convex if it is the boundary of a convex domain. A closed surface* $\mathbf{p}$ *is star-shaped if there exists a point* P *in* R^3 *such that* $(\mathbf{p} - P) \cdot \mathbf{N}(\mathbf{p}) > 0$ *for all* $\mathbf{p}$.

$\angle(\mathbf{p} - P, \mathbf{N}(\mathbf{p})) < \pi/2$ on a star-shaped surface. A ray issued from P can meet the surface only once; otherwise there must be some points where the normal is perpendicular to the radius vector $\mathbf{p} = P$. We always choose the normals of (orientable) closed surfaces pointing *outward.* Therefore we change, *for this section only,* the expressions k_i, l_{ij}, H into $-k_i$, $-l_{ij}$, $-H$ in order to retain positive curvatures for convex surfaces.

The study of closed surfaces belongs to global differential geometry. Compactness is not a local property. Nevertheless, we are able to treat here a few problems connected with closed surfaces. The main tool is Stokes's theorem in the following form: An *orientable* closed C^1 surface is covered by a finite number of simply connected closed neighborhoods $\bar{U}_i$ such that two adjacent neighborhoods $\bar{U}_i$ and $\bar{U}_j$ have only a smooth arc $\mathbf{c}_{ij}$ in common. $\bar{U}_i$ is a one-to-one image of I^2 such that its interior U_i is differentiably homeomorphic to the interior of the unit square and its boundary ∂U_i is a piecewise differentiable image of ∂I^2. Formula (9-24) applies to this case. For any linear differential form, $\iint_{U_i} d\omega = \int_{\partial U_i} \omega$; hence $\iint_{\mathbf{p}} d\omega = \Sigma \int_{\mathbf{c}_{ij}} \omega$. In the sum on the right-hand side of the last equation each $\mathbf{c}_{ij}$ appears twice, once in the boundary of U_i and once in that of U_j. If the surface is orientable, the two orientations induced on $\mathbf{c}_{ij}$ by the local coordinates in U_i and U_j are opposite to one another, and all integrals cancel out pairwise. It is easy to prove this statement. If $\bar{U}_i$ and $\bar{U}_j$ are imbedded in bigger neighborhoods V_i and V_j (Fig. 10-5), the local coordinates in V_i and V_j induce the same orientation on smooth curvilinear triangles built on $\mathbf{c}_{ij}$, since the surface areas must have equal signs if computed in different systems of coordinates.

Lemma 10-28. *On an orientable closed surface* $\mathbf{p}$,

$$\iint_{\mathbf{p}} d\omega = 0$$

The *support function* of a surface

$$h(\mathbf{p}) = \mathbf{p} \cdot \mathbf{N}(\mathbf{p}) \tag{10-74}$$

is the oriented distance from the origin of the tangent plane at $\mathbf{p}$. If the origin is an interior point of the convex domain bounded by a convex surface, or if the origin is a point P for which the surface is star-shaped, then $h(\mathbf{p}) > 0$ for all points on the surface. The support function appears in some important integral formulas, due to Minkowski, that

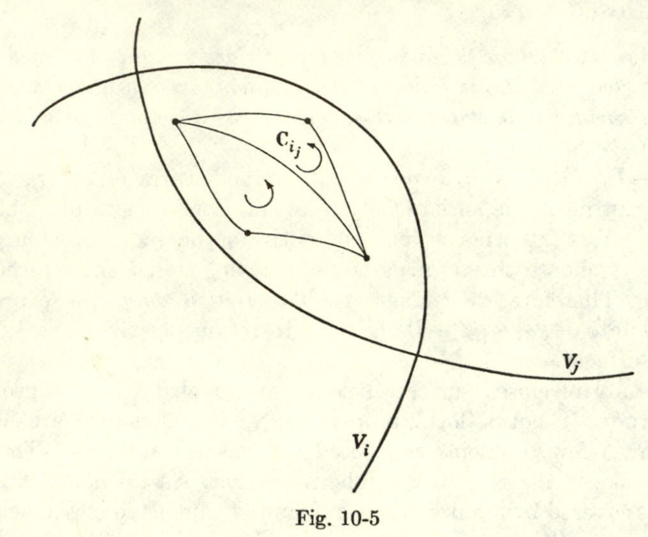

Fig. 10-5

may be derived from the equations of Hopf and Voss [(10-23) to (10-25)]. Since

$$\begin{aligned} d\text{``}\|\text{''}\mathbf{p},\mathbf{N},d\mathbf{p}\text{``}\|\text{''} &= \text{``}\|\text{''}\, d\mathbf{p},\mathbf{N},d\mathbf{p}\text{``}\|\text{''} + \text{``}\|\text{''}\, \mathbf{p},d\mathbf{N},d\mathbf{p}\text{``}\|\text{''} \\ &= -2(1 - hH)\alpha \\ d\text{``}\|\text{''}\mathbf{p},\mathbf{N},d\mathbf{N}\text{``}\|\text{''} &= \text{``}\|\text{''}\, d\mathbf{p},\mathbf{N},d\mathbf{N}\,\text{``}\|\text{''} + \text{``}\|\text{''}\, \mathbf{p},d\mathbf{N},d\mathbf{N}\,\text{``}\|\text{''} \\ &= 2(hK - H)\alpha \end{aligned}$$

hence, by lemma 10-28,

$$\iint hH\alpha = \iint \alpha \tag{10-75}$$

$$\iint hK\alpha = \iint H\alpha \tag{10-76}$$

Double integrals in which the domain of integration is not indicated are taken over the whole surface.

Theorem 10-29. *A closed surface always has elliptic points.*

A closed surface, being compact, may be enclosed in a sphere with a certain center O and radius R. Let S' be the smallest closed spherical ball with center O that contains the surface. The boundary sphere of S' and the surface have at least one point $\mathbf{x}$ in common. At $\mathbf{x}$, the Gauss curvature of the surface is at least that of the sphere; otherwise there would be points of the sphere outside S' (see page 212); hence $K(\mathbf{x}) > 0$.

This argument can be greatly refined. For a closed convex surface Σ, let Σ^* be its *convex hull*, i.e., the boundary of the smallest convex domain whose interior contains the interior of Σ. Σ^* will be composed of certain parts of Σ and of plane regions. The points of Σ which are on Σ^* have non-negative curvature, as shown by the preceding argument. The points on Σ^* which are not on Σ are parabolic. Hence

$$\iint_{\Sigma^*} K_{\Sigma^*} \alpha_{\Sigma^*} \leq \iint_{K_\Sigma \geq 0} K_\Sigma \alpha_\Sigma \tag{10-77}$$

The Gauss curvature is the determinant of the jacobian of the map $\Sigma^* \to S^2$ which associates its spherical image with $\mathbf{x} \in \Sigma^*$. If the origin is in the interior of Σ^*, the spherical image of Σ^* covers the unit sphere S^2 just once,

$$\iint_{\Sigma^*} K_{\Sigma^*} \alpha_{\Sigma^*} = \iint_{S^2} \alpha_{S^2} = 4\pi \tag{10-78}$$

The plane regions contribute only a set of surface area zero on the sphere (jacobian determinant 0). The combination of Eqs. (10-77) and (10-78) shows the following:

Theorem 10-30 (Voss). *For a closed C^2 surface,*

$$\iint_{\{\mathbf{x} \mid K(\mathbf{x}) \geq 0\}} K\alpha \geq 4\pi$$

This theorem has an interesting application on closed space curves. A *canal surface* is obtained by putting a circle of fixed radius r and center

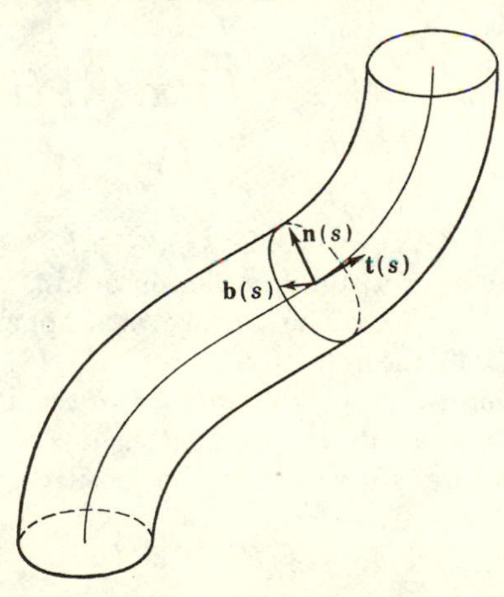

Fig. 10-6

$\mathbf{x}(s)$ in each normal plane $(\mathbf{n}(s),\mathbf{b}(s))$ of a closed space curve $\mathbf{x}(s)$ of length l. The canal surface is

$$\mathbf{X}(s,\phi) = \mathbf{x}(s) + r(\mathbf{n} \cos \phi + \mathbf{b} \sin \phi) = \mathbf{x}(s) + r\mathbf{N}(s,\phi)$$

Its surface elements result in

$$\alpha = r(1 - rk \cos \phi)\, ds \wedge d\phi$$

We must take $r < \min |k(s)|^{-1}$ to make positive the coefficient of $ds \wedge d\phi$. The coordinate lines $s =$ const and $\phi =$ const are curvature lines. The Gauss curvature is computed from (10 = 25)

$$K\alpha = k_1\pi^1 \wedge k_2\pi^2 = -k \cos \phi\, ds \wedge d\phi$$

or

$$\iint K\alpha = -\int_0^l \int_0^{2\pi} k(s) \cos \phi\, ds \wedge d\phi$$

The curvature of a space curve is a non-negative function; the integral of theorem 10-30 becomes

$$\iint_{K \geq 0} K\alpha = -\int_0^l \int_{\pi/2}^{3\pi/2} k(s) \cos \phi\, ds \wedge d\phi = 2 \int_0^l k(s)\, ds$$

Theorem 10-31 (Fenchel). *The integral curvature* $\int_0^l k(s)\, ds$ *of a closed space curve is* $\geq 2\pi$.

The integral curvature is also the length of the closed spherical curve $\mathbf{t}(s)$, the tangent indicatrix of $\mathbf{x}(s)$.

We explore next the implications of formulas (10-75) and (10-76).

Theorem 10-32 (Liebmann). *The only closed oriented star-shaped surface with constant mean curvature is the sphere.*

$H = 0$ only in hyperbolic points; by theorem 10-29, $H = c \neq 0$ in our case. Equations (10-75) and (10-76) become

$$c \iint h\alpha = \iint \alpha \qquad \iint hK\alpha = c \iint \alpha$$

or

$$\iint h\left(1 - \frac{K}{c^2}\right)\alpha = 0$$

For any surface $1 - K/H^2 = 1 - 2k_1k_2/(k_1 + k_2)^2 \geq 0$. Since $h > 0$ on star-shaped surfaces for a suitably chosen origin, the integral cannot vanish unless $K = H^2 = c^2$. The surface is a sphere since all points are umbilics and $c^2 > 0$ (theorem 10-8).

Theorem 10-33 (Liebmann). *The only closed oriented star-shaped surface with constant Gauss curvature is the sphere.*

The Gauss curvature $K = c$ must be positive. Formula (10-76) implies

$$c\iint h\alpha = \iint H\alpha \geq c^{\frac{1}{2}} \iint \alpha$$

Hence, by Eq. (10-75),

$$\iint h\alpha \geq c^{\frac{1}{2}} \iint hH\, \alpha$$

$h > 0$ implies either that $H = c^{\frac{1}{2}} = K^{\frac{1}{2}}$ or that in some region $H < c^{\frac{1}{2}}$. The inequality is impossible; all points are umbilics.

Theorem 10-33 is not the best possible one. One shows by topological arguments that a closed surface with everywhere positive Gauss curvature is convex (*Hadamard's theorem*). The words "oriented star-shaped" may be eliminated from theorem 10-33.

Some theorems on closed surfaces are based on properties of the *principal curvature radii* $R_i = 1/k_i$. On a local surface without parabolic points one may write $d\mathbf{N} = \sigma^1\mathbf{e}_1 + \sigma^2\mathbf{e}_2$, $(\omega^1,\omega^2) = (\sigma^1,\sigma^2)(\lambda_{ij})$. The R_i are the eigenvalues of the matrix (λ_{ij}) since the eigenvalues of the inverse of a matrix are the inverses of the eigenvalues. In curvature-line parameters, $\omega^i = R_i\sigma^i$ (no summation). The invariants of the matrix (λ_{ij}) are its trace $2P_1 = \lambda_{11} + \lambda_{22} = R_1 + R_2 = 2H/K$ and its determinant $P_2 = \lambda_{11}\lambda_{22} - (\lambda_{12})^2 = R_1R_2 = 1/K$.

A surface $\mathbf{p}$ is *strictly* convex if it is convex and has no parabolic points. The spherical image of a strictly convex surface is a one-to-one mapping. A closed strictly convex surface may be given as a function of the points on the unit sphere. If two surfaces $\mathbf{p}$ and $\mathbf{p}^*$ are given in this way, points with equal parameter values have parallel normals $\mathbf{N} = \mathbf{N}^*$. The integrals over $\mathbf{p}$ and $\mathbf{p}^*$ may be written as integrals over the unit sphere with the surface area element $\alpha_S = \sigma^1 \wedge \sigma^2$. In addition to the functions P_i on $\mathbf{p}$ and the corresponding ones P_i^* on $\mathbf{p}^*$, we introduce $Q = \frac{1}{2}(R_1R_2^* + R_1^*R_2) = \frac{1}{2}(\lambda_{11}\lambda_{22}^* + \lambda_{11}^*\lambda_{22} - 2\lambda_{12}\lambda_{12}^*)$, the *polar form of the determinant*, which makes sense as a function on the unit sphere. There are some integral formulas parallel to (10-75) and (10-76):

$$\begin{aligned} d\text{“}\|\text{”}\,\mathbf{p},\mathbf{N},d\mathbf{p}\,\text{“}\|\text{”} &= \text{“}\|\text{”}\,d\mathbf{p},\mathbf{N},d\mathbf{p}\,\text{“}\|\text{”} + \text{“}\|\text{”}\,\mathbf{p},d\mathbf{N},d\mathbf{p}\,\text{“}\|\text{”} \\ &= -\text{“}\|\text{”}\,\mathbf{N},d\mathbf{p},d\mathbf{p}\,\text{“}\|\text{”} + h\,\text{“}\|\text{”}\,\mathbf{N},d\mathbf{N},d\mathbf{p}\,\text{“}\|\text{”} \\ &= 2(hP_1 - P_2)\alpha_S \end{aligned}$$

Hence
$$\iint (hP_1 - P_2)\alpha_S = 0 \tag{10-79}$$

The next determinants are best computed in a family of frames on S^2 which diagonalizes simultaneously the second fundamental forms of both $\mathbf{p}$ and $\mathbf{p}^*$.

$$\begin{aligned} d\text{“}\|\text{”}\,\mathbf{p},\mathbf{N},d\mathbf{p}^*\,\text{“}\|\text{”} &= \text{“}\|\text{”}\,d\mathbf{p},\mathbf{N},d\mathbf{p}^*\,\text{“}\|\text{”} + \text{“}\|\text{”}\,\mathbf{p},d\mathbf{N},d\mathbf{p}^*\,\text{“}\|\text{”} \\ &= -\text{“}\|\text{”}\,\mathbf{N},d\mathbf{p},d\mathbf{p}^*\,\text{“}\|\text{”} + h\,\text{“}\|\text{”}\,\mathbf{N}^*,d\mathbf{N}^*,d\mathbf{p}^*\,\text{“}\|\text{”} \\ &= 2(hP_1^* - Q)\alpha_S \end{aligned}$$

or
$$\iint (hP_1^* - Q)\alpha_S = 0 \tag{10-80}$$

$$\begin{aligned} d\text{“}\|\text{”}\,\mathbf{p},\mathbf{p}^*,d\mathbf{p}^*\,\text{“}\|\text{”} &= \text{“}\|\text{”}\,d\mathbf{p},\mathbf{p}^*,d\mathbf{p}^*\,\text{“}\|\text{”} + \text{“}\|\text{”}\,\mathbf{p},d\mathbf{p}^*,d\mathbf{p}^*\,\text{“}\|\text{”} \\ &= 2(hP_2^* - h^*Q)\alpha_S \end{aligned}$$

that is,
$$\iint (hP_2^* - h^*Q)\alpha_S = 0 \tag{10-81}$$

The following theorem is due to E. B. Christoffel, A. Hurwitz, H. Minkowski, A. D. Aleksandrov, W. Fenchel, and B. Jessen. The proof presented here is that of S. S. Chern.

Theorem 10-34. *Two closed strictly convex C^2 surfaces* $\mathbf{p}$ *and* $\mathbf{p}^*$ *are congruent and obtained from one another by a translation if either* P_1 *or* P_2 *takes the same values on both surfaces at points with parallel normals.*

We have to prove that $P_1 = P_1^*$ or $P_2 = P_2^*$ (as functions on S^2) implies $R_1 = R_1^*$ and $R_2 = R_2^*$. These equations, together with

$$d\mathbf{N} = d\mathbf{N}^*$$

give $\omega^i = \omega^{i*}$ and $\omega_i^{\ j} = \omega_i^{\ j*}$; by theorem 10-18, the two surfaces are congruent (this notion of congruence includes symmetry). Since corresponding points have identical normals the congruence is a translation. The theorem says that a closed strictly convex surface is defined (up to congruence) by either $R_1 + R_2$ or R_1R_2. The result is *much* stronger than the local one obtained in Sec. 10-3. It implies that a strictly convex closed surface is determined by its Gauss curvature or by the ratio H/K.

First we assume that $P_1(\mathbf{u}) = P_1^*(\mathbf{u})$, $\mathbf{u}$ being a point on S^2. Equations (10-79) to (10-81) and the corresponding ones obtained by permutation of $\mathbf{p}$ and $\mathbf{p}^*$ give for the function

$$\begin{aligned} F &= (R_1 - R_1^*)(R_2 - R_2^*) = P_2 + P_2^* - 2Q \\ \iint F\alpha_S &= \iint[(P_2 - Q) + (P_2^* - Q)]\alpha_S \\ &\qquad = \iint[h(P_1 - P_1^*) + h^*(P_1^* - P_1)]\alpha_S \\ &= 0 \end{aligned}$$

On the other hand,

$$2F = 4(P_1 - P_1^*)^2 - (R_1 - R_1^*)^2 - (R_2 - R_2^*)^2$$

Hence for $P_1 = P_1^*$ the integral over F cannot vanish unless $R_1 = R_1^*$, $R_2 = R_2^*$.

We assume next that $P_2(\mathbf{u}) = P_2^*(\mathbf{u})$. From (10-81) and its corresponding formula for h^*P_2 it follows that

$$\iint h(P_2^* - Q)\alpha_S = \iint h^*(Q - P_2)\alpha_S$$

or

$$2\iint h(P_2^* - Q)\alpha_S = \iint[h^*(Q - P_2) - h(Q - P_2^*)]\alpha_S$$

The right-hand side of this equation is an antisymmetric function of $\mathbf{p}$ and $\mathbf{p}^*$; hence it is zero: The Gauss curvature of a strictly convex closed surface is positive. If $P_2 = P_2^* > 0$, one may write

$$Q = \tfrac{1}{2}P_2(R_1/R_1^* + R_1^*/R_1)$$

For positive x, the function $f(x) = x + 1/x$ has a minimum for $x = 1$; hence $Q \geq P_2$ and $Q = P_2$ only for $R_1 = R_1^*$, $R_2 = R_2^*$. Since the integral on the left-hand side of the equation must vanish and we may choose the origin such that $h(\mathbf{p}) > 0$, the principal curvature radii are equal on both surfaces.

Corollary 10-35. *Two closed strictly convex surfaces are congruent (or symmetric) if they are isometric.*

PROOF: Theorems 10-22 and 10-34.

Exercise 10-5

1. (*a*) Prove that a convex surface is always star-shaped.
(*b*) Give an example of a nonconvex star-shaped surface.

2. Show that, if a closed surface is star-shaped with respect to a point P, every ray through P has one point in common with the surface.

3. Prove that, on a closed orientable surface, $\iint f\,d\omega = \iint \omega \wedge df$ for all differentiable functions f and all C^1 linear differential forms ω.

4. Prove that no surface with $K \leq 0$ throughout can be compact.

5. Show that, if $\int_0^l k(s)\,ds = 2\pi$ for a closed curve, that curve is plane convex.

6. Prove that a closed space curve with curvature $k(s) \leq 1/R$ has length $l \geq 2\pi R$ (Schwarz).

7. A closed space curve is *knotted* if the spherical image of the canal surface of the curve covers the unit sphere at least four times. Prove that $\oint k(s)\,ds \geq 4\pi$ for a knotted curve (Milnor-Voss). (HINT: On the canal surface, $K \geq 0$ for $\pi/2 \leq \phi \leq 3\pi/2$. Prove that the integral of theorem 10-30 is $\geq 8\pi$.)

8. Let $n(\mathbf{g})$ be the number of intersections of the tangent indicatrix $\mathbf{t}(s)$ of a space curve with a great circle $\mathbf{g}$ on S^2. Show that theorem 10-31 implies that $n(\mathbf{g}) \geq 2$.

9. Let $\mathbf{r}$ be a constant vector and w a function of u^1,u^2. With a surface $\mathbf{p}$ we associate $\mathbf{p}^* = \mathbf{p} + w\mathbf{r}$. Prove that

$$\begin{aligned} d\text{“}\|\text{”}\,\mathbf{N},w\mathbf{r},d\mathbf{p}\,\text{“}\|\text{”} &= 2Hw\mathbf{r}\cdot\mathbf{N}\alpha + \mathbf{N}\cdot\mathbf{N}^*\alpha^* - \alpha \\ d\text{“}\|\text{”}\,\mathbf{N}^*,w\mathbf{r},d\mathbf{p}\,\text{“}\|\text{”} &= 2H^*w\mathbf{r}\cdot\mathbf{N}\alpha + \alpha^* - \mathbf{N}\cdot\mathbf{N}^*\alpha \end{aligned}$$

and if $\mathbf{p}$ is closed oriented,

$$2\iint(H^* - H)w\mathbf{r}\cdot\mathbf{N}\alpha + \iint(1 - \mathbf{N}\cdot\mathbf{N}^*)(\alpha + \alpha^*) = 0$$

all integrals taken over $\mathbf{p}$ (Hopf and Voss).

10. Use the last formula of Prob. 9 to prove that if two closed oriented surfaces $\mathbf{p}$ and $\mathbf{p}^* = \mathbf{p} + w\mathbf{r}$, $\mathbf{r}$ const, are such that $H(\mathbf{p}) = H^*(\mathbf{p}^*)$, then w is const (Hopf and Voss).

11. $\mathbf{p}^*$ is said to be obtained from $\mathbf{p}$ by *central projection* if $\mathbf{p}^* = f(\mathbf{p})\mathbf{p}$, where f is a positive C^2 function on $\mathbf{p}$. Compute

$$d\text{“}\|\text{”}\,\mathbf{N} - \mathbf{N}^*,\ \mathbf{p},d\mathbf{p}\,\text{“}\|\text{”}$$

and prove that

$$\iint \left[(\mathbf{N} - \mathbf{N}^*)^2 + 2(|\mathbf{p}|H - |\mathbf{p}^*|H^*) \frac{\mathbf{p} \cdot \mathbf{N}}{|\mathbf{p}|} \right] \alpha = 0 \quad \text{(Aeppli)}$$

12. If in a one-to-one central projection of a closed oriented surface $\mathbf{p}$ onto a closed oriented surface $\mathbf{p}^*$ the function $|\mathbf{p}|H$ is invariant, show that the projection is a homothety, that is, $f(\mathbf{p}) = \text{const}$ (Aeppli). Use Prob. 11.

13. If two closed oriented surfaces are mapped onto one another by a one-to-one central projection, prove that

$$\iint (|\mathbf{p}|H - |\mathbf{p}^*|H^*)\mathbf{p} \cdot \mathbf{N}\alpha \leq 0$$

See Prob. 11 (Aeppli).

14. Show that, in the infinitesimal deformation (Sec. 10-4) of a surface with positive Gauss curvature, $ac - b^2 \leq 0$. [Hint: Use a frame which makes II and the matrix introduced in (10-71) diagonal. Then use the first equation (10-72).]

15. (Sequel to Prob. 14.) With the notations introduced in Sec. 10-4 for infinitesimal deformations, show that $d\|\mathbf{p},\mathbf{y},d\mathbf{y}\| = 2h(ac - b^2)\alpha$. Show also that the integral of the right-hand side of this equation over a closed oriented surface vanishes.

16. Use Probs. 14 and 15 to show that *a strictly convex closed surface is infinitesimally rigid* (Blaschke).

♦17. Find a formula $\iint \cdots = 0$ on a closed oriented surface from $d(\mathbf{p} \cdot \mathbf{e}_1\omega_2{}^3 + \mathbf{p} \cdot \mathbf{e}_2\omega_3{}^1) = ?$.

18. Show that theorem 10-34 remains true for open surfaces with boundary curves if the Gauss curvature is positive on the surface, zero on the boundary, and $\iint K\alpha = 4\pi$ [i.e., the spherical image covers the unit sphere exactly once, perhaps with the exception of some lines (of surface area 0)].

10-6. Line Congruences

In this section we use the notations and results of Sec. 8-2.

A *line congruence* is a function $\mathbf{A}(u^1,u^2)$ of two variables with values on the dual unit sphere, or a two-parameter family of lines. We assume throughout that $d\mathbf{A}$ "$\times$" $d\mathbf{A} \neq 0$. The Frenet frame attached to a curve on a unit sphere is its tangent normal frame (if $\mathbf{A}_3$ is replaced by $\mathbf{A}_1 = \mathbf{A}$, and $\mathbf{A}_1$ by $-\mathbf{A}_3$) since on a sphere the normal is the radius $\mathbf{A}$. The treatment of surfaces cannot be transposed immediately to line congruences since $\mathbf{A} = \mathbf{N}$ means $\text{II} = ds^2$; all points on the dual unit sphere are umbilics just like the points on the cartesian sphere. The deeper reason for this is the fact that $ds^2 = d\mathbf{A} \cdot d\mathbf{A} = d\mathbf{a} \cdot d\mathbf{a} + 2\tau\, d\mathbf{a} \cdot d\bar{\mathbf{a}}$

contains in itself two quadratic forms that are sufficient to develop the whole theory, viz., the first

$$d\sigma^2 = d\mathbf{a} \cdot d\mathbf{a} = g_{11}(du^1)^2 + 2g_{12}\, du^1\, du^2 + g_{22}(du^2)^2$$

and the second fundamental form

$$\mathrm{II} = d\mathbf{a} \cdot d\bar{\mathbf{a}} = l_{11}(du^1)^2 + 2l_{12}\, du^1\, du^2 + l_{22}(du^2)^2$$

The integrability conditions for given fundamental forms may be obtained from the fact that the Gauss curvature computed formally for the metric ds^2 must be $+1$ since the dual unit sphere is its own spherical image.

We choose a family of frames $\{\mathbf{E}_i\}$ on the dual sphere such that $\mathbf{E}_1 = \mathbf{A}$ and $\mathbf{E}_2$, $\mathbf{E}_3$ are tangent to the sphere. The moving frame is obtained from a fixed one by an orthogonal dual transformation. The Frenet equation is

$$d\begin{pmatrix} \mathbf{E}_1 \\ \mathbf{E}_1 \\ \mathbf{E}_3 \end{pmatrix} = \begin{pmatrix} 0 & \Omega_1{}^2 & \Omega_1{}^3 \\ -\Omega_1{}^2 & 0 & \Omega_2{}^3 \\ -\Omega_1{}^3 & -\Omega_2{}^3 & 0 \end{pmatrix} \begin{pmatrix} \mathbf{E}_1 \\ \mathbf{E}_2 \\ \mathbf{E}_3 \end{pmatrix}$$

$$\Omega_i{}^j = \omega_i{}^j + \tau\bar{\omega}_i{}^j$$

A rotation of the frames by a real matrix

$$\begin{pmatrix} 1 & 0 & 0 \\ 0 & \cos\phi & \sin\phi \\ 0 & -\sin\phi & \cos\phi \end{pmatrix}$$

transforms the Cartan matrix into

$$\begin{pmatrix} 0 & \Omega_1{}^2\cos\phi + \Omega_1{}^3\sin\phi & -\Omega_1{}^2\sin\phi + \Omega_1{}^3\cos\phi \\ -\Omega_1{}^2\cos\phi - \Omega_1{}^3\sin\phi & 0 & \Omega_2{}^3 + d\phi \\ \Omega_1{}^2\sin\phi - \Omega_1{}^3\cos\phi & -\Omega_2{}^3 - d\phi & 0 \end{pmatrix} \tag{10-82}$$

The two fundamental forms

$$d\sigma^2 = (\omega_1{}^2)^2 + (\omega_1{}^3)^2 \qquad \text{and} \qquad \mathrm{II} = \omega_1{}^2\bar{\omega}_1{}^2 + \omega_1{}^3\bar{\omega}_1{}^3$$

will both be diagonal if $\bar{\omega}_1{}^2 = \kappa_1\omega_1{}^2$, $\bar{\omega}_1{}^3 = \kappa_2\omega_1{}^3$; this can be achieved by a rotation determined from

$$\bar{\omega}_1{}^2\cos\phi + \bar{\omega}_1{}^3\sin\phi = \kappa_1(\omega_1{}^2\cos\phi + \omega_1{}^3\sin\phi)$$
$$-\bar{\omega}_1{}^2\sin\phi + \bar{\omega}_1{}^3\cos\phi = \kappa_2(-\omega_1{}^2\sin\phi + \omega_1{}^3\cos\phi)$$

As for surfaces, we call the two directions so obtained the *principal directions* of the congruence in $\mathbf{E}_1$. Through every line in the congruence there pass two *principal surfaces* whose images on the dual sphere we take as parameter lines. For this special system we denote the frame vectors again by $\mathbf{A}_i$ and the linear forms in the Cartan matrix by $\pi_i{}^j$ and $\bar{\pi}_i{}^j$.

The dual arc lengths on the principal surfaces may be taken as parameters. The Frenet equations are now

$$
\begin{aligned}
d\begin{pmatrix}\mathbf{a}_1\\ \mathbf{a}_2\\ \mathbf{a}_3\end{pmatrix} &= \begin{pmatrix}0 & \pi_1{}^2 & \pi_1{}^3\\ -\pi_1{}^2 & 0 & \pi_2{}^3\\ -\pi_1{}^3 & -\pi_2{}^3 & 0\end{pmatrix}\begin{pmatrix}\mathbf{a}_1\\ \mathbf{a}_2\\ \mathbf{a}_3\end{pmatrix}\\
d\begin{pmatrix}\bar{\mathbf{a}}_1\\ \bar{\mathbf{a}}_2\\ \bar{\mathbf{a}}_3\end{pmatrix} &= \begin{pmatrix}0 & \varkappa_1\pi_1{}^2 & \varkappa_2\pi_1{}^3\\ -\varkappa_1\pi_1{}^2 & 0 & \bar{\pi}_2{}^3\\ -\varkappa_2\pi_1{}^3 & -\bar{\pi}_2{}^3 & 0\end{pmatrix}\begin{pmatrix}\mathbf{a}_1\\ \mathbf{a}_2\\ \mathbf{a}_2\end{pmatrix} + \begin{pmatrix}0 & \pi_1{}^2 & \pi_1{}^3\\ -\pi_1{}^2 & 0 & \pi_2{}^3\\ -\pi_1{}^3 & -\pi_2{}^3 & 0\end{pmatrix}\begin{pmatrix}\bar{\mathbf{a}}_1\\ \bar{\mathbf{a}}_2\\ \bar{\mathbf{a}}_3\end{pmatrix}
\end{aligned}
\tag{10-83}
$$

$\varkappa_1$ and $\varkappa_2$ are the invariants $k_2(\sigma)$ on the principal surfaces. The frame is attached to the point of striction of the two principal surfaces which pass through a given ruling; this is the *central point* $\mathbf{m}$ of the line in the congruence. For its change along a ruled surface one differentiates the equations $\mathbf{m} \times \mathbf{a}_i = \bar{\mathbf{a}}_i$ [see page 167 for $\mathbf{s}(u)$] to obtain

$$d\mathbf{m} = \bar{\pi}_2{}^3\mathbf{a}_1 - \varkappa_2\pi_1{}^3\mathbf{a}_2 + \varkappa_1\pi_1{}^2\mathbf{a}_3 \tag{10-84}$$

A dual curve through a certain ruling, making an angle ϕ with the principal surface $\pi_1{}^3 = 0$, has its frame defined by the vector

$$\mathbf{e}_2 = \mathbf{a}_2 \cos \phi + \mathbf{a}_3 \sin \phi$$

A comparison of Eq. (8-19) with (10-82) shows that for this curve

$$k_2(\phi) = \varkappa_1 \cos^2 \phi + \varkappa_2 \sin^2 \phi \tag{10-85}$$

This formula of *Hamilton* is analogous to Euler's formula (10-17) in the theory of surfaces. The point of striction in a ruled surface of the congruence may be found from its distance ρ from the central point

$$\mathbf{s} = \mathbf{m} + \rho\mathbf{e}_1 = \mathbf{m} + \rho\mathbf{a}_1$$

By (8-22), $d\mathbf{s} \cdot d\mathbf{e}_1 = 0$; hence (in symmetric products)

$$\rho[(\pi_1{}^2)^2 + (\pi_1{}^3)^2] + (\varkappa_1 - \varkappa_2)\pi_1{}^2\pi_1{}^3 = 0$$

For $\pi_1{}^2 = \cos \phi \, d\sigma$, $\pi_1{}^3 = \sin \phi \, d\sigma$ we obtain an analogue to the second equation (10-17):

$$\rho = (\varkappa_2 - \varkappa_1) \cos \phi \sin \phi \tag{10-86}$$

Exercise 10-6

◆ **1.** Compute the frame $\{\mathbf{A}_i\}$ and the invariants for the congruence $\mathbf{A} = (\sin u^1 \cos \alpha + \tau u^2 \sin u^1 \sin \alpha,$
$\sin u^1 \sin \alpha - \tau(u^2 \sin u^1 \cos \alpha + u^2 \cos u^1), \cos u^1 + \tau u^2 \sin u^1 \sin \alpha)$
(α const).

2. Write the integrability conditions for equations (10-83) and state the necessary differentiability assumptions.

3. Explain how $\pi_2{}^3$ and $\bar{\pi}_2{}^3$ can be computed if $\pi_1{}^2$, $\pi_1{}^3$, $\varkappa_1$, $\varkappa_2$ are given. (It is necessary first to solve Prob. 2.)

4. Is it possible to find a developable through a given ruling of any congruence? (Discuss hyperbolic and elliptic regions separately. The equations of Prob. 2 are needed.)

5. Prove that a congruence **A** is that of normals to a surface $\mathbf{p} = \mathbf{m} + \rho\mathbf{e}_1$ if and only if $\varkappa_1 + \varkappa_2 = 0$. (HINT: $d\mathbf{N} = d\mathbf{e}_1$.)

6. If **A** is a congruence of normals and **B** is a congruence for which $\mathbf{A}(u^1,u^2) \cdot \mathbf{B}(u^1,u^2) = \cos\phi_0 = \text{const}$, show that **B** also is a congruence of normals (Malus and Dupin).

7. State a necessary and sufficient condition for a congruence to be invariant in a two-parameter group of euclidean motions in R^3.

8. If $\varkappa_1 = \varkappa_2$, the ruling is *umbilical*. A congruence is *isotropic* if all its rulings are umbilical. Show that for such a congruence all lines of striction of ruled surfaces in the congruence are on the *central surface* **m**.

9. Prove that a congruence is isotropic (Prob. 8) if and only if the invariants k_2 of all its ruled surfaces are functions only of u^1, u^2, not of du^1, du^2.

10. $d\mathbf{m} \cdot d\mathbf{e}_1 = 0$ for an isotropic congruence. Try to establish this result without computation.

11. A congruence is *of revolution* if it is generated by the rotation of a ruled surface about an axis. It always has a parametric representation

$$\begin{aligned} \mathbf{a} &= (\sin u^1 \sin u^2, -\cos u^1 \sin u^2, \cos u^2) \\ \bar{\mathbf{a}} &= (f(u^2)\cos u^1 \sin u^2 + g(u^2)\sin u^1 \cos u^2, \\ &\qquad f(u^2)\sin u^1 \sin u^2 - g(u^2)\sin u^1 \cos u^2, -g(u^2)\sin u^2) \end{aligned}$$

Show that a congruence of revolution is isotropic if and only if $f' = 0$ and $g' \sin u^2 - g \cos u^2 = 0$.

12. A *focal plane* is spanned by two directions for which $\mathrm{II}(du^1,du^2) = 0$. Show that an isotropic congruence has no focal planes.

References

The great source book on the euclidean geometry of surfaces is the following:

Darboux, G.: "Leçons sur la théorie générale des surfaces," 4 vols., Gauthier-Villars, Paris, 1887, 1889, 1896.

Other standard introductions are:

Bianchi, L.: "Lezioni di geometria differenziale," 4 vols., 2d ed., Niccolo Zanichelli, Bologna, 1930.

Eisenhart, L. P.: "An Introduction to Differential Geometry," Princeton University Press, Princeton, N.J., 1947.

Willmore, T. J.: "An Introduction to Differential Geometry," Oxford University Press, London, 1959.

A direct approach to the study of surfaces in euclidean space which is not based on the theory of space curves and which often allows a considerable weakening of differentiability assumptions has been developed by A. D. Aleksandrov. See, for example:

Aleksandrov, A. D.: "Vnutrennyaya geometriya vypuklyh poverhnosteĭ," Gostehisdat, Moscow and Leningrad, 1948; also, in German translation, "Die innere Geometrie der konvexen Flächen," Akademie-Verlag VeB, Berlin, 1955.

SEC. 10-2

For the circular sections of the ellipsoid, compare the following:

McCrea, W. H.: "Analytical Geometry of Three Dimensions," 2d ed., pp. 109–112, Oliver & Boyd Ltd., Edinburgh and London, 1953.

SECS. 10-3 AND 10-4

Cartan, E.: "La théorie des groupes finis et continus et la géométrie différentielle," Gauthier-Villars, Paris, 1937.

The modern (global) theory of integration of systems of differential forms is based on the work of G. Reeb in the following:

Wu, W. T., and G. Reeb: "Sur les espaces fibrés et les variétés feuilletées," Hermann & Cie, Paris, 1952.

The classical work on mappings and deformations is summarized in the following:

Voss, A.: Abbildung und Abwicklung zwier Flächen aufeinander, Enz. Math. Wiss., vol. III, part 3, pp. 355–440, B. G. Teubner Verlagsgesellschaft, mbH, Leipzig, 1903.

The standard reference on the modern developments is:

Efimow, N. W.: "Flächenverbiegung im Grossen," Akademie-Verlag GmbH, Berlin, 1957.

SEC. 10-5

Chern, S. S.: Integral Formulas for Hypersurfaces in Euclidean Space and Their Applications to Uniqueness Theorems, *J. Math. Mech.*, **8**: 947–956 (1959).
Hopf, H.: Selected Topics in Differential Geometry in the Large, Lecture Notes, New York University, New York, 1955.
Voss, K.: Eine Bemerkung über die Totalkrümmung geschlossener Raumkurven, *Arch. Math.*, **6**: 259–263 (1955).

SEC. 10-6

See Sec. 8-2.

11
INNER GEOMETRY OF SURFACES

11-1. Geodesics

It was shown in the preceding chapter that certain properties of local surfaces depend only on the metric, not on the imbedding of the surface in three space. In this chapter we study in a systematic way those properties of surfaces that can be obtained by measurements on the surface itself, without reference to the surrounding space. It is the kind of geometry which a two-dimensional being living on a curved surface would develop. The simplest setting for the theory is obtained if we identify the space R^2 of the coordinates and the point set of the local surface. The inner geometry of surfaces then is simply the study of a neighborhood in the cartesian plane in which a positive-definite quadratic form $g_{ij}(u^1,u^2)$ is given. The metric is $ds^2 = g_{ij}\,du^i\,du^j$, and the arc length of any curve $u^i = u^i(t)$ in the neighborhood is $s = \int_0^t (g_{ij}u^{i\cdot}u^{j\cdot})^{\frac{1}{2}}\,dt$. Here arises a most interesting global problem: Given a collection of local surfaces, what kinds of surfaces have this collection as their atlas, provided the mapping functions are defined in suitable regions of R^2 such that the functions (9-27) have all the required properties? This is a topological problem to be solved by topological methods; although it is intimately connected with some of the most interesting applications of inner geometry, we cannot treat it in this book.

The inner geometry of surfaces is radically different from the problems treated in the preceding chapters. In the discussion of isometric surfaces (Sec. 10-4) we saw that only two special classes of metrics exist that admit a differentiable group of metric-preserving transformations. In general, there exists no local transformation of a local surface other than the identity that maps arcs into arcs of equal length. The inner geometry of surfaces is not a Klein geometry; the general arguments of the existence of differential invariants, invariant frames, and natural orders of differentiability do not apply. Nevertheless, we shall see that the methods of group theory may be extended to cover this case also, but in a manner different from Klein geometry.

Any nonsingular symmetric matrix may be brought into diagonal form. This means that there exists a nonsingular matrix $D = (d_i{}^j)$ such that $G = (g_{ij}) = DD^t$. The algebraic formulas show that it is possible to find D as a differentiable function of $\mathbf{u}$ if G is differentiable. At each point $\mathbf{u}$ (which now serves both as a parameter and as a point on the local surface) a tangent plane T_u is defined on the set of equivalence classes of curves through $\mathbf{u}$. It will help to understand this if each T_u is visualized as a separate plane. If the metric (first fundamental form) is defined in a neighborhood V in the plane, we introduce a four-dimensional space $T(V)$. A point in $T(V)$ is a couple $(\mathbf{u},\mathbf{x})$, $\mathbf{u} \in V$, $\mathbf{x} \in T_u$. (See Sec. 9-4.) The map π: $(\mathbf{u},\mathbf{x}) \to \mathbf{u}$ is the *projection* of $T(V)$ onto V. A differentiable curve $\mathbf{c}(t)$ defines a map $\hat{\mathbf{c}}$: $R^1 \to T(V)$ by $\hat{\mathbf{c}}(t) = (\mathbf{c}(t),\mathbf{c}^{\cdot}(t))$. $\hat{\mathbf{c}}$ has the special property that (at least for an interval $I = [t - \epsilon, t + \epsilon]$) the intersection of $\hat{\mathbf{c}}(I)$ with any set $(\mathbf{u},T_u)$ consists of one point at most, and $\pi\hat{\mathbf{c}}(t) = \mathbf{c}(t)$. $\hat{\mathbf{c}}$ is a *section* in $T(V)$. In each tangent space T_u we have a natural basis formed by the vectors $\mathbf{a}_i$ which represent the two coordinate lines through $\mathbf{u}$. We define the scalar product of two vectors $\mathbf{x} = x^i\mathbf{a}_i$ and $\mathbf{y} = y^i\mathbf{a}_i$ by $\mathbf{x} \circ \mathbf{y} = g_{ij}x^iy^j$ (or $\mathbf{x}G\mathbf{y}^t$ in coordinate vectors). The length of a vector in T_u is $|\mathbf{x}| = (\mathbf{x} \circ \mathbf{x})^{\frac{1}{2}}$; the angle of two vectors is defined by $\cos \angle(\mathbf{x},\mathbf{y}) = \mathbf{x} \circ \mathbf{y}/|\mathbf{x}|\,|\mathbf{y}|$. The vectors $\{\mathbf{e}_i\} = D^{-1}\{\mathbf{a}_i\}$ are orthonormal in the $\circ$ product, $(\mathbf{e}_i \circ \mathbf{e}_j) = D^{-1}G(D^t)^{-1} = D^{-1}DD^t(D^t)^{-1} = U$. In the basis $\{\mathbf{e}_i\}$ the $\circ$ product reduces to the usual dot product. If $\mathbf{x} = x^i\mathbf{e}_i$, then $\mathbf{x} = x^id_i{}^j\mathbf{a}_j$ and $\mathbf{x} \circ \mathbf{y} = x^id_i{}^jg_{jk}d_l{}^ky^l = x^i\delta_{il}y^l = \mathbf{x} \cdot \mathbf{y}$. Any other orthonormal system $\{\mathbf{e}_{i'}\}$ is obtained from $\{\mathbf{e}_i\}$ by an orthogonal matrix, $\{\mathbf{e}_{i'}\} = A\{\mathbf{e}_i\}$, $A \in O_2$. The diagonalizing matrix D itself is defined only up to an orthogonal matrix, since $(DA)(DA)^t = DD^t$ for $AA^t = U$. If a family of frames $\{\mathbf{e}_i\}$ is fixed in V, the metric appears as $ds^2 = (\omega^1)^2 + (\omega^2)^2$, where $(\omega^1,\omega^2) = (du^1,du^2)D^{-1}$.

The tangent normal frame of a curve on a surface in R^3 defines a map of the tangent space T_{x_0} of a point $\mathbf{x}_0$ on the curve onto the tangent space T_x of a nearby point $\mathbf{x}$ on the same curve. Let ϕ be the angle of the normal projection of $\mathbf{t}(\mathbf{x}_0)$ onto the plane T_x with the tangent $\mathbf{t}(\mathbf{x})$. The two tangents serve as basis vectors in the tangent spaces. T_{x_0} is mapped onto T_x by a normal projection followed by a rotation of angle ϕ. Equation (10-2) shows that $k_g = d\phi/ds$. The same mapping can be defined in inner geometry. Given a curve $\mathbf{c}(t)$, the vector

$$\mathbf{e}_1 = d\mathbf{c}/ds = \mathbf{c}^{\cdot}(g_{ij}c^{\cdot i}c^{\cdot j})^{-\frac{1}{2}}$$

is a unit vector. A vector $\mathbf{e}_2$ orthogonal to $\mathbf{e}_1$ in the $\circ$ product gives a frame belonging to $\mathbf{c}(s)$ in $T_{c(s)}$. This frame is now introduced in all tangent spaces along $\mathbf{c}$. The tangent spaces are hitherto unrelated. If we define a map $T_{c_0} \to T_c$ through the identification of $\mathbf{e}_i \in T_{c_0}$ with $\mathbf{e}_i + \int_{s_0}^{s} d\mathbf{e}_i(s) \in T_c$, where

$$d\begin{pmatrix}\mathbf{e}_1\\ \mathbf{e}_2\end{pmatrix} = \begin{pmatrix}0 & \omega_1{}^2\\ -\omega_1{}^2 & 0\end{pmatrix}\begin{pmatrix}\mathbf{e}_1\\ \mathbf{e}_2\end{pmatrix} \tag{11-1}$$

and

$$d\omega^1 = \omega_1{}^2 \wedge \omega^2 \qquad d\omega^2 = \omega^1 \wedge \omega_1{}^2 \tag{11-2}$$

a comparison with the developments of Sec. 10-1 shows that this definition of maps of tangent spaces along curves coincides for surfaces in R^3 with the map induced by the movement of the tangent normal frame along the curve. Therefore formulas (11-1) and (11-2) hold also for frames not directly connected with the curve in which

$$d\mathbf{c} = \omega^1\mathbf{e}_1 + \omega^2\mathbf{e}_2 \tag{11-3}$$

If all tangent spaces are identified with one fixed cartesian plane R^2 referred to an orthonormal basis $\{\mathbf{i}_i\}$, the matrix

$$\begin{pmatrix} 0 & \dfrac{\omega_1{}^2}{ds} \\ -\dfrac{\omega_1{}^2}{ds} & 0 \end{pmatrix}$$

is the Cartan matrix of an orthogonal matrix function $A(s)$. The curve determined in R^2 by $A(s)$ is the *development of* $\mathbf{c}$ *on* R^2. The situation encountered here, that each tangent space may be given the structure of a euclidean plane and any differentiable curve defines a map of tangent spaces by integration of a matrix function in $\mathfrak{L}(O_2)$, is summed up by saying that inner geometry is *locally euclidean.* We have shown that the formalism of Cartan matrices applies also to local geometries. The group-theoretical background of local geometry will be discussed in Chap. 14.

Definition 11-1. *A geodesic is a curve whose development on R^2 is a straight line.*

For a geodesic $\omega_1{}^2 = 0$, its geodesic curvature vanishes. The new definition concides with that of corollary 10-23. Inner geometry, to a great extent, reduces to the study of geodesics, so much so that a property of a local surface in R^3 which belongs to inner geometry is said to be a *geodesic* property.

The integration theorem 10-18 assures that for every point $\mathbf{u}_0$ it is possible to find a unique geodesic starting from $\mathbf{u}_0$ in any given direction. These geodesics together with their orthogonal trajectories (in the sense of the $\circ$ angle) form an intrinsic system of coordinates in some ring domain about $\mathbf{u}_0$.

Definition 11-2. *A family of curves forms a field in a domain V if through every point of V there passes exactly one curve of the family.*

The geodesics and their $\circ$-orthogonal trajectories are a system of coordinate lines in any domain in which the geodesics issuing from $\mathbf{u}_0$ form a *field.* We introduce parameters r and ϕ such that $\phi = \text{const}$ are geodesics through $\mathbf{u}_0$ and $r = \text{const}$ are orthogonal trajectories. By orthogonality,

$$ds^2 = E\,dr^2 + G\,d\phi^2$$

Hence $$\omega^1 = \sqrt{E}\,dr \qquad \omega^2 = \sqrt{G}\,d\phi$$

For a geodesic, $\omega_1{}^2 = 0$; hence also $d\omega^1 = 0$, that is,

$$\frac{\partial E}{\partial \phi} = 0$$

It is possible to change the r scale to normalize $E = E(r) = 1$ by $dr^* = \sqrt{E}\,dr$. If we again replace r^* by r, the metric appears as

$$ds^2 = dr^2 + G(r,\phi)\,d\phi^2 \qquad G(r,\phi) > 0 \tag{11-4}$$

(r,ϕ) is the system of *geodesic polar coordinates* in $\mathbf{u}_0$.

The length of the arcs of geodesics between two fixed orthogonal trajectories is constant:

$$\int_{\mathbf{u}_1}^{\mathbf{u}_2} ds = \int_{\mathbf{u}_1}^{\mathbf{u}_2} dr = r(2) - r(1)$$

For any other arc $\mathbf{p}$ joining $\mathbf{u}_1$ to $\mathbf{u}_1$ we have

$$s = \int_{\mathbf{u}_1}^{\mathbf{u}_2} (dr^2 + G\,d\phi^2)^{\frac{1}{2}} > \int_{\mathbf{u}_1}^{\mathbf{u}_2} dr$$

Theorem 11-3. *If an arc of a geodesic is part of a field of geodesics for its whole length, it is the shortest connection between its endpoints.*

The theorem follows from the preceding inequality since any field of geodesics with its orthogonal trajectories may be given a metric [Eq. (11-4)]. The theorem is important because it settles the question of the sufficient conditions in the calculus of variations for all problems of integrals over the square root of a positive quadratic form in the differentials. We proceed to show in which sense the condition of the theorem is also necessary.

Any given geodesic in a field may be taken to be $\phi = 0$. The distance of the point (r,a) on the geodesic $\phi = a$ from the corresponding point

$(r,0)$ on $\phi = 0$ is $\int_0^a \sqrt{G(r,\phi)}\, d\phi$. Since $\sqrt{G} \geq 0$ (in a field, $\sqrt{G} > 0$) it follows that the geodesics issuing from (0,0) in an angle $-\epsilon < \phi < +\epsilon$ form a field at least as long as $\sqrt{G(r,0)} > 0$. In the metric (11-4) we have $\omega_1{}^2 = (d\sqrt{G(r,0)}/dr)\, d\phi$. The Gauss equation $d\omega_1{}^2 = -K\omega^1 \wedge \omega^2$ implies that $\sqrt{G(r,0)}$ is a solution of the *Jacobi equation*

$$\frac{d^2\sqrt{G(r,0)}}{dr^2} + K(r,0)\sqrt{G(r,0)} = 0 \qquad \sqrt{G(0,0)} = 0 \qquad (11\text{-}5)$$

Definition 11-4. *The point* $\mathbf{u}^* = (r^*,0)$ *corresponding to the smallest positive root of a solution of the Jacobi equation is the conjugate point to* $\mathbf{u}_0 = (0,0)$ *on the geodesic* $\phi = 0$.

Theorem 11-3 may now be made more definite:

Theorem 11-5. *An arc of a geodesic is the shortest connection between its endpoints if it does not contain a point and its conjugate point.*

The locus of the conjugate points to (0,0) on all geodesics belongs to the point set defined by $\|J_{(r,\phi)}(u^1,u^2)\| = 0$ (or $= \infty$). If the geodesics $u^i = u^i(r,a)$ have an envelope, this envelope must be a part of the locus

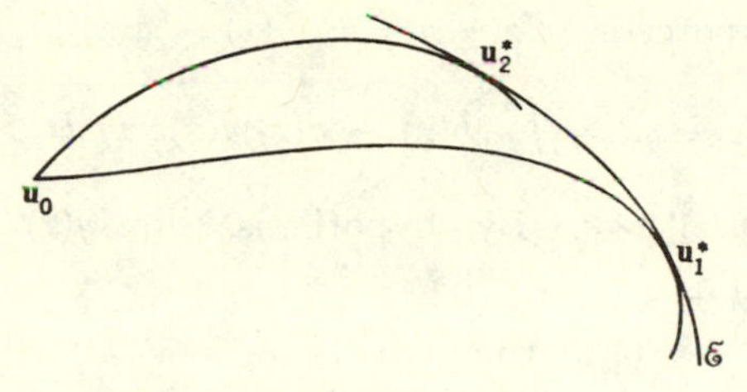

Fig. 11-1

$\|J_{(r,\phi)}(u^1,u^2)\| = 0$. The line element of the envelope $\mathcal{E}$ is always identical with the line element of some geodesic; hence $G = 0$ on $\mathcal{E}$. This leads to the following situation (Fig. 11-1): Let $\mathbf{u}_1^*$, $\mathbf{u}_2^*$ be two conjugate points to $\mathbf{u}_0$ situated on $\mathcal{E}$. Then

$$\underset{\text{on geodesic}}{\int_{\mathbf{u}_0}^{\mathbf{u}_1^*} ds} = \underset{\text{on geodesic}}{\int_{\mathbf{u}_0}^{\mathbf{u}_2^*} ds} + \underset{\text{on envelope}}{\int_{\mathbf{u}_2^*}^{\mathbf{u}_1^*} ds}$$

There can be only one geodesic through one point in a given direction; the envelope cannot be the shortest connection between $\mathbf{u}_1^*$ and $\mathbf{u}_2^*$ since it is not a geodesic. Therefore it is possible to find an arc connecting $\mathbf{u}_1^*$ and $\mathbf{u}_2^*$ of shorter length; the geodesic arc from $\mathbf{u}_0$ to $\mathbf{u}_1^*$ is the shortest. This argument shows more or less that a geodesic arc is not the shortest if it contains two conjugate points. Because of the complications of the theory of envelopes, it does not seem that any published proof along these lines takes care of all possible exceptional cases. (For example, on a sphere there exists a unique conjugate point to any point, and any meridian between a point and its antipodal point is shortest; see theorem 11-8.) A rigorous proof that a geodesic arc is not shortest if it contains two mutually conjugate points, one of them as an interior point, is given in exercise 11-1, Prob. 6. The proof, due to Bliss, is based on a supplementary variational problem and uses exercise 4-1, Prob. 9.

Theorem 11-5 shows that the fundamental problem in the geometry of geodesics is to obtain as much information as possible on conjugate points. The main tool is the following famous theorem on differential equations:

Lemma 11-6 (Sturm). *If two differential equations*

$$\text{I}: y'' + a(x)y = 0 \qquad \text{II}: z'' + b(x)z = 0$$

are given in an interval $[x_0,x_1]$ *and if* $a(x) < b(x)$ *in that interval, then for any two solutions* $y(x)$ *of* I *and* $z(x)$ *of* II *such that* $y(x_0) = z(x_0) = 0$ *one has*

$$y(x) > z(x)$$

for $x > x_0$ *and as long as both functions are non-negative.*

PROOF: By hypothesis, $y''z - yz'' = [b(x) - a(x)]yz$; hence

$$y'z - yz' = \int_{x_0}^{x} [b(x) - a(x)]y(x)z(x)\,dx > 0$$

This means $(y/z)' > 0$. By hypothesis, $\lim_{x \downarrow x_0} y(x)/z(x) = 1$; hence $y(x)/z(x) > 1$ for $x > x_0$.

If $K(r,\phi) < -\epsilon < 0$, we compare (11-5),

$$(\sqrt{G})'' - |K|\sqrt{G} = 0 \qquad \sqrt{G(0)} = 0$$

and

$$z'' - \epsilon z = 0 \qquad z(0) = 0$$

By Sturm's theorem, $\sqrt{G} > z(r) = \operatorname{Sinh}\sqrt{\epsilon}\,r > 0$; $\sqrt{G}$ has no zero for $r > 0$.

Theorem 11-7. *On a local surface with negative Gauss curvature bounded away from zero there are no conjugate points; every geodesic arc is the shortest connection between its endpoints.*

If $K(r,\phi) > \epsilon > 0$, we compare (11-5) and $y'' + \epsilon y = 0$, $y(0) = 0$. In this case, $\sqrt{G} < \sin\sqrt{\epsilon}\,r$ has a zero for some $r \leq \pi/\sqrt{\epsilon}$.

Theorem 11-8. *On a surface with positive curvature bounded away from zero the maximal distance of two mutually conjugate points (measured on the geodesic) is* $\leq \pi(\inf K)^{-\frac{1}{2}}$.

The Gauss curvature of a sphere of radius R is R^{-2}. Here conjugate points are antipodal points; their geodesic distance is πR.

A practical method for the computation of the geodesics is the following: For given ds^2 and curvature lines we know how to compute ω^1, ω^2, and $\omega_1{}^2$. If $\theta = \arctan \omega^2/\omega^1$ is the angle of the tangent of the parameter line $u^2 = \text{const}$ with the tangent of the geodesic, Eqs. (10-5) show that the differential equation of the geodesics is

$$\omega_1{}^2 + d\theta = 0 \tag{11-6}$$

***Example* 11-1.** The ds^2 of a surface of revolution may be brought into the form $ds^2 = dz^2 + r^2(z)\, d\phi^2$. The geodesics are the solutions of (11-6): $r'\, d\phi + d\theta = 0$. Since $\tan\theta = r\, d\phi/dz$, the equation becomes $(r'/r) \tan\theta\, dz + d\theta = 0$, or $d(r(z) \sin\theta(z)) = 0$ (Fig. 11-2).

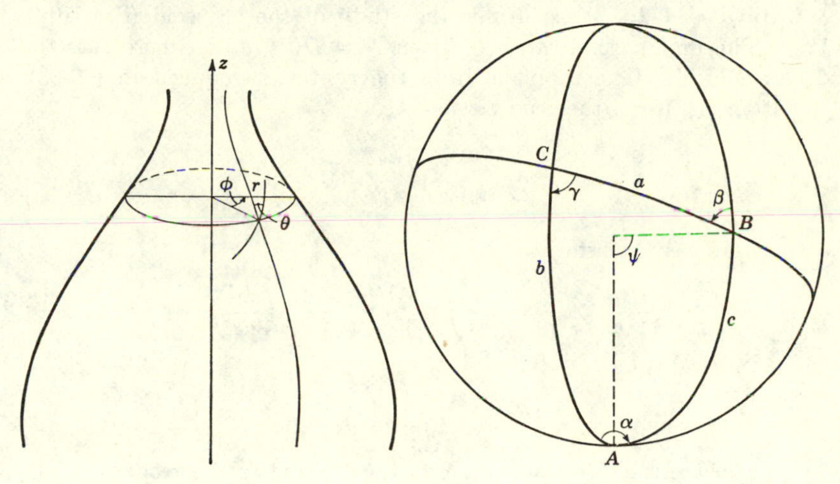

Fig. 11-2 Fig. 11-3

Theorem 11-9 (Clairaut). *On a surface of revolution the product of the sine of the angle of the geodesic and meridian with the distance from the axis of revolution is constant.*

On a unit sphere, $r = \sin\psi$. We take a spherical triangle with one vertex at the south pole and the two other vertices on meridians from the south to the north pole. In the notation of Fig. 11-3, $r(C) = \sin b$, $r(B) = \sin c$, and for the arc of the great circle joining B to C, $\theta(C) = \gamma$ and $\theta(B) = \beta$. Hence $\sin b \sin\gamma = \sin c \sin\beta$ or

$$\frac{\sin b}{\sin\beta} = \frac{\sin c}{\sin\gamma} = \frac{\sin a}{\sin\alpha} \tag{11-7}$$

Clairaut's theorem on the sphere is the sine theorem of spherical trigonometry. (Compare exercise 11-1, Prob. 22.)

The geodesics of a surface in R^3 are sometimes more easily found from

$$k_g = \left\| \mathbf{t}, \frac{d\mathbf{t}}{ds}, \mathbf{N} \right\| \tag{11-8}$$

a formula which follows directly from (10-2). The geodesics are the curves for which $\|\mathbf{t},d\mathbf{t},\mathbf{N}\| = 0$. Also from (10-2) it follows that the geodesics are solutions of

$$d\mathbf{t} \times \mathbf{N} = k_g\mathbf{t}\, ds = 0 \tag{11-9}$$

and

$$d\mathbf{t} \cdot d\mathbf{N} = -k_g t_r\, ds^2 = 0 \tag{11-10}$$

The last equation may be used only on elliptic surfaces. For the sphere it follows from (11-9) that $d(\mathbf{t} \times \mathbf{N}) = 0$, the binormal is $\mathbf{t} \times \mathbf{N}$ and constant, and the geodesics are the great circles.

***Example* 11-2.** We continue the study of the ellipsoid (Sec. 10-2, IV). The unit normal can be written $\mathbf{N} = D(x^i/(a^i)^2)$, where D is the distance of the tangent plane from the center introduced in (10-33). Equation (11-10) of the geodesics is

$$\sum_{i=1}^{3} \frac{d^2x^i}{ds^2} \frac{d}{ds} \frac{Dx^i}{(a^i)^2} = 0$$

or, after an easy transformation,

$$\frac{D}{2} \frac{d}{ds} \sum \frac{1}{(a^i)^2} \left(\frac{dx^i}{ds}\right)^2 = \frac{dD}{ds} \sum \frac{1}{(a^i)^2} \left(\frac{dx^i}{ds}\right)^2$$

whose integral is

$$D^2 = c \sum \frac{1}{(a^i)^2} \left(\frac{dx^i}{ds}\right)^2$$

The dx^i/ds are the direction cosines of the tangent to the geodesic. The line $(\lambda dx^i/ds)$ through the origin is parallel to that tangent and meets the ellipsoid at points of parameter λ determined by $\lambda^2\Sigma(a^i)^{-2}\,(dx^i/ds)^2 = 1$. This proves the following theorem.

Theorem 11-10 (Joachimsthal). *The product of the distance D of the tangent plane to the center of the ellipsoid with the length 2λ of the diameter of the ellipsoid parallel to the tangent of the geodesic is constant along any geodesic.*

All four umbilics of the ellipsoid have the same D. The central sections of the umbilics are circles. This shows that *all geodesics passing through the umbilics belong to the same value of $D\lambda$.* Consider now two geodesics through umbilics meeting at a point $\mathbf{x}_0$. D depends only on $\mathbf{x}_0$, and the value of $D\lambda$ is the same on both curves; hence λ has the same value on both geodesics. The central section through $\mathbf{x}_0$ is an ellipse. Diameters of equal length in an ellipse make equal angles with the axes. The directions of the axes of the central section are the principal directions on the ellipsoid. *The principal directions are the angle bisectors of the tangents to the geodesics joining a point to two umbilics.* The same situation is found also in the plane. The tangents to an ellipse and hyperbola bisect the angle made by the two lines which connect a point on the conic with its foci. Curvature lines appear here as geodesic conics with foci at the umbilics (see exercise 11-1, Prob. 25).

Exercise 11-1

1. Prove that $|\cos \angle(\mathbf{x},\mathbf{y})| \leq 1$ in the $\circ$ product.
2. Instead of the set $T(V)$ one may use the set

$$V_1 = \{(\mathbf{u},\mathbf{x}); \mathbf{x} \in T_u, |\mathbf{x}| = 1\}$$

Formulate the equations for the map of $V_1(\mathbf{u}_0)$ onto $V_1(\mathbf{u})$ corresponding to that of T_{u_0} onto T_u along a curve.
3. Characterize the geodesics of local surfaces in R^3 by the relative positions of their tangent normal and their space-curve frames.
4. Show that the geodesics are solutions of the Euler equations for $\int ds = \min$. [HINT: It is sufficient to treat the problem for orthogonal parameter lines, $ds^2 = E(du^1)^2 + G(du^2)^2$.]
5. *Geodesic cartesian coordinates* are obtained in the following manner: A geodesic is taken as a parameter line $v = 0$. The parameter lines $u = \text{const}$ are the geodesics orthogonal to $v = 0$. The parameter lines $v = \text{const}$ are the orthogonal trajectories of the geodesics $u = \text{const}$. Show that it is possible to choose the parameters so as to bring the metric into the form $ds^2 = E\,du^2 + dv^2$, where $E(u,0) = 1$, $\partial E/\partial v(u,0) = 0$. (HINT: Take u to be the arc length on $v = 0$.)
6. In geodesic cartesian coordinates (Prob. 5) we compare the geodesic $v = 0$ with curves $(u,v^*(u))$, $v^*(u) = v(u)$, $v(0) = v(a) = 0$. The length of such a curve between 0 and a is

$$s(\epsilon) = \int_0^a (E + v^{*\prime 2})^{\frac{1}{2}}\,du = s(0) + \epsilon s'(0) + \tfrac{1}{2}\epsilon^2 s''(0) + \cdots$$

From Prob. 5 we have a Taylor expansion

$$E^{\frac{1}{2}} = 1 + \tfrac{1}{2}v^{*2}(E^{\frac{1}{2}})_{vv}(u,0) + R_3$$

This implies for $s(\epsilon)$ the Euler equation $s'(0) = 0$ and

$$s''(0) = \int_0^a (v^2(E^{\frac{1}{2}})_{vv} + v'^2)\,du$$

The geodesic will be the shortest in the field if $s''(0) > 0$. By (11-5), $(E^{\frac{1}{2}})_{vv} = -K(u,0)$.

The variational problem $\int_0^a (v'^2 - K(s)v^2)\,ds = \min$ has the Jacobi equation (11-5) as its Euler equation. $v = 0$ is a solution of the problem for $s''(0) \geq 0$. On the geodesic $v = 0$ let A be the conjugate point of O. On an arc OB which contains A we take a nonzero solution of (11-5) from O to A and $v = 0$ from A to B. This is a solution of a variational problem with a cusp. Exercise 4-1, Prob. 9, then shows that $v'(A) = 0$. By hypothesis, $v(A) = 0$; hence $v(s) = 0$ for all s. This contradicts the hypothesis $s''(0) > 0$. Prove that the geodesic OB is not the shortest and analyze the differentiability assumptions (Bliss).

7. Show that, if a metric has the form $ds^2 = du^2 + G\,dv^2$, the curves $v = \text{const}$ must be geodesics.

8. What are the images of the geodesics on developable surfaces in an isometric mapping onto the plane? Find the geodesics of cylinders and of cones without computation.

◆ **9.** Find the geodesics for $ds^2 = (dx^2 + dy^2)/y$, $y > 0$. (Use Euler equations.)

◆ **10.** Find the geodesics for

$$ds^2 = \frac{dx^2 + dy^2}{1 - x^2 - y^2} \qquad x^2 + y^2 < 1$$

Determine K for this metric.

11. $ds^2 = [f(u^1) + g(u^2)][(du^1)^2 + (du^2)^2]$ is a *Liouville* metric. Show that the geodesics of a Liouville metric admit an integral

$$\int (f(u^1) + a)^{-\frac{1}{2}}\,du^1 + \int (g(u^2) - a)^{-\frac{1}{2}}\,du^2 = \text{const}$$

12. Show that, if a surface admits two conjugate families (exercise 10-1, Prob. 22) of geodesics, it is of constant curvature (Liouville).

13. Prove that a surface has two mutually orthogonal families of geodesics if and only if $K = 0$.

14. Show that, if $\mathbf{p}^*$ is a point conjugate to $\mathbf{p}$, then $\mathbf{p}$ is conjugate to $\mathbf{p}^*$.

15. Two intersecting families of geodesics may be used as parameter lines in some neighborhood. The angle of the coordinate lines at any point is given by $\cos\omega = g_{12}(g_{11}g_{22})^{-\frac{1}{2}}$. Show that the ds^2 in these coordinates is

$$ds^2 = \frac{(du^1)^2 + 2\cos\omega\,du^1\,du^2 + (du^2)^2}{\sin^2\omega}$$

(Weingarten).

16. Show that, if two families of geodesics meet at a constant angle the surface is developable (Liouville). (Use Prob. 15.)

17. With the notations of Prob. 15, prove that

$$\frac{\partial^2\omega}{\partial u^1\,\partial u^2} = K(g_{11}g_{22} - g_{12}{}^2)^{\frac{1}{2}}$$

(Crudeli).

18. Prove that an asymptotic line is a geodesic if and only if it is straight.

19. The metric $ds^2 = x(dx^2 + dy^2)$ has been discussed in example 4-3. Compute the curvature and explain the phenomena of that example in the light of the theorems of Sec. 11-1.

20. What are the conjugate points to (0,0) in the problem of the brachistochrone (example 4-2)?

21. Find the geodesics for $ds^2 = dz^2 + r^2\, d\phi^2$.

22. On a sphere, let x be the longitude, y the latitude, and s the length of a great circle; $ds^2 = dy^2 + \sin^2 y\, dx^2$. The geodesics are solutions of the Euler equations for $\int ds = \min$. Show that they may be integrated [for $x(0) = y(0) = 0$] by $\sin x = \cos A \sin y/\sin A \cos y$. From this equation it follows that

$$\sin s = \sin y/\sin A \qquad \cos s = \cos x \cos y$$

The last equation is the cosine theorem of a right-angle spherical triangle of legs x and y and hypothenuse s (Euler).

23. A *geodesic circle of the first kind* is a curve of constant geodesic curvature. A *geodesic circle of the second kind* is the locus of points of constant geodesic distance from a fixed point. Geodesic circles of the second kind are curves $u = \text{const}$ in geodesic polar coordinates. Show that these curves are also geodesic circles of the first kind if the metric is of constant curvature.

24. Show that theorem 11-10 holds also for curvature lines.

25. In curvature-line parameters on an ellipsoid (exercise 10-2, Prob. 20) the equation of a geodesic is

$$E\, d\sigma^2 + G\, d\sigma^3 + 2\sqrt{EG}\,(\sigma^2 - \sigma^3)\, d\theta = 0$$

Since $\sqrt{E}\, d\sigma^2 = ds \cos\theta$, $\sqrt{G}\, d\sigma^3 = ds \sin\theta$, the geodesics are the curves $\sigma^2 \cos^2\theta + \sigma^3 \sin^2\theta = C$. The geodesic has contact with the two curvature lines $\sigma^2 = C$ and $\sigma^3 = C$. Show that

$$D^{-2}\lambda^{-2} = (a^1)^{-2}(a^2)^{-2}(a^3)^{-2}\{[(\sigma^1)^2 - (\sigma^2)^2]\cos^2\theta + [(\sigma^1)^2 - (\sigma^3)^2]\sin^2\theta\}$$

and that the locus of the points at which two geodesics belonging to the same value of C intersect is the spherical conic $(\sigma^2)^2 + (\sigma^3)^2 = 2C$.

26. Show that the central sections through two of the axes of an ellipsoid are geodesics.

27. The conjugate point to an umbilic on an ellipsoid is the opposite umbilic. Prove this theorem in the case of the plane section $x^3 = 0$ (see Prob. 26).

28. Discuss our theorems on the ellipsoid for an ellipsoid of revolution.

11-2. Clifford-Klein Surfaces

In this section we study transformation groups of isometries on surfaces (see Sec. 10-4, example 10-6).

Two curves $\mathbf{x}(s)$ and $\mathbf{y}(s)$ issuing from a point $\mathbf{u}_0 = \mathbf{x}(0) = \mathbf{y}(0)$ are represented in T_{u_0} by their initial vectors $\boldsymbol{\xi} = \mathbf{x}'(0)$ and $\boldsymbol{\eta} = \mathbf{y}'(0)$. The scalar product in T_{u_0} is $\boldsymbol{\xi} \circ \boldsymbol{\eta} = \boldsymbol{\xi} G(u_0)\boldsymbol{\eta}^t$. The action of a differentiable map F of the surface into itself induces an action on the scalar product by

$$(\boldsymbol{\xi} \circ \boldsymbol{\eta})_F = \boldsymbol{\xi}_F G(Fu_0)\boldsymbol{\eta}_F{}^t$$

where

$$\boldsymbol{\xi}_F = \left[\frac{d}{ds} F(\mathbf{x}(s))\right]_{s=0} = \xi^j \frac{\partial F^i}{\partial u^j} = \boldsymbol{\xi}[J_u F]_{\mathbf{u}=\mathbf{u}_0} \qquad \boldsymbol{\eta}_F = \boldsymbol{\eta}[J_u F]_{\mathbf{u}=\mathbf{u}_0}$$

are the initial vectors of the transformed curves. F is an isometry if it leaves the scalar product invariant:

$$(\boldsymbol{\xi} \circ \boldsymbol{\eta})_F = \boldsymbol{\xi} \circ \boldsymbol{\eta}$$

If $F = F(\mathbf{u},t)$ is a one-parameter group of isometries (or *motions*) with vector field $\Xi(\mathbf{u}) = dF(\mathbf{u},t)/dt_{t=0}$, then naturally

$$\pounds(\boldsymbol{\xi} \circ \boldsymbol{\eta}) = -\left[\frac{d}{dt}(\boldsymbol{\xi} \circ \boldsymbol{\eta})_F\right]_{t=0} = 0$$

The vector field Ξ of a group of motions is called a *Killing* vector. $\pounds$ obeys the Leibniz rule as a directional derivative,

$$\begin{aligned}\pounds(\boldsymbol{\xi} \circ \boldsymbol{\eta}) &= \pounds\boldsymbol{\xi} G\boldsymbol{\eta}^t + \boldsymbol{\xi}\pounds G\boldsymbol{\eta}^t + \boldsymbol{\xi} G\pounds\boldsymbol{\eta}^t \\ &= \left(\xi^j\left[\frac{\partial \Xi^i}{\partial u^j} g_{ik} + \Xi^i \frac{\partial g_{jk}}{\partial u^i} + g_{ji}\frac{\partial \Xi^i}{\partial u^k}\right]\eta^k\right) = 0\end{aligned}$$

This equation must be an identity in $\boldsymbol{\xi}$ and $\boldsymbol{\eta}$.

Theorem 11-11 (Killing). *The vector field Ξ of a one-parameter group of isometries satisfies*

$$\Xi^s \frac{\partial g_{jk}}{\partial u^s} + g_{sk}\frac{\partial \Xi^s}{\partial u^j} + g_{js}\frac{\partial \Xi^s}{\partial u^k} = 0 \qquad (11\text{-}11)$$

for all j,k.

We choose a local system of coordinates in which the parameter lines are the trajectories of the transformation group and their orthogonal trajectories. Then $g_{12} = g_{21} = 0$ and $\Xi = (0,1)$. The Killing equations (11-11) are

$$\frac{\partial g_{11}}{\partial u^2} = \frac{\partial g_{22}}{\partial u^2} = 0$$

It is possible to choose the parameters on the orthogonal trajectories such that $g_{11}(u^1) = 1$ [see (11-4)]; the metric then appears as

$$ds^2 = (du^1)^2 + G(u^1)(du^2)^2$$

Theorem 11-12. *A surface admits a one-parameter group of motions if and only if it is isometric to a surface of revolution.*

(Compare example 10-6, case $B2$.) The one-parameter group is $(u^1,u^2) \to (u^1,\ u^2 + t)$. We assume now that on the surface there exists another one-parameter group of motions whose vector field H is independent of Ξ. We write the metric as one of a surface of revolution $ds^2 = (du^1)^2 + r^2(u^1)(du^2)^2$. The Killing equations are

$$\frac{\partial \mathrm{H}^1}{\partial u^1} = 0 \qquad \frac{\partial \mathrm{H}^2}{\partial u^1} + r^{-2}\frac{\partial \mathrm{H}^1}{\partial u^2} = 0 \qquad \frac{\partial \mathrm{H}^2}{\partial u^2} + \frac{r'}{r}\mathrm{H}^1 = 0 \tag{11-12}$$

It follows that $H^1 = f(u^2)$ is a function of u^2 only. The symmetry condition $\partial^2\mathrm{H}^2/\partial u^1\,\partial u^2 = \partial^2\mathrm{H}^2/\partial u^2\,\partial u^1$ is

$$\frac{f''}{r^2} = \frac{r''r - r'^2}{r^2} f$$

or

$$\frac{f''(u^2)}{f(u^2)} = r''(u^1)r(u^1) - r'^2(u^1) = \text{const} \tag{11-13}$$

This equation has several important consequences. First, $(r''/r)' = (rr''' - r'r'')/r^2 = (rr'' - r'^2)'/r^2 = 0$. Hence $K = -r''/r = \text{const}$.

Theorem 11-13. *If a surface admits more than one one-parameter group of motions, it is of constant curvature.*

Second, the group of motions is obtained from

$$f''(u^2) = cf(u^2) \tag{11-13a}$$

The general solution of this second-order differential equation depends on two parameters; the full group of motions on the surface will be of dimension three. It will turn out in the discussion that in every case the group acts transitively on the surface and that the greatest subgroup leaving a point $\mathbf{u}_0$ fixed $[F(\mathbf{u}_0 t) = \mathbf{u}_0]$ induces in T_{u_0} the action of the rotation group. Such a group may bring any orthonormal frame in $T(V)$ on any other such frame; it is *transitive* on the set V_1 of orthonormal frames in $T(V)$. This leads to the following definition:

Definition 11-14. *A surface V is a Clifford-Klein surface if its inner geometry is the Klein geometry of a group transitive on V_1.*

Like most problems in group theory, the problem of finding all the Clifford-Klein surfaces has two parts. There is a local, differential-geometric problem to characterize all metrics that may appear on such a surface. We solve this problem here. The next question is to find all the surfaces admitting a given group and a given metric. This problem is topological; it belongs to global geometry, a subject outside the scope of this book. The topological problem is far from being trivial. For example, we shall treat $K = 0$ as just one case but the geometry may be realized on the euclidean plane, a straight circular cylinder, or a torus. In the last two cases, plane "euclidean" geometry on the surface is not induced by the euclidean geometry of the surrounding space (see theorem 10-9). The action of the rotation group on the cylinder cannot be described globally. Even more complicated is the search for nonorientable Clifford-Klein surfaces or for nonmanifold Clifford-Klein spaces.

We determine all groups that may act on a Clifford-Klein surface, i.e., all two-dimensional metric geometries. They are distinct according to the sign of K.

I. $K = 0$, *euclidean geometry.* Here $r'' = 0$, and the metric is $ds^2 = (du^1)^2 + (au^1 + b)^2(du^2)^2$.

A. If $a = 0$, a simple change of coordinates reduces the ds^2 to the euclidean form $ds^2 = (du^1)^2 + (du^2)^2$. The transformation group has the vector fields $\Xi = (0,1)$ and $\mathrm{H} = (c_1u^2 + c_2, -c_1u^1 + c_3)$. Ξ appears as a field H for $c_1 = c_2 = 0$, $c_3 = 1$. The algebra of the transformation group is generated by

$$\pounds_1 = \frac{\partial}{\partial u^1} \qquad \pounds_2 = \frac{\partial}{\partial u^2} \qquad \pounds_3 = u^2\frac{\partial}{\partial u^1} - u^1\frac{\partial}{\partial u^2}$$

corresponding to the parameter vectors (0,1,0), (0,0,1), and (1,0,0). It is the algebra of the group of euclidean motions in the plane (see example 7-1).

B. If $a \neq 0$, one may obtain $a = 1$, $b = 0$ by a change in the parametrization of the curves $u^2 = \text{const}$. Equation (11-13*a*) is $f'' + f = 0$; $H = (c_1 \cos (u^2 + c_2), (-c_1/u^1) \sin (u^2 + c_2) + c_3)$. The group, acting on a surface isometric with a right circular cone, is generated by

$$\pounds_1 = \sin u^2\frac{\partial}{\partial u^1} + \frac{\cos u^2}{u^2}\frac{\partial}{\partial u^2} \qquad \pounds_2 = \cos u^2\frac{\partial}{\partial u^1} - \frac{\sin u^2}{u^2}\frac{\partial}{\partial u^2}$$

$$\pounds_3 = \frac{\partial}{\partial u^2}$$

[parameter vectors $(-1,\pi/2,0)$, (1,0,0), and (0,0,1)]. This group is isomorphic to the group discussed under *A* since

$$[\pounds_1,\pounds_2] = 0 \qquad [\pounds_1,\pounds_3] = -\pounds_2 \qquad [\pounds_2,\pounds_3] = \pounds_1$$

as in example 7-1. A comparison of cases *A* and *B* is an illustration for theorem 10-24.

II. $K > 0$, *elliptic geometry.* By theorem 10-24, all surfaces with the same K are isometric. Therefore we may discuss the general case based on a particular solution of $r'' + Kr = 0$. We choose

$$ds^2 = K^{-1}[(du^1)^2 + \sin^2 u^1(du^2)^2]$$

The surface is isometric with a sphere of radius $\sqrt{K}$. Integration of Eqs. (11-13a) and (11-12) gives the transformation group

$$\mathfrak{L}_1 = \sin u^2 \frac{\partial}{\partial u^1} + \cot u^1 \cos u^2 \frac{\partial}{\partial u^2}$$

$$\mathfrak{L}_2 = \cos u^2 \frac{\partial}{\partial u^1} - \cot u^1 \sin u^2 \frac{\partial}{\partial u^2} \qquad \mathfrak{L}_3 = \frac{\partial}{\partial u^2}$$

As expected, the group represents the action of O_3 since

$$[\mathfrak{L}_1,\mathfrak{L}_2] = \mathfrak{L}_3 \qquad [\mathfrak{L}_2,\mathfrak{L}_3] = \mathfrak{L}_1 \qquad [\mathfrak{L}_3,\mathfrak{L}_1] = \mathfrak{L}_2$$

is the multiplication table of $\mathfrak{L}(O_3)$ (example 7-2).

III. $K < 0$, *hyperbolic geometry*. This geometry does not correspond to anything encountered in elementary geometry; therefore we study it in some detail. All local surfaces with the same value of K are isometric, by theorem 10-24. Hence we may use any particular solution of $R^2r'' - r = 0$, the equation obtained for $K = -1/R^2$. We choose $ds^2 = (du^1)^2 + e^{2u^1/R}(du^2)^2$. By the substitution $x = u^2$, $y = Re^{-u^1/R}$, the metric becomes

$$ds^2 = R^2 \frac{dx^2 + dy^2}{y^2} \qquad y > 0 \tag{11-14}$$

This is an isothermal metric (exercise 10-4, Probs. 1 and 3) in the upper cartesian half plane. The angle measure in this geometry (page 262) coincides with the euclidean one since $\mathbf{a} \circ \mathbf{b} = (R^2/y^2)\mathbf{a} \cdot \mathbf{b}$. The geodesics, solutions of

$$\frac{d}{dy} \frac{dx/dy}{y(1 + (dx/dy)^2)^{\frac{1}{2}}} = 0$$

are the euclidean circles with their center on the x axis $(x - a)^2 + y^2 = h^2$. In u^1,u^2 coordinates, hyperbolic geometry is defined in the whole cartesian plane. It satisfies all the axioms of euclidean geometry except the parallel postulate: For a line l there exists an infinity of lines through a point P not on l which do not intersect l (Fig. 11-4). Hyperbolic geometry is a *non-euclidean geometry*. The length of an arc of a geodesic is easily computed in polar coordinates. For $x = a + h \cos \theta$, $y = h \sin \theta$, one has

$$s(AB) = R \int_{\theta(A)}^{\theta(B)} \csc \theta \, d\theta = R \ln \frac{\tan \theta(B)/2}{\tan \theta(A)/2}$$

The logarithm tends to infinity for $\theta \to 0$ and $\theta \to \pi$; a geodesic has infinite hyperbolic length and the line $y = 0$ is at an infinite distance from any point in the geometry [by (11-4) the metric is not defined on the x axis].

The group of motions may be found from the Killing equations

$$-\Xi^2 + y\frac{\partial \Xi^1}{\partial x} = 0 \qquad \frac{\partial \Xi^2}{\partial x} + \frac{\partial \Xi^1}{\partial y} = 0 \qquad -\Xi^2 + y\frac{\partial \Xi^2}{\partial y} = 0$$

By a straightforward integration $\Xi = (\frac{1}{2}a(x^2 - y^2) + bx + c, (ax + b)y)$. The group of motions is generated by the Lie derivatives

$$\pounds_1 = \frac{\partial}{\partial x} \qquad \pounds_2 = x\frac{\partial}{\partial x} + y\frac{\partial}{\partial y} \qquad \pounds_3 = (x^2 - y^2)\frac{\partial}{\partial x} + 2xy\frac{\partial}{\partial y}$$

whose multiplication table is

$$[\pounds_2,\pounds_1] = \pounds_1 \qquad [\pounds_3,\pounds_1] = 2\pounds_2 \qquad [\pounds_3,\pounds_2] = \pounds_3$$

This algebra has been studied in example 7-3.

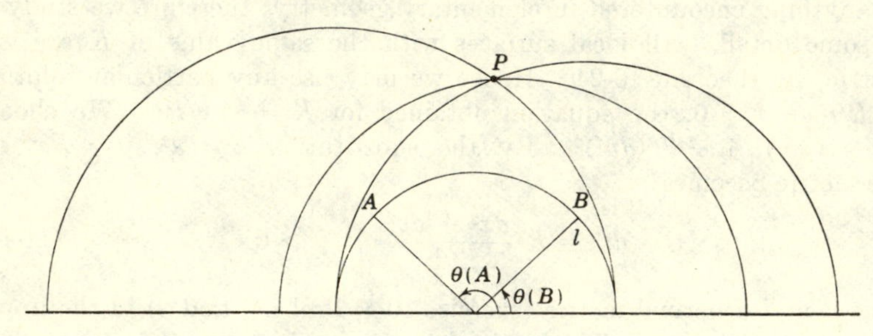

Fig. 11-4

Theorem 11-15. *Plane hyperbolic geometry is the geometry of the group SL_2 of projectivities of P^1 acting as a group of motions in a plane.*

The trajectories of the transformation group are the integral curves of

$$\frac{dx}{dt} = a_1 + a_2x + a_3(x^2 - y^2) \qquad \frac{dy}{dt} = a_2y + 2a_3xy$$

that is, $\quad a_1 + a_2x + a_3(x^2 - y^2) = y\dfrac{d}{dy}[a_1 + a_2x + a_3(x^2 + y^2)]$

They are the circles

$$a_1 + a_2x + a_3(x^2 + y^2) = ky \tag{11-15}$$

By definition, these curves must have constant geodesic curvature; they are the circles of hyperbolic geometry. Equation (11-15) may be rewritten as

$$\left(x + \frac{a_2}{2a_3}\right)^2 + \left(y - \frac{k}{2a_3}\right)^2 = \frac{a_2{}^2 - 4a_1a_3 + k^2}{4a_3{}^2} \tag{11-15a}$$

I. If $a_2^2 - 4a_1a_3 < 0$, there exists a fixed point $\mathbf{x}_0$ of the transformation of coordinates $x_0 = -a_2/2a_3$, $y_0 = +(4a_1a_3 - a_2^2)^{\frac{1}{2}}/2a_3$. The orbits are the orthogonal trajectories to the family of geodesics represented by euclidean circles through (x_0,y_0) and $(x_0,-y_0)$. We may call

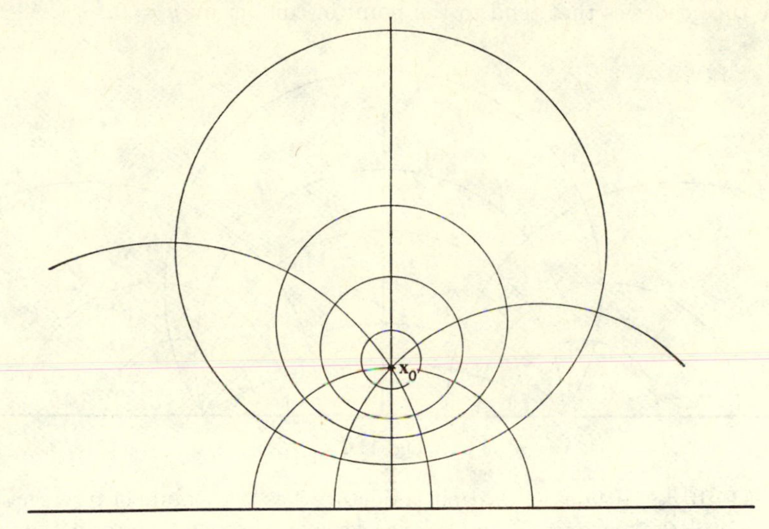

Fig. 11-5

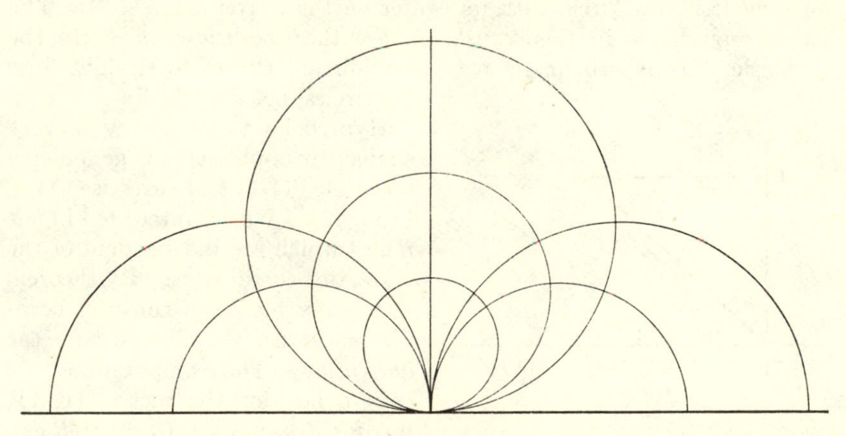

Fig. 11-6

the movement a *rotation* about x_0. Figure 11-5 shows the trajectories of a one-parameter family of rotations about x_0.

II. If $a_2^2 - 4a_1a_3 = 0$ and $a_3 = a_2 = 0$, the trajectories are euclidean lines $y = c$; they may be considered as the circles orthogonal to the geodesics $x = c$.

If $a_3 \neq 0$, the trajectories are *horocycles*. These are hyperbolic circles with hyperbolic radius ∞ which touch the "line at infinity" $y = 0$. The movement, a kind of rotation about a point at infinity, has no analogue in euclidean geometry (Fig. 11-6). The orthogonal trajectories are the geodesics that tend to the point of contact on $y = 0$.

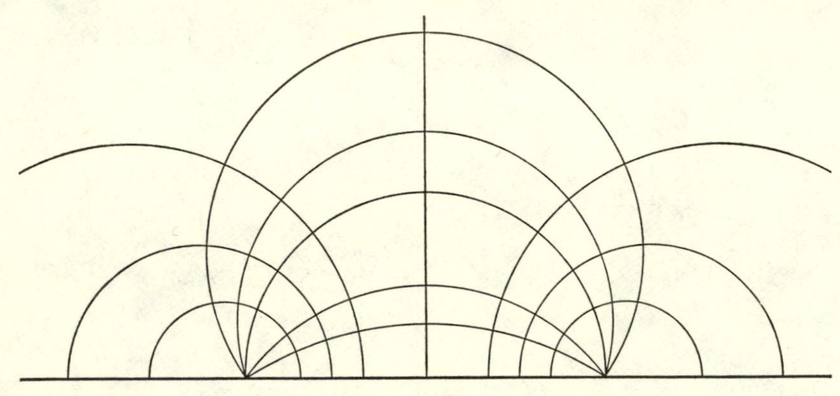

Fig. 11-7

III. If $a_2^2 - 4a_1a_3 < 0$, each trajectory has two points of intersection with $y = 0$. Among all trajectories of a one-parameter group there is one geodesic, the circle with its center on the x axis (Fig. 11-7). The other trajectories in this *translation* are the *equidistant curves* to the geodesic. In hyperbolic geometry, equidistant curves to straight lines are circles, not straight lines.

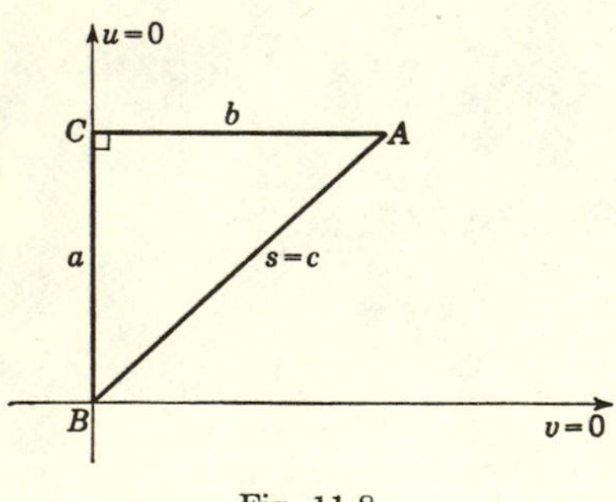

Fig. 11-8

Hyperbolic trigonometry is very similar to spherical trigonometry (example 11-1 and exercise 11-1, Prob. 22). We are interested in formulas which are independent of the system of coordinates. By theorem 10-24, any metric of constant negative curvature may be used for the derivation. The computations are easiest not for the metric (11-14) but for $ds^2 = du^2 + \operatorname{Cosh}^2 u/R\, dv^2$. This metric belongs to a cartesian geodesic system of parameter lines. The curves $v = \text{const}$ and $u = 0$ are geodesic. A right triangle may be brought into the position indicated in Fig. 11-8 by a hyperbolic motion; the vertex B is in the origin, and the side a is on the coordinate line $u = 0$. If v is taken as the independent variable, the geodesic BA is the integral curve of

$$\frac{d}{dv}\frac{u'}{(u'^2 + \operatorname{Cosh}^2 u/R)^{\frac{1}{2}}} - \frac{1}{R}\frac{\operatorname{Sinh} u/R \operatorname{Cosh} u/R}{(u'^2 + \operatorname{Cosh}^2 u/R)^{\frac{1}{2}}} = 0$$

$$u(0) = 0 \qquad \frac{dv}{ds}(0) = \cos B \tag{11-16}$$

If a variational problem $\int F(y,y')\,dx = \min$ does not contain the independent variable explicitly, a first integral of the Euler equation is $F - y'\,\partial F/\partial y' = \text{const}$; in our case

$$\left(u'^2 + \operatorname{Cosh}^2 \frac{u}{R}\right)^{\frac{1}{2}} - \frac{u'^2}{(u'^2 + \operatorname{Cosh}^2 u/R)^{\frac{1}{2}}} = k$$

that is,
$$dv = \frac{k\,du}{\operatorname{Cosh} u/R\,(\operatorname{Cosh}^2 u/R - k^2)^{\frac{1}{2}}}$$

and
$$ds = \frac{\operatorname{Cosh}^2 u/R}{k}\,dv$$

The initial conditions show that $k = \cos B$. A straightforward integration gives

$$\operatorname{Sinh}\frac{v}{R} = \frac{\cos B \operatorname{Sinh} u/R}{\sin B \operatorname{Cosh} u/R} \qquad \operatorname{Sinh}\frac{s}{R} = \frac{\operatorname{Sinh} u/R}{\sin B}$$

The sides of the triangle are $u = a$, $v = b$, and $s = c$. The last equation may be written

$$\frac{\operatorname{Sinh} c/R}{\sin C} = \frac{\operatorname{Sinh} b/R}{\sin B} \tag{11-17}$$

This is the sine theorem of hyperbolic geometry. The equation for v gives

$$\operatorname{Cosh}\frac{a}{R} = \left(1 + \operatorname{Sinh}^2\frac{a}{R}\right)^{\frac{1}{2}} = \left(\sin^2 B + \operatorname{Sinh}^2\frac{b}{R}\right)^{\frac{1}{2}}\Big/\sin B \operatorname{Cosh}\frac{b}{R}$$

and that for s, $\operatorname{Cosh} c/R = (\sin^2 B + \operatorname{Sinh}^2 b/R)^{\frac{1}{2}}/\sin B$; hence

$$\operatorname{Cosh}\frac{c}{R} = \operatorname{Cosh}\frac{a}{R}\operatorname{Cosh}\frac{b}{R} \tag{11-18}$$

This is the cosine formula of hyperbolic geometry which replaces Pythagoras's theorem in euclidean geometry.

Although hyperbolic geometry may be studied as an inner geometry in the cartesian plane, it is impossible to obtain a complete hyperbolic geometry as a geometry of a surface in euclidean three space. For example, the hyperbolic metrics

$$ds_1^2 = dq^2 + \operatorname{Cosh}^2\frac{q}{R}\,d\phi^2$$

$$ds_2^2 = dq^2 + \operatorname{Sinh}^2\frac{q}{R}\,d\phi^2$$

$$ds_3^2 = dq^2 + e^{-2q/R}\,d\phi^2$$

may be realized on the surfaces of revolution:

$$x_1{}^3 = \int \left(1 - \frac{1}{R^2}\,\text{Sinh}^2\,\frac{q}{R}\right)^{\frac{1}{2}} dq \qquad r(x^3) = \text{Cosh}\,\frac{q}{R}$$

$$x_2{}^3 = \int \left(1 - \frac{1}{R^2}\,\text{Cosh}^2\,\frac{q}{R}\right)^{\frac{1}{2}} dq \qquad r(x^3) = \text{Sinh}\,\frac{q}{R}$$

$$x_3{}^3 = \int \left(1 - \frac{1}{R^2}\,e^{-2q/R}\right)^{\frac{1}{2}} dq \qquad r(x^3) = e^{-q/R}$$

The surfaces are shown in Fig. 11-9. In the first two cases we obtain only a strip of surface since x^3 has to be real, and dx^3/dq vanishes at the boundary. The third surface has a singularity at $x^3 = 0$. Hilbert has shown that no analytic imbedding of the hyperbolic plane in R^3

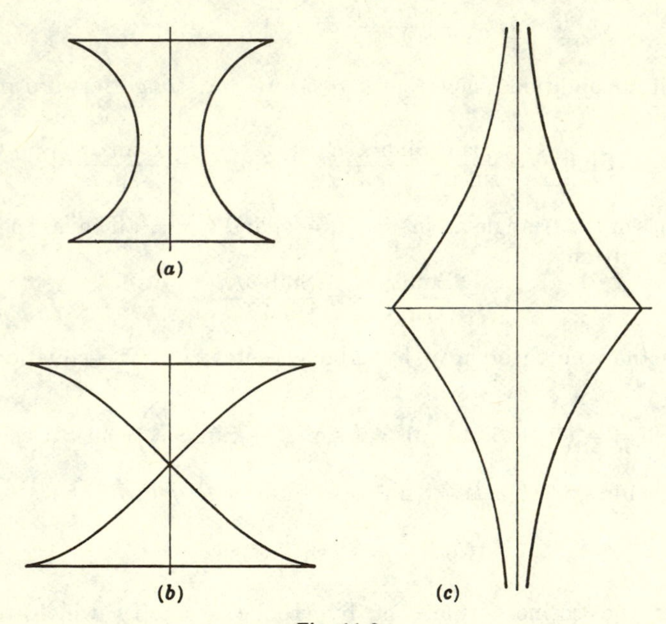

Fig. 11-9

exists; the proof holds good also for C^2 imbeddings. A clear and correct presentation of the proof may be found in Willmore's text cited in the references of Chap. 10. Recently, Kuiper has shown that there exist C^1 global imbeddings of the hyperbolic plane in R^3; the Gauss curvature cannot be everywhere defined on such a surface.

Exercise 11-2

1. Show that the definitions of isometry of Secs. 11-2 and 10-4 coincide.

2. A torus is the surface

$$(R \cos \phi + r \cos \phi \cos \psi,\ R \sin \phi + r \sin \phi \cos \psi,\ r \sin \psi)$$

Find the conditions for R and r to ensure that the surface is a closed manifold. Discuss its inner geometry (compare exercise 11-3, Prob. 1).

3. Show that in a motion of a surface into itself geodesics are mapped into geodesics.

4. Prove that two distinct one-parameter groups of motions cannot have the same trajectories.

5. A group of motions is a *translation* if all its trajectories are geodesics. Show that if a surface admits a translation it is isometric with the euclidean plane (Bianchi).

6. Compare the results of this section with exercise 6-2, Prob. 9, to show that a three-dimensional group may act as a transitive group on V_1 only if it is abelian or if its derived algebra is isomorphic to its Lie algebra.

◆ **7.** Find the Killing equations for the metric

$$ds^2 = \lambda(x^1,x^2)[(dx^1)^2 + (dx^2)^2]$$

8. The most general metric of revolution which belongs to elliptic geometry is $ds^2 = (du^{1'})^2 + c_1 \sin^2 (\sqrt{K}\, u^{2'} + c_2)(du^{2'})^2$. Find an explicit formula for the change of coordinates $(u^{1'},u^{2'}) \to (u^1,u^2)$ which reduces the ds^2 to the form used in the text.

9. Use the Taylor expansion of the hyperbolic functions to show that the cosine theorem [Eq. (11-18)] becomes the pythagorean theorem for $R \to \infty$.

10. Prove the sine theorem [Eq. (11-17)] for oblique triangles.

11. Prove that $\cos A = \operatorname{Cosh} a/R \sin B$ for a hyperbolic right triangle. Since $\operatorname{Cosh} a/R > 1$, $A + B + \pi/2 < \pi$, *the sum of the angles of a hyperbolic right triangle is less than* π. Establish the same result for an oblique triangle.

12. Find the explicit formulas for hyperbolic movements. [Integrate (7-7) and combine the transformations obtained.]

13. Prove that for a hyperbolic oblique triangle

$$\operatorname{Cosh}\frac{c}{R} = \operatorname{Cosh}\frac{a}{R}\operatorname{Cosh}\frac{b}{R} - \operatorname{Sinh}\frac{a}{R}\operatorname{Sinh}\frac{b}{R}\cos C$$

and $$\cos C = -\cos A \cos B + \sin A \sin B \operatorname{Cosh}\frac{c}{R}$$

It follows from the last formula that the angles of a hyperbolic triangle uniquely determine its sides; *there are no similitudes in hyperbolic geometry.* [Hint: Split the triangle into two right ones by a height (why does one exist?) and use (11-17), (11-18), and Prob. 11.]

14. Show that, if the equidistant curves of a line are lines, the Klein geometry is euclidean (Posidonius).

15. What are the circles of elliptic geometry?

16. Write the Frenet formulas corresponding to (2-3) and (2-5) in hyperbolic geometry.

17. Show that the projective plane is a nonorientable elliptic Clifford-Klein surface.

18. Three-dimensional hyperbolic geometry may be based on the metric

$$ds^2 = R^2 \frac{dx^2 + dy^2 + dz^2}{z^2} \qquad z > 0$$

In this geometry, "planes" are euclidean half spheres over the (x,y) plane. The metric is conformal with the euclidean one. Use the results of Sec. 10-2 to find the group of hyperbolic motions in space without computation.

◆**19.** Elliptic geometry in the projective space P^3 of homogeneous coordinates $x^1{:}x^2{:}x^3{:}x^4$ $[\Sigma(x^i)^2 = 1]$ may be based on the metric $ds^2 = R^2\Sigma(dx^i)^4$. Use example 5-5 and Eq. (5-11) to find the matrix group of three-dimensional elliptic movements. In a one-parameter group of such movements the trajectories are skew lines equidistant in an elliptic metric (*Clifford parallels*) discussed in exercise 5-2, Prob. 12.

11-3. The Bonnet Formula

It has been shown in Sec. 10-5 that Stokes's theorem holds in the form

$$\oint_{\partial U} \omega = \iint_U d\omega$$

where U is a differentiable homeomorphic image of the interior of the unit square I^2 and ∂U a piecewise differentiable image of ∂I^2. By Eq. (10-20), $\oint_{\partial U} \omega_1{}^2 = -\iint_U K\omega^1 \wedge \omega^2$. Let θ be the angle of rotation *from* a given frame $\{\mathbf{e}_i\}$ *to* the tangent normal frame of ∂U at those points where the tangent normal frame exists. Then $\omega_1{}^2 = k_g\,ds - d\theta$, that is,

$$\iint_U K\omega^1 \wedge \omega^2 + \oint_{\partial U} k_g\,ds = \oint_{\partial U} d\theta \qquad (11\text{-}19)$$

This is the famous *Bonnet formula*, one of the strongest tools of global differential geometry.

θ is defined only up to a multiple of 2π; if ∂U is a differentiable curve the tangent normal frame rotates exactly once in the given continuous field of frames, $\oint d\theta = 2\pi$.† If ∂U is only piecewise differentiable, the integral over $d\theta$ is taken only on the smooth arcs. At a vertex P_i of the curve ∂U the jump of θ is the *exterior* angle θ_i counted in the sense of increasing arc length (Fig. 11-10). In terms of the *interior* angles, $\phi_i = \pi - \theta_i$, $\oint d\theta = 2\pi - \sum_i (\pi - \phi_i)$; hence

$$\iint_U K\alpha + \oint k_g\, ds = 2\pi + \sum_i (\phi_i - \pi) \tag{11-19a}$$

We give some applications of this formula.

On a surface of *constant* curvature K let ∂U be a *geodesic polygon*, i.e., a curve composed of a finite number k of geodesic arcs which bounds a local surface (a simply connected domain). The *angle defect* of the

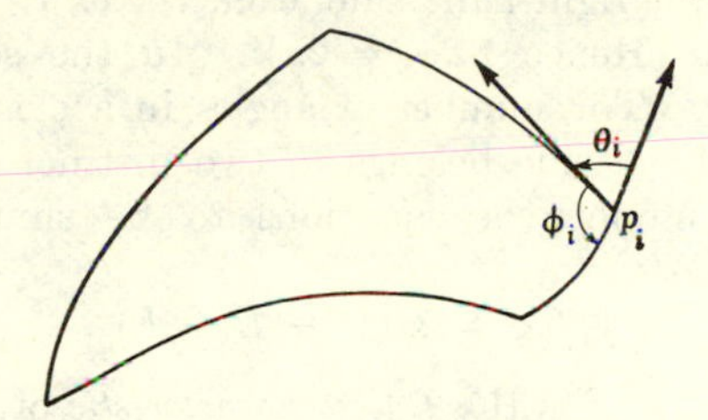

Fig. 11-10

polygon is $\Delta = \Sigma\phi_i - (k - 2)\pi$. The area A of the polygon is proportional to the angle defect, since $k_g = 0$ and, by (11-19a),

$$\Delta = KA \tag{11-20}$$

By definition, the area is always positive.

Theorem 11-16 (Gauss). *The sum of the angles of a geodesic triangle is* $>\pi$ *in elliptic and* $<\pi$ *in hyperbolic geometry.*

In Sec. 10-5 it was shown that a closed C^2 surface may always be divided into a finite number of simply connected local surfaces bounded by piecewise differentiable closed curves. The same construction applies to closed C^2 two-dimensional manifolds. An inner geometry is defined on such a manifold if each of its neighborhoods U_α is given a metric ds^2 and if all maps (9-27) are isometries. There exist interesting 2-manifolds whose metric cannot be realized on surfaces in three space, like closed manifolds with everywhere negative Gauss curvature (theorem 10-29).

† This proof is incomplete. One should show that a continuous field of vectors $\mathbf{e}_1$ in a simply connected domain never has a closed trajectory; i.e., the movement of $\mathbf{e}_1$ along the curve has no influence on $\oint d\theta$. This is a fundamental theorem in algebraic topology. See P. Alexandroff and H. Hopf, "Topologie," chap. XII, §3, Springer-Verlag OHG, Berlin, 1935.

Each $\mathbf{c}_{ij}$ will appear in two (or more) charts (U_α, p_α); for metric purposes all representations of one arc are equivalent.

In a covering of an orientable closed surface by local surfaces U_i with boundary curves $\mathbf{c}_{ij}$ let V be the number of vertices of the polygons ∂U_i, E the number of arcs $\mathbf{c}_{ij}$, and F the number of local surfaces (= simply connected domains) on the surface. For each U_i we have a Bonnet formula (11-19*a*):

$$\iint_{U_i} K\alpha + \sum_j \oint_{\mathbf{c}_{ij}} k_g\, ds = \Sigma\phi_s - \Sigma\pi + 2\pi$$

If we take the sum of all these equations there appears on the left-hand side the *integral curvature* $\iint K\alpha$ of the surface. The sum $\Sigma \oint k_g\, ds = 0$ for an orientable surface, since every arc $\mathbf{c}_{ij}$ appears twice with opposite orientations.† On the right-hand side, each vertex P_i contributes a sum of inner angles 2π. Hence $\Sigma\Sigma\phi_s = 2\pi V$. In the second term, each angle contributes π. The number of angles in a closed polygon is the number of sides. Each side belongs to two distinct polygons. Hence $\Sigma\Sigma\pi = 2\pi E$. The number of equations to be summed is F. This shows that

$$\iint K\alpha = 2\pi(V - E + F) \tag{11-21}$$

The number $V - E + F$ is the *Euler characteristic* of the polygonal net on the surface.

Theorem 11-17. *The Euler characteristic of a closed, oriented surface is independent of the polygonal net chosen on the surface; it is* $(2\pi)^{-1}$ *times the integral curvature.*

A polygonal net is transformed into another such net in any differentiable homeomorphism.

Corollary 11-18. *The integral curvature is invariant in any differentiable homeomorphism.*

The *genus* g of an orientable, closed surface is defined by

$$V - E + F = 2(1 - g) \tag{11-22}$$

We do not go into the proof here that g is a non-negative integer. Some simple cases are treated in the exercises.

A closed convex surface is homeomorphic to the unit sphere for which $\iint K\alpha = \iint \alpha = 4\pi$. *A closed convex surface is of genus zero.* The Bonnet formula written as

$$\iint K\alpha = 4\pi(1 - g) \tag{11-23}$$

may be used to deduce some statements about the impossibility of certain inner geometries on 2-manifolds: *A metric with everywhere positive Gauss curvature is impossible on surfaces of genus* $g > 0$. *A metric with nowhere*

† Compare page 249.

positive Gauss curvature is impossible on surfaces of genus zero. One may combine (11-23) with theorem 10-30 to obtain an estimate for the *absolute integral curvature*

$$\iint |K|\alpha = \iint_{K>0} K\alpha - \iint_{K<0} K\alpha \geq 4\pi(g+1) \qquad \text{(11-24)}$$

We proceed to the study of some global properties of geodesics. Most of the arguments will suppose that the surface is a *complete* metric space. For any two points P and Q on the surface we take all rectifiable curves $\mathbf{c}$ from P to Q, of length l_c. The *distance* between P and Q is defined as $d(P,Q) = \inf l_c$. A sequence of points P_i is a *Cauchy sequence* if $d(P_n,P_m) < \epsilon$ for $n,m \geq N(\epsilon)$. A space is *complete* if in it every Cauchy sequence converges.

A compact space is complete. Not every complete surface is compact (for example, the plane). An example of a noncomplete surface is obtained if a point is deleted on any complete surface.

Theorem 11-19 (Hopf-Rinow). *Any geodesic on a complete surface may be produced to any given length. Any two points P, Q on a complete surface may be joined by an arc of a geodesic whose length is the distance between P and Q.*

In the first statement of the theorem one has to include closed geodesics on which one goes around a number of times. The second statement has been proved if $d(P,Q)$ is less than the distance from P to its nearest conjugate point. This distance is positive; therefore there is a neighborhood of P in which both statements are true.

If the first statement were not true for all lengths, there would exist a geodesic γ through P on which it would be possible to report any length $l < L$, but no length $l' > L$. On γ we take two sequences P_i and P_i' such that (on γ) $\int_P^{P_i} ds = L - \epsilon_i$, $\int_P^{P_i'} ds = L - \epsilon_i'$, $\lim \epsilon_i = \lim \epsilon_i' = 0$. Both sequences are Cauchy sequences. Therefore $Q = \lim P_i$ and $Q' = \lim P_i'$ exist. But $P_1, P_1', P_2, \ldots, P_n, P_n', \ldots$ is also a Cauchy sequence; hence $Q = Q'$ is independent of the sequence chosen, and $\int_P^Q ds = L$. γ has a tangent $\mathbf{t}$ in Q. The geodesic in Q with tangent $-\mathbf{t}$ may be produced to a certain length. This geodesic is a smooth prolongation of γ to a total length $> L$. No number L may exist on any geodesic.

We assume now that the second statement is true for P and all Q such that $d(P,Q) < L$ but that there exist points R with $d(P,R) = L + \epsilon$ for which it is false. ϵ may be taken to be less than the distance from R to its nearest conjugate point. As before, it follows that the statement is true for all Q such that $d(P,Q) \geq L$. The theorem will be proved if we show that there exists a Q such that $d(P,Q) = L$, $d(Q,R) = d(P,R) - L$;

in this case all three points must be on one geodesic. We know that there exist curves $\mathbf{c}_n(P,R)$ from P to R of length $\leq d(P,R) + 1/n$ for all integers n. On $\mathbf{c}_n(P,R)$ let S_n be the last point at distance L from P. To each point S in the set $\{S|d(P,S) = L\}$ there corresponds a *unique* vector in T_p, the initial vector of the shortest geodesic from P to S. The set $\{S\}$ is compact because it is homeomorphic to the compact unit circle in T_P. Therefore the sequence S_n has a subsequence which converges to a point S^*. By hypothesis,

$$d(P,S^*) = L \qquad d(S^*,R) \leq d(P,R) - L + \frac{1}{n}$$

for all n; hence $d(P,R) = d(P,S^*) + d(S^*,R)$.

A typical application of the notion of completeness is the following:

Theorem 11-20 (Hopf-Rinow). *A complete surface of positive curvature bounded away from zero is closed.*

PROOF: Two points on the surface may always be connected by a shortest geodesic arc (theorem 11-19); their distance is $\leq \pi(\inf K)^{-\frac{1}{2}}$ (theorem 11-8). Therefore the point set of the surface is compact.

A closed geodesic without a double point on a closed convex surface divides the surface into two local surfaces. By Bonnet's formula (11-19), $\iint_\Sigma K\alpha = 2\pi$, where Σ denotes one of the two local surfaces.

Theorem 11-21 (Hadamard). *The spherical image of a simple closed geodesic on a closed convex surface bisects the unit sphere.*

Since a closed convex surface is homeomorphic with the unit sphere, the theorem implies the following:

Corollary 11-22. *Two simple closed geodesics on a convex closed surface always intersect.*

We do not prove here the existence of closed geodesics on convex surfaces. This is a subject that belongs to the calculus of variations in the large.

The example of the sphere shows that on a surface with positive curvature there may exist *geodesic biangles;* these are local surfaces bounded by *two* geodesic arcs. Any two meridians on the sphere bound a geodesic biangle. No such configuration exists on surfaces with nonpositive Gauss curvature.

Theorem 11-23 (Hadamard). *On a surface with nonpositive curvature there cannot exist a geodesic biangle.*

The Bonnet formula for biangles is $\iint_\Sigma K\alpha = -\iint_\Sigma |K|\alpha = \phi_1 + \phi_2$. The ϕ_i are positive—a contradiction! The theorem does *not* say that two geodesics cannot intersect twice on a surface with nonpositive curvature, only that two geodesics with two points of intersection cannot bound a local surface (simply connected domain). For $\phi_1 = \phi_2 = \pi$ we have the following interesting conclusion:

Corollary 11-24. *On a surface with nonpositive curvature there cannot exist a closed geodesic which bounds a simply connected domain.*

Exercise 11-3

◆ **1.** Show that the surface in R^4 $(R \cos \phi, R \sin \phi, r \cos \theta, r \sin \theta)$ is a torus (exercise 11-2, Prob. 2) with the metric $ds^2 = R^2\, d\phi^2 + r^2\, d\theta^2$. Can this metric be realized on a torus in R^3? (What is K?)

◆ **2.** Draw a polygonal net and compute the genus for the sphere, the torus, and the pretzel (Fig. 11-11).

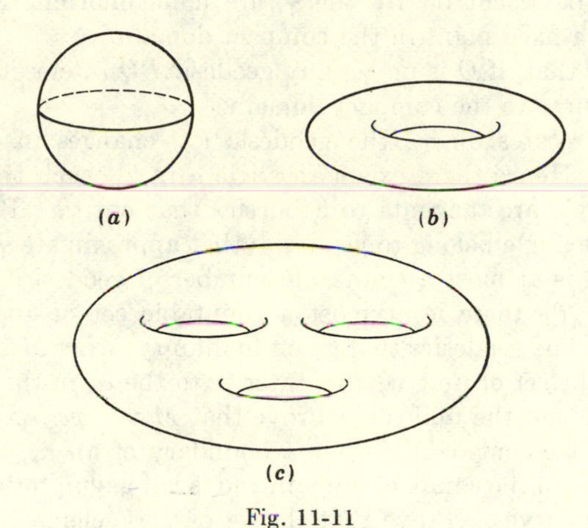

Fig. 11-11

3. Prove that the greatest lower bound of the Gauss curvature on a complete open (= not closed) surface is ≤ 0.

4. Show on the example of the elliptic paraboloid that the hypothesis "bounded away from zero" is essential in theorem 11-20.

5. Show that, if a geodesic is the shortest connection between all its points, the curvature of the surface cannot be positive at all points of the geodesic.

6. Show that there exists no complete metric of strictly positive curvature on a cylinder. (Use Prob. 5.)

7. Find a metric in the plane for which the plane is *not* complete. (HINT: The plane may be given the metric of the interior of the unit circle. No length then is ≥ 2.)

8. Show that on

$$\left[\frac{(x-d)^2}{a^2} + \frac{y^2}{b^2} - \frac{z^2}{c^2} - 1\right]\left[\frac{(x+d)^2}{a^2} + \frac{y^2}{b^2} - \frac{z^2}{c^2} - 1\right] = \epsilon$$

two geodesics may intersect twice.

9. Show that two geodesics in a local surface of nonpositive curvature intersect, at most, once.

10. Prove that the line of striction of the hyperboloid of one sheet is a geodesic and that it is the only closed geodesic of the surface. (Assume that there is another one and derive a contradiction by Bonnet's formula.)

11. Show that, if a cylinder is given a metric with nonpositive curvature, it has, at most, one closed geodesic.

***12.** Given a complete surface of nonpositive curvature such that a finite number of closed geodesics γ_i bounds a compact domain and that the parts cut off by the γ_i are homeomorphic to a cylinder. Let P be a fixed point in the compact domain.

(*a*) Show that, if Q is on γ_i, any geodesic PQ intersects γ_i once and never returns to the compact domain.

(*b*) If Q rotates on γ_i, the geodesic PQ changes in a continuous manner. Hence there exists an angle α in T_P such that all vectors in the angle are tangents to geodesics that cut γ_i. Prove that the legs of the angle belong to geodesics that approximate γ_i indefinitely.

(*c*) There is at most a countable number of geodesics from P to Q. Hence in T_P there is at most a countable set of angles α_n whose vectors define geodesics that go off to infinity. Let M be the complement of the set of unit vectors interior to the α_n in the unit circle in T_P. M is not the null set. Prove that M is a *perfect* set.

(*d*) If an element of M is not a boundary of an α_n, its geodesic is completely in the compact domain and is not asymptotic to one of the boundary curves. Prove that the set of these elements is uncountable (Hadamard).

13. On a complete surface there exists a shortest curve from a point P to a geodesic γ. Show that the shortest curve is perpendicular to γ.

14. Show that, on a complete surface with nonpositive curvature, two shortest geodesics PQ_0 and PQ_1 from a point P to a geodesic γ together with the arc of geodesic Q_0Q_1 cannot bound a local surface (Hadamard).

15. The rulings of the hyperboloid of one sheet are geodesics. Prove that two rulings of one family have, at most, one ruling of the second family as a common perpendicular.

16. For a closed space curve $\mathbf{c}$ with curvature k and torsion t define $w = (k^2 + t^2)^{\frac{1}{2}}$, $\phi = \arctan t/k$. The arc length of the normal indicatrix $\mathbf{n}(s)$ is obtained from $ds_n/ds = w$. Prove that the geodesic curvature of $\mathbf{n}(s_n)$ on S^2 is $k_g = d\phi/ds_n$; hence

$$\oint k_g \, ds_n = \oint d\phi = 2\pi\nu$$

Show that $\nu = 0$ if $k \neq 0$ on the whole of $\mathbf{c}$. In this case $\mathbf{n}$ bisects the unit sphere (Jacobi).

17. (See exercise 10-1, Prob. 17.) Show that the surface area and volume of a closed convex surface and its parallel surfaces are related by

$$A(c) = A_0 + 2c\iint H\alpha + 4\pi c^2$$
$$V(c) = V_0 + A_0 c + c^2\iint H\alpha + (4\pi/3)c^3$$

18. (Sequel to Prob. 17.) Prove that $M = \iint H\alpha \geq 2\sqrt{\pi}$ and that $V^2 - 36\pi A^3 \geq 0$, where equality holds only for $H^2 = K$. What is the closed convex surface of minimal surface area which bounds a given volume?

References

The general references to Chap. 10 are fundamental also for Chap. 11.

SEC. 11-1

Gauss, C. F.: "Werke," vol. 4, Königliche Gesellschaft der Wissenschaften, Göttingen, 1894.

SEC. 11-2

Euler, L.: "Principes de la trigonometrie sphérique tirés de la méthode des plus grands et plus petits (E 214)." Opera Omnia, Ser. I, vol. 27, pp. 277–308. Opera Omnia, sub auspiciis Societatis Scientiarum Naturalium Helvetica, Lausannae et Turici, 1954.

Killing, W.: Ueber die Grundlagen der Geometrie, *J. Reine Angew. Math.*, **109**: 121–186 (1892).

For the global theory compare the following:

Kuiper, N. H.: On C^1 isometric imbeddings, *Proc. Neth. Acad. Sci., Ser.* A, **58**: 545–556, 683–689 (1955).

Mostow, George Daniel: The Extensibility of Local Lie Groups of Transformations and Groups on Surfaces, *Ann. of Math.*, **52**(2): 606–636 (1950).

SEC. 11-3

Hadamard, J.: Les surfaces à courbures opposées et leurs lignes géodésiques, *J. Math. Pures Appl.* (Liouville), Ser. 5, **4**: 27–73 (1898).

Modern amplifications of Hadamard's work use strong methods of statistics and go under the names of "topological dynamics" and "geodesic flow."

Hopf, H., and W. Rinow: Ueber den Begriff der vollständigen differentialgeometrischen Fläche, *Comment. Math. Helv.*, **3**: 209–225 (1931).

An investigation of the Bonnet formula on *open* surfaces is found in the following:

Cohn-Vossen, S.: Kürzeste Wege und Totalkrümmung auf Flächen, *Compositio Math.*, **2**: 69–133 (1935).

12
AFFINE GEOMETRY OF SURFACES

12-1. Frenet Formulas

In this chapter we study the properties of surfaces invariant under the group of unimodular affine transformations of R^3:

$$\mathbf{x}^* = \mathbf{x}A + \mathbf{a} \qquad \|A\| = 1$$

As always, the main problem is the construction of invariant frames on a surface. The group is of dimension 11; we expect one invariant of order three and three independent invariants of order four. The integrability conditions then are of order five. We develop the theory for C^5 surfaces.

Having fixed a cartesian system of coordinates in R^3, on the local surface $\mathbf{p}(u^1,u^2)$ we take a family of frames $\{\mathbf{e}_i\}$ such that $\mathbf{e}_1$ and $\mathbf{e}_2$ span T_p,

$$d\mathbf{p} = \omega^1\mathbf{e}_1 + \omega^2\mathbf{e}_2 \tag{12-1}$$

and the frame is a unimodular affine image of the fixed orthonormal frame of R^3:

$$\|\mathbf{e}_1,\mathbf{e}_2,\mathbf{e}_3\| = 1 \tag{12-2}$$

Equation (12-2) replaces the orthonormality condition of metric geometry. The Frenet formula is

$$d\{\mathbf{e}_i\} = (\omega_i{}^j)\{\mathbf{e}_j\} \tag{12-3}$$

The Cartan matrix $(\omega_i{}^j)$ is in $\mathfrak{L}(SL_3)$; hence its trace vanishes:

$$\omega_i{}^i = 0 \tag{12-4}$$

The integrability conditions $dd\mathbf{p} = dd\mathbf{e}_i = 0$ are

$$d\omega^\alpha = \omega^\beta \wedge \omega_\beta{}^\alpha \tag{12-5a}$$
$$0 = \omega^1 \wedge \omega_1{}^3 + \omega^2 \wedge \omega_2{}^3 \tag{12-5b}$$
$$d\omega_i{}^j = \omega_i{}^k \wedge \omega_k{}^j \tag{12-5c}$$

Equation (12-5*b*) implies that there exists a symmetric matrix $(l_{\alpha\beta})$

$$(\omega_1{}^3,\omega_2{}^3) = (\omega^1,\omega^2)\begin{pmatrix} l_{11} & l_{12} \\ l_{12} & l_{22} \end{pmatrix}$$

If the frames are changed in a unimodular transformation which leaves the tangent plane invariant,

$$\{\mathbf{e}_{i'}\} = \begin{pmatrix} e_{1'}{}^1 & e_{1'}{}^2 & 0 \\ e_{2'}{}^1 & e_{2'}{}^2 & 0 \\ e_{3'}{}^1 & e_{3'}{}^2 & e_{3'}{}^3 \end{pmatrix}\{\mathbf{e}_i\} \qquad \|e_{i'}{}^j\| = 1 \tag{12-6}$$

the linear forms are transformed by [see Eq. (1-5)]

$$\begin{aligned} (\omega^1,\omega^2) &= (\omega^{1'},\omega^{2'})\begin{pmatrix} e_{1'}{}^1 & e_{1'}{}^2 \\ e_{2'}{}^1 & e_{2'}{}^2 \end{pmatrix} = (\omega^{\alpha'})E_{3'}{}^3\dagger \\ (\omega_{1'}{}^{3'},\omega_{2'}{}^{3'}) &= (\omega_1{}^3,\omega_2{}^3)E_{3'}{}^{3t} \\ &= (\omega^{1'},\omega^{2'})E_{3'}{}^3(l_{\alpha\beta})E_{3'}{}^{3t} \end{aligned} \tag{12-7}$$

The other forms are changed into complicated expressions which we do not need at this moment. Since $\|E_{3'}{}^3\| \neq 0$, it follows from the second equation (12-7) that

$$\operatorname{sign} \|l_{\alpha\beta}\| = \operatorname{sign} \|l_{\alpha'\beta'}\|$$

is an affine invariant. By analogy with the nomenclature of euclidean geometry, we classify points on a surface according to the sign of this determinant:

$$\text{Elliptic point } \|l_{\alpha\beta}\| > 0$$
$$\text{Hyperbolic point } \|l_{\alpha\beta}\| < 0$$
$$\text{Parabolic point } \|l_{\alpha\beta}\| = 0$$

The Frenet formulas differ from one region to the other.

I. *Elliptic region.* The transformation matrix may be found such that

$$(l_{\alpha'\beta'}) = E_{3'}{}^3(l_{\alpha\beta})E_{3'}{}^{3t} = U$$

that is,

$$\omega_{1'}{}^{3'} = \omega^{1'} \qquad \omega_{2'}{}^{3'} = \omega^{2'} \tag{12-8}$$

The most general unimodular transformation (12-6) which leaves (12-8) invariant is one in which $E_{3'}{}^3$ is orthogonal and hence $e_{3'}{}^3 = 1$. In order to simplify the computations, we study first the special trans-

† $A_i{}^j$ is the matrix of the minor of the element $a_i{}^j$ in the matrix $(a_i{}^j)$.

formation

$$\{\mathbf{e}_{i''}\} = \begin{pmatrix} 1 & 0 & 0 \\ 0 & 1 & 0 \\ e_{3''}{}^{1'} & e_{3''}{}^{2'} & 1 \end{pmatrix} \{\mathbf{e}_{i'}\} \tag{12-9}$$

which will serve to define a unique *affine normal.* Here

$$\omega_{3''}{}^{3''} = \omega_{3'}{}^{3'} + e_{3''}{}^{1'}\omega^{1'} + e_{3''}{}^{2'}\omega^{2'}$$

Since the ω^α are a basis in T_p^*, we may write

$$\omega_i{}^j = w_{i\alpha}{}^j\omega^\alpha$$

in any system of coordinates. The *affine normal* $\mathbf{e}_{3''}$ is the unique field of vectors for which

$$\omega_{3''}{}^{3''} = 0$$

This is obtained for $e_{3''}{}^{1'} = -w_{3'1'}{}^{3'}$, $e_{3''}{}^{2'} = -w_{3'2'}{}^{3'}$. In the new frame, $\omega_{1''}{}^{1''} = -\omega_{2''}{}^{2''}$, by Eq. (12-4). By Eq. (12-8), $d\omega_{\alpha''}{}^{3''} = d\omega^{\alpha''}$ ($\alpha'' = 1'',2''$). A comparison of the integrability conditions (12-5*a*) and (12-5*c*) shows that

$$\begin{aligned} 2w_{1''2''}{}^{1''} - (w_{1''1''}{}^{2''} + w_{2''1''}{}^{1''}) &= 0 \\ 2w_{1''1''}{}^{1''} + w_{1''2''}{}^{2''} + w_{2''2''}{}^{1''} &= 0 \end{aligned} \tag{12-10}$$

The only remaining admissible transformations are

$$\{\mathbf{a}_i\} = \begin{pmatrix} \cos\theta & \sin\theta & 0 \\ -\sin\theta & \cos\theta & 0 \\ 0 & 0 & 1 \end{pmatrix} \{\mathbf{e}_{i''}\} \tag{12-11}$$

for which we get, by (1-5) and (12-10),

$$\begin{aligned} \pi_1{}^1 = (w_{1''1''}{}^{1''} \cos 3\theta + \tfrac{1}{2}w_{1''2''}{}^{1''} \sin 3\theta)\pi^1 \\ + (w_{1''2''}{}^{1''} \cos 3\theta - \tfrac{1}{2}\, w_{1''1''}{}^{1''} \sin 3\theta)\pi^2 \end{aligned}$$

if we denote again by π^α and $\pi_i{}^j = p_{i\alpha}{}^j\pi^\alpha$ the linear forms which belong to the invariant frame $\{\mathbf{a}_i\}$. A possible choice for the unique definition of the $\{\mathbf{a}_i\}$ frame is

$$\tan 3\theta = 2\,\frac{w_{1''2''}{}^{1''}}{w_{1''1''}{}^{1''}}$$

which leads to

$$\pi_1{}^1 = I\pi^1$$

The *elliptic affine curvature* I vanishes, and the frame $\{\mathbf{a}_i\}$ is not defined if $w_{1''2''}{}^{1''} = w_{1''1''}{}^{1''} = 0$; such a point is a *Darboux umbilic.* This case is treated in detail in the next section. In points other than Darboux umbilics, θ is determined up to a multiple of $\pi/3$. There are three

Darboux directions for which the Cartan matrix appears in reduced form. These Darboux directions take the place of the principal directions in euclidean geometry. Equation (12-10) now becomes $\pi_1^2 + \pi_2^1 = -2I\pi^2$. The last equation (12-5*c*), $d\pi_3^3 = 0 = \pi_3^1 \wedge \pi^1 + \pi_3^2 \wedge \pi^2$, shows that $p_{32}^1 = p_{31}^2$. Since the frame is now defined in an invariant way, the functions I, p_{11}^2, p_{12}^2, p_{31}^1, p_{31}^2, p_{32}^2 are invariants.

Theorem 12-1. *At any point of an elliptic region there exist a unique affine normal and, except in Darboux umbilics, three Darboux directions two of which, together with the affine normal, determine an invariant frame whose Frenet equations are*

$$d\{\mathbf{a}_i\} = \begin{pmatrix} I\pi^1 & \pi_1^2 & \pi^1 \\ \pi_2^1 & -I\pi^1 & \pi^2 \\ \pi_3^1 & \pi_3^2 & 0 \end{pmatrix} \{\mathbf{a}_i\}$$

$$\pi_1^2 + \pi_2^1 = -2I\pi^2 \qquad \pi_3^1 \wedge \pi^1 + \pi_3^2 \wedge \pi^2 = 0$$

The integrability conditions (12-5) may be written in invariant derivatives as

$$d\pi^1 = p_{11}^2\pi^1 \wedge \pi^2 \qquad d\pi^2 = (p_{12}^2 + I)\pi^1 \wedge \pi^2$$

$$\begin{aligned} p_{31}^2 &= p_{32}^1 = 3p_{11}^2 I - I_{,2} \\ p_{32}^2 &= p_{12}^2(p_{12}^2 - I) + (p_{11}^2)^2 + p_{12}^2{}_{,1} - p_{11}^2{}_{,2} \\ p_{31}^1 &= (p_{12}^2 + 3I)(p_{12}^2 + 2I) + p_{12}^2{}_{,1} - p_{11}^2{}_{,2} + 2I_{,1} \end{aligned} \tag{12-12}$$

$$p_{11}^2(p_{31}^1 - p_{32}^2) + 2p_{32}^1(p_{12}^2 + 2I) + p_{32}^1{}_{,1} - p_{31}^1{}_{,2} = 0$$
$$p_{12}^2(p_{31}^1 - p_{32}^2) - 2p_{32}^1 p_{11}^2 + p_{32}^1{}_{,2} - p_{32}^2{}_{,1} = 0$$

These equations show that all invariants may be computed if I and the π^α are known. Only the last two equations then are genuine compatibility conditions. We shall come back later to the question of the actual determination of the invariant frame and the invariants.

II. *Hyperbolic region.* The matrix $(l_{\alpha\beta})$ may be normalized either by $E_{3'}{}^3(l_{\alpha\beta})E_{3'}{}^{3t} = \begin{pmatrix} 1 & 0 \\ 0 & -1 \end{pmatrix}$ or by $E_{3'}{}^3(l_{\alpha\beta})E_{3'}{}^{3t} = \begin{pmatrix} 0 & 1 \\ 1 & 0 \end{pmatrix}$. The first case corresponds to the treatment of the elliptic region, but it leads to serious complications (see exercise 12-1, Prob. 11). Therefore we choose the second way which is characterized by

$$\omega_{1'}{}^{3'} = \omega^{2'} \qquad \omega_{2'}{}^{3'} = \omega^{1'} \tag{12-13}$$

As before, the *affine normal* is defined by a transformation (12-9) and the condition $\omega_{3''}{}^{3''} = 0$. From (12-13) and the integrability conditions (12-5) one obtains

$$w_{1''2''}{}^{2''} = w_{2''1''}{}^{1''} = 0$$

The only unimodular matrices for which (12-13) and $\mathbf{e}_{3''}$ are invariant are those which appear in

$$\{\mathbf{a}_i\} = \begin{pmatrix} t & 0 & 0 \\ 0 & \frac{1}{t} & 0 \\ 0 & 0 & 1 \end{pmatrix} \{\mathbf{e}_{i''}\}$$

The Cartan matrix is changed into

$$\begin{pmatrix} \pi_1{}^1 & \pi_1{}^2 & \pi^2 \\ \pi_2{}^1 & -\pi_1{}^1 & \pi^1 \\ \pi_3{}^1 & \pi_3{}^2 & 0 \end{pmatrix} = \begin{pmatrix} t & 0 & 0 \\ 0 & \frac{1}{t} & 0 \\ 0 & 0 & 1 \end{pmatrix} \begin{pmatrix} \omega_{1''}{}^{1''} & \omega_{1''}{}^{2''} & \omega^{2''} \\ \omega_{2''}{}^{1''} & -\omega_{1''}{}^{1''} & \omega^{1''} \\ \omega_{3''}{}^{1''} & \omega_{3''}{}^{2''} & 0 \end{pmatrix} \begin{pmatrix} t & 0 & 0 \\ 0 & \frac{1}{t} & 0 \\ 0 & 0 & 1 \end{pmatrix} + \frac{dt}{t} \begin{pmatrix} 1 & 0 & 0 \\ 0 & -1 & 0 \\ 0 & 0 & 0 \end{pmatrix}$$

We do not try to normalize $\pi_1{}^1$ since this would involve a differential equation. From

$$\pi_1{}^2 = t^3 w_{1''1''}{}^{2''}\pi^1 \qquad \pi_2{}^1 = t^{-3} w_{2''2''}{}^{1''}\pi^2$$

it follows that

$$p_{11}{}^2 p_{22}{}^1 = w_{1''1''}{}^{2''} w_{2''2''}{}^{1''}$$

is an invariant. It is possible by a unimodular transformation to change the direction of either $\mathbf{a}_1$ or $\mathbf{a}_2$ (and $\mathbf{a}_3$) into its opposite; hence it is possible to choose the frame so that

$$p_{11}{}^2 p_{22}{}^1 = J^2 > 0 \tag{12-14}$$

If $J \neq 0$, we take $t = (J/p_{11}{}^2)^{\frac{1}{3}}$ to obtain $\pi_1{}^2 = J\pi^1$, $\pi_2{}^1 = J\pi^2$. The case $J = 0$ will be discussed in the next section.

Theorem 12-2. *At any point of a hyperbolic region with nonzero hyperbolic affine curvature J there exists a unique frame for which the Frenet equations are*

$$d\{\mathbf{a}_i\} = \begin{pmatrix} \pi_1{}^1 & J\pi^1 & \pi^2 \\ J\pi^2 & -\pi_1{}^1 & \pi^1 \\ \pi_3{}^1 & \pi_3{}^2 & 0 \end{pmatrix} \{\mathbf{a}_i\}$$

The integrability conditions (12-5) are

$$d\pi^1 = p_{12}{}^1\pi^1 \wedge \pi^2 \qquad d\pi^2 = p_{11}{}^1\pi^1 \wedge \pi^2$$

$$\begin{aligned} p_{31}{}^1 &= p_{32}{}^2 = J^2 - 2p_{11}{}^1p_{12}{}^1 + p_{11}{}^1{}_{,2} - p_{12}{}^1{}_{,1} \\ p_{31}{}^2 &= -3Jp_{12}{}^1 + J_{,2} \\ p_{32}{}^1 &= 3Jp_{11}{}^1 + J_{,1} \end{aligned} \tag{12-15}$$

$$\begin{aligned} p_{31}{}^2J - 2p_{32}{}^1p_{11}{}^1 + p_{31}{}^1{}_{,2} - p_{32}{}^1{}_{,1} &= 0 \\ -p_{32}{}^1J - 2p_{31}{}^2p_{12}{}^1 + p_{31}{}^2{}_{,1} - p_{32}{}^2{}_{,1} &= 0 \end{aligned}$$

Here again the frame and all invariants are determined if π^1, π^2, and J are given, subject to the integrability conditions. The directions of $\mathbf{a}_1$ and $\mathbf{a}_2$ are the *asymptotic directions* on the surface (Sec. 12-3).

III. *Parabolic region.* We assume now that $\|l_{\alpha\beta}\| = 0$ on some open set of the surface. We do not treat the case of parabolic curves on a surface. Since elliptic and hyperbolic regions are so completely different in affine geometry, it is not to be expected that it may be possible to connect elliptic and hyperbolic frames over a parabolic curve in a continuous manner.

There are two cases to consider.

A. If all $l_{\alpha\beta}$ vanish, $\omega_1{}^3 = \omega_2{}^3 = 0$. This means that

$$d\mathbf{e}_1 = \omega_1{}^1\mathbf{e}_1 + \omega_1{}^2\mathbf{e}_2 \qquad d\mathbf{e}_2 = \omega_2{}^1\mathbf{e}_1 + \omega_2{}^2\mathbf{e}_2$$

The tangent plane is constant for all points, and *the surface is the plane* of $\mathbf{e}_1$ and $\mathbf{e}_2$.

B. If the rank of $(l_{\alpha\beta})$ is one, we may normalize the matrix by

$$E_{3'}{}^3(l_{\alpha\beta})E_{3'}{}^{3t} = \begin{pmatrix} 0 & 0 \\ 0 & 1 \end{pmatrix}$$

that is, $\omega_{1'}{}^{3'} = 0$ and $\omega_{2'}{}^{3'} = \omega^{2'}$. The integrability condition

$$d\omega_{1'}{}^{3'} = 0 = \omega_{1'}{}^{2'} \wedge \omega^{2'}$$

shows that $\omega_{1'}{}^{2'} = w_{1'2'}{}^{2'}\omega^{2'}$. The equations

$$\begin{aligned} d\mathbf{p} &= \omega^{1'}\mathbf{e}_{1'} + \omega^{2'}\mathbf{e}_{2'} \\ d\mathbf{e}_{1'} &= \omega_{1'}{}^{1'}\mathbf{e}_{1'} + w_{1'2'}{}^{2'}\omega^{2'}\mathbf{e}_{2'} \\ d\mathbf{e}_{2'} &= \omega_{2'}{}^{1'}\mathbf{e}_{1'} + \omega_{2'}{}^{2'}\mathbf{e}_{2'} + \omega^{2'}\mathbf{e}_{3'} \end{aligned}$$

along the curves $\omega^{2'} = 0$ reduce to

$$d\mathbf{p} = \omega^{1'}\mathbf{e}_{1'}$$

$$d\mathbf{e}_{1'} = \omega_{1'}{}^{1'}\mathbf{e}_{1'} \qquad d\mathbf{e}_{2'} = \omega_{2'}{}^{1'}\mathbf{e}_{1'} + \omega_{2'}{}^{2'}\mathbf{e}_{2'}$$

The curves $\omega^{2'} = 0$ are straight lines since their tangents $\mathbf{e}_{1'}$ have constant direction. The tangent plane $\mathbf{e}_{1'},\mathbf{e}_{2'}$ is constant along such a ruling.

Theorem 12-3. *Parabolic surfaces are developable.*

1. If $w_{1'2'}{}^{2'} = 0$, the direction of $\mathbf{e}_1$ is constant on the surface. All rulings are parallel, and the surface is a *cylinder*.

2. If $w_{1'2'}{}^{2'} \neq 0$, the admissible transformations of frames are unimodular matrices

$$\begin{pmatrix} e_{1''}{}^{1'} & 0 & 0 \\ e_{2''}{}^{1} & e_{2''}{}^{2'} & 0 \\ e_{3''}{}^{1'} & e_{3''}{}^{2'} & e_{3''}{}^{3'} \end{pmatrix}$$

We study first the influence of

$$\{\mathbf{e}_{i''}\} = \begin{pmatrix} t & 0 & 0 \\ 0 & \frac{1}{t} & 0 \\ 0 & 0 & 1 \end{pmatrix} \{\mathbf{e}_i'\}$$

for which $\omega_{1''}{}^{2''} = t^3 w_{1'2'}{}^{2'} \omega^{2''}$. t is chosen as

$$t = (w_{1'2'}{}^{2'})^{-\frac{1}{3}}$$

for the normalization

$$\omega_{1''}{}^{2''} = \omega^{2''}$$

The integrability conditions for $d\omega^{2''} = d\omega_{1''}{}^{2''} = d\omega_{2''}{}^{3''}$ imply

$$(\omega_{1''}{}^{1''} - \omega^{1''}) \wedge \omega^{2''} = 0$$
$$(\omega_{1''}{}^{1''} - 2\omega_{2''}{}^{2''} + \omega_{3''}{}^{3''}) \wedge \omega^{2''} = 0$$

The last equation subtracted from the trivial one $(\omega_{i''}{}^{i''}) \wedge \omega^{2''} = 0$ means that

$$\omega_{2''}{}^{2''} \wedge \omega^{2''} = 0$$

These equations tell us that

$$\omega_{1''}{}^{1''} = \omega^{1''} + w_{1''2''}{}^{1''}\omega^{2''} \qquad \omega_{2''}{}^{2''} = w_{2''2''}{}^{2''}\omega^{2''}$$
$$\omega_{3''}{}^{3''} = -\omega^{1''} - (w_{1''2''}{}^{1''} + w_{2''2''}{}^{2''})\omega^{2''}$$

A transformation

$$\{\mathbf{e}_{i^*}\} = \begin{pmatrix} 1 & 0 & 0 \\ e_{2^*}{}^{1''} & 1 & 0 \\ 0 & 0 & 1 \end{pmatrix} \{\mathbf{e}_i''\}$$

conserves the previous relations and gives $\omega_{2^*}{}^{2^*} = \omega_{2''}{}^{2''} + e_{2^*}{}^{1''}\omega^{2''}$. Therefore it is possible to choose the frame so that

$$\omega_{2^*}{}^{2^*} = 0$$

The integrability condition $d\omega_{2^*}{}^{2^*} = 0 = (\omega_{2^*}{}^{1^*} - \omega_{3^*}{}^{2^*}) \wedge \omega^{2^*}$ shows that $w_{2^*1^*}{}^{1^*} = w_{3^*1^*}{}^{2^*}$.

a. If $w_{1^*2^*}{}^{1^*} = 0$, then $d\mathbf{e}_{1^*} = \omega^{1^*}\mathbf{e}_{1^*} + \omega^{2^*}\mathbf{e}_{2^*} = d\mathbf{p}$; the point $\mathbf{p} - \mathbf{e}_{1^*}$ is fixed, and the surface is a *cone with vertex* $\mathbf{p} - \mathbf{e}_{1^*}$.

b. If $w_{1^*2^*}{}^{1^*} \neq 0$, the only admissible transformations are

$$\{\mathbf{a}_i\} = \begin{pmatrix} 1 & 0 & 0 \\ 0 & 1 & 0 \\ e_3{}^{1^*} & 0 & 1 \end{pmatrix} \{\mathbf{e}_{i^*}\}$$

for which $\pi_2{}^1 = \omega_{2^*}{}^{1^*} - e_3{}^{1^*}\pi^2$. An invariant frame therefore may be obtained from the condition $p_{22}{}^1 = 0$. The Frenet equations are

$$d\{\mathbf{a}_i\} = \begin{pmatrix} \pi^1 + p_{12}{}^1\pi^2 & \pi^2 & 0 \\ p_{21}{}^1\pi^1 & 0 & \pi^2 \\ p_{31}{}^1\pi^1 + p_{32}{}^1\pi^2 & p_{21}{}^1\pi^1 + p_{32}{}^2\pi^2 & -\pi^1 - p_{12}{}^1\pi^2 \end{pmatrix} \{\mathbf{a}_i\}$$

It is left to the reader to compute the integrability conditions in terms of invariant derivatives. They show that all invariants may be determined if π^1, π^2, and $p_{12}{}^1$ are known.

Having determined the form of the invariant affine Cartan matrices at all points of a local surface, we turn to the actual computation of the invariants by means of *symmetric* differential forms. The *first fundamental form* of affine geometry is

$$\begin{aligned} \Phi = \|\mathbf{a}_1,\mathbf{a}_2,d^2\mathbf{p}\| &= (\pi^1)^2 + (\pi^2)^2 && \text{elliptic region} \\ &= 2\pi^1\pi^2 && \text{hyperbolic region} \\ &= (\pi^2)^2 && \text{parabolic region other than plane} \end{aligned} \tag{12-16}$$

Only elliptic and hyperbolic points will be considered here; developables are better treated as tangent surfaces to space curves (Sec. 8-3).

For the computation of Φ as a function of a parametric representation $\mathbf{p}(u^1,u^2)$ we start from the fact that there exists a nonsingular matrix $(\lambda_\alpha{}^\beta)$ such that

$$\left\{\frac{\partial \mathbf{p}}{\partial u^\alpha}\right\} = (\lambda_\alpha{}^\beta)\{\mathbf{a}_\beta\} \qquad \alpha, \beta = 1, 2$$

It follows that $(\pi^\alpha) = (du^\beta)(\lambda_\beta{}^\alpha)$ and

$$\left\|\frac{\partial \mathbf{p}}{\partial u^1},\frac{\partial \mathbf{p}}{\partial u^2},d^2\mathbf{p}\right\| = L_{\alpha\beta}\, du^\alpha\, du^\beta = \|\lambda_\alpha{}^\beta\|\Phi$$

If we replace Φ by its expression in terms of the π^α and these again by the $\lambda_\alpha{}^\beta$, it follows that

$$\|L_{\alpha\beta}\| = \begin{cases} \|\lambda_\alpha{}^\beta\|^4 & \text{in the elliptic case} \\ -\|\lambda_\alpha{}^\beta\|^4 & \text{in the hyperbolic case} \end{cases}$$

The sign of $\|L_{\alpha\beta}\|$ can be used to distinguish between elliptic and hyperbolic regions. The fundamental form

$$\Phi = |\,\|L_{\alpha\beta}\|\,|^{-\frac{1}{4}} L_{\alpha\beta}\, du^\alpha\, du^\beta = |\,\|L_{\alpha\beta}\|\,|^{-\frac{1}{4}} \left\|\frac{\partial \mathbf{p}}{\partial u^1},\frac{\partial \mathbf{p}}{\partial u^2},d^2\mathbf{p}\right\| \tag{12-17}$$

may be computed if the parametric representation is given. Its (symmetric) differential is

$$\begin{aligned} d\Phi &= \|\mathbf{a}_1,\mathbf{a}_2,d^3\mathbf{p}\| + \|\mathbf{a}_1,d\mathbf{a}_2,d^2\mathbf{p}\| + \|d\mathbf{a}_1,\mathbf{a}_2,d^2\mathbf{p}\| \\ &= \|\mathbf{a}_1,\mathbf{a}_2,d^3\mathbf{p}\| - I[(\pi^1)^3 - 3\pi^1(\pi^2)^2] - [\pi^1 d\pi^1 + \pi^2 d\pi^2] \quad \text{elliptic} \\ &= \|\mathbf{a}_1,\mathbf{a}_2,d^3\mathbf{p}\| - J[(\pi^1)^3 + (\pi^2)^3] - [\pi^1 d\pi^2 + \pi^2 d\pi^1] \quad \text{hyperbolic} \end{aligned}$$

In both cases the cubic form

$$\begin{aligned}\Psi &= \|\mathbf{a}_1,\mathbf{a}_2,d^3\mathbf{p}\| - \frac{3}{2}\,d\Phi = |\,\|L_{\alpha\beta}\|\,|^{-\frac{1}{4}} \left\| \frac{\partial \mathbf{p}}{\partial u^1},\frac{\partial \mathbf{p}}{\partial u^2},d^3\mathbf{p} \right\| - \frac{3}{2}\,d\Phi \\ &= I[(\pi^1)^3 - 3\pi^1(\pi^2)^2] \quad \text{elliptic} \\ &= J[(\pi^1)^3 + (\pi^2)^3] \quad \text{hyperbolic}\end{aligned} \tag{12-18}$$

is invariant. Ψ is the second fundamental form of affine geometry.

In the elliptic case, π^1 can be taken as a multiple of any of the *three* linear factors θ of Ψ, $\pi^1 = a\theta$. The function a is found by the condition that $\Phi - a^2\theta^2 = (\pi^2)^2$ is a complete square. The sign of a is determined by the condition $I > 0$, and the vectors $\mathbf{a}_\alpha$ are found from $d\mathbf{p} = \pi^\alpha \mathbf{a}_\alpha$.

In the hyperbolic case it follows from (12-16) and (12-18) that Ψ may be diagonalized at the same time in which the matrix of Φ is reduced to the second diagonal; this problem has a unique solution (up to a change in terms).

The two forms Φ and $\Psi = b_{\alpha\beta\gamma}\, du^\alpha\, du^\beta\, du^\gamma$ are not independent of one another. It is easily checked from the definitions that

$$\begin{aligned}L_{22}b_{111} - 2L_{12}b_{112} + L_{11}b_{122} &= 0 \\ L_{22}b_{112} - 2L_{12}b_{122} + L_{11}b_{222} &= 0\end{aligned}$$

By virtue of these conditions, Φ and Ψ are said to be *apolar*. Since apolarity of forms of different degrees has no simple geometric interpretation, we shall not study it further.

The invariants I and J describe the shape of the surface in the neighborhood of a point. We take the frame $\mathbf{a}_i(\mathbf{p}_0)$ as the basis of an (oblique) system of coordinates x^i. The first terms of the Taylor expansion of $\mathbf{p}(u^1,u^2)$ in the neighborhood of $\mathbf{p}_0$,

$$\mathbf{p} = \mathbf{p}_0 + (d\mathbf{p})_0 + \tfrac{1}{2}(d^2\mathbf{p})_0 + \tfrac{1}{6}(d^3\mathbf{p})_0 + R_4$$

either are fixed by the coordinates, like $\mathbf{p}_0 = 0$, $(d\mathbf{p})_0 = x^1\mathbf{a}_1 + x^2\mathbf{a}_2$ [hence $(\pi^\alpha)_0 = x^\alpha$], or may be computed from the fundamental forms. It follows that

$$\begin{aligned}x^3 &= \|\mathbf{a}_1(\mathbf{p}_0),\mathbf{a}_2(\mathbf{p}_0),\mathbf{p}\| \\ &= \frac{1}{2}\,[(x^1)^2 + (x^2)^2] + \frac{I}{6}\,[(x^1)^3 - 3x^1(x^2)^2] + R_4 \quad \text{elliptic} \\ x^3 &= x^1x^2 + \frac{J}{6}\,[(x^1)^3 + (x^2)^3] + R_4 \quad \text{hyperbolic}\end{aligned} \tag{12-19}$$

It is easily seen from these formulas that the words "elliptic, hyperbolic, parabolic" points have the same meaning in euclidean and in affine geometry.

Exercise 12-1

◆ **1.** Determine the elliptic and the hyperbolic regions on the surface $(x^1)^3 + (x^2)^3 + (x^3)^3 = 1$.

◆ **2.** Find the affine curvature for the surface of Prob. 1.

3. Find the affine curvature at (0,0,1) of the surface $x^3 e^{x^1 x^2} = 1$. [Use Eq. (12-19).]

4. Show that the affine curvature vanishes for a quadric.

5. Discuss (in cartesian coordinates) the surfaces described by formulas (12-19) for constant affine curvature and $R_4 = 0$.

◆ **6.** Find the affine normal and the affine curvature at (0,0,0) for the surface $x^3 = 4x^1x^2 - 2(x^1)^2 - (x^2)^4$. [Hint: Find oblique axes to obtain a development (12-19).]

7. With the affine normal $\mathbf{a}_3$ we associate a *covariant* vector by $\boldsymbol{\alpha}(\mathbf{a}_3) = 1$.

(*a*) Show that $\boldsymbol{\alpha}(\mathbf{p}_{u^1}) = \boldsymbol{\alpha}(\mathbf{p}_{u^2}) = (\partial\boldsymbol{\alpha}/\partial u^\beta)(\mathbf{p}_{u^\gamma}) = 0$.

(*b*) Show that the tangent plane is $\alpha_i(X^i - p^i) = 0$.

(*c*) Why is the distinction between covariant and contravariant vectors essential in this problem?

8. Find a formula corresponding to (12-19) for the parabolic surface in case 2*b*.

9. Formulate the existence and unicity theorems that follow from theorem 10-18 for affine geometry.

10. Write the integrability conditions for the invariants of parabolic surfaces.

11. If in the hyperbolic case one chooses the normalization

$$E_{3'}{}^{3}(l_{\alpha\beta})E_{3'}{}^{3t} = \begin{pmatrix} 1 & 0 \\ 0 & -1 \end{pmatrix}$$

show that the admissible transformations are

$$\begin{pmatrix} \operatorname{Cosh} t & \operatorname{Sinh} t & 0 \\ \operatorname{Sinh} t & \operatorname{Cosh} t & 0 \\ 0 & 0 & 1 \end{pmatrix}$$

and that $w_{1''2''}{}^{1''} = 0$ for $\operatorname{Tanh} 3t = -w_{1'2'}{}^{1'}/w_{1'1'}{}^{1'}$. When does this equation have a real root? What is the condition for

$$w_{1''1''}{}^{1''} = 0$$

Are both normalizations possible on one surface?

12. Find the formulas for the change of the dual coordinates of a line in a unimodular transformation of R^3.

13. In the hyperbolic case, study the invariant frame defined by a transformation from $\{\mathbf{e}_{i''}\}$ to $\{\mathbf{a}_i\}$ with $t_{,2}/t^3 = -w_{1''2''}{}^{1''}$.

♦14. Find the behavior of the affine curvatures in a homothety.

12-2. Special Surfaces

I. $I = 0$. We begin with a study of the surfaces all of whose points are Darboux umbilics. One may start from

$$d\{\mathbf{e}_{i''}\} = \begin{pmatrix} 0 & \omega_{1''}{}^{2''} & \omega^{1''} \\ -\omega_{1''}{}^{2''} & 0 & \omega^{2''} \\ \omega_{3''}{}^{1''} & \omega_{3''}{}^{2''} & 0 \end{pmatrix} \{\mathbf{e}_i''\}$$

The integrability conditions for $d\omega_{1''}{}^{1''}$ and $d\omega_{3''}{}^{3''}$ imply

$$w_{3''2''}{}^{1''} = w_{3''1''}{}^{2''} = 0$$

The only transformations admissible on the frame $\{\mathbf{e}_{i''}\}$ are given by (12-11); they result in

$$\begin{aligned}
\pi_1{}^2 &= \omega_{1''}{}^{2''} + d\theta \\
\pi_3{}^1 &= (w_{3''1''}{}^{1''} \cos^2\theta + w_{3''2''}{}^{2''} \sin^2\theta)\pi^1 \\
&\qquad\qquad + (w_{3''2''}{}^{2''} - w_{3''1''}{}^{1''}) \sin\theta\cos\theta\,\pi^2 \\
\pi_3{}^2 &= (w_{3''2''}{}^{2''} - w_{3''1''}{}^{1''}) \sin\theta\cos\theta\,\pi^1 \\
&\qquad\qquad + (w_{3''1''}{}^{1''} \sin^2\theta + w_{3''2''}{}^{2''} \cos^2\theta)\pi^2
\end{aligned}$$

Since the integrability conditions must hold also after the transformation, one has $p_{32}{}^1 = p_{31}{}^2 = 0$, that is, $w_{3''2''}{}^{2''} = w_{3''1''}{}^{1''}$ and also $p_{32}{}^2 = p_{31}{}^1$. The integrability conditions for $d\pi_3{}^1$ and $d\pi_3{}^2$ give $dp_{31}{}^1 = 0$; $p_{31}{}^1 = c$ is a constant. The Frenet equations are, finally,

$$d\{\mathbf{a}_i\} = \begin{pmatrix} 0 & \pi_1{}^2 & \pi^1 \\ -\pi_1{}^2 & 0 & \pi^2 \\ c\pi^1 & c\pi^2 & 0 \end{pmatrix} \{\mathbf{a}_i\} \tag{12-20}$$

$$d\pi_1{}^2 = c\pi^1 \wedge \pi^2$$

A. $c = 0$. In this case, $\pi_1{}^2$ is a total differential $-d\theta(u^1,u^2)$; hence there exists a transformation (12-11) that gives a frame $\{\mathbf{a}_i\}$ for which $\pi_1{}^2 = 0$. Then also $d\pi^\alpha = 0$, and one may find new coordinates (locally) such that $\pi^\alpha = dv^\alpha$. The Frenet equations

$$d\mathbf{p} = dv^1\,\mathbf{a}_1 + dv^2\,\mathbf{a}_2$$
$$d\mathbf{a}_1 = dv^1\,\mathbf{a}_3 \qquad d\mathbf{a}_2 = dv^2\,\mathbf{a}_3 \qquad d\mathbf{a}_3 = 0$$

are easily integrated:

$$\mathbf{p} = \mathbf{p}_0 + v^1\mathbf{a}_1(0) + v^2\mathbf{a}_2(0) + \frac{(v^1)^2 + (v^2)^2}{2}\,\mathbf{a}_3(0)$$

In the oblique system defined by the vectors $\mathbf{a}_i(0)$ the surface has the equation

$$x^3 = \tfrac{1}{2}[(x^1)^2 + (x^2)^2]$$

It is an *elliptic paraboloid.*

B. $c \neq 0$. Here $d\mathbf{a}_3 = c(\pi^1\mathbf{a}_1 + \pi^2\mathbf{a}_2) = c\,d\mathbf{p}$. If the origin is chosen such that $\mathbf{a}_3(0) - \mathbf{p}(0) = 0$, the affine normal is parallel to the radius vector: $\mathbf{a}_3 = c\mathbf{p}$. Such a surface is called an *affine sphere* ($\mathbf{N} = c\mathbf{p}$ characterizes euclidean spheres). We try to change the length of the vectors $\{\mathbf{a}_i\}$ by

$$\{\mathbf{a}_{i'}\} = \begin{pmatrix} k & 0 & 0 \\ 0 & k & 0 \\ 0 & 0 & \dfrac{1}{k^2} \end{pmatrix} \{\mathbf{a}_i\}$$

in order to make skew-symmetric the Cartan matrix in (12-20). The result is

$$d\begin{pmatrix} k\mathbf{a}_1 \\ k\mathbf{a}_2 \\ \dfrac{\mathbf{a}_3}{k^2} \end{pmatrix} = \begin{pmatrix} 0 & \pi_1{}^2 & k^3\pi^1 \\ -\pi_1{}^2 & 0 & k^3\pi^2 \\ \dfrac{c\pi^1}{k^3} & \dfrac{c\pi^2}{k^3} & 0 \end{pmatrix} \begin{pmatrix} k\mathbf{a}_1 \\ k\mathbf{a}_2 \\ \dfrac{\mathbf{a}_3}{k^2} \end{pmatrix}$$

Hence we should want to choose $k^6 = -c$.

1. $c < 0$. The Cartan matrix is in $\mathfrak{L}(O_3)$ if we choose $k^6 = -c$. If the frame $\{k\mathbf{a}_1(0), k\mathbf{a}_2(0), \mathbf{a}_3(0)/k^2\}$ were orthonormal, the endpoint of $\mathbf{a}_3$ would describe a sphere of radius k^2 if the origin of the frame is kept fixed. In the general case, then, the locus of $\mathbf{a}_3(u^1,u^2)$ is an *ellipsoid,* the affine image of a sphere. $\mathbf{p}(u^1,u^2)$ describes a homothetic ellipsoid.

2. $c > 0$. If we choose $k^6 = c$, the Cartan matrix is in the Lie algebra of the group of matrices A for which

$$A^t\begin{pmatrix} -1 & 0 & 0 \\ 0 & -1 & 0 \\ 0 & 0 & 1 \end{pmatrix} A = \begin{pmatrix} -1 & 0 & 0 \\ 0 & -1 & 0 \\ 0 & 0 & 1 \end{pmatrix}$$

since that group is characterized by the symmetry property

$$\begin{pmatrix} -1 & 0 & 0 \\ 0 & -1 & 0 \\ 0 & 0 & 1 \end{pmatrix} C(A) = -\left(\begin{pmatrix} -1 & 0 & 0 \\ 0 & -1 & 0 \\ 0 & 0 & 1 \end{pmatrix} C(A)\right)^t$$

If the frame is "orthonormal" in the sense of this group, viz.,

$$\{a_i{}^k(0)\}^t\begin{pmatrix} -1 & 0 & 0 \\ 0 & -1 & 0 \\ 0 & 0 & 1 \end{pmatrix}\{a_i{}^k(0)\} = \begin{pmatrix} -1 & 0 & 0 \\ 0 & -1 & 0 \\ 0 & 0 & 1 \end{pmatrix} \qquad (12\text{-}21)$$

it remains so for all values of the parameters. This shows that the endpoint of $\mathbf{a}_3/k^2$ is always on the equilateral hyperboloid of two sheets $-(x^1)^2 - (x^2)^2 + (x^3)^2 = 1$. The surfaces $\mathbf{p}$ obtained from it by a general affine transformation are again *hyperboloids of two sheets.*

II. $J = 0$. It follows from Eq. (12-14) that we have to distinguish two cases.

A. $p_{22}{}^1 \neq 0$. The Frenet formulas are

$$d\{\mathbf{a}_i\} = \begin{pmatrix} \pi_1{}^1 & 0 & \pi^2 \\ p_{22}{}^1\pi^2 & -\pi_1{}^1 & \pi^1 \\ p_{31}{}^1\pi^1 & p_{32}{}^2\pi^2 & 0 \end{pmatrix} \{\mathbf{a}_i\}$$

The curves $\pi^2 = 0$ are straight lines ($d\mathbf{a}_1 = \pi_1{}^1\mathbf{a}_1$). This case comprises all *ruled surfaces* that are not developables.

B. $p_{22}{}^1 = 0$.

1. $p_{11}{}^2 \neq 0$. A change of terms $\mathbf{a}_1 \to \mathbf{a}_2$, $\mathbf{a}_2 \to \mathbf{a}_1$, $\mathbf{a}_3 \to -\mathbf{a}_3$ makes this case identical with *A*.

2. $p_{11}{}^2 = 0$. The Frenet formulas are

$$d\{\mathbf{a}_i\} = \begin{pmatrix} \pi_1{}^1 & 0 & \pi^2 \\ 0 & -\pi_1{}^1 & \pi^1 \\ c\pi^1 & c\pi^2 & 0 \end{pmatrix} \{\mathbf{a}_i\}$$
$$dc = 0$$

by the same arguments as in case I. Both families of parameter lines are straight , and the surfaces are *ruled quadrics.* For $c = 0$, one has the *hyperbolic paraboloid*

$$\mathbf{p} = \mathbf{p}_0 + x^1\mathbf{a}_1(0) + x^2\mathbf{a}_2(0) + x^1x^2\mathbf{a}_3(0)$$

For $c \neq 0$, an argument like that of I*B* shows that the surface is a *one-sheeted hyperboloid*, the affine image of

$$(x^1)^2 + (x^2)^2 - (x^3)^2 = 1$$

III. *Surfaces admitting a transitive group of unimodular affine transformations of* R^3. These surfaces are characterized by the property that *all* their invariants are constant. We determine all such surfaces of elliptic and of hyperbolic type. The parabolic surfaces are dealt with in the exercises.

A. *Elliptic surfaces.* The last two equations (12-12) become

$$\begin{aligned} (p_{31}{}^1 - p_{32}{}^2)p_{12}{}^2 - 2p_{32}{}^1\, p_{11}{}^2 \qquad\qquad &= 0 \\ (p_{31}{}^1 - p_{32}{}^2)p_{11}{}^2 + 2p_{32}{}^1(p_{12}{}^2 + 2I) &= 0 \end{aligned}$$

Either this homogeneous system has the trivial solution

$$p_{31}{}^1 - p_{32}{}^2 = 0 \qquad p_{32}{}^1 = 0 \tag{12-22a}$$

or its determinant vanishes:

$$p_{12}{}^2(p_{12}{}^2 + 2I) + (p_{11}{}^2)^2 = 0 \tag{12-22b}$$

1. $p_{31}{}^1 = p_{32}{}^2,\ p_{32}{}^1 = 0$. Equations (12-12) imply

$$p_{11}{}^2 I = 0 \qquad 6I(p_{12}{}^2 + I) = (p_{11}{}^2)^2$$

a. $I = 0$. The elliptic surfaces of vanishing affine curvature are quadrics. All quadrics admit transitive groups of unimodular affine transformations. We determine here the group of the elliptic paraboloid (case I*A*). The method applies to all quadrics.

The equation of the paraboloid may be written in matrix form

$$(x^1,x^2,x^3,1)\begin{pmatrix}1 & 0 & 0 & 0\\ 0 & 1 & 0 & 0\\ 0 & 0 & 0 & -1\\ 0 & 0 & -1 & 0\end{pmatrix}\begin{pmatrix}x^1\\ x^2\\ x^3\\ 1\end{pmatrix} = 0$$

This equation should be invariant in a transformation

$$(x^{1'},x^{2'},x^{3'},1) = (x^1,x^2,x^3,1)\begin{pmatrix}a_1{}^{1'} & a_1{}^{2'} & a_1{}^{3'} & 0\\ a_2{}^{1'} & a_2{}^{2'} & a_2{}^{3'} & 0\\ a_3{}^{1'} & a_3{}^{2'} & a_3{}^{3'} & 0\\ b^{1'} & b^{2'} & b^{3'} & 1\end{pmatrix} \qquad \|a_i{}^{j'}\| = 1$$

Hence

$$\begin{pmatrix}a_1{}^{1'} & a_1{}^{2'} & a_1{}^{3'} & 0\\ a_2{}^{1'} & a_2{}^{2'} & a_2{}^{3'} & 0\\ a_3{}^{1'} & a_3{}^{2'} & a_3{}^{3'} & 0\\ b^{1'} & b^{2'} & b^{3'} & 1\end{pmatrix}\begin{pmatrix}1 & 0 & 0 & 0\\ 0 & 1 & 0 & 0\\ 0 & 0 & 0 & -1\\ 0 & 0 & -1 & 0\end{pmatrix}\begin{pmatrix}a_1{}^{1'} & a_2{}^{1'} & a_3{}^{1'} & b^{1'}\\ a_1{}^{2'} & a_2{}^{2'} & a_3{}^{2'} & b^{2'}\\ a_1{}^{3'} & a_2{}^{3'} & a_3{}^{3'} & b^{3'}\\ 0 & 0 & 0 & 1\end{pmatrix} = \begin{pmatrix}1 & 0 & 0 & 0\\ 0 & 1 & 0 & 0\\ 0 & 0 & 0 & -1\\ 0 & 0 & -1 & 0\end{pmatrix}$$

The only matrices compatible with this condition are

$$\begin{pmatrix}\cos\phi & -\sin\phi & b^{1'}\cos\phi - b^{2'}\sin\phi & 0\\ \sin\phi & \cos\phi & b^{1'}\sin\phi + b^{2'}\cos\phi & 0\\ 0 & 0 & 1 & 0\\ b^{1'} & b^{2'} & \frac{1}{2}(b^{1'})^2 + (b^{2'})^2 & 1\end{pmatrix}$$

The group has three parameters.

b. $I \neq 0$. In this case $p_{11}{}^2 = 0$ and $p_{12}{}^2 = -I$. By Eqs. (12-12), $d\pi^\alpha = 0$, we may use parameters u^α such that $\pi^\alpha = du^\alpha$. The Frenet equations are

$$d\mathbf{p} = \mathbf{a}_1\,du^1 + \mathbf{a}_2\,du^2$$

$$d\{\mathbf{a}_i\} = \begin{pmatrix} I\,du^1 & -I\,du^2 & du^1 \\ -I\,du^2 & -I\,du^1 & du^2 \\ 2I^2\,du^1 & 2I^2\,du^2 & 0 \end{pmatrix} \{\mathbf{a}_i\}$$

Since $d\mathbf{a}_3 = 2I^2\,d\mathbf{p}$, the surface is an *affine sphere.* $\mathbf{p}$ is an integral of the system of partial differential equations

$$\mathbf{p}_{u^1} = \mathbf{a}_1 \qquad \mathbf{p}_{u^2} = \mathbf{a}_2$$

$$\mathbf{p}_{u^1u^1} = I\mathbf{p}_{u^1} + 2I^2\mathbf{p} \qquad \mathbf{p}_{u^1u^2} = -I\mathbf{p}_{u^2} \qquad \mathbf{p}_{u^2u^2} = -I\mathbf{p}_{u^1} + 2I^2\mathbf{p}$$

if the origin is chosen so that $\mathbf{a}_3 = 2I^2\mathbf{p}$. The equation for $\mathbf{p}_{u^1u^2}$ shows that $\mathbf{p} = e^{-Iu^1}\mathbf{x}(u^2) + \mathbf{y}(u^1)$; the other ones give

$$\mathbf{p} = \mathbf{c}_1 e^{2Iu^1} + e^{-Iu^1}(\mathbf{c}_2 e^{I\sqrt{3}u^2} + \mathbf{c}_3 e^{-I\sqrt{3}u^2})$$

The condition $\|\mathbf{a}_1,\mathbf{a}_2,\mathbf{a}_3\| = 1$ becomes for $u^1 = u^2 = 0$

$$12\sqrt{3}\,I^4\|\mathbf{c}_1,\mathbf{c}_2,\mathbf{c}_3\| = 1$$

If the linearly independent vectors $\mathbf{c}_i$ are used as coordinate vectors, $\mathbf{p} = x^i\mathbf{c}_i$, the equation of the surface is

$$x^1x^2x^3 = 1$$

It admits the two-parameter group

$$\mathbf{x}^* = \mathbf{x}\begin{pmatrix} a & 0 & 0 \\ 0 & b & 0 \\ 0 & 0 & \dfrac{1}{ab} \end{pmatrix}$$

Since there exist three families of Darboux frames, it is possible to compose each transformation of the group with a rotation of $\pi/3$ or $2\pi/3$. The full group of transformations cannot be obtained from its group germ; it has three "connected components." This situation obtains for all elliptic surfaces with nonvanishing curvature.

2. $p_{12}{}^2(p_{12}{}^2 + 2I) + (p_{11}{}^2)^2 = 0$. As explained before, we are interested only in the case $I \neq 0$.

a. $p_{11}{}^2 = 0$.

(1) $p_{12}{}^2 = 0$. Since $d\pi^1 = 0$, $d\pi^2 = I\pi^1 \wedge \pi^2$, it is possible to find parameters such that $\pi^1 = du^1$ and $\pi^2 = e^{Iu^1}\,du^2$. The equations to integrate are

$$d\mathbf{p} = du^1\,\mathbf{a}_1 + e^{Iu^1}\,du^2\,\mathbf{a}_2$$

$$d\{\mathbf{a}_i\} = \begin{pmatrix} I\,du^1 & 0 & du^1 \\ -2Ie^{Iu^1}\,du^2 & -I\,du^1 & e^{Iu^1}\,du^2 \\ 6I^2\,du^1 & 0 & 0 \end{pmatrix} \{\mathbf{a}_i\}$$

Since $\mathbf{p}_{u^1u^2} = 0$, it follows that

$$\mathbf{p} = \mathbf{x}(u^1) + \mathbf{y}(u^2) \tag{12-23}$$

A surface that can be written in the form (12-23) is called a *surface of translation*. The equation easiest to integrate is

$$d^2\mathbf{a}_1 = (I\,d\mathbf{a}_1 + 6I^2\mathbf{a}_1\,du^1)\,du^1$$

from which we have

$$\mathbf{a}_1 = \mathbf{c}_1 e^{-2Iu^1} + \mathbf{c}_2 e^{3Iu^1}$$

and

$$\mathbf{a}_3 = -3I\mathbf{c}_1 e^{-2Iu^1} + 2I\mathbf{c}_2 e^{3Iu^1}$$

$$d(e^{Iu^1}\mathbf{a}_2) = -5I\mathbf{c}_1\,du^2$$

Hence

$$\mathbf{a}_2 = (\mathbf{c}_3 - 5I\mathbf{c}_1 u^2)e^{-Iu^1}$$

If we choose the origin so as not to introduce new constants in the integration of $d\mathbf{p}$, the final result is

$$\mathbf{p} = -\left[\frac{1}{2I}e^{-2uI^1} + \frac{5}{2}I(u^2)^2\right]\mathbf{c}_1 + \frac{1}{3I}e^{3Iu^1}\mathbf{c}_2 + u^2\mathbf{c}_3$$

subject to

$$\|\mathbf{a}_1,\mathbf{a}_2,\mathbf{a}_3\| = 5I\|\mathbf{c}_1,\mathbf{c}_2,\mathbf{c}_3\| = 1$$

If $5I\mathbf{c}_1$, $\mathbf{c}_2$, $\mathbf{c}_3$ are taken as basis vectors of a unimodular system of coordinates, the equation of the surface is

$$[(x^3)^2 + 2x^1]^3(x^2)^2 = -\frac{1}{1{,}125I^8}$$

on which acts the group

$$\mathbf{x}^* = \mathbf{x}\begin{pmatrix} a^{-\frac{1}{2}} & 0 & 0 \\ 0 & a & 0 \\ ba^{-\frac{1}{2}} & 0 & a^{-\frac{1}{2}} \end{pmatrix} + \left(\frac{b^2}{2},0,b\right)$$

for each one of the possible Darboux frames.

(2) $p_{12}{}^2 = -2I$. The third equation (12-12) reduces to $12I^2 = 0$. No surface exists for $I \neq 0$.

b. $p_{11}{}^2 \neq 0$. Equations (12-12) have a unique solution

$$p_{11}{}^2 = \tfrac{3}{2}I \qquad p_{12}{}^2 = -\tfrac{3}{2}I$$

It is easily checked that the Cartan matrix of case $a(1)$ is brought into the matrix for case b by a rotation (12-11) by $\pm 2\pi/3$. The two cases are identical, up to a different choice of the Darboux frames.

B. Hyperbolic surfaces. Formulas (12-15) become

$$\begin{aligned} p_{31}{}^1 &= p_{32}{}^2 = J^2 - 2p_{11}{}^1p_{12}{}^1 \\ p_{31}{}^2 &= -3Jp_{12}{}^1 \\ p_{32}{}^1 &= 3Jp_{11}{}^1 \end{aligned} \tag{12-24}$$

$$Jp_{31}{}^2 - 2p_{11}{}^1p_{32}{}^1 = 0$$
$$2p_{12}{}^1p_{31}{}^2 + Jp_{32}{}^1 = 0$$

1. $J = 0$. If we look for surfaces other than quadrics, we have to start from the Cartan matrix given under II*A*. The equation

$$d\pi_2{}^1 = \pi_2{}^i \wedge \pi_i{}^1$$

becomes $p_{22}{}^1 p_{11}{}^1 = 0$. Since we must have $p_{22}{}^1 \neq 0$, it follows that $p_{11}{}^1 = 0$, and the Frenet equations reduce to

$$d\mathbf{p} = du^1\,\mathbf{a}_1 + du^2\,\mathbf{a}_2$$

$$d\{\mathbf{a}_i\} = \begin{pmatrix} 0 & 0 & du^2 \\ du^2 & 0 & du^1 \\ 0 & 0 & 0 \end{pmatrix} \{\mathbf{a}_i\}$$

Such a surface with constant affine normal ($d\mathbf{a}_3 = 0$) is called an *improper affine sphere*. A simple integration gives

$$\mathbf{p} = (u^1 + \tfrac{1}{2}(u^2)^2)\mathbf{c}_1 + u^2\mathbf{c}_2 + (u^1u^2 + \tfrac{1}{6}(u^2)^3)\mathbf{c}_3$$

subject to $\|\mathbf{c}_1,\mathbf{c}_2,\mathbf{c}_3\| = 1$. This surface, known as *Cayley's surface*, is in coordinates to the basis $\{\mathbf{c}_i\}$,

$$x^3 = x^1x^2 - \tfrac{1}{3}(x^2)^3$$

Its group is

$$(x^{1'},x^{2'},x^{3'}) = (x^1,x^2,x^3)\begin{pmatrix} 1 & 0 & a \\ a & 1 & b \\ 0 & 0 & 1 \end{pmatrix} + (b,a,ab - \tfrac{1}{3}b^3)$$

2. $J \neq 0$. The last two equations (12-24) either have only the trivial solution $p_{31}{}^2 = p_{32}{}^1 = 0$ or else their determinant vanishes.

a. $p_{31}{}^2 = p_{32}{}^1 = 0$. By (12-24), the equations to integrate are

$$d\mathbf{p} = du^1\,\mathbf{a}_1 + du^2\,\mathbf{a}_2$$

$$d\{\mathbf{a}_i\} = \begin{pmatrix} 0 & J\,du^1 & du^2 \\ J\,du^2 & 0 & du^1 \\ J^2\,du^1 & J^2\,du^2 & 0 \end{pmatrix} \{\mathbf{a}_i\}$$

The surface is an affine sphere. We choose the origin at the point $\mathbf{a}_3(0) - J^2\mathbf{p}(0)$ common to all affine normals. The partial differential equations of the surface then become

$$\mathbf{p}_{u^1} = \mathbf{a}_1 \qquad \mathbf{p}_{u^2} = \mathbf{a}_2$$

$$\mathbf{p}_{u^1u^1} = J\mathbf{p}_{u^2} \qquad \mathbf{p}_{u^1u^2} = J^2\mathbf{p} \qquad \mathbf{p}_{u^2u^2} = J\mathbf{p}_{u^1}$$

$$\mathbf{p}_{u^1u^1u^1} = J^3\mathbf{p} = \mathbf{p}_{u^2u^2u^2}$$

From the last line it follows that

$$\mathbf{p} = \mathbf{x}_1(u^2)e^{Ju^1} + e^{-Ju^1/2}\left[\mathbf{x}_2(u^2)\cos\frac{\sqrt{3}}{2}Ju^1 + \mathbf{x}_3(u^2)\sin\frac{\sqrt{3}}{2}Ju^1\right]$$
$$= \mathbf{y}_1(u^1)e^{Ju^2} + e^{-Ju^2/2}\left[\mathbf{y}_2(u^1)\cos\frac{\sqrt{3}}{2}Ju^2 + \mathbf{y}_3(u^1)\sin\frac{\sqrt{3}}{2}Ju^2\right]$$

It is then seen that

$$\mathbf{p} = \frac{2}{3^{\frac{3}{2}}J^4}e^{J(u^1+u^2)}\mathbf{c}_1 + e^{-\frac{1}{2}J(u^1+u^2)}\left[\mathbf{c}_2\cos\frac{\sqrt{3}}{2}J(u^1+u^2) + \mathbf{c}_3\sin\frac{\sqrt{3}}{2}J(u^1+u^2)\right]$$

with $\|\mathbf{c}_1,\mathbf{c}_2,\mathbf{c}_3\| = 1$. The surface

$$x^1[(x^2)^2 + (x^3)^2] = 2\cdot 3^{-\frac{3}{2}}J^{-4}$$

is invariant under the action of

$$(x^{1'},x^{2'},x^{3'}) = (x^1,x^2,x^3)\begin{pmatrix} \frac{1}{c^2} & 0 & 0 \\ 0 & c\cos\theta & c\sin\theta \\ 0 & -c\sin\theta & c\cos\theta \end{pmatrix}$$

b. $J^2 + 4p_{11}{}^1p_{12}{}^1 = 0$. Equations (12-24) have a unique solution:

$$p_{11}{}^1 = \frac{J}{2} \qquad p_{12}{}^1 = -\frac{J}{2} \qquad p_{31}{}^1 = p_{32}{}^1 = p_{31}{}^2 = p_{32}{}^2 = \frac{3J^2}{2}$$

By Eqs. (12-15), $d(\pi^1 + \pi^2) = 0$, $d(\pi^1 - \pi^2) = -J\pi^1 \wedge \pi^2$. Instead of the frame $\mathbf{a}_i$ we use the (not unimodular) frame $\mathbf{b}_1 = \frac{1}{2}(\mathbf{a}_1 + \mathbf{a}_2)$, $\mathbf{b}_2 = \frac{1}{2}(\mathbf{a}_1 - \mathbf{a}_2)$, $\mathbf{b}_3 = \mathbf{a}_3$. We have to integrate

$$d\mathbf{p} = du^1\,\mathbf{b}_1 + du^2\,\mathbf{b}_2$$

$$d\{\mathbf{b}_i\} = \begin{pmatrix} \frac{J}{2}\,du^1 & 0 & \frac{1}{2}\,du^1 \\ Je^{Ju^1/2}\,du^2 & -\frac{J}{2}\,du^1 & -\frac{1}{2}e^{Ju^1/2}\,du^2 \\ 3J^2 & 0 & 0 \end{pmatrix}\{\mathbf{b}_i\}$$

From

$$2\frac{d^2\mathbf{b}_1}{(du^1)^2} - J\frac{d\mathbf{b}_1}{du^1} - 3J^2\mathbf{b}_1 = 0$$

one has immediately

$$\mathbf{b}_1 = +\frac{1}{10J}e^{-Ju^1}\mathbf{c}_1 + e^{\frac{3}{2}Ju^1}\mathbf{c}_2$$
$$\mathbf{b}_2 = -\tfrac{1}{2}u^2\mathbf{c}_1 + \mathbf{c}_3$$
$$\mathbf{b}_3 = \tfrac{3}{5}e^{-Ju^1}\mathbf{c}_1 + 2Je^{\frac{3}{2}Ju^1}\,\mathbf{c}_2$$

Hence

$$\mathbf{p} = \left[\frac{1}{8}(u^2)^2 - \frac{1}{10J^2}e^{-Ju^1}\right]\mathbf{c}_1 + \frac{2}{3J}e^{\frac{1}{2}Ju^1}\mathbf{c}_2 + u^2\mathbf{c}_3$$

$\mathbf{p}$ is a surface of translation whose equation in the unimodular system $\mathbf{c}_i$ is

$$(x^2)^2[(x^3)^2 - 8x^1]^3 = \frac{2^8}{3^2 5^3 J^8}$$

invariant under

$$(x^{1'},x^{2'},x^{3'}) = (x^1,x^2,x^3)\begin{pmatrix} a^2 & 0 & 0 \\ 0 & \dfrac{1}{a^3} & 0 \\ \dfrac{ab}{4} & 0 & a \end{pmatrix} + \left(\frac{b^2}{8},0,b\right)$$

The following theorem summarizes the results:

Theorem 12-4. *The surfaces admitting a transitive group of unimodular affine transformations are the quadrics, the elliptic surfaces*

$$x^1x^2x^3 = 2^{-2}3^{-\frac{1}{3}}I^{-4} \qquad \textit{and} \qquad [(x^3)^2 + 2x^1]^3(x^2)^2 = -3^{-2}5^{-3}I^{-8}$$

the hyperbolic surfaces

$$x^1[(x^2)^2 + (x^3)^2] = 2^1 3^{-\frac{1}{3}}J^{-4} \qquad \textit{and} \qquad (x^2)^2[(x^3)^2 - 8x^1]^3 = 2^8 3^{-2}5^{-3}J^{-8}$$

and the ruled surface $x^3 = x^1x^2 - \frac{1}{3}(x^2)^3$.

Exercise 12-2

1. Find the three-parameter groups of unimodular affinities
   (*a*) Of the ellipsoid
   (*b*) Of the hyperboloid of two sheets
   (*c*) Of the hyperboloid of one sheet
   (*d*) Of the hyperbolic paraboloid
2. Find the invariant affine frame of the right helicoid.
3. Show that a euclidean sphere is an affine sphere.
4. An affine sphere is defined by $d\mathbf{a}_3 = \lambda\, d\mathbf{p}$. Show that $\lambda = \text{const.}$
5. Show that the only convex closed surface with constant affine curvature is the ellipsoid (Blaschke). HINT: Show that

$$d``\|"\mathbf{a}_1,d\mathbf{p},\mathbf{a}_3``\|" = \pi^1 \wedge \pi_1{}^2 + I\pi^1 \wedge \pi^2$$

and that $d``\|"\mathbf{a}_2,d\mathbf{p},\mathbf{a}_3``\|" = \pi_2{}^1 \wedge \pi^2$. From the first equation and Stokes's theorem it follows that $I\iint\pi^1 \wedge \pi^2 = -\iint p_{12}{}^2\pi^1 \wedge \pi^2$ and from the second, $\iint p_{21}{}^1\pi^1 \wedge \pi^2 = 0 = -\iint p_{12}{}^2\pi^1 \wedge \pi^2$. It is possible to conclude $I = 0$ and to use the list in I of this section.

6. Show that, if in every point of a surface there exists a quadric having contact of order >3 with the surface, the surface is a quadric. [HINT: By (12-19) the affine curvature must vanish. Then there remains only one case to be excluded.]

◆7. Find the rulings of Cayley's surface.

◆8. Show that in the parabolic case the only surface with a transitive group of affinities which is not a quadric, a cylinder, or a cone has the Frenet equation

$$d\{\mathbf{a}_i\} = \begin{pmatrix} \pi^1 & \pi^2 & 0 \\ p_{22}{}^1\pi^2 & 0 & \pi^2 \\ -2p_{22}{}^1\pi^1 & -p_{22}{}^1\pi^1 & -\pi^1 \end{pmatrix} \{\mathbf{a}_i\}$$

(*a*) Integrate the equations for $p_{22}{}^1 = 0$ and find the group of the surface.

(*b*) In the case $p_{22}{}^1 \neq 0$ identify the surface among the parabolic surfaces studied in Sec. 12-1.

9. Find all cylinders and all cones with transitive groups of unimodular affinities.

12-3. Curves on a Surface

I. *Elliptic case.* Just as in metric geometry, we study curves on surfaces by their affine tangent normal frames. Let θ be the angle between the tangent $\boldsymbol{\tau}$ to a curve and the Darboux direction $\mathbf{a}_1$ at the same point. The *tangent normal frame* $\{\boldsymbol{\tau},\mathbf{v},\mathbf{a}_3\}$ to the curve is the frame obtained from $\{\mathbf{a}_i\}$ by a transformation (12-11). $\boldsymbol{\tau}$ is the *affine tangent* and $\mathbf{v}$ the *affine tangent normal.* In the elliptic case there are two possible choices for the tangent normal, depending on the Darboux frames. We may define an *affine surface arc length* s by

$$\pi^1 = \cos\theta\, ds \qquad \pi^2 = \sin\theta\, ds$$

that is,

$$\Phi = ds^2$$

The surface arc length has no connection whatsoever with the affine arc length introduced in the theory of curves (Secs. 7-3 and 8-3). The derivation formulas are

$$d\begin{pmatrix} \boldsymbol{\tau} \\ \mathbf{v} \\ \mathbf{a}_3 \end{pmatrix} = \begin{pmatrix} I\cos 3\theta\, ds & (-I\sin 3\theta + k_g)\, ds & ds \\ (-I\sin 3\theta - k_g)\, ds & -I\cos 3\theta\, ds & 0 \\ \pi_3{}^1\cos\theta + \pi_3{}^2\sin\theta & -\pi_3{}^1\sin\theta + \pi_3{}^2\cos\theta & 0 \end{pmatrix} \begin{pmatrix} \boldsymbol{\tau} \\ \mathbf{v} \\ \mathbf{a}_3 \end{pmatrix}$$

where

$$k_g\, ds = d\theta + p_{11}{}^2\pi^1 + (p_{12}{}^2 + I)\pi^2$$

is the geodesic curvature of the curve for the metric Φ. An *affine geodesic* is defined by

$$-I\sin 3\theta + k_g = 0$$

Its osculating plane contains the affine normal $\mathbf{a}_3$. The curves

$$I \sin 3\theta + k_g = 0$$

are the *contour lines.* The boundaries between light and shadow on surfaces exposed to parallel light rays are contour lines since the affine tangent normal has a constant direction along a contour line. The affine tangent of a contour line is the solution of

$$\frac{d\boldsymbol{\tau}}{ds} = I \cos 3\theta\, \boldsymbol{\tau} - 2I \sin 3\theta\, \mathbf{v} + \mathbf{a}_3$$

If a curve is plane, its tangent is a linear combination of, at most, two linearly independent vectors. If $I \neq 0$, not all contour lines on a local surface may be plane.

Theorem 12-5 (Maschke). *An (elliptic) surface all of whose contour lines are plane is an (elliptic) quadric.*

The *affine curvature lines* are the integral curves of Rodrigues's equation

$$d\mathbf{e}_3 + k\, d\mathbf{p} = 0$$

Their tangents are in the *affine principal directions* defined by

$$\frac{\pi_3{}^1}{ds} \cos \theta + \frac{\pi_3{}^2}{ds} \sin \theta = -k \cos \theta$$

$$-\frac{\pi_3{}^1}{ds} \sin \theta + \frac{\pi_3{}^2}{ds} \cos \theta = -k \sin \theta$$

There are, in general, two *affine principal curvatures* k_1, k_2, the solutions of

$$\begin{Vmatrix} p_{31}{}^1 + k & p_{32}{}^1 \\ p_{32}{}^1 & p_{32}{}^2 + k \end{Vmatrix} = 0$$

The k_α are affine invariants, as are also the *affine Gauss curvature*

$$K = k_1k_2 = p_{31}{}^1p_{32}{}^2 - (p_{32}{}^1)^2$$

and the *affine mean curvature*

$$H = \tfrac{1}{2}(k_1 + k_2) = -\tfrac{1}{2}(p_{31}{}^1 + p_{32}{}^2)$$

If $k_1 = k_2$, the point on the surface is an *affine umbilic.* (Affine umbilics and Darboux umbilics are distinct notions.) A surface is an affine sphere if all its points are affine umbilics. If the affine principal curvatures are all zero, the surface is an improper affine sphere. Many of the arguments of Sec. 10-5 apply to affine geometry (see exercise 12-3, Probs. 14 and 15).

II. *Hyperbolic case.* The hyperbolic metric $ds^2 = 2\pi^1\pi^2$ has two asymptotic directions which divide each tangent plane into four regions according to the sign of ds^2. The sign must be conserved in the transformation of the $\{\mathbf{a}_i\}$ frame into the tangent normal frame of a curve.

A. If sign π^1 = sign π^2 for the tangent directions of a curve on the surface, we define the affine surface arc length by

$$\pi^1 = \frac{ds}{\sqrt{2}} e^t \qquad \pi^2 = \frac{ds}{\sqrt{2}} e^{-t}$$

and the affine tangent normal frame by

$$\begin{pmatrix} \boldsymbol{\tau} \\ \mathbf{v} \\ \mathbf{a}_3 \end{pmatrix} = \begin{pmatrix} \dfrac{e^t}{\sqrt{2}} & \dfrac{e^{-t}}{\sqrt{2}} & 0 \\ \dfrac{-e^t}{\sqrt{2}} & \dfrac{e^{-t}}{\sqrt{2}} & 0 \\ 0 & 0 & 1 \end{pmatrix} \begin{pmatrix} \mathbf{a}_1 \\ \mathbf{a}_2 \\ \mathbf{a}_3 \end{pmatrix}$$

Its Frenet equation becomes

$$\frac{d}{ds}\begin{pmatrix} \boldsymbol{\tau} \\ \mathbf{v} \\ \mathbf{a}_3 \end{pmatrix}$$

$$= \begin{pmatrix} \dfrac{J}{\sqrt{2}} \operatorname{Cosh} 3t & \dfrac{J}{\sqrt{2}} \operatorname{Sinh} 3t - \dfrac{\pi_1{}^1 + dt}{ds} & 1 \\ -\dfrac{J}{\sqrt{2}} \operatorname{Sinh} 3t - \dfrac{\pi_1{}^1 + dt}{ds} & -\dfrac{J}{\sqrt{2}} \operatorname{Cosh} 3t & 0 \\ \dfrac{\pi_3{}^1}{ds}\dfrac{e^{-t}}{\sqrt{2}} + \dfrac{\pi_3{}^2}{ds}\dfrac{e^t}{\sqrt{2}} & -\dfrac{\pi_3{}^1}{ds}\dfrac{e^{-t}}{\sqrt{2}} + \dfrac{\pi_3{}^2}{ds}\dfrac{e^t}{\sqrt{2}} & 0 \end{pmatrix} \begin{pmatrix} \boldsymbol{\tau} \\ \mathbf{v} \\ \mathbf{a}_3 \end{pmatrix}$$

The same families of special curves may be investigated as in the elliptic case. *Affine geodesics* are integral curves of

$$\pi_1{}^1 + dt - \frac{J}{\sqrt{2}} \operatorname{Sinh} 3t\, ds = 0$$

and *contour lines* integral curves of

$$\pi_1{}^1 + dt + \frac{J}{\sqrt{2}} \operatorname{Sinh} 3t\, ds = 0$$

For a contour line,

$$\frac{d\boldsymbol{\tau}}{ds} = \frac{J}{\sqrt{2}} \operatorname{Cosh} 3t\, \boldsymbol{\tau} + \sqrt{2}\, J \operatorname{Sinh} 3t\, \mathbf{v} + \mathbf{a}_3$$

If all contour lines of a surface are plane, it follows that

$$J = p_{11}{}^2 = p_{22}{}^1 = 0$$

The surface is a quadric. Theorem 12-5 holds also in the hyperbolic case.

The affine principal directions should be solutions of

$$\begin{aligned} \pi_3{}^1 e^{-t} + \pi_3{}^2 e^t &= -ke^t\, ds \\ -\pi_3{}^1 e^{-t} + \pi_3{}^2 e^t &= -ke^{-t}\, ds \end{aligned}$$

The principal curvatures should be solutions of

$$\begin{aligned} p_{31}{}^1 + p_{32}{}^2 + \sqrt{2}\, ke^t + 2p_{32}{}^1 \operatorname{Cosh} 2t &= 0 \\ -p_{31}{}^1 + p_{32}{}^2 - \sqrt{2}\, ke^t + 2p_{32}{}^1 \operatorname{Sinh} 2t &= 0 \end{aligned}$$

independent of t. This is impossible; hyperbolic surfaces do not have affine curvature lines.

B. If $\pi^1\pi^2 < 0$, the surface arc length and tangent normal frame are defined by

$$\pi^1 = \frac{ds}{\sqrt{2}} e^t \qquad \pi^2 = -\frac{ds}{\sqrt{2}} e^{-t}$$

and

$$\begin{pmatrix} \mathbf{v} \\ \boldsymbol{\tau} \\ \mathbf{a}_3 \end{pmatrix} = \begin{pmatrix} \dfrac{e^t}{\sqrt{2}} & \dfrac{-e^{-t}}{\sqrt{2}} & 0 \\ \dfrac{e^t}{\sqrt{2}} & \dfrac{e^{-t}}{\sqrt{2}} & 0 \\ 0 & 0 & 1 \end{pmatrix} \begin{pmatrix} \mathbf{a}_1 \\ \mathbf{a}_2 \\ \mathbf{a}_3 \end{pmatrix}$$

All formulas may be established analogous to case *A*.

Exercise 12-3

1. The *Lie quadric* at a point of a surface is defined (*a*) in the elliptic case as the quadric with center $\mathbf{p} + \frac{1}{2}(k_1 + k_2)\mathbf{a}_3$ which has contact of order ≥ 3 with the surface in $\mathbf{p}$ and (*b*) in the hyperbolic case as the osculating quadric of the ruled surface defined by the directions $\mathbf{a}_2$ along an asymptotic line $\pi^2 = 0$ through $\mathbf{p}$. Find the equation of the Lie quadric [use (12-19)].
2. Prove that in the hyperbolic case the Lie quadric is the integral surface of

$$d\{\mathbf{a}_i\} = \pi^1 \begin{pmatrix} p_{11}{}^1(0) & J(0) & 0 \\ 0 & -p_{11}{}^1(0) & 1 \\ p_{31}{}^1(0) & p_{32}{}^1(0) & 0 \end{pmatrix} \{\mathbf{a}_i\}$$

3. Show that in the hyperbolic case the two Lie quadrics defined by the two asymptotic lines through one point coincide. (Use the answer to Prob. 1; the formula must be symmetric in x^1 and x^2.)
4. Prove that the center of the Lie quadric is constant if and only if the surface is an affine sphere.
5. Show that all Lie quadrics of a surface coincide if the cubic fundamental form is identically zero.

6. Show that, if $H = 0$, both $\mathbf{a}_2$ and $\mathbf{a}_3$ remain parallel to a fixed plane along any asymptotic line $\pi^\alpha = 0$.

7. Discuss the metric and the contour lines of $(x^1)^3 + 3x^1x^2 - 2x^3 = 0$.

8. Show that no proper affine sphere is a surface of translation (Reidemeister).

9. The inner geometry of $ds^2 = dx^1\,dx^2$ admits a transitive group of transformations $(\partial/\partial x^1,\ \partial/\partial x^2,\ -x^1\partial/\partial x^1 - x^2\partial/\partial x^2)$. Find its geodesics.

10. Determine the transformation behavior of the invariants Φ, Ψ, I, J, and $p_{ij}{}^k$ in a *general* (not unimodular) affine transformation of R^n. Use your result to construct invariants of general affine geometry. What are the differentiability assumptions?

11. State and prove an affine analogue to the theorem of Joachimsthal.

12. Find all parabolic affine spheres.

13. Find all hyperbolic improper affine spheres.

14. Show that formulas (10-23) to (10-25) are valid also in affine geometry if $\alpha = \pi^1 \wedge \pi^2$ and $\mathbf{N} = \mathbf{a}_3$.

15. Use the results of Prob. 14 to prove formulas (10-75) and (10-76) as well as theorems 10-32 and 10-33 in affine geometry.

16. Show that the euclidean normal $\mathbf{N}$ and the elliptic affine normal $\mathbf{a}_3$ are defined *uniquely* by the following properties:

(*a*) The normal in $\mathbf{p}$ is not in T_p.

(*b*) Its definition is geometrically invariant.

(*c*) It is invariant in the group under consideration.

(*d*) The developables contained in the congruence of normal lines define a double system of "curvature lines" on the surface.

(*e*) The definition depends only on derivatives up to the natural order of the group geometry in question (not counting the additional order for the integrability condition) (Sannia).

17. Show that the Darboux directions are the only directions for which the circle of curvature of the euclidean normal section has contact of order ≥ 3 with the section (Transon, 1841).

References

See References for Chap. 8.

13 RIEMANNIAN GEOMETRY

13-1. Parallelism and Curvature

Riemannian geometry is a generalization of inner geometry to n dimensions. A *Riemann space* V is an open set in a cartesian space R^n in which is defined a positive-definite symmetric metric $ds^2 = g_{ij}(\mathbf{x})\, dx^i\, dx^j$. A *Riemann manifold* is a manifold on which is defined a positive-definite symmetric twice-covariant tensor field $G = g_{ij}(\mathbf{x})$. The metric tensor is assumed to be C^4; coordinate transformations are admissible if they are at least C^5. It will be left to the reader to check that the necessary differentiability assumptions are, in fact, always satisfied. In the first three sections of this chapter, only Riemann spaces are considered.

If we are given the metric tensor field $g_{ij}(\mathbf{x})$, we are looking for another system of coordinates $(x^{i'})$ such that, at *some* point $\mathbf{x}_0$, $g_{i'j'}(\mathbf{x}_0) = \delta_{i'j'}$. In terms of the jacobian $J = (\partial x^i/\partial x^{i'})$, this means

$$U = J(\mathbf{x}_0)G(\mathbf{x}_0)J(\mathbf{x}_0)^t \tag{13-1}$$

Since G is symmetric and positive-definite, a real nonsingular matrix $J(\mathbf{x}_0)$ may be found. By the implicit-function theorem coordinate transformations with the desired property exist in some neighborhood of $\mathbf{x}_0$. Condition (13-1) cannot, in general, be enforced in the neighborhood outside $\mathbf{x}_0$.

In the tangent space $T_{\mathbf{x}_0}$ the scalar product $\mathbf{a} \circ \mathbf{b} = \mathbf{a}U\mathbf{b}^t$ induced by the $(x^{i'})$ coordinates is the euclidean one. Covariant and contravariant vectors may be identified in euclidean geometry. A covariant vector

(α_i) and a contravariant one (a^i) represent the same geometric object in T_{x_0} if the transformed vector $\{\alpha_{i'}\} = J\{\alpha_i\}$ is the transpose of

$$(a^{i'}) = (a^i)J^{-1}$$

that is, if

$$(a^i)^t = J^tJ\{\alpha_i\}$$

By (13-1),
$$G^{-1}(x_0) = J(x_0)^tJ(x_0)$$

It is customary to denote the matrix G^{-1} by (g^{ij}). Then

$$a^i = g^{ij}\alpha_j \tag{13-2}$$

is the relation between the *covariant coordinates* $\{\alpha_j\}$ and the *contravariant coordinates* (a^i) of a vector in the euclidean geometry of T_{x_0}. In general, we *raise* and *lower* indices by

$$T_{i_1i_2\cdots i_k}{}^{j_1\cdots j_l} = g_{i_ks}T_{i_1\cdots i_{k-1}}{}^{sj_1\cdots j_l}$$
$$T_{i_1\cdots i_{k-1}}{}^{j_1\cdots j_{l+1}} = g^{j_1s}T_{i_1\cdots i_{k-1}s}{}^{j_2\cdots j_{l+1}}$$

This is justified by the identification of tensors of different types in euclidean geometry and by

$$g^i_j = g^{is}g_{sj} = \delta^i_j$$

In riemannian geometry a tensor is characterized only by the number of its indices and eventual symmetry properties. A k tensor has representations by i times covariant, j times contravariant coordinates for all i and j such that $i + j = k$. It is important to indicate exactly the place of the indices. This is achieved usually by dots in the void places. Unless the tensor is symmetric, one will have, for example,

$$T_{i_1i_2\cdot i_4}{}^{\;\;\;i_3} \neq T^{i_1}{}_{\cdot i_2i_3i_4}$$

In each tangent space T_x we choose two frames of reference. The vectors $\mathbf{i}_j$ of the first frame are the tangent vectors to the parameter curves $x^k = \text{const}$ $(k \neq j)$. The frame $\{\mathbf{i}_j\}$ defines in T_x an oblique system of cartesian coordinates with unequal unit lengths. The scalar product in T_x is

$$\mathbf{i}_j \circ \mathbf{i}_k = g_{jk} \tag{13-3a}$$

The other frame $\{\mathbf{e}_\alpha\}$ will be an orthonormal one,

$$\mathbf{e}_\alpha \circ \mathbf{e}_\beta = \delta_{\alpha\beta} \tag{13-3b}$$

chosen as a differentiable function of $\mathbf{x}$. In general, *the vectors* $\{\mathbf{e}_\alpha\}$ *cannot be chosen as tangents to parameter lines.* Hence the index α in $\mathbf{e}_\alpha$ does not refer to any system of *coordinates* in the space but to a system of *frames*. We shall reserve Greek indices for this kind of coordinates in

T_x not derived from coordinates in V (often called *non-holonomic* coordinates). Take now a differentiable curve $\mathbf{x}(t)$ through the point $\mathbf{x}_0 = \mathbf{x}(0)$. The vectors in T_{x_0}, by definition, represent classes of curves at $\mathbf{x}_0$. The tangent vector to our curve will be

$$d\mathbf{x} = dx^i\, \mathbf{i}_i \tag{13-4a}$$

or

$$d\mathbf{x} = \omega^\alpha \mathbf{e}_\alpha \tag{13-4b}$$

The arc length of the curve, by definition, is the integral over ds, where

$$ds^2 = g_{ij}\, dx^i\, dx^j = \Sigma(\omega^\alpha)^2 \tag{13-5}$$

We want to associate with the curve $\mathbf{x}(t)$ a linear map $T_{x_0} \to T_{x(t)}$ that should be obtained by integration over t of a Frenet formula

$$d\{\mathbf{e}_\alpha\} = (\omega^\beta{}_\alpha)\{\mathbf{e}_\beta\} \tag{13-6a}$$

or

$$d\{\mathbf{i}_i\} = (\Gamma^j{}_i)\{\mathbf{i}_j\} \tag{13-6b}$$

If we use the orthonormal frames, we are dealing with orthogonal geometry; hence

$$\omega^\beta{}_\alpha + \omega^\alpha{}_\beta = 0 \tag{13-7}$$

It must be underscored that, a priori, nothing is known about the relations between the tangent spaces except the neighborhood structure introduced in Chap. 9. The existence of formulas (13-6) must be considered as an *axiom* of riemannian geometry, as also the continued validity of

$$dd\mathbf{x} = 0 \tag{9-22}$$

a formula that has been established only in the framework of euclidean geometry. We justify the two postulates by showing that they uniquely define the matrices $\omega = (\omega^\beta{}_\alpha)$ and $\Gamma = (\Gamma^j{}_i)$. The condition (9-22) implies

$$d\omega^\alpha = \omega^\beta \wedge \omega^\alpha{}_\beta \tag{13-8}$$

and

$$dx^j \wedge \Gamma^i{}_j = 0$$

If we write

$$\Gamma^i{}_j = \Gamma^i{}_{jk}\, dx^k \tag{13-9}$$

Cartan's theorem shows that the functions $\Gamma^i{}_{jk}$ are symmetric in their lower indices,

$$\Gamma^i{}_{jk} = \Gamma^i{}_{kj} \tag{13-10}$$

It is customary to write

$$\Gamma_{i,jk} = g_{is}\Gamma^s{}_{jk} \tag{13-11}$$

If as a last axiom we postulate the Leibniz rule

$$d(\mathbf{a} \circ \mathbf{b}) = d\mathbf{a} \circ \mathbf{b} + \mathbf{a} \circ d\mathbf{b} \tag{13-12}$$

for the $\circ$ product (which varies from point to point), it follows from (13-3*a*) that

$$\Gamma_{i,jk} + \Gamma_{j,ik} = \frac{\partial g_{ij}}{\partial x^k}$$

If we write the two equations obtained from this one by cyclic permutation of the indices and add two equations, subtracting the third one, then, by (13-10),

$$\Gamma_{i,jk} = \frac{1}{2}\left(\frac{\partial g_{ik}}{\partial x^j} + \frac{\partial g_{ij}}{\partial x^k} - \frac{\partial g_{jk}}{\partial x^i}\right) \tag{13-13}$$

From this formula and from

$$\{\mathbf{e}_\alpha\} = A\{\mathbf{i}_i\}$$

the matrix ω is immediately obtained:

$$\omega = A\Gamma A^{-1} + C(A) \tag{13-14}$$

where $\qquad C(A) = dAA^{-1}$

The functions $\Gamma^i{}_{jk}$ and $\Gamma_{i,jk}$ are called the *Christoffel symbols.* The functions $\gamma^\alpha{}_{\beta\gamma}$ appearing in $\omega^\alpha{}_\beta = \gamma^\alpha{}_{\beta\gamma}\omega^\gamma$ are the *Ricci coefficients.* Often the notation

$$\Gamma^i{}_{jk} = \begin{Bmatrix} i \\ j\ k \end{Bmatrix} \qquad \Gamma_{i,jk} = [i,jk]$$

is used in order to distinguish the Christoffel symbols from the more general objects to be introduced in Chap. 14.

Theorem 13-1. *For any* C^1 *metric* G *and frames* $\{\mathbf{i}_i\}$ *and* $\{\mathbf{e}_\alpha\}$ *there exist unique matrices* ω *and* Γ *such that* (13-3), (9-22), *and* (13-12) *hold.*

The Christoffel symbols are no tensor fields since they transform by the rule of Cartan matrices:

$$(\Gamma^{j'}{}_{i'}) = J(\Gamma^j{}_i)J^{-1} + C(J) \tag{13-15}$$

If we choose the coordinate transformation such that $C(J^{-1}) = \Gamma$ at a certain point $\mathbf{x}_0$, then $(\Gamma^{j'}{}_{i'}(\mathbf{x}_0)) = 0$. If a tensor field vanishes at $\mathbf{x}_0$ in *some* system of coordinates, it vanishes at $\mathbf{x}_0$ in *any* admissible system of coordinates. This is not true for Cartan matrices and quantities derived from them. On the other hand, it is impossible to find a coordinate system in which Γ vanishes in a whole neighborhood unless the equation $\Gamma = C(J^{-1})$ is totally integrable. The matrices ω and Γ are functions [with values in $T \otimes T^* \otimes T^*(V)$] of the *frames* in $T(V)$, *not* of the points on V.

Along the curve $\mathbf{x}(t)$ the matrix ω is a function of the parameter t alone. We may introduce a standard euclidean space R^n referred to an orthonormal basis $\{\mathbf{e}_{\alpha 0}\}$. Each tangent space T_x is the image of R^n under the orthogonal map

$$M(\mathbf{x})\colon \{\mathbf{e}_{\alpha 0}\} \to \{\mathbf{e}_\alpha(\mathbf{x})\}$$

M is a solution of the ordinary differential equation

$$\omega = C(M)$$

Two vectors $\mathbf{v} \in T_x$ and $\mathbf{v}^* \in T_{x^*}$ are the images of the same vector $\mathbf{v}_0 \in R^n$ if and only if

$$\mathbf{v}^* = M(\mathbf{x}^*)M(\mathbf{x})^{-1}\mathbf{v} \qquad (13\text{-}16)$$

Definition 13-2. *Two vectors* $\mathbf{v}$ *and* $\mathbf{v}^*$ *in tangent spaces* T_x *and* T_{x^*} *are parallel with respect to a curve joining* $\mathbf{x}$ *and* $\mathbf{x}^*$ *if they are images of one another in a mapping* (13-16).

If the metric G is constant, M is constant and $\mathbf{v}^* = \mathbf{v}$. Parallel vectors of equal length are identified in vector algebra. The notion of (*Levi-Civita*) parallelism introduced in this definition is a natural generalization of euclidean parallelism. The vectors of a vector field $\mathbf{v}(\mathbf{x})$ remain parallel to a vector $\mathbf{v}_0 = \mathbf{v}(\mathbf{x}_0)$ if

$$\mathbf{v}(\mathbf{x}(t)) = \mathbf{v}_0 + \int_{t_0}^{t} d\mathbf{v}(\mathbf{x}(t)) = M(\mathbf{x}(t))M(\mathbf{x}_0)^{-1}\mathbf{v}_0$$

that is, if $\quad d\mathbf{v}(\mathbf{x}(t)) = \omega(\mathbf{x}(t))\mathbf{v}(\mathbf{x}(t))$

The operation

$$\nabla\mathbf{v} = d\mathbf{v} - \omega\mathbf{v} \qquad (13\text{-}17)$$

therefore measures the deviation of a vector field from parallelism.

Definition 13-3. $\nabla\mathbf{v}$ *is the covariant differential of the vector field* $\mathbf{v}(\mathbf{x})$.

Let $E(\mathbf{x})$ be a differentiable matrix function of automorphisms of T_x. The covariant derivative of the vector $E\mathbf{v}$ relative to the family of frames $E\{\mathbf{e}_\alpha\}$ will be

$$\nabla_{E\mathbf{e}_\alpha}E\mathbf{v} = C(E)E\mathbf{v} + E\,d\mathbf{v} - (E\omega E^{-1} + C(E))E\mathbf{v} = E\nabla_{\mathbf{e}_\alpha}\mathbf{v} \qquad (13\text{-}18)$$

where the frame is indicated as a subscript of the differentiation symbol. This equation shows that the covariant differential of a covariant vector is a covariant tensor field. In components one works better with the *covariant derivatives* defined by

$$\nabla\{v_\iota\} = v_{\iota|\kappa}\omega^\kappa$$

or

$$\nabla\{v_i\} = v_{i;k}\,dx^k$$

The explicit formula then is

$$v_{\iota|\kappa} = v_{\iota,\kappa} - \gamma^\sigma{}_{\iota\kappa}v_\sigma \qquad (13\text{-}19)$$

and, by (13-18) (for $E = A^{-1}$),

$$v_{i;k} = \frac{\partial v_i}{\partial x^k} - \Gamma^s{}_{ik} v_s \tag{13-20}$$

The contravariant coordinates of a vector (v^i) are transformed by the inverse matrix $M^{-1} = M^t$. Hence

$$\nabla(v^\iota) = v^\iota{}_{|\kappa}\omega^\kappa = dv^\iota - (v^\sigma)C(M^t) = (dv^\iota) + (v^\sigma)(\omega^\iota{}_\sigma) \tag{13-21}$$

and also, again by (13-18),

$$\nabla(v^i) = v^i{}_{;k}\, dx^k = dv^i + v^s\Gamma^i{}_s \tag{13-22}$$

As a rule, we use a semicolon to indicate covariant derivatives with respect to coordinates and a vertical line for covariant derivatives with respect to frames that do not define parameter curves. The comma has the meaning defined in Sec. 10-3. Tensor product spaces such as $T \otimes T^*$ are transformed by the tensor product of the transformations in question. A tensor $t_i{}^j$ will be mapped by parallelism into

$$(t_\iota{}^\kappa) + \int_{t_0}^{t} d(t_\iota{}^\kappa) = (M(\mathbf{x}))(M(\mathbf{x}_0)^{-1})(t_\iota{}^\kappa)(M(\mathbf{x}_0))(M(\mathbf{x})^{-1})$$

From this one confirms immediately that

$$t_\iota{}^\kappa{}_{|\lambda} = t_\iota{}^\kappa{}_{,\lambda} + t_\iota{}^\sigma\gamma^\kappa{}_{\sigma\lambda} - \gamma^\sigma{}_{\iota\lambda}t_\sigma{}^\kappa \tag{13-23a}$$

and

$$t_i{}^j{}_{;k} = \frac{\partial t_i{}^j}{\partial x^k} + t_i{}^s\Gamma^j{}_{ks} - t_s{}^j\Gamma^s{}_{ik} \tag{13-23b}$$

For the terms additional to the partial derivative one considers each index by itself and uses either (13-19) and (13-20) or (13-21) and (13-22). The same rule holds in more complicated cases:

$$t_{i_1\cdots i_a}{}^{j_1\cdots j_b}{}_{;k} = \frac{\partial t_{i_1\cdots i_a}{}^{j_1\cdots j_b}}{\partial x^k} + \sum_{r=1}^{b} \Gamma^{j_r}{}_{ks} t_{i_1\cdots i_a}{}^{j_1\cdots j_{r-1} s j_{r+1}\cdots j_b} - \sum_{r=1}^{a} \Gamma^s{}_{i_r k} t_{i_1\cdots i_{r-1} s i_{r+1}\cdots i_a}{}^{j_1\cdots j_b} \tag{13-24}$$

By the definition of the **i** frames, *the metric tensor is parallel to itself along any curve,*

$$g_{ij;k} = 0 \tag{13-25}$$

This result may be confirmed by computation from (13-24) and (13-13). The formula is equivalent to the following statement: *If a vector remains parallel to itself,*

$$\nabla \mathbf{v} = 0 \qquad \text{or} \qquad d\mathbf{v} = \Gamma\mathbf{v}$$

then its length remains unchanged:

$$\begin{aligned} d(g_{ij}v^iv^j) &= (dg_{ij})v^iv^j + g_{ij}\,dv^iv^j + g_{ij}v^i\,dv^j \\ &= (dg_{ij} - g_{sj}\Gamma^s{}_i - g_{is}\Gamma^s{}_j)v^iv^j \\ &= (\nabla g_{ij})v^iv^j = 0 \end{aligned}$$

A simple, but less geometric, explanation of covariant differentiation is the following: For any vector $\mathbf{v} = v^i\mathbf{i}_i$ we have $d\mathbf{v} = (dv^i + v^j\Gamma^i{}_j)\mathbf{i}_i$. This shows that $\nabla v^i = dv^i + v^j\Gamma^i{}_j$ are the components of a differential vector.

Covariant derivation may be combined with exterior derivation. We define

$$D(a_{i_1\ldots i_k}\,dx^{i_1} \wedge \cdots \wedge dx^{i_k}) = (\nabla a_{i_1\ldots i_k}) \wedge dx^{i_1} \wedge \cdots \wedge dx^{i_k} \qquad (13\text{-}26)$$

D, ∇, and d coincide for scalars (functions). D and d coincide for vectors because of the symmetry (13-10) of the Γ's.

Let us now take a closed curve $\mathbf{c}\colon I \to V$ that begins and ends at $\mathbf{x}_0$. Any vector $\mathbf{v} \in T_{x_0}$ may be moved along $\mathbf{c}$ parallel to itself. This results in a linear map of T_{x_0} into itself:

$$R(\mathbf{c})\colon \mathbf{v} \to \mathbf{v} + \int_0^1 \Gamma\mathbf{v} \qquad (13\text{-}27)$$

R is orthogonal because it conserves lengths. Let $\mathbf{c}_P{}^Q(t)$ be any curve beginning at P and ending at $Q[\mathbf{c}_P{}^Q(0) = P,\ \mathbf{c}_P{}^Q(1) = Q]$. It is possible to define a multiplication in the set of curves. The product is defined if the endpoint of the first curve is the starting point of the second one, by

$$\hat{\mathbf{c}}_Q{}^R \cdot \mathbf{c}_P{}^Q(t) = \begin{cases} \mathbf{c}_P{}^Q(2t) & 0 \le t \le \frac{1}{2} \\ \hat{\mathbf{c}}_Q{}^R(2t-1) & \frac{1}{2} \le t \le 1 \end{cases} \qquad (13\text{-}28)$$

The *inverse* of a curve $\mathbf{c}_P{}^Q(t)$ from P to Q is the curve $\mathbf{c}_P{}^Q(1-t) = [\mathbf{c}_P{}^Q]^{-1}$. It follows from (13-27) that $R(\hat{\mathbf{c}}_{x_0}{}^{x_0} \cdot \mathbf{c}_{x_0}{}^{x_0})$ maps $\mathbf{v}$ into $R(c)\mathbf{v} + \int_0^1 \Gamma R(c)\mathbf{v}$, that is,

$$R(\hat{\mathbf{c}} \cdot \mathbf{c})\mathbf{v} = R(\hat{\mathbf{c}})R(\mathbf{c})\mathbf{v} \qquad (13\text{-}29)$$

(Since we consider here only curves whose endpoints coincide, we may suppress the indices.) The product is associative. This proves the following theorem:

Theorem 13-4. *The Levi-Civita parallelism defines a map of the set of all closed C^1 curves attached to a point $\mathbf{x}_0$ into a subgroup $H(\mathbf{x}_0)$ of the orthogonal group O_n. $H(\mathbf{x}_0)$ is the homogeneous holonomy group of V at $\mathbf{x}_0$.*

As stated at the beginning of this section, we deal only with Riemann *spaces.* Any two points in R^n may be connected by a smooth arc.

Theorem 13-5. *The homogeneous holonomy groups at any two points are isomorphic.*

Let $\mathbf{c}_1$ be any closed curve beginning and ending at $\mathbf{x}_1$. Choose a fixed curve $\mathbf{c}^* = \mathbf{c}_{x_0}{}^{x_1}$. Then $R(\mathbf{c}^{*-1} \cdot \mathbf{c}_1 \cdot \mathbf{c}^*) = R(\mathbf{c}_1)$ since the integrations

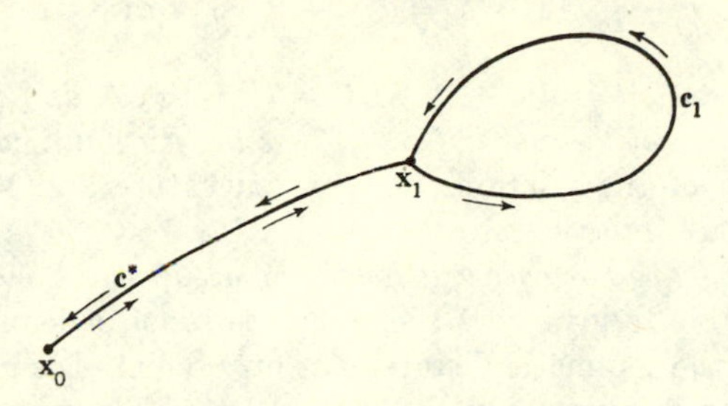

Fig. 13-1

along $\mathbf{c}^*$ destroy one another. The equation defines an isomorphism of $H(\mathbf{x}_1)$ into $H(\mathbf{x}_0)$ (Fig. 13-1). The same reasoning shows that there exists an isomorphism of $H(\mathbf{x}_0)$ into $H(\mathbf{x}_1)$ and that the composition of the two maps is the identity. Therefore $H(\mathbf{x}_0)$ is isomorphic to $H(\mathbf{x}_1)$.

The holonomy group vanishes in euclidean space. The "size" of H in a Riemann space is a certain measure of "non-euclidicity." The holonomy group does not vanish even for Klein geometries other than the euclidean. An example is given by the geometry of the sphere. It follows from (13-25) that the angle of two vectors defined by

$$\cos \angle(\mathbf{a},\mathbf{b}) = \frac{g_{ij}a^i b^j}{(a_i a^i)^{\frac{1}{2}}(b_i b^i)^{\frac{1}{2}}} \qquad (13\text{-}30)$$

is constant if both vectors are displaced by parallelism along a given curve. On the sphere (Fig. 13-2) we draw a triangle with three right angles. If the tangent vector to AB in A is moved first along AB, then along BC, and finally back along CA, it follows that $R(A \to B \to C \to A)$ is a rotation by $\pi/2$. (It is shown in the next section that tangents along great circles are parallel.)

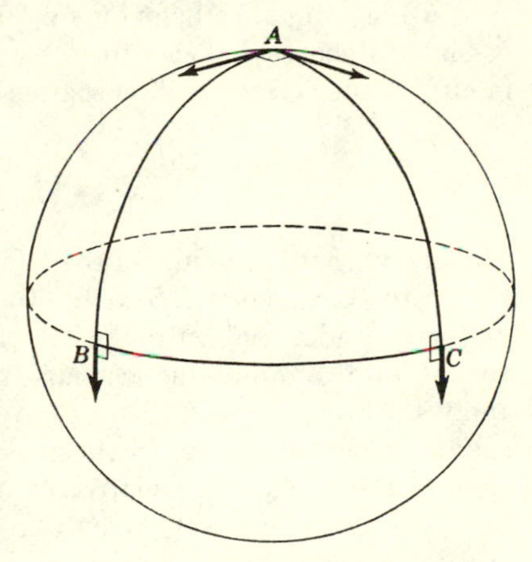

Fig. 13-2

To any point $\mathbf{c}(t)$ is attached a vector $\mathbf{v}(t) = \mathbf{v}(0) + \int_0^t \Gamma\mathbf{v}$ as a parallel translation of the vector $\mathbf{v}(0)$. The points $s\mathbf{c}(t)$ $(0 \leq t \leq 1, 0 \leq s \leq 1)$ define a two-dimensional subspace (not necessarily a local one) in V.

With the point $s\mathbf{c}(t)$ we associate the vector $\mathbf{v}(s,t) = s\mathbf{v}(t)$ in the $\mathbf{i}$ coordinate system of $T_{c(s,t)}$. The functions $v^i(s,t)$ are defined and are differentiable at all points of the subspace. By Stokes's theorem,

$$R(\mathbf{c})\mathbf{v} = U\mathbf{v} + \int_0^1 \Gamma(t)\mathbf{v}(t) = U\mathbf{v} + \int_0^1 \int_0^1 (d\Gamma - \Gamma \text{ ``}\cdot\text{'' } \Gamma)\mathbf{v}(s,t)$$

Definition 13-6. $\mathrm{P} = d\Gamma - \Gamma \text{ ``}\cdot\text{'' } \Gamma = (R_{i\cdot\cdot kl}^{\ j} dx^k \wedge dx^l)$ *is the Riemann (curvature) matrix of the metric G.* $R_{i\cdot\cdot kl}^{\ i}$ *is the Riemann curvature tensor.*

A discussion of the relations between curvature and holonomy rests on the following fundamental:

Lemma 13-7. *The homogeneous holonomy group is a Lie group.*

It is impossible to give here the complete proof of lemma 13-7. The reason is that the lemma is a statement on (global) Lie groups whereas we have at our disposal only the apparatus of Lie group germs. We know that $H(\mathbf{x}_0)$ is a subgroup of O_n. Each element of $H(\mathbf{x}_0)$ is obtained from a certain closed curve $\mathbf{c}$ attached to $\mathbf{x}_0$. For two such curves $\mathbf{c}(s)$ and $\mathbf{c}^*(s)$, $H(\mathbf{x}_0)$ contains also all elements belonging to the curves $\mathbf{c}_t(s)$ defined by (10-44). This shows that any two elements in $H(\mathbf{x}_0)$ can be connected by a continuous image of the unit interval $0 \leq t \leq 1$. One expresses this fact by saying that $H(\mathbf{x}_0)$ is *arcwise connected.* An arcwise-connected subgroup of a Lie group is a Lie group.†

As a Lie group, the holonomy group has a Lie algebra $\mathfrak{L}(H(\mathbf{x}_0))$. It is possible to choose the coordinates in such a way that the tangent plane in $\mathbf{c}(0)$ to the surface $s\mathbf{c}(t)$ is spanned by $\mathbf{i}_1$ and $\mathbf{i}_2$. Then

$$\left(\frac{dR(\mathbf{c}(s,t))}{ds}\right)_{s=0} = \mathrm{P}(\mathbf{i}_1 \wedge \mathbf{i}_2)$$

This shows that the values of the Riemann matrix on the bivectors $\mathbf{a} \wedge \mathbf{b}$ belong to the algebra $\mathfrak{L}(H(\mathbf{x}_0))$. But this reasoning does not prove that $\mathfrak{L}(H(\mathbf{x}_0))$ is identical with the set of these matrices, since the curves $\mathbf{c}$ do not form a finite-dimensional space. This makes it impossible to apply the reasonings of Chap. 6. Also it is possible that P vanishes at one point but not in the whole space; theorem 13-5 then shows that the values of the Riemann matrix cannot span $\mathfrak{L}(H(\mathbf{x}_0))$. The reason is that the homogeneous holonomy group is defined on all closed curves at $\mathbf{x}_0$, whereas in the definition of the Riemann matrix only curves in a small neighborhood of $\mathbf{x}_0$ enter.

If we take a nested sequence of neighborhoods $U_i(\mathbf{x}_0)$ $(i = 1,2, \ldots)$ whose intersection reduces to $\mathbf{x}_0$ and restrict our Riemann space to one defined only in $U_i(\mathbf{x}_0)$, we obtain for each one a holonomy group $H_{U_i}(\mathbf{x}_0)$.

† H. Yamabe, On an Arcwise Connected Subgroup of a Lie Group, *Osaka Math. J.*, **2:** 14–15 (1950).

From the hypothesis $U_{i+1} \subset U_i$, it follows that closed curves in U_{i+1} are also in U_i; hence $H_{U_{i+1}}(\mathbf{x}_0) \subset H_{U_i}(\mathbf{x}_0)$, and, by the same reason,

$$\mathfrak{L}(H_{U_{i+1}}(\mathbf{x}_0)) \subset \mathfrak{L}(H_{U_i}(\mathbf{x}_0))$$

The Lie algebras are finite-dimensional vector spaces. Their dimensions may decrease only a finite number of times. This means that there exists an index n such that $\mathfrak{L}(H_{U_{n+k}}(\mathbf{x}_0)) = \mathfrak{L}(H_{U_n}(\mathbf{x}_0))$. The same also must hold for the Lie groups themselves.

Definition 13-8. *The local holonomy group $H^*(\mathbf{x}_0)$ is the group of elements common to all homogeneous holonomy groups $H_{U(\mathbf{x}_0)}(\mathbf{x}_0)$.*

Then we have shown the following:

Theorem 13-9 (Nijenhuis). *For each $\mathbf{x}_0$ there exists $U(\mathbf{x}_0)$ such that $H^*(\mathbf{x}_0) = H_{U(\mathbf{x}_0)}(\mathbf{x}_0)$. $\mathfrak{L}(H^*(\mathbf{x}_0))$ is spanned by the Riemann matrices at $\mathbf{x}_0$.*

The rank of the matrices P need not be constant over the space; neither does the dimension of $\mathfrak{L}(H^*(\mathbf{x}_0))$. No theorem similar to 13-5 can hold for local holonomy groups which are important for local geometry. The determination of properties of $H(\mathbf{x}_0)$ from those of $H^*(\mathbf{x}_0)$ is one of the main problems of global differential geometry. (There are some additional complications in the definition of holonomy groups on manifolds.)

Although we have used $dd\mathbf{x} = 0$ to define ω and Γ in a unique fashion, we are not able to control the choice of the matrices to such an extent as to ensure a Poincaré relation $dd\mathbf{i}_k = 0$, since $dd\{\mathbf{i}_k\} = \mathrm{P}\{\mathbf{i}_k\}$. An intuitive interpretation may be given to this as the result of moving the frame by parallelism about an "infinitesimal parallelogram." The value of such interpretations is questionable (but see exercise 13-1, Prob. 5).

Again let $E(\mathbf{x})$ be a differentiable automorphism of the tangent spaces. Under the action of E, the $\mathbf{i}$ frame is transformed into $\{\mathbf{e}_\alpha\} = E\{\mathbf{i}_j\}$, Γ into $\omega = E\Gamma E^{-1} + C(E)$, and P into

$$\begin{aligned}\Omega &= \mathrm{P}(\omega) = \mathrm{P}(E\Gamma E^{-1} + C(E)) \\ &= E\Gamma\text{``}\cdot\text{''}\Gamma E^{-1} + E\Gamma E^{-1}\text{``}\cdot\text{''}C(E) + C(E)\text{``}\cdot\text{''}E\Gamma E^{-1} + C(E)\text{``}\cdot\text{''}C(E) \\ &\qquad - C(E)\text{``}\cdot\text{''}E\Gamma E^{-1} - E\,d\Gamma E^{-1} - E\Gamma E^{-1}\text{``}\cdot\text{''}C(E) - dC(E) \\ &= E\mathrm{P}(\Gamma)E^{-1} + \mathrm{P}(C(E))\end{aligned}$$

The tensor field character of the curvature matrices P and

$$\Omega = d\omega - \omega\text{``}\cdot\text{''}\omega = (\Omega_{\lambda\cdot\cdot\mu\nu}^{\ \ \kappa}\,\omega^\mu \wedge \omega^\nu)$$

that is, the formula

$$\Omega = E\mathrm{P}(\Gamma)E^{-1} \tag{13-31}$$

now follows from this fundamental statement:

Theorem 13-10. $\mathrm{P} = 0$ *in a neighborhood if and only if Γ is a Cartan matrix.*

If $\Gamma = dAA^{-1}$, then

$$d\Gamma - \Gamma``\cdot"\Gamma = dAA^{-1}``\cdot" dAA^{-1} - dAA^{-1}``\cdot" dAA^{-1} = 0$$

If $d\Gamma = \Gamma``\cdot"\Gamma$, we have to find a matrix $A$ such that $\Theta = dA - \Gamma A = 0$. Since $d\Theta = -d\Gamma A + \Gamma``\cdot" dA = \Gamma``\cdot"\Theta$, the system $\Theta_i{}^j = 0$ is totally integrable (theorem 10-25) and A may be found nonsingular in some neighborhood of any given point.

Theorem 13-10 and formula (13-31) have several important consequences.

If the matrices E are jacobians, (13-31) indicates

$$\begin{aligned} \mathrm{P}(\Gamma') &= R_{i'..k'l'}^{j'}\, dx^{k'} \wedge dx^{l'} \\ &= R_{i..kl}^{j}(J^{-1})_j^{j'} J_{i'}^{i}\, dx^k \wedge dx^l \\ &= R_{i..kl}^{j}(J^{-1})_j^{j'} J_{i'}^{i} J_{k'}^{k} J_{l'}^{l}\, dx^{k'} \wedge dx^{l'} \end{aligned}$$

Corollary 13-11. *The Riemann curvature* $R_{i..kl}^{j}$ *is a tensor field.*

If $\mathrm{P} = 0$, we know that $\Gamma = dAA^{-1}$. The frame $\{\mathbf{k}_\alpha\} = A^{-1}\{\mathbf{i}_j\}$ is constant over the whole space,

$$\{d\mathbf{k}_\alpha\} = -A^{-1}\Gamma\{\mathbf{i}_j\} + A^{-1}\Gamma\{\mathbf{i}_j\} = 0$$

Since $\|\mathbf{k}_1, \ldots, \mathbf{k}_n\| \neq 0$, the curves $\mathbf{x}_{i'}(t)$, $\mathbf{x}'_{i'} = \mathbf{k}_i$ define a system of parameter curves in which the metric $G' = (\mathbf{k}_i \circ \mathbf{k}_j)$ is constant; the Riemann space is a euclidean space in curvilinear coordinates.

Corollary 13-12. *A metric* $g_{ij}dx^i\, dx^j$ *can be brought into the euclidean form* $\delta_{i'j'}\, dx^{i'}\, dx^{j'} = \Sigma(dx^{i'})^2$ *if and only if its Riemann curvature vanishes.*

We want to restrict the notation $\Omega = d\omega - \omega``\cdot"\omega$ to the curvature expressed in *orthonormal* frames $\{\mathbf{e}_\alpha\}$. Ω is often simpler to use than P since by (13-7) it has the property

$$\Omega_{\lambda..\mu\nu}^{\kappa} + \Omega_{\kappa..\mu\nu}^{\lambda} = 0 \tag{13-32}$$

Exterior derivation of (13-8) gives

$$d(\omega^\beta \wedge \omega^\alpha{}_\beta) = \omega^\kappa \wedge \Omega_\kappa{}^\lambda = 0 \tag{13-33}$$

This identity, commonly named after G. Ricci-Curbastro, shows that

$$\Omega_{\kappa..\mu\nu}^{\lambda} + \Omega_{\mu..\nu\kappa}^{\lambda} + \Omega_{\nu..\kappa\mu}^{\lambda} = 0 \tag{13-34a}$$

Hence also, by (13-31),

$$R_{j..kl}^{i} + R_{k..lj}^{i} + R_{l..jk}^{i} = 0 \tag{13-34b}$$

Since from $\mathbf{i}_j = A_j^\alpha \mathbf{e}_\alpha$ it follows that $G = A^tA$ and

$$R_{j..kl}^{i} = A_j^\kappa (A^{-1})_\lambda^i A_k^\mu A_l^\nu \Omega_{\kappa..\mu\nu}^{\lambda}$$

(13-32) becomes

$$R_{kl,mn} + R_{lk,mn} = 0 \tag{13-35}$$

Another important identity (*Bianchi*'s identity) results from

$$dP = -d\Gamma``\cdot"\Gamma + \Gamma``\cdot" \, d\Gamma = P``\cdot"\Gamma - \Gamma``\cdot"P$$

that is,

$$DP = D\Omega = 0 \tag{13-36a}$$

or, explicitly,

$$R^{\;i}_{j..kl;m} + R^{\;i}_{j..lm;k} + R^{\;i}_{j..mk;l} = 0 \tag{13-36b}$$

Formula (13-13) and Def. 13-6 give a formula for the computation of the curvature tensor from the metric,

$$R_{ij,kl} = \frac{1}{2}\left(\frac{\partial^2 g_{jk}}{\partial x^i\,\partial x^l} + \frac{\partial^2 g_{il}}{\partial x^j\,\partial x^k} - \frac{\partial^2 g_{ik}}{\partial x^j\,\partial x^l} - \frac{\partial^2 g_{jl}}{\partial x^i\,\partial x^k}\right) + g^{st}([t,il][s,jk] - [t,ik][s,jl]) \tag{13-37}$$

from which the last identity for the curvature tensor

$$R_{ij,kl} = R_{kl,ij} \tag{13-38}$$

is verified by inspection.

The curvature tensor appears in the relations between the second covariant derivatives. For vectors,

$$v^i{}_{;j;k} = \frac{\partial^2 v^i}{\partial x^k\,\partial x^j} - \frac{\partial \Gamma^i{}_{js}}{\partial x^k} v_s - \Gamma^i{}_{js}\frac{\partial v^s}{\partial x^k} - \Gamma^i{}_{ks}\frac{\partial v^s}{\partial x^j} + \Gamma^i{}_{ks}\Gamma^s{}_{jt}v^t + \Gamma^s{}_{jk}\frac{\partial v^i}{\partial x^s} - \Gamma^s{}_{jk}\Gamma^i{}_{st}v^t$$

or

$$v^i{}_{;j;k} - v^i{}_{;k;j} = R^{\;i}_{t..jk}v^t \tag{13-39}$$

The second covariant derivatives are not symmetric unless the curvature tensor vanishes. Analogous relations for all other types of tensors may be derived from (13-39).

Since

$$R^{\;s}_{s..kl} = g^{st}R_{st,kl} = \sum_{s<t} g^{st}(R_{st,kl} + R_{ts,kl}) = 0$$

there exists only one type of nontrivial contracted curvature tensor, the *Ricci tensor* $R_{ik} = R^{\;j}_{i..kj}$. Equation (13-35) shows that the Ricci tensor is symmetric,

$$R_{ik} = R_{ki}$$

Another contraction yields the *curvature scalar* $R = R^{\;i}_{i.} = g^{rs}R_{rs} = R^{\;j\;i}_{i..j}$.

In the two-dimensional case, the metric tensor

$$G = \begin{pmatrix} g_{11} & g_{12} \\ g_{12} & g_{22} \end{pmatrix}$$

gives the contravariant metric

$$G^{-1} = \begin{pmatrix} g^{11} & g^{12} \\ g^{21} & g^{22} \end{pmatrix} = \begin{pmatrix} \dfrac{g_{22}}{\|G\|} & -\dfrac{g_{12}}{\|G\|} \\ -\dfrac{g_{12}}{\|G\|} & \dfrac{g_{11}}{\|G\|} \end{pmatrix}$$

The Christoffel symbols are

$$[1,11] = \frac{1}{2}\frac{\partial g_{11}}{\partial x^1} \qquad [1,12] = \frac{1}{2}\frac{\partial g_{11}}{\partial x^2} \qquad [1,22] = \frac{\partial g_{12}}{\partial x_2} - \frac{1}{2}\frac{\partial g_{22}}{\partial x^1}$$

$$[2,11] = \frac{\partial g_{12}}{\partial x^1} - \frac{1}{2}\frac{\partial g_{11}}{\partial x^2} \qquad [2,12] = \frac{1}{2}\frac{\partial g_{22}}{\partial x^1} \qquad [2,22] = \frac{1}{2}\frac{\partial g_{22}}{\partial x^2}$$

In the $\{\mathbf{e}_\alpha\}$ frame,

$$\omega = \begin{pmatrix} 0 & \omega_1{}^2 \\ -\omega_1{}^2 & 0 \end{pmatrix} \qquad \Omega = \begin{pmatrix} 0 & K\omega^1 \wedge \omega^2 \\ -K\omega^1 \wedge \omega^2 & 0 \end{pmatrix}$$

and since $\omega^1 \wedge \omega^2 = \|G\|\, dx^1 \wedge dx^2$, the only nonzero coordinate of the curvature tensor is

$$R_{12,12} = K\|G\| \tag{13-40}$$

For the curvature scalar one obtains

$$R = 2\|G^{-1}\|R_{12,12} = 2K \tag{13-41}$$

Exercise 13-1

1. A manifold is postulated to have a covering in which each neighborhood has a nonzero intersection only with a finite number of neighborhoods. Prove that every such manifold may be given the structure of a Riemann manifold. (HINT: Show first that each neighborhood U_α may be turned into a Riemann space with a metric $ds_\alpha{}^2$ which is zero outside $\bar{U}_\alpha$. Then define $ds^2 = \Sigma\, ds_\alpha{}^2$. Why does this sum always make sense?)
2. Show that a *constant* metric G is always isometric with the euclidean one U. (HINT: Consider G as a euclidean metric in an oblique system of coordinates.)
3. Prove (13-7) from (13-12).
4. Find the covariant coordinates of the contravariant vector $(0,r)$ for the metric $\begin{pmatrix} 1 & 0 \\ 0 & r^2 \end{pmatrix}$.
5. The *development* of a curve $\mathbf{x}(t)$ in R^n is the curve defined by $\mathbf{x}^*(0) = 0$, $\dot{\mathbf{x}}^*(0) = \mathbf{e}_0$, and the Cartan matrix $\omega(t)$. If $\mathbf{x}(t)$ is closed,

$\mathbf{x}(0) = \mathbf{x}(1) = \mathbf{x}_0$, the development $\mathbf{x}^*(t)$ will, in general, not be closed. But since $\omega(t)$ is skew-symmetric, there exists a euclidean movement which brings $\mathbf{x}^*(0)$ onto $\mathbf{x}^*(1)$ and the Frenet frame in $\mathbf{x}^*(0)$ onto the Frenet frame in $\mathbf{x}^*(1)$. All sufficiently differentiable closed curves at $\mathbf{x}_0$ define the *nonhomogeneous holonomy group* $\mathcal{H}(\mathbf{x}_0)$ as the group of these euclidean motions. Prove that $\mathcal{H}(\mathbf{x}_0)$ is isomorphic to any $\mathcal{H}(\mathbf{x}_1)$ and that it is a Lie subgroup of $O_n \times R_n$.

6. Find all groups that may be nonhomogeneous holonomy groups of two-dimensional Riemann spaces. (Use the last statement of Prob. 5.)

◆ **7.** Find all groups that may be (*a*) homogeneous and (*b*) nonhomogeneous holonomy groups of three-dimensional Riemann spaces. (Use the last statement of Prob. 5.)

8. Assume that the curvature matrix splits on the whole Riemann space: $\mathrm{P} = \begin{pmatrix} \mathrm{P}_1 & 0 \\ 0 & \mathrm{P}_2 \end{pmatrix}$, where 0 stands for rectangular zero matrices. What does this imply for the structure of the homogeneous holonomy group?

9. Define a product of closed curves at $\mathbf{x}_0$ such that its image in the holonomy group is the matrix product.

10. Give all the details of the proof that the map $H(\mathbf{x}_0) \to H(\mathbf{x}_1)$ defined on page 321 is linear.

◆**11.** Compute P for

(*a*) $G = \begin{pmatrix} 1 + (x^1)^2 & 1/\sqrt{2} & 0 \\ 1/\sqrt{2} & 1 + (x^2)^2 & 0 \\ 0 & 0 & 1 \end{pmatrix}$

(*b*) $G = \begin{pmatrix} 1 & 0 & 0 \\ 0 & (x^1)^2 & 0 \\ 0 & 0 & 1 \end{pmatrix}$

12. Compute Γ for a two-dimensional space. Obtain a formula for K as a function of the g_{ij}.

13. Find Γ for $ds^2 = dr^2 + r^2\,d\phi^2$. Why must we have $\mathrm{P} = 0$?

14. Show that $dd\mathbf{e}_\alpha = \Omega_\alpha{}^\beta \mathbf{e}_\beta$.

15. Use formula (13-37) to prove the Ricci and the Bianchi identities.

16. Use (13-39) to find an expression for $v_{i;j;k} - v_{i;k;j}$.

17. Compute $T^i_{.jk;l;m} - T^i_{.jk;m;l}$ as a function of the curvature tensor and $T^i_{.jk}$.

18. Prove that the tensor field $(\delta^i_{.j})$ is covariant constant (that is, $\delta^i_{.j;k} = 0$) for any metric.

19. Prove that

$$\Gamma^i{}_{ij} = \frac{1}{\|G\|} \frac{\partial \|G\|^{\frac{1}{2}}}{\partial x^j}$$

20. Prove that

$$R_{i.;j}^{\ j} = \frac{1}{2}\frac{\partial R}{\partial x^i}$$

21. Prove that

$$v^i{}_{;i} = \frac{1}{\sqrt{\|G\|}}\frac{\partial \sqrt{\|G\|}\, v^i}{\partial x^i}$$

22. Prove that $t^i{}_{;j} = g^{is}t_{s;j}$.

23. Prove that for any quantity in $\overset{p}{\wedge}\, T(V)$

$$\begin{aligned}(t^{ii_2\cdots i_p}{}_{;j}t^j{}_{i_2\ldots i_p})_{;i} &- (t^{ii_2\cdots i_p}{}_{;i}t^j{}_{i_2\ldots i_p})_{;j} \\ &= -t^{i_1\cdots i_p}g^{jk}t_{i_1\ldots i_p;j;k} + t^{ii_2\cdots i_p}{}_{;j}t^j{}_{i_2\ldots i_p;i} - t^{ii_2\cdots i_p}{}_{;i}\, t^j{}_{i_2\ldots i_p;j}\end{aligned}$$

(Yano).

24. A Riemann space is *recurrent* if $R_{ij,kl;m} = v_m R_{ij,kl}$, where (v_m) is a nonzero vector. Prove that in this case

$$R_{ij,kl}v_m + R_{ik,lm}v_j + R_{il,mj}v_k = 0$$

25. Prove that

$$G = \begin{pmatrix} 1 & 0 & 0 \\ 0 & g_{22} & g_{23} \\ 0 & g_{32} & g_{33} \end{pmatrix}$$

is a metric with recurrent curvature. (See Prob. 24.)

26. Prove that

$$\begin{aligned}R_{ij,kl;m;n} - R_{ij,kl;n;m} = R_{i..,mn}^{\ s}R_{sj,kl} &+ R_{j..,mn}^{\ s}R_{si,kl} \\ &+ R_{k..,mn}^{\ s}R_{ij,sl} + R_{l..,mn}^{\ s}R_{ij,sk}\end{aligned}$$

27. Let $H_{ijkl,mn}$ denote the right-hand side of the equation of Prob. 26. Prove that $R^s{}_{.j,kl;i;s} = R_{jl;k;i} - R_{jk;l;i} + H^s{}_{.jkl,si}$.

28. An *Einstein* space is a Riemann space for which

$$R_{ij} = \frac{R}{n}g_{ij}$$

Show that $R = \text{const}$ in an Einstein space of dimension $n \geq 3$. (Use the Bianchi identities.)

29. The *Einstein* tensor is $G_{ij} = R_{ij} - \frac{1}{2}Rg_{ij}$. Prove that $G^i_{.j;i} = 0$. (Use the Bianchi identities.)

30. Prove that $\delta^{abcd}_{ijkl}R_{ab,cd} = 0$. (The left-hand side is often referred to as $4!R_{[ij,kl]}$.)

31. Use the Bianchi identities to prove $R_{i..,kl;s}^{\ s} - 2\delta^{st}_{ik}R_{sl;t} = 0$.

32. Find the formula which expresses the second covariant exterior differential $DD\alpha$ of a k-form α in terms of the curvature and of α.

33. The star operator [formula (9-17)] in a Riemann space is

$$*(a_{i_1\ldots i_k}\,dx^{i_1}\wedge\cdots\wedge dx^{i_k}) = \sqrt{\|G\|}\,\delta^{1\ \cdots\ n}_{i_1\cdots i_n}a^{i_1\cdots i_k}\,dx^{i_{k+1}}\wedge\cdots\wedge dx^{i_n}$$

Prove that the star of a tensor field is again a tensor field and compute the *differential parameter* $\nabla(f,g) = *(*\,df\wedge dg)$ for two functions f and g on a two-dimensional Riemann space.

34. The *Laplacean* of a function $f(\mathbf{x})$ is $\Delta f = g^{ij}f_{;i;j}$. Show that $\Delta(v^iv_i) = 2(g^{jk}v^i{}_{;j;k}v_i + v^i{}_{;}{}^{j}v_{i;j})$.

35. Find the formula for

$$\Delta T^{i_1\cdots i_p}T_{i_1\ldots i_p}$$

analogous to the one of Prob. 34.

36. The *codifferential* of an exterior form $\alpha = \alpha_{i_1\ldots i_p}\,dx^{i_1}\wedge\cdots\wedge dx^{i_p}$ is $\delta\alpha = \alpha^{\ j}_{.i_2\cdots i_p;j}\,dx^{i_2}\wedge\cdots\wedge dx^{i_p}$.

(*a*) Compute explicitly $\delta\alpha$ for a one form and a three form in R^3.

(*b*) Show that the definition of Prob. 34 is identical with $\Delta = d\delta + \delta d$.

(*c*) Show that $\delta\delta\alpha = 0$.

37. Which of the developments of this chapter remain true for C^2 metrics?

38. Define

$$\Phi_k = \Sigma\delta^{\alpha_1\cdots\alpha_{n-1}}_{1\cdots n-1}\Omega^{\alpha_1}{}_{\alpha_2}\wedge\Omega^{\alpha_3}{}_{\alpha_4}\wedge\cdots\wedge\Omega^{\alpha_{2k-1}}{}_{\alpha_{2k}}\wedge\omega^{\alpha_{2k+1}}{}_n\wedge\cdots\wedge\omega^{\alpha_{n-1}}{}_n$$

$$\Psi_k = 2(k+1)\Sigma\delta^{\alpha_1\cdots\alpha_{n-1}}_{1\cdots n-1}\Omega^{\alpha_1}{}_{\alpha_2}\wedge\cdots\wedge\Omega^{\alpha_{2k-1}}{}_{\alpha_{2k}}\wedge\Omega^{\alpha_{2k+1}}{}_n\wedge\omega^{\alpha_{2k+2}}{}_n\wedge\cdots\wedge\omega^{\alpha_{n-1}}{}_n$$

$$\Psi_{-1} = 0$$

$$\Theta_k = \sum_{s=0}^{k}(-1)^{k-s}\frac{(2k+2)\cdots(2s+2)}{(n-2s-1)\cdots(n-2k-1)}\Phi_s$$

Show that $$d\Phi_k = \Psi_{k-1} + \frac{n-2k-1}{2(k+1)}\Psi_k$$

and $$d\Theta_{k-1} = \Phi_k \qquad n \text{ odd}$$
$$d\Theta_{k-1} = \Psi_{k-1} \qquad n \text{ even}$$

(Chern).

39. Show that the product defined in (13-28) is associative.

13-2. Geodesics

A straight line in euclidean geometry is characterized by the fact that its unit tangent is a constant vector. In riemannian geometry, the corresponding curves are the *geodesics* or *autoparallel curves* whose tangents $d\mathbf{x}/ds$ are all parallel to one another, i.e.,

$$\frac{\nabla \mathbf{x}'(s)}{ds} = 0$$

or

$$\frac{d^2x^i}{ds^2} + \Gamma^i{}_{jk}\frac{dx^j}{ds}\frac{dx^k}{ds} = 0 \tag{13-42}$$

where s is the arc length of the geodesic. The matrix Γ is C^2. The existence theorems on ordinary differential equations show that (13-42) has a unique solution, in some neighborhood, for any initial conditions $\mathbf{x}(0) = \mathbf{x}_0$, $\mathbf{x}'(0) = \mathbf{a}$.

Geodesics may serve to define several families of invariant frames.

Any point P in some neighborhood of a point P_0 is on a unique geodesic through P_0. At P_0 we choose a frame $\{\mathbf{n}_i(0)\}$ of orthonormal vectors. Let $\mathbf{a}(P) = a^i\mathbf{n}_i(0)$ be the unit tangent at P_0 to the geodesic P_0P, and let $s(P)$ be the arc length P_0P measured on that geodesic. Then

$$x^i = s(P)a^i \tag{13-43}$$

are the *normal coordinates* of P. P_0 is the origin of the system of normal coordinates whose **i** frame is $\{\mathbf{n}_i(P)\}$. In this frame,

$$g_{ij}(P_0) = \delta_{ij} \tag{13-44}$$

The geodesics issuing from P_0 have equations

$$x^i(s) = a^is \tag{13-45}$$

Hence, by (13-42),

$$\Gamma^i{}_{jk}a^ja^k = 0 \tag{13-46}$$

along any such geodesic. At P_0, the equation must hold for an *arbitrary* vector $\mathbf{a}$. Hence

$$\Gamma^i{}_{jk}(P_0) = 0 \tag{13-47}$$

Normal coordinates in n dimensions depend on $n + 1$ parameters $a^1, \ldots, a^n, s$.

As a first example of the application of normal coordinates, we prove the fundamental existence theorem of riemannian geometry.

Theorem 13-13. *The metric tensor field g_{ij} is uniquely defined if the Riemann curvature tensor is given as a function of normal coordinates of origin P_0 and if at each point P an orthonormal frame in T_P is given which is assigned to be the frame obtained from $\{\mathbf{n}_i(P_0)\}$ by parallel transport along the geodesic PP_0.*

Let $\{\mathbf{m}_i(P)\}$ be the orthonormal frame assigned in T_P. This means that along the geodesic P_0P

$$dP = a^\iota\mathbf{m}_\iota\, ds \qquad d\mathbf{m}_\iota = 0$$

by (13-47). If we take this family of orthonormal frames as **e** frames, the corresponding linear forms may be written as

$$\omega^\iota(s,\mathbf{a};ds,d\mathbf{a}) = a^\iota\, ds + \omega^{*\iota}(s,\mathbf{a};d\mathbf{a})$$

and, by (13-8),

$$\omega^\kappa{}_\iota(s,\mathbf{a};ds,d\mathbf{a}) = \omega^{*\kappa}{}_\iota(s,\mathbf{a};d\mathbf{a})$$

Equation (13-8) becomes

$$da^\iota \wedge ds + ds \wedge \frac{\partial \omega^{*\iota}}{\partial s} - a^\kappa\, ds \wedge \omega^{*\iota}{}_\kappa = \omega^{*\kappa} \wedge \omega^{*\iota}{}_\kappa$$

Both sides of this equation must vanish since the left-hand side depends on ds, whereas the right-hand side does not. Hence

$$\frac{\partial \omega^{*\iota}}{\partial s} = da^\iota + a^\kappa \omega^{*\iota}{}_\kappa$$

In the same way we obtain from the definition of the curvature matrix that

$$\frac{\partial \omega^{*\iota}{}_\kappa}{\partial s} = \Omega_{\kappa\cdot,\lambda\mu}^{\cdot\iota} a^\lambda \omega^{*\mu}$$

Since, by definition,

$$\frac{\partial}{\partial s}\, da^\iota = 0$$

we see that we may take the a^ι, da^ι as fixed parameters. The differential forms $\omega^{*\iota}$ are the unique solution of the system of linear differential equations

$$\frac{d^2\omega^{*\iota}}{ds^2} = \Omega_{\kappa\cdot,\lambda\mu}^{\cdot\iota} a^\kappa a^\lambda \omega^{*\mu} \qquad \omega^{*\iota}(0) = 0 \qquad \frac{d\omega^{*\iota}}{ds}(0) = da^\iota$$

Therefore, $ds^2 = \Sigma(\omega^\iota)^2$ is uniquely determined in some neighborhood of P_0.

Another important system of coordinates may be obtained as follows: A local hypersurface defines in the tangent space T_p of any of its points **p** an $(n-1)$-dimensional subspace T' spanned by the curves in the local hypersurface. In the euclidean geometry of T_p there exists a unique normal to T'. This normal we call *the normal to the hypersurface* at **p**. We fix a local hypersurface and denote by x^1 the arc length on any geodesic normal to the hypersurface, i.e., any geodesic whose class in T_p is the normal to the hypersurface at **p**. Let $x^2, \ldots, x^n$ be differentiable coordinates on the local hypersurface $x^1 = 0$. A geodesic normal is defined by the coordinates $(x^2, \ldots, x^n)$ of its point of intersection with the given hypersurface; the $x^1, \ldots, x^n$ are differentiable coordinates in some neighborhood of the local hypersurface. In this *geodesic cartesian* system, $g_{11}(x^1, \ldots, x^n) = 1$. If the Christoffel symbols in (13-42) are replaced by their expressions (13-13) in terms of the metric, the equations of the geodesics become

$$\frac{d}{ds}\left(g_{il}\frac{dx^i}{ds}\right) - \frac{1}{2}\frac{\partial g_{jk}}{\partial x^l}\frac{dx^j}{ds}\frac{dx^k}{ds} = 0 \tag{13-48}$$

The curve $(s,0, \ldots, 0)$ is a geodesic only if

$$\frac{\partial g_{1k}}{\partial x^1} = 0$$

By hypothesis, this geodesic is orthogonal to the hypersurface $x^1 = 0$, that is, $g_{1k}(0,x^2, \ldots, x^n) = 0$. The above differential equation shows that $g_{1k}(x^1, \ldots, x^n) = 0$ always; the metric is of the form

$$ds^2 = (dx^1)^2 + \sum_{r,t=2}^{n} g_{rt}\, dx^r\, dx^t \tag{13-49}$$

As in Sec. 11-1, it follows from (13-49) that a geodesic is a curve of minimal length as long as it can appear as a curve $(s,0, \ldots, 0)$ in a geodesic cartesian system.

Theorem 13-14. *A geodesic is a curve of minimal length as long as it can be imbedded in a field of geodesic normals to some local hypersurface.*

Two vectors $\mathbf{a}_1$ and $\mathbf{a}_2$ in the tangent space T_0 of a point P_0 define a local surface, which is the locus of the geodesics tangent at P_0 to the plane spanned by $\mathbf{a}_1$ and $\mathbf{a}_2$. We introduce a system of *normal* coordinates (x^i) with the origin at P_0. In these coordinates

$$\mathbf{a}_A = a^i_A \mathbf{n}_i(0) \qquad A = 1, 2$$

The plane spanned by $\mathbf{a}_1$ and $\mathbf{a}_2$ is the set of vectors

$$\boldsymbol{\xi} = \xi^1\mathbf{a}_1 + \xi^2\mathbf{a}_2 = \xi^A\mathbf{a}_A$$

The Riemann metric of the space induces an inner metric of the surface by

$$\tilde{g}_{AB}\, d\xi^A\, d\xi^B = g_{ij}\frac{\partial x^i}{\partial \xi^A}\frac{\partial x^j}{\partial \xi^B}\, d\xi^A\, d\xi^B$$

that is,

$$\tilde{g}_{AB} = g_{ij}a^i_A a^j_B \tag{13-50}$$

A simple computation shows that

$$\|\tilde{G}\| = (g_{ij}g_{kl} - g_{ik}g_{jl})a^i_1 a^j_1 a^k_2 a^l_2$$

Equation (13-47) implies that also

$$\tilde{\Gamma}_{A,BC}(P_0) = \Gamma_{i,jk}a^i_A a^j_B a^k_C = 0$$

Hence

$$\tilde{R}_{12,12}(P_0) = \left(\frac{\partial \Gamma_{i,jk}}{\partial x^l} - \frac{\partial \Gamma_{i,jl}}{\partial x^k}\right) a_2^i a_1^j a_1^k a_2^l = R_{ji,kl}(P_0) a_2^i a_1^j a_1^k a_2^l$$

By (13-40), the Gauss curvature of the surface of geodesics is

$$K = \frac{\tilde{R}_{12,12}(P_0)}{\|\tilde{G}(P_0)\|} = \frac{R_{ij,kl} a_1^i a_2^j a_1^k a_2^l}{(g_{ik}g_{jl} - g_{il}g_{jk}) a_1^i a_2^j a_1^k a_2^l} \tag{13-51}$$

Definition 13-15. *$K = K(P_0, \mathbf{a}_1 \wedge \mathbf{a}_2)$ is the Riemann curvature at P_0 in the direction of the bivector $\mathbf{a}_1 \wedge \mathbf{a}_2$.*

Formula (13-51) shows that K indeed depends only on the bivector $\mathbf{a}_1 \wedge \mathbf{a}_2$ since both numerator and denominator are bilinear alternating functions of the vectors. The necessary and sufficient condition that K be the same for *all* directions at P_0 is that there exists a function $K(P_0)$ such that

$$R_{ij,kl} = K(g_{ik}g_{jl} - g_{il}g_{jk}) \tag{13-52}$$

In this case we say that K is *constant at P_0*. If (13-52) holds in some neighborhood of P_0, Bianchi's identity (13-36) is

$$K_{;m}(g_{ik}g_{jl} - g_{il}g_{jk}) + K_{;l}(g_{im}g_{jk} - g_{ik}g_{jm}) + K_{;k}(g_{il}g_{jm} - g_{im}g_{jl}) = 0$$

Contracting with g^{ik}, we obtain, because of $g^{ik}g_{il}g_{jk} = 2g_{jl}$,

$$(n - 2)(K_{;m}g_{jl} - K_{;l}g_{jk}) = 0$$

By the hypothesis $\|G\| \neq 0$, no two columns of the matrix G may be proportional. Therefore, if $n \neq 2$, we deduce that $K_{;m} = \partial K/\partial x^m = 0$ for all m, K constant.

Theorem 13-16 (F. Schur). *If the Riemann curvature of a space of dimension $n > 2$ is constant at all points, it is a constant.*

If K is a constant, the space is said to be of *constant curvature.* The following corollary is immediate:

Corollary 13-17. *If in a Riemann space there exists a group of motions that is transitive on every tangent space T_P, the space is of constant curvature.*

One may ask in what measure the geometry of geodesics determines the metric. To put it in a more definite way: Is it possible that two different metrics $ds^2 = g_{ij}\, dx^i\, dx^j$ and $d\hat{s}^2 = \hat{g}_{ij}\, dx^i\, dx^j$ have identical solutions of the respective equations (13-42)

$$\frac{d^2x^i}{ds^2} + \Gamma^i{}_{jk} \frac{dx^j}{ds}\frac{dx^k}{ds} = 0$$

$$\frac{d^2x^i}{d\hat{s}^2} + \hat{\Gamma}^i{}_{jk} \frac{dx^j}{d\hat{s}}\frac{dx^k}{d\hat{s}} = 0$$

that is, that

$$(\Gamma^i{}_{jk} - \hat{\Gamma}^i{}_{jk}) \frac{dx^j}{ds} \frac{dx^k}{ds} = \frac{d^2 s}{d\hat{s}^2} \left(\frac{d\hat{s}}{ds}\right)^2 \frac{dx^i}{ds}$$

for *all* vectors $d\mathbf{x}/ds$? Obviously the quantity $\Delta^i{}_{jk} = \Gamma^i{}_{jk} - \hat{\Gamma}^i{}_{jk}$ has the property that for all vectors $\mathbf{a}$

$$\Delta^i{}_{jk} a^j a^k = c(\mathbf{a}) a^i \qquad \Delta^i{}_{jk} = \Delta^i{}_{kj}$$

where $c(\mathbf{a})$ is a real-valued function of $\mathbf{a}$. If we take the special vector $a^i = \delta^i_j + \delta^i_k$ (for fixed j,k), it follows that

$$\Delta^i{}_{jk} = \delta^i_j \phi_k + \delta^i_k \phi_j \tag{13-53}$$

where $\{\phi_j\}$ is an arbitrary vector. If (13-53) holds, then

$$\Delta^i{}_{jk} a^j a^k = \delta^i_j \phi_k a^j a^k + \delta^i_k \phi_j a^j a^k = 2(\phi_j a^j) a^i$$

Therefore, two matrices Γ and $\hat{\Gamma}$ define the same geodesics if and only if

$$\Gamma = \hat{\Gamma} + (\delta^i_j \phi_k \, dx^k + \phi_j \, dx^i) \tag{13-54}$$

For the curvature matrix it follows that

$$\mathrm{P} = \hat{\mathrm{P}} + (\phi_{j;k} \, dx^k \wedge dx^i + \delta^i_j \phi_{l;k} \, dx^k \wedge dx^l - \phi_j \phi_k \, dx^k \wedge dx^i)$$

or

$$R^{\;i}_{j.,kl} = \hat{R}^{\;i}_{j.,kl} + \delta^i_j(\phi_{lk} - \phi_{kl}) + \delta^i_l \phi_{jk} - \delta^i_k \phi_{jl} \tag{13-55}$$

where we have put

$$\phi_{jk} = \phi_{j;k} - \phi_j \phi_k \tag{13-56}$$

It is possible to compute the ϕ_{jk} as functions of the curvature matrices. The relation between the curvature tensors gives, by contraction,

$$R_{jk} = \hat{R}_{jk} - \phi_{kj} + n\phi_{jk}$$

The symmetry of the Ricci tensor implies the symmetry of the ϕ_{jk}; hence

$$\phi_{jk} = \frac{1}{n-1} (R_{jk} - \hat{R}_{jk}) \tag{13-57}$$

In Eq. (13-55) the ϕ_{jk} may now be replaced by their values (13-57). The result is

$$R^{\;i}_{j.,kl} + \frac{1}{n-1} (\delta^i_k R_{jl} - \delta^i_l R_{jk}) = \hat{R}^{\;i}_{j.,kl} + \frac{1}{n-1} (\delta^i_k \hat{R}_{jl} - \delta^i_l \hat{R}_{jk})$$

Theorem 13-18 (H. Weyl). *Two metrics G and $\hat{G}$ may define the same geodesics only if they have the same Weyl tensor*

$$W^{\;i}_{j.,kl} = R^{\;i}_{j.,kl} + \frac{1}{n-1} (\delta^i_k R_{jl} - \delta^i_l R_{jk})$$

Of special interest are the metrics whose geodesics are straight. These metrics have the same Weyl tensor as the euclidean metric with vanishing curvature; they satisfy

$$(n-1)R^{i}_{j..kl} = \delta^i_l R_{jk} - \delta^i_k R_{jl}$$

or

$$(n-1)R_{ji,kl} = g_{il}R_{jk} - g_{ik}R_{jl} \tag{13-58}$$

For $j = i$ it follows that $g_{il}R_{ik} - R_{il}g_{ik} = 0$; hence

$$\frac{R_{ik}}{g_{ik}} = \frac{R_{il}}{g_{il}} = \text{const}$$

and

$$R_{ji,kl} = \frac{c}{n-1}(g_{il}g_{jk} - g_{ik}g_{jl})$$

The space is of constant curvature. On the other hand, if a metric is of constant curvature, (13-58) holds and the pfaffian system derived from (13-56) and (13-57),

$$\theta_l = d\phi_l - \frac{1}{n-1}R_{kl}\,dx^k - \phi_l\phi_k\,dx^k = 0$$

is totally integrable since $d\theta_l = \phi_l\theta_m \wedge dx^m$. For $n = 2$, the Weyl tensor vanishes identically.

Theorem 13-19 (Schläfli). *A Riemann space of dimension $n \geq 3$ admits a system of coordinates in which the geodesics are straight lines if and only if it is of constant curvature.*

Exercise 13-2

1. Show that a curve is a geodesic if and only if its development (exercise 13-1, Prob. 5) is a straight line.
2. Show that the jacobian of a given coordinate system and the normal (or the geodesic cartesian) coordinates is really a nonsingular matrix in some neighborhood.
3. Define geodesic polar coordinates in analogy to the developments of Sec. 11-1. Discuss their connection with geodesic cartesian coordinates. Define conjugate points in n dimensions.
4. Given a geodesic $\mathbf{x}(s)$, the coordinate x^1 will be the arc length along that geodesic. The other coordinates will be the arc lengths of geodesics defined by the remaining $n-1$ vectors of an orthonormal frame moved along $\mathbf{x}(s)$ by parallelism. Each point in some neighborhood of $\mathbf{x}$ will be in a unique hypersurface $x^1 = c$ generated by the geodesics trough $(c,0, \ldots ,0)$ and normal to the geodesic $\mathbf{x}$. Therefore this system of *Fermi coordinates* is well defined. Show that $\Gamma(x^1,0, \ldots ,0) = 0$. Find the form of the matrix $G(x^1,0, \ldots ,0)$ (Fermi).

◆ **5.** Find the equation of the geodesics in a parameter t which is not the arc length.

6. If the equation of the geodesics is of the form (13-42), s is called an *affine parameter.* The discussion of the Weyl tensor shows that there exist affine parameters other than the arc length. Show that the geodesics and the assignment of some affine parameter as arc length uniquely define the connection matrix Γ (Thomas).

◆ **7.** Find the geodesics for the metric

$$\frac{1}{1+(x^1)^2+(x^2)^2+(x^3)^2}\begin{pmatrix} 1+(x^2)^2+(x^3)^2 & -x^1x^2 & -x^1x^3 \\ -x^2x^1 & 1+(x^1)^2+(x^3)^2 & -x^2x^3 \\ -x^3x^1 & -x^3x^2 & 1+(x^1)^2+(x^2)^2 \end{pmatrix}$$

(See Prob. 10.)

8. Use normal coordinates to derive the symmetry properties of the curvature tensor from the defining equation (13-37).

9. Find the ds^2 of a space of constant curvature in normal coordinates.

10. Find the equations of the geodesics in covariant coordinates.

11. Let $\{\mathbf{e}_\alpha\}$ be an orthonormal frame and let K_i be the Riemann curvature in the direction $\mathbf{e}_1 \wedge \mathbf{e}_i$. Express K_i in terms of $\Omega_{\alpha\beta}$, and of R_{ij}. The *point curvature* is the mean of the Riemann curvatures of the $\frac{1}{2}n(n-1)$ bivectors in a frame $\{\mathbf{e}_\alpha\}$. Prove that the point curvature is a function of R and is independent of the frame.

12. Show that $\Delta^i{}_{jk}$ is a tensor field [use (13-15)] and hence that ϕ_i must be a tensor field in any change of coordinates.

13. Compute $W_{ji,kl}$ for the metrics of exercise 13-1, Prob. 11.

14. Prove that the "projective connection"

$$\overset{p}{\Gamma}{}^i{}_{jk} = \Gamma^i{}_{jk} - \left(\frac{1}{n}+1\right)(\delta^i_k\Gamma^l{}_{jl} + \delta^i_j\Gamma^l{}_{kl})$$

is invariant in any change of connection $\Gamma \to \hat{\Gamma}$ [(13-54)]. Find the transformation law of $\overset{p}{\Gamma}{}^i{}_{jk}$ under a linear map of the tangent spaces (Thomas).

15. Find $\overset{p}{\Gamma}{}^k{}_{jk}$ (Prob. 14) as a function of the $\Gamma^k{}_{jk}$.

16. Formulate theorem 13-18 as a statement on the existence of maps of Riemann spaces which carry geodesics onto geodesics.

17. Find the integrability conditions for the system $\theta_l = 0$ (page 335) for spaces of nonconstant curvature in order to obtain a sufficient condition in theorem 13-18.

18. Prove that a two-dimensional space admits locally straight geodesics only if it is of constant Gauss curvature.

♦**19.** Suppose $n > 3$. Two Riemann metrics G and $\hat{G}$ are said to be *conformally equivalent* if $G = e^{2\phi}\hat{G}$. Show that

$$\Gamma^i{}_{jk} = \hat{\Gamma}^i{}_{jk} + \delta^i_j\phi_{;k} + \delta^i_k\phi_{;j} - g_{jk}\phi_{;}{}^i$$

and

$$R^{\ \ i}_{j..kl} = \hat{R}^{\ \ i}_{j..kl} + \delta^i_l\phi_{jk} - \delta^i_k\phi_{jl} + g^{is}(g_{jk}\phi_{sl} - g_{jl}\phi_{sk}) + (\delta^i_l g_{jk} - \delta^i_k g_{jl})g^{rs}\phi_{;r}\phi_{;s}$$

where $\phi_{jk} = \phi_{;j;k} - \phi_{;j}\phi_{;k}$.

Deduce from these results that two metrics can be conformally equivalent only if they define the same *conformal Weyl tensor*

$$C_{ji,kl} = R_{ji,kl} - \frac{1}{n-2}(g_{ik}R_{jl} - g_{il}R_{jk} + g_{jl}R_{ik} - g_{jk}R_{il}) - \frac{R}{(n-1)(n-2)}(g_{il}g_{jk} - g_{ik}g_{jl})$$

What are the conformally euclidean metrics (Weyl)?

13-3. Subspaces

If a local variety V^M of dimension M is given in a Riemann space V^n of dimension $n = M + \mu$, the metric in the ambient space induces one in the subspace and turns V^M also into a Riemann space.

Indices running from 1 to M will be denoted by capitals, those from $M + 1$ to n by Greek letters, and those from 1 to n by lower case. If $u^1, \ldots, u^M$ are local coordinates in V^M and $x^1, \ldots, x^n$ coordinates in V^n, the local subspace V^M may be given by a set of equations

$$x^i = x^i(u^1, \ldots, u^M)$$

The metric induced in V^M by that of V^n is

$$\mathfrak{g}_{AB} = g_{ij}\frac{\partial x^i}{\partial u^A}\frac{\partial x^j}{\partial u^B} \tag{13-59}$$

The tangent space to V^n at a point P is a vector space of classes of curves through P. The curves in V^M define a subspace of the tangent space which may be identified with the tangent space to V^M at P. The most simple treatment of subspaces is through a special $\mathbf{e}$ frame in which the vectors $\mathbf{e}_A$ form an orthonormal frame in the tangent space to V^M and the remaining unit vectors $\mathbf{e}_\alpha$ are orthogonal in the tangent space of V^M to that of V^n. (In the same way a frame $\mathbf{i}_j$ would be suitable if V^M is given locally by a system of equations $x^\alpha = 0$.) The Frenet equations of the geometry of V^n are

$$\begin{aligned} d\mathbf{p} &= \omega^A \mathbf{e}_A + \omega^\alpha \mathbf{e}_\alpha \\ d\mathbf{e}_A &= \omega^B{}_A \mathbf{e}_B + \omega^\alpha{}_A \mathbf{e}_\alpha \\ d\mathbf{e}_\alpha &= \omega^B{}_\alpha \mathbf{e}_B + \omega^\beta{}_\alpha \mathbf{e}_\beta \end{aligned} \tag{13-60}$$

The geometry of V^M is obtained from these equations and

$$\omega^\alpha = 0$$

Equations (13-8) and Def. 13-6 split into

$$\begin{aligned} d\omega^A &= \omega^B \wedge \omega^A{}_B \\ 0 &= \omega^B \wedge \omega^\alpha{}_B \\ d\omega^B{}_A &= \omega^C{}_A \wedge \omega^B{}_C + \omega^\gamma{}_A \wedge \omega^B{}_\gamma + \Omega_A{}^B \\ d\omega^\beta{}_A &= \omega^C{}_A \wedge \omega^\beta{}_C + \omega^\gamma{}_A \wedge \omega^\beta{}_\gamma + \Omega_A{}^\beta \\ d\omega^B{}_\alpha &= \omega^C{}_\alpha \wedge \omega^B{}_C + \omega^\gamma{}_\alpha \wedge \omega^B{}_\gamma + \Omega_\alpha{}^B \\ d\omega^\beta{}_\alpha &= \omega^C{}_\alpha \wedge \omega^\beta{}_C + \omega^\gamma{}_\alpha \wedge \omega^\beta{}_\gamma + \Omega_\alpha{}^\beta \end{aligned} \tag{13-61}$$

The second equation of this set shows by Cartan's theorem that

$$\gamma^\beta{}_{AB} = \gamma^\beta{}_{BA}$$

The remaining equations and those obtained from them by exterior differentiation contain the whole geometry of subspaces of Riemann spaces. In this theory, only coordinate transformations are admitted that at any point $\mathbf{p}$ induce linear maps of the tangent space $T[V^n]$ onto itself in which both the tangent space $T[V^M]$ to V^M and its orthogonal complement are invariant. The matrix of such a transformation in the $\mathbf{e}$ frame is

$$E = \begin{pmatrix} a_A{}^B & 0 \\ 0 & a_\alpha{}^\beta \end{pmatrix}$$

$C(E)$ splits in the same way into two nonzero and two zero matrices; the formula (13-15) [i.e., (1-5)] shows that $\gamma^\alpha{}_{AB}$ and $\gamma^A{}_{\alpha B}$ behave like tensor fields in this geometry.

A *hypersurface* ($M = n - 1$) has a unique normal $\mathbf{e}_n$ at each point. Hence we may define its *second fundamental form*

$$\text{II} = -d\mathbf{e}_n \cdot d\mathbf{p} = \omega^A \omega^n{}_A = \gamma^n{}_{AB} \omega^A \omega^B$$

II is a symmetric quadratic form; it is possible to choose the frame $\{\mathbf{e}_A\}$ so as to bring it into diagonal form:

$$\text{II} = \Sigma k_A (\omega^A)^2 \tag{13-62}$$

The k_A are the *principal curvatures* of the hypersurface. Curvature lines and asymptotic lines may be defined just as in two dimensions. A frame for which (13-62) holds is a *principal frame*, characterized by $\gamma^n{}_{AB} = 0$ for $A \neq B$. Along the curvature line in direction $\mathbf{e}_A$,

$$d\mathbf{e}_n = k_A \omega^A \mathbf{e}_A \quad \text{no summation}$$

This is a generalization of Rodrigues's formula: *A curve on a hypersurface is a curvature line if and only if* $d\mathbf{e}_n = \lambda \mathbf{t}\, ds$.

As a last application, we study the theory of curves. Since the $\mathbf{e}$ frames are orthonormal, we may attach to a curve $\mathbf{x}(s)$ in V^n a family of frames for which the Frenet equation is

$$d\{\mathbf{e}_i\} = \begin{pmatrix} 0 & k_1 & 0 & 0 & \cdot & 0 \\ -k_1 & 0 & k_2 & 0 & \cdot & 0 \\ \cdots & \cdots & \cdots & \cdots & \cdots & \cdots \\ 0 & 0 & 0 & \cdot & -k_{n-1} & 0 \end{pmatrix} \{\mathbf{e}_i\} \qquad (13\text{-}63)$$

[if we are ready to make the necessary differentiability assumptions for $\mathbf{x}(s)$]. If $\mathbf{x}(s)$ is a curve in V^M, we are no longer in a position freely to choose the frame $\mathbf{e}_i$. A frame generalizing the tangent normal one must be defined in this case. Starting from $\mathbf{e}_1 = d\mathbf{x}/ds$, we fix $\mathbf{e}_2$ and $\mathbf{e}_{M+1}$ by

$$\frac{d\mathbf{e}_1}{ds} = k_{1g}\mathbf{e}_2 + k_{1e}\mathbf{e}_{M+1} \qquad \mathbf{e}_2 \in T(V^M)$$

k_{1g} is the *first geodesic curvature* and k_{1e} the *first enforced curvature.* The Cartan matrix of this frame appears as shown.

$$\left(\begin{array}{ccccc:cccc} 0 & k_{1g} & 0 & \cdots & 0 & k_{1e} & 0 & \cdots & 0 \\ -k_{1g} & & & & & & & & \\ 0 & & a_{AB} & & & & a_{A\beta} & & \\ \vdots & & & & & & & & \\ 0 & & & & & & & & \\ \hdashline -k_{1e} & & & & & & & & \\ 0 & & & & & & & & \\ \vdots & & a_{\alpha B} & & & & a_{\alpha\beta} & & \\ 0 & & & & & & & & \end{array}\right)$$

Proceeding now with

$$\frac{d\mathbf{e}_2}{ds} = -k_{1g}\mathbf{e}_1 + k_{2g}\mathbf{e}_3 + a_2\mathbf{e}_{M+1} + k_{2e}\mathbf{e}_{M+2}$$

we see that there exists a unique frame for which the Cartan matrix is as follows:

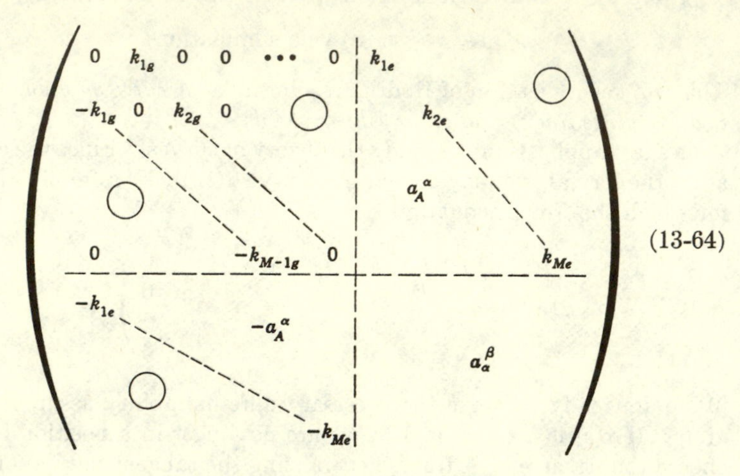

A geodesic in V^n [$k_1 = 0$ in (13-63)] is also a geodesic in V^M [$k_{1g} = 0$]. On the other hand, a geodesic in V^M is a geodesic in V^n only if $k_{1e} = 0$. V^M is called *totally geodesic* if all geodesics of V^M are also geodesics of V^n, that is, if k_{1e} vanishes identically on V^M. A comparison of (13-64) and (13-62) shows that this is possible only if

$$\gamma^{\alpha}{}_{AB} = 0$$

for all A, B, and α. This condition obviously is also sufficient for V^M to be totally geodesic in V^n. Since the $\gamma^{\alpha}{}_{AB}$ is a tensor field in the geometry of V^M, it is, in general, impossible to find a totally geodesic subspace of a V^n.

Exercise 13-3

1. Prove that a subspace V^M of a euclidean space R^n has a curvature matrix $\Omega = (\omega_A{}^{\alpha} \wedge \omega_{\alpha}{}^{B})$.
2. Define $T_r = \Sigma\Omega_{A_1}{}^{A_2} \wedge \Omega_{A_2}{}^{A_3} \wedge \cdots \wedge \Omega_{A_r}{}^{A_1}$, the sum taken over all combinations $A_1 \neq A_2 \neq \cdots \neq A_r$. If V^M is a subspace of R^{M+2}, show that $2T_{2k} = (-1)^{k-1} \overset{k}{\wedge} T_2$. (Use Prob. 1.)
3. Show that any given tensor $R_{ij,kl}$ $(1 \leq i,j,k,l \leq n)$ with the symmetry properties of the Riemann curvature tensor may be realized as the curvature tensor of a subspace V^n of $R^{n(n-1)/2}$ (Cartan). (Use Prob. 1.)
4. If V^{n-1} is a hypersurface in R^n $(n > 3)$, show that the curvature (Prob. 1) uniquely defines the tensor $(\gamma^n{}_{AB})$ if the rank of this matrix is ≥ 3. Deduce from this result that a hypersurface in R^n is rigid

(has no isometric mapping which is not a congruence) if it has more than two nonvanishing principal curvatures (Killing).

5. An *umbilic* is a point of a hypersurface at which all directions are principal. Show that a point is an umbilic if and only if at it

$$\gamma^n{}_{AB} = c\mathfrak{g}_{AB}$$

6. For a hypersurface V^N in R^{N+1} define

$$A_r = \text{“}\|\text{”}\ \mathbf{x},\mathbf{e}_{N+1},\underbrace{d\mathbf{e}_{N+1},\ .\ .\ .\ ,d\mathbf{e}_{N+1}}_{N-1-r\text{ times}},\underbrace{d\mathbf{x},\ .\ .\ .\ ,d\mathbf{x}}_{r\text{ times}}\ \text{“}\|\text{”}$$

$$C_r = \text{“}\|\text{”}\ \mathbf{x},\underbrace{d\mathbf{e}_{N+1},\ .\ .\ .\ ,d\mathbf{e}_{N+1}}_{N-r\text{ times}},\underbrace{d\mathbf{x},\ .\ .\ .\ ,d\mathbf{x}}_{r\text{ times}}\ \text{“}\|\text{”}$$

$$D_r = \text{“}\|\text{”}\ \mathbf{e}_{N+1},\underbrace{d\mathbf{e}_{N+1},\ .\ .\ .\ ,d\mathbf{e}_{N+1}}_{N-r\text{ times}},\underbrace{d\mathbf{x},\ .\ .\ .\ ,d\mathbf{x}}_{r\text{ times}}\ \text{“}\|\text{”}$$

$h(\mathbf{x}) = \mathbf{x} \cdot \mathbf{e}_{N+1}$ is the *support function* of the hypersurface $\mathbf{x}$. Prove that $C_r = hD_r$ and that $dA_r = C_r - D_{r+1}$ (Chern).

7. The elementary symmetric function of degree r of the principal curvatures defines the rth *mean curvature* of a hypersurface V^N in V^{N+1} by

$$\binom{N}{r} H_r = \Sigma\, k_1 k_2 \cdot \cdot \cdot k_r$$

Prove that for a hypersurface in R^{N+1} (notations of Prob. 6)

$$D_r = (-1)^{N-r} N!\, H_{N-r} D_N$$

(Voss).

8. Prove that the point curvature (exercise 13-2, Prob. 11) in the induced metric in V^N is

$$\frac{1}{N(N-1)} \sum_{A \neq B} \Omega_{AB,AB} - \frac{1}{N(N-1)} \sum_{\alpha} \sum_{A \neq B} (\gamma^\alpha{}_{AA}\gamma^\alpha{}_{BB} - \gamma^\alpha{}_{AB}{}^2)$$

(Ricci).

9. Find the formulas corresponding to (13-61) for the $\mathbf{i}$ frame adapted to V^N in V^n.

10. Write the integrability condition (13-33) for the curvatures of V^N in V^n.

11. Let V^N be a hypersurface of V^{N+1} and suppose that all points of V^N are umbilics. All principal curvatures are equal; let $s(\mathbf{x})$ be this common value. Show that the Riemann curvature for a bivector in $\overset{2}{\wedge} T(V^N)$, computed for the induced metric in V^N, is equal to the Riemann curvature for the same bivector computed in V^{N+1} plus $s^2(\mathbf{x})$ (Voss).

12. Show that a totally geodesic hypersurface is a hypersurface all of whose points are umbilics (Voss).

13. For a surface V^2 in a V^3 with principal curvatures k_1 and k_2, show that the *induced Gauss curvature* K_i defined by $d\omega^2{}_1 = -K_i\omega^1 \wedge \omega^2$ is $K_i = k_1k_2 - R_{12,12}$ in any **e** frame adapted to the surface.

14. The Frenet formula for a curve on a V^2 in a V^3 may be written in the form of Eq. (10-2). (Why?) Show that

$$\gamma^3{}_{11;1} = \frac{dk_n}{ds} - 2t_rk_g$$

along the curve. It follows that the expression on the right-hand side of this equation is the same for all curves tangent to the same $\mathbf{e}_1$ at a given point (Laguerre).

15. Compute the induced metric and the curvature $\hat{R}_{AB,CD}$ for the hypersurface $x^3 = 0$ in the space of the metric of exercise 13-2, Prob. 7.

16. Prove an analogue of the first equation (10-17) for curves in V^M.

17. Prove that Dupin's theorem holds for the intersection of V^M and V^{n-M+1} in V^n.

18. Discuss the relation of $H^*(\mathbf{x}_0)$ for V^M to $H^*(\mathbf{x}_0)$ for V^n.

13-4. Groups of Motions

A group of motions on a Riemann space is a pseudogroup of transformations all of whose infinitesimal transformations satisfy

$$\mathfrak{L}(d\mathbf{x}\, G\, d\mathbf{x}^t) = 0$$

The vectors Ξ defining a one-parameter group of motions are the *Killing vectors,* solutions of the Killing equation (11-11). The argument of Sec. 11-2 applies here without change. The Killing equations may be rewritten in the very simple form

$$\Xi_{i;k} + \Xi_{k;i} = 0 \tag{13-65}$$

The literature on groups of motions is very vast. In this section we restrict our attention to the generalization of results obtained for surfaces in Sec. 11-2. In addition, we show how simple group-theoretic arguments lead to strong results on transformation groups. These results may indicate the power of modern direct methods.

If the coordinate system is chosen such that $\Xi = \mathbf{i}_1$, Eq. (11-11) becomes

$$\frac{\partial g_{ij}}{\partial x^1} = 0$$

The following theorem generalizes theorem 11-12 and follows at once:

Theorem 13-20. *A Riemann space admits a one-parameter group of motions if and only if a coordinate system may be found such that the metric* (g_{ij}) *does not depend on one of the coordinates.*

In this case, the transformation is

$$(x^1,x^2, \ldots ,x^n) \to (x^1 + t, x^2, \ldots , x^n)$$

The trajectories of the group are the parameter lines $x^i = \text{const}$ $(i = 2, \ldots ,n)$.

Definition 13-21. *A one-parameter group of motions is a translation if its Killing vector is of constant length along any trajectory.*

We retain the system of coordinates in which $\Xi = \mathbf{i}_1$. The motion is a translation if and only if $g_{11} = \text{const}$. This implies $\Gamma^i{}_{11} = 0$; all curves $x^i = \text{const}$ $(i = 2, \ldots ,n)$ are geodesics.

On the other hand, if the trajectories $x^i = \text{const}$ $(i = 2, \ldots ,n)$ of a one-parameter group of motions are geodesics, then $\Gamma^i{}_{11}(dx^1/ds)^2 = 0$ for all i. But in this case

$$\frac{1}{2}\frac{\partial g_{11}}{\partial x^j} = -\Gamma_{j,11} = -g_{ji}\Gamma^i{}_{11} = 0$$

g_{11} is constant.

Theorem 13-22 (Bianchi). *A one-parameter group of motions is a translation if and only if its trajectories are geodesics.*

Another important result is the following:

Theorem 13-23. *Two distinct one-parameter groups of motions cannot have the same trajectories in some neighborhood.*

If two Killing vector fields $\Xi_{(1)}$ and $\Xi_{(2)}$ define the same trajectories, we may choose the coordinates such that

$$\Xi_{(1)}{}^j = \delta^j_1 \qquad \Xi_{(2)}{}^j = f(x^1, \ldots ,x^n)\delta^j_1$$

The Killing equations for the second field are

$$g_{1j}\frac{\partial f}{\partial x^i} + g_{i1}\frac{\partial f}{\partial x^j} = 0$$

In this equation we change j into k and eliminate g_{i1}:

$$\frac{\partial f}{\partial x^i}\left(g_{1j}\frac{\partial f}{\partial x^k} - g_{1k}\frac{\partial f}{\partial x^j}\right) = 0$$

This means that either $\partial f/\partial x^i = 0$ or

$$g_{1j}\frac{\partial f}{\partial x^k} = g_{1k}\frac{\partial f}{\partial x^j}$$

A special case of this relation is

$$g_{11}\frac{\partial f}{\partial x^k} = g_{1k}\frac{\partial f}{\partial x^1} \tag{13-66}$$

and we know that the coordinates may be chosen so that $g_{11} \neq 0$. This relation may be inserted into the Killing equation

$$g_{1j}\frac{\partial f}{\partial x^i} + g_{1i}\frac{\partial f}{\partial x^j} = 2\frac{\partial f}{\partial x^1}\frac{g_{1j}g_{1i}}{g_{11}} = 0$$

From this equation and (13-66) it follows that $\partial f/\partial x^1 = \partial f/\partial x^k = 0$, f is constant, and $\Xi_{(2)}$ determines the same group as does $\Xi_{(1)}$.

It follows from theorem 13-23 that the group of motions that do not leave a certain point P fixed is of dimension $\leq n$ since any ray in T_P belongs to, at most, one field of Killing vectors. The motions that leave P fixed induce in T_P a subgroup of the orthogonal group O_n of dimension $n(n-1)/2$.

Theorem 13-24. *The dimension of the group of motions of a V^n is, at most, $n + n(n-1)/2 = n(n+1)/2$.*

If the dimension of the group is the maximum, the group induced in T_P must be O_n, which is transitive in T_P. By corollary 13-17, the space is of constant curvature. We proceed to show that any space of constant curvature is, in fact, a *Clifford-Klein space*, i.e., admits a group of motions of dimension $n(n+1)/2$, transitive in $T(V)$.

We show that for two Riemann spaces of the same dimension and constant and equal curvatures there always exists an isometric mapping which brings a given point $\mathbf{p}$ onto a given point $\mathbf{q}$ and some assigned $\mathbf{e}$ frame in T_p into a given $\mathbf{e}$ frame in T_q. If the two spaces coincide, this shows the existence of a differentiable group of motions transitive in $T(V)$.

In the space V the Riemann structure with respect to a family of $\mathbf{e}$ frames is given by pfaffian forms $\omega^\alpha(x^1, \ldots, x^n; dx^1, \ldots, dx^n)$ and $\omega^\alpha{}_\beta(x^i, dx^i)$. In a space $\bar{V}$ we have forms $\bar{\omega}^\alpha(\bar{x}^i, d\bar{x}^i)$ and $\bar{\omega}^\alpha{}_\beta(\bar{x}^i, d\bar{x}^i)$. If both spaces have the same constant curvature K, the system

$$\theta^\alpha = \omega^\alpha - \bar{\omega}^\alpha = 0$$
$$\theta^\alpha{}_\beta = \omega^\alpha{}_\beta - \bar{\omega}^\alpha{}_\beta = 0$$

is completely integrable since

$$d\theta^\alpha = \theta^\beta \wedge \omega^\alpha{}_\beta + \bar{\omega}^\beta \wedge \theta^\alpha{}_\beta$$
$$d\theta^\alpha{}_\beta = \theta^\gamma{}_\beta \wedge \omega^\alpha{}_\gamma + \bar{\omega}^\gamma{}_\beta \wedge \theta^\alpha{}_\gamma + K(\delta_{\alpha\beta}\delta_{\gamma\epsilon} - \delta_{\alpha\epsilon}\delta_{\beta\gamma})(\theta^\gamma \wedge \omega^\epsilon + \bar{\omega}^\gamma \wedge \theta^\epsilon)$$

The proof is a simple copy of that of theorem 10-24.

Theorem 13-25. *A Riemann space V admits a group of motions transitive over $T(V)$ if and only if it is of constant curvature.*

Naturally, the theorem is not true for Riemann manifolds.

It was shown in Sec. 11-2 that no group of motions on a *local* surface can be of dimension two. A generalization of this fact may be based on the following theorem on Lie groups, due to Montgomery and Samelson: *For $n \neq 4$, there exists no proper subgroup of O_n of dimensions $>(n-1)(n-2)/2$.*

Theorem 13-26. *If a Riemann space of dimension $n \neq 4$ is not of constant curvature, it admits no group of motions of dimension*

$$> \binom{n-1}{2} + n = 1 + \frac{n(n-1)}{2}$$

Any group of dimension $n(n-1)/2 + 1$ is transitive on V.

A differentiable homeomorphism of a Riemann space onto itself, $\mathbf{q} = \phi(\mathbf{p})$, induces a map

$$\phi^J = \left(\frac{\partial \phi^i}{\partial x^j}\right)$$

of T_p onto T_q. ϕ is a motion if

$$\mathbf{x} \circ \mathbf{y} = \phi^J \mathbf{x} \circ \phi^J \mathbf{y}$$

It is a *homothety* if

$$\phi^J \mathbf{x} \circ \phi^J \mathbf{y} = \text{const } \mathbf{x} \circ \mathbf{y}$$

Given a curve $\mathbf{c}$ which joins $\mathbf{p}$ to $\mathbf{p}_1$, let $\tau_c \colon T_p \to T_{p_1}$ be the map of tangent spaces defined by parallel translation of the vectors of T_p along $\mathbf{c}$. The map ϕ will transform $\mathbf{c}$ into a curve $\mathbf{c}^\phi$. Let $\tau_c{}^\phi$ be the map by parallelism along $\mathbf{c}^\phi$. ϕ is *affine* if it commutes with parallelism, i.e., if

$$\phi^J[\tau_c(\mathbf{x})] = \tau_c{}^\phi[\phi^J(\mathbf{x})] \tag{13-67}$$

for all $\mathbf{x} \in T_p$. Motions, homotheties, and affine maps on a space V form groups $M(V)$, $S(V)$, and $A(V)$, respectively. It is clear that

$$M(V) \subset S(V) \subset A(V)$$

The following theorem is a far-reaching generalization of the fact that there are no homotheties (hence no similitudes other than congruences) in non-euclidean geometry.

Theorem 13-27 (Kobayashi). *If on a complete Riemann manifold the curvature is not identically zero, then all homotheties are motions,*

$$S(V) = M(V)$$

The notion of a complete space has been discussed in Sec. 11-3.

Let ϕ be a homothety but not a motion. If the constant c of the homothety is >1, ϕ^{-1} has constant $c^{-1} < 1$. Therefore we may assume $c < 1$. c surely is positive. Consider the sequence P_0, $P_1 = \phi(P_0)$,

$\ldots, P_\nu = \phi(P_{\nu-1}), \ldots$ for arbitrary P_0. If the length of an arc $\mathbf{c}$ is $s = \int\sqrt{d\mathbf{x} \circ d\mathbf{x}}$, the length of $\mathbf{c}^\phi$ is cs. It follows for the distance (see page 285) between two points that $d(P_\nu,P_{\nu+1}) = c^\nu d(P_0,P_1)$; hence

$$d(P_\nu,P_\mu) \leq \sum_{i=0}^{\mu-\nu-1} d(P_{\nu+i},P_{\nu+i+1}) < \frac{c^\nu}{1-c}\, d(P_0,P_1)$$

The sequence $\{P_\mu\}$ is a Cauchy sequence; $P = \lim_{\mu\to\infty} P_\mu$ exists. By definition, $d(P,P_\nu) < \epsilon$ for $\nu > \nu_0(\epsilon)$. Hence $d(P,\phi(P)) \leq d(P,P_{\nu+1}) + d(P_{\nu+1},\phi(P)) < 2\epsilon$ for all ϵ. This means that $P = \phi P$; *P is a fixed point for ϕ.*

Now let $\mathbf{c}$ be a closed C^2 curve through the fixed point P. $\mathbf{c}$ is referred to some parameter t, $\mathbf{c}(0) = \mathbf{c}(1) = P$. Since $\mathbf{c}$ is compact, it has a finite length l and $\max_c |d\mathbf{c}/dt|$ exists. The curves $\mathbf{c}^\phi$ all are closed curves through P and

$$\max\left|\frac{d\mathbf{c}^{\phi^\nu}}{dt}\right| = c^\nu \max\left|\frac{d\mathbf{c}}{dt}\right| \to 0$$

as well as

$$l(\mathbf{c}^{\phi^\nu}) = c^\nu l \to 0$$

for $\nu \to \infty$. The maps $\tau_c{}^\phi$ are rotations in T_P. It follows from the limit relations and (13-27) that

$$\lim_{\nu\to\infty} \tau_c{}^{\phi^\nu} = U$$

In our case both ϕ^J and τ_c are nonsingular linear transformations in T_P, and (13-67) may be written

$$\tau_c = (\phi^J)^{-1}\tau_c{}^\phi\phi^J$$

Hence also

$$\tau_c = (\phi^J)^{-\nu}\tau_c{}^{\phi^\nu}(\phi^J)^\nu$$

It is a well-known theorem of algebra† that for any matrix A and any nonsingular matrix B the *characteristic polynomial*

$$p(A) = \|U - \lambda A\| = p(B^{-1}AB) = \|U - \lambda B^{-1}AB\|$$

is not changed in a "similarity" $A \to B^{-1}AB$. In our case this means that $p(\tau_c) = p(\tau_c{}^{\phi^\nu})$ for all ν. Hence also

$$p(\tau_c) = \lim_{\nu\to\infty} p(\tau_c{}^{\phi^\nu}) = p(\lim_{\nu\to\infty} \tau_c{}^{\phi^\nu}) = p(U) = (1-\lambda)^n$$

No matrix other than U has this characteristic polynomial; hence $\tau_c = U$ for all c. By Def. 13-6 this means that $\mathrm{P} = 0$, and the space is *locally*

† See, for example, C. C. MacDuffee, "The Theory of Matrices," Theorem 38.1, Chelsea Publishing Company, New York, 1946 (reprint); or G. de B. Robinson, "Vector Geometry," Theorem 9.35, Allyn and Bacon, Inc., Englewood Cliffs, N.J., 1962.

euclidean. In all other spaces the constant of any homothety must be 1; the homotheties are isometries.

The relation between $S(V)$ and $A(V)$ depends on another, purely local notion.

Definition 13-28. *A Riemann space is reducible if it is possible to find a system of coordinates in which the metric splits:*

$$G = \begin{pmatrix} g_{11} & \cdots & g_{1k} & 0 & \cdots & 0 \\ \cdot & \cdots & \cdot & \cdot & \cdots & \cdot \\ g_{k1} & \cdots & g_{kk} & 0 & \cdots & 0 \\ 0 & \cdots & 0 & g_{k+1\,k+1} & \cdots & g_{k+1\,n} \\ \cdot & \cdots & \cdot & \cdot & \cdots & \cdot \\ 0 & & 0 & g_{n\,k+1} & \cdots & g_{nn} \end{pmatrix}$$

Otherwise it is irreducible.

If a space is reducible, the definitions show that $P_i^j = 0$ for $i \leq k$, $j > k$ or $i > k$, $j \leq k$. If the matrix P is reduced, it is possible to find **e** frames such that Ω also is reduced and vice versa. If all matrices of a group are of the form

$$\begin{pmatrix} A & 0 \\ 0 & B \end{pmatrix}$$

where A and B are square matrices of a constant number of columns and the 0's are rectangular zero matrices, the A's and the B's form groups themselves. The given group is the *direct product* of the two groups. The group is said to be *irreducible* if it is not a direct product. The previous argument shows that the local holonomy group is reducible if the Riemann space is reducible. If the local holonomy group is reducible, the proof of theorem 13-13 shows that there exists a system of normal coordinates in which the space is reducible.

Theorem 13-29. *A Riemann space is reducible if and only if its homogeneous holonomy group is reducible.*

The inner product defined by G is a quadratic form in T_p which is invariant under the action of the local holonomy group. (A tensor is invariant under that group if it is covariant constant.) If the space is reducible, G surely is not the only invariant form, since all matrices

$$\begin{pmatrix} ag_{ij} & 0 \\ 0 & bg_{ij} \end{pmatrix}$$

with constant a,b are covariant constant. However, G is essentially the only invariant quadratic form on an irreducible space.

Theorem 13-30 (Eisenhart). *In an irreducible Riemann space every covariant constant symmetric tensor $A = (a_{ij})$ is a constant multiple of the metric tensor g_{ij}.*

We have to show only that a_{ij} is a multiple of g_{ij}. The ratio must be constant over the space since its covariant derivatives, which are its partial derivatives, all vanish. Therefore we may work in *one* tangent space T_P.

It is shown in linear algebra that all roots $\lambda_{(h)}$ $(h = 1, \ldots, n)$ of $\|A - \lambda G\| = 0$ are real and that there exists an orthonormal basis of T_P formed of vectors $\mathbf{x}_{(i)}$ $(i = 1, \ldots, n)$, solutions of

$$\mathbf{x}(A - \lambda G) = 0 \tag{13-68}$$

$(\lambda = \lambda_{(h)})$. Assume now that not all $\lambda_{(h)}$ are equal. Equation (13-68) is invariant under the action of the homogeneous holonomy group, i.e., under parallel translation along closed curves. This means that the linear space spanned in T_P by the vectors $\mathbf{x}_{(h)}$ belonging to one value of $\lambda_{(h)}$ is transformed into itself by the local holonomy group. The vectors $\mathbf{x}_{(h)}$ may be taken as the vectors of an $\mathbf{e}$ frame, and the corresponding matrix Ω is reduced; hence the space is reducible.

Eisenhart's theorem is a very strong tool. Its force is shown in the proof of the next theorem.

Theorem 13-31 (Nomizu). *All affine maps of irreducible Riemann spaces are homotheties,*

$$S(V) = A(V)$$

If ϕ is affine, we define a new scalar product in T_P by $(\mathbf{x},\mathbf{y}) = \phi^J\mathbf{x} \circ \phi^J\mathbf{y}$. Let $\mathbf{c}$ be a closed curve through P; then

$$(\tau_c\mathbf{x},\tau_c\mathbf{y}) = \phi^J\tau_c\mathbf{x} \circ \phi^J\tau_c\mathbf{y} = \tau_{c^\phi}\phi^J\mathbf{x} \circ \tau_{c^\phi}\phi^J\mathbf{y} = \phi^J\mathbf{x} \circ \phi^J\mathbf{y} = (\mathbf{x},\mathbf{y}) \tag{13-69}$$

Hence the symmetric quadratic form $(\mathbf{x},\mathbf{y}) = a_{ij}x^iy^j$ is invariant under the action of the homogeneous holonomy group. By Eisenhart's theorem, $(\mathbf{x},\mathbf{y}) = c^2\mathbf{x} \circ \mathbf{y}$ with constant c^2.

Kobayashi's and Nomizu's theorems are of a kind which cannot be obtained by the local theory of transformation groups. They deal with single transformations; hence they are valid also if the groups in question are discrete.

Exercise 13-4

◆ **1.** Integrate the Killing equations for

(*a*) $ds^2 = (dx^1)^2 + f^2(x^1)[(dx^2)^2 + (dx^3)^2 + (dx^4)^2]$

(*b*) $ds^2 = (dx^1)^2 + f^2(x^1)[(dx^2)^2 + \sin^2 x^2\,(dx^3)^2]$

What are the conditions on f?

2. Find the Killing equations for

$$ds^2 = \frac{4(dx^2 + dy^2) + \{[1 + \frac{1}{4}(x^2 + y^2)]\,dz + (x\,dy - y\,dx)\}^2}{[1 + \frac{1}{4}(x^2 + y^2)]^2}$$

3. Show that geodesics are transformed into geodesics in any motion of a Riemann space.

4. Show that a field of unit vectors Ξ defines a one-parameter group of translations if and only if the vectors Ξ make a constant angle with the tangents along *any* geodesic.

5. Show that the trajectories of two one-parameter groups of translations meet under a constant angle. (Use Prob. 4.)

6. If a space admits a group of translations, show that any local surface generated by a one-parameter family of trajectories must be of zero curvature (Bianchi). (The surface admits a group of translations.)

7. Find the transformation groups of dimension six that may act on a three-dimensional Riemann space.

8. For any point P on a Riemann space and a nearby point M let M' be the point on the geodesic MP whose distance from P is equal to the distance of M from P. If the map $M \to M'$ is an isometry for all P, the space is *symmetric*. In normal coordinates with origin P the symmetry is $(x^i) \to (-x^i)$. Use these coordinates to show that a space is symmetric if and only if its curvature tensor is covariant constant (Cartan).

9. For a transformation group G acting on a space V let G_P be the set of all transformations $\phi \in G$ such that $\phi(P) = P$.

(*a*) Show that G_P is a group (the *stability group* of P).

(*b*) For two points P and Q in an arcwise-connected space, show that G_P is isomorphic to G_Q.

(*c*) If G is a global group, show that G_P is a closed set in G.

10. The stability group G_P of a group of motions induces in T_P a rotation group $G_P{}^J$. Let $S \subset T_P$ be the linear space of vectors of T_P invariant under the action of $G_P{}^J$, and let **S** be the local subspace generated by the geodesics tangent to vectors of S. Show that **S** is totally geodesic (Cartan). (Use Prob. 9 to show that **S** is invariant under any G_Q, $Q \in$ **S**.)

♦**11.** Compute $\pounds\Gamma^i{}_{jk}$ for any one-parameter group of a vector field Ξ.

12. Show that a transformation group is one of affine maps if and only if $\pounds\Gamma = 0$.

13. If a space is reducible, the g_{ij} $(i,j \leq k)$ are independent of the x^s $(s > k)$. (Show that $g_{ij;s} = 0$ implies $\partial g_{ij}/\partial x^s = 0$.) A space is *totally reducible* if its metric is a sum of squares,

$$ds^2 = \Sigma h_i^2(x^1, \ldots, x^n)(dx^i)^2$$

Show that h_i depends only on x^i.

♦**14.** Find the conditions for a transformation group (not of motions) to carry geodesics onto geodesics. [Use Eq. (13-54).]

15. If both partial metrics of a reducible metric are of constant curvature, show that the scalar R of the total metric is constant and that its Weyl tensor vanishes (Tachibana).

16. Show that, if a reducible metric splits into two Einstein metrics, R is constant for the total metric (Tachibana).

13-5. Integral Theorems

As in the theory of surfaces, there are some simple and interesting applications of Stokes's formula to global riemannian geometry that do not presuppose a study of topology. The method explained in this section is due mainly to Kentaro Yano.

A compact Riemann manifold V^n is the union of a finite number of Riemann spaces. In each of these spaces the Stokes formula (9-24) holds. If the manifold is orientable, each ∂I^n itself is covered by a finite number of $(n-1)$-dimensional spaces, each of which appears on the boundaries of two distinct n-dimensional spaces but with different orientations. Therefore

$$\int_{V^n} d\alpha^{(n-1)} = 0 \tag{13-70}$$

for a *tensor field* form $d\alpha^{(n-1)}$ on an *orientable, compact* manifold V^n. [If $d\alpha^{(n-1)}$ is defined on the whole manifold, its coordinate tensor must be given by a map of the point on the manifold into the tangent tensor space; this means that the coordinate tensor is a tensor field.]

Let v^i be a vector *field* in V^n. In each of the spaces covering V^n we choose an **e** frame to which belong forms ω^α. For

$$\beta = \sum_{\sigma=1}^{n} (-1)^{\sigma-1} v^\sigma \omega^1 \wedge \cdots \hat{\omega}^\sigma \cdots \wedge \omega^n$$

one has $d\beta = (v^\sigma{}_{;\sigma})\omega^1 \wedge \cdots \wedge \omega^n = (v^i{}_{;i})\|G\|^{\frac{1}{2}}\, dx^1 \wedge \cdots \wedge dx^n$. We denote by

$$\alpha = \omega^1 \wedge \cdots \wedge \omega^n = \|G\|^{\frac{1}{2}}\, dx^1 \wedge \cdots \wedge dx^n \tag{13-71}$$

the *element of volume* of the Riemann space. In any coordinate transformation $(dx^{i'}) = (dx^i)J_i^{i'}$ one has

$$\|g_{i'j'}\|^{\frac{1}{2}}\, dx^{1'} \wedge \cdots \wedge dx^{n'} = \|JGJ^t\|^{\frac{1}{2}}\|J^{-1}\|\, dx^1 \wedge \cdots \wedge dx^n = \alpha$$

This shows that α is a map $V^n \to \overset{n}{\wedge}\, T(V^n)$. Stokes's formula holds and

$$\int_{V^n} v^i{}_{;i}\alpha = 0 \tag{13-72}$$

Equation (13-72) is often referred to as *Green's formula* in analogy with the three-dimensional euclidean case in which div $\mathbf{v} = v^i{}_{,i}$. It follows from the formula also that

$$\int_{V^n} (v^i{}_{;j}v^j)_{;i}\alpha = \int_{V^n} (v^i{}_{;i}v^j)_{;j}\alpha = 0$$

since the expressions in parentheses are vector fields if v^i is one. But

$$(v^i{}_{;j}v^j)_{;i} = v^i{}_{;j;i}v^j + v^i{}_{;j}v^j{}_{;i} = v^i{}_{;i;j} + v^i{}_{;j}v^j{}_{;i} + R_{ij}v^iv^j$$

and

$$(v^i{}_{;i}v^j)_{;j} = v^i{}_{;i;j}v^j + v^i{}_{;i}v^j{}_{;j}$$

Hence we obtain *Yano's formula*

$$\int(R_{ij}v^iv^j + v^i{}_{;}{}^jv_{j;i} - v^i{}_{;i}v^j{}_{;j})\alpha = 0 \tag{13-73}$$

All integrals are taken over V^n.

If v^i is a field of Killing vectors, it follows from (13-65) that

$$v_{i;}{}^i + v^i{}_{;i} = 2v^i{}_{;i} = 0$$

For Killing vector fields one has the stronger relation

$$\int(R_{ij}v^iv^j - v^i{}_{;}{}^jv_{i;j})\alpha = 0$$

The change in sign of the second term is also a consequence of (13-65) since we have permuted indices. The following theorem is an immediate consequence of the preceding formula:

Theorem 13-32 (Bochner). *There cannot exist a one-parameter group of motions on a compact orientable Riemann manifold if the quadratic form $R_{ij}v^iv^j$ is everywhere negative-definite.*

A vector field v^i determines the corresponding Lie derivative $\pounds_v\Gamma^i{}_{jk}$. The dragging of the points of V^n along the integral curves of the vector field v^i defines a change of coordinates $x^{i'}(t) = x^i(0)$ (see Sec. 7-1). The jacobian of this transformation is

$$J = \left(\delta^i_j + \frac{\partial v^i}{\partial x^j}t + \cdots\right)$$

It follows from the transformation formula (13-15) that

$$\begin{aligned}\pounds_v\Gamma^i{}_{jk} &= \lim_{t\to 0}\frac{1}{t}\,(\Gamma^i{}_{jk} - \Gamma^{i'}{}_{j'k'}(t)) \\ &= v^s\frac{\partial\Gamma^i{}_{jk}}{\partial x^s} + \Gamma^i{}_{sk}\frac{\partial v^s}{\partial x^j} + \Gamma^i{}_{js}\frac{\partial v^s}{\partial x^k} - \Gamma^s{}_{jk}\frac{\partial v^i}{\partial x^s} + \frac{\partial^2 v^i}{\partial x^j\,\partial x^k} \\ &= v^i{}_{;j;k} + R_j{}^i{}_{.kl}v^l\end{aligned} \tag{13-74}$$

The field v^i defines a one-parameter group of affine maps [see Eq. (13-67)] if $\pounds\nabla = \nabla\pounds$ for the action on arbitrary geometric objects; this is possible only if $\pounds\Gamma^i_{jk} = 0$. For such a group of affine maps, then,

$$v^i{}_{;j;k} + R^{\ i}_{j.,kl}v^l = 0 \tag{13-75}$$

As a consequence

$$v^i{}_;{}^k{}_{;k} + R^i{}_l v^l = 0 \tag{13-76}$$

and also $v^i{}_{;i;k} = 0$. This implies $v^i{}_{;i} =$ const, by Green's formula (13-72),

$$v^i{}_{;i} = 0 \tag{13-77}$$

The identity

$$(v_k v^k)_;{}^i{}_{;j} = g^{ij}(v_k v^k)_{;i;j} = 2v_k g^{ij} v^k{}_{;i;j} + 2v^i{}_;{}^j v_{i;j}$$

gives by Green's formula and a slight change in terms

$$\int (v_i v^i{}_;{}^j{}_{;j} + v^i{}_;{}^j v_{i;j})\alpha = 0 \tag{13-78}$$

If we add this to (13-73) we obtain another of Yano's formulas:

$$\int (v_i v^i{}_;{}^j{}_{;j} + R_{ij}v^i v^j + v^i{}_;{}^j v_{i;j} + v^i{}_;{}^j v_{j;i} - v^i{}_{;i} v^j{}_{;j})\alpha$$
$$= \int [v_i(v^i{}_;{}^k{}_{;k} + R^i{}_l v^l) + \tfrac{1}{2}(v^i{}_;{}^j + v^j{}_;{}^i)(v_{i;j} + v_{j;i})]\alpha = 0$$

If v^i is the vector field of a group of affine maps,

$$\int (v^i{}_;{}^j + v^j{}_;{}^i)(v_{i;j} + v_{j;i})\alpha = 0 \tag{13-79}$$

by (13-76). Since G is a positive-definite matrix,

$$(v^i{}_;{}^j + v^j{}_;{}^i)(v_{i;j} + v_{j;i}) = g_{ir}g_{js}(v^i{}_;{}^j + v^j{}_;{}^i)(v^r{}_;{}^s + v^s{}_;{}^r)$$

is nowhere negative. Equation (13-79) is possible only if

$$v_{i;j} + v_{j;i} = 0 \tag{13-65}$$

Theorem 13-33 (Yano). *A one-parameter group of affine maps on a compact orientable Riemann manifold is a group of motions.*

This theorem and Nomizu's (theorem 13-31) do not agree as to hypotheses and to results.

The *Laplacean* ΔF of a real-valued function $F(x)$ is defined as

$$\Delta F = F_;{}^i{}_{;i} = (g^{ij}F_{;i})_{;j}$$

(compare exercise 13-1, Prob. 34). By Green's formula $\int \Delta F \alpha = 0$. This shows that $\Delta F \geq 0$ on a compact orientable Riemann manifold only if $\Delta F = 0$. In this case

$$\int \Delta(F^2)\alpha = 2\int (F\,\Delta F + g^{ij}F_{;i}F_{;j})\alpha = \int F_;{}^i F_{;i}\alpha = \int |F_{;i}|^2\alpha = 0$$

Hence $F_{;i} = \partial F/\partial x^i = 0$, F constant.

Theorem 13-34 (E. Hopf-Bochner). *If a function F satisfies $\Delta F \geq 0$ on a compact orientable Riemann manifold, F is a constant.*

This theorem is a very strong tool in the theory of *harmonic* vectors. A vector field w_i on V^n is said to be harmonic if it satisfies

$$w_{i;j} - w_{j;i} = 0 \qquad w^i{}_{;i} = 0 \tag{13-80}$$

The contraction of a harmonic field w_i with a Killing vector field v^i is $F = w_iv^i = g^{ij}w_iv_j$. It follows easily from Eqs. (13-80) and (13-65) that $\Delta F = g^{st}w_{i;s;t}v^i + 2w_{i;j}v^i{}_{;}{}^j + w_ig^{st}v^i{}_{;s;t} = 0$.

Corollary 13-35 (Bochner). *In a compact orientable Riemann manifold, the contraction of a harmonic and a Killing vector is constant.*

If a manifold admits a transitive group of motions, there exists a nonzero Killing vector in every tangent space. Hence:

Corollary 13-36 (Constantinescu). *If a compact orientable Riemann manifold admits a transitive group of motions, a harmonic vector vanishes identically if it is zero at one point.*

The importance of harmonic vectors can be explained only within the framework of algebraic topology.

Exercise 13-5

All problems in this exercise deal with compact orientable Riemann manifolds.

1. Prove that

$$\int [R_{ij}v^iv^j + v^i{}_{;}{}^jv_{i;j} - \tfrac{1}{2}(v^i{}_{;}{}^j - v^j{}_{;}{}^i)(v_{i;j} - v_{j;i}) - v^j{}_{;j}v^i{}_{;i}]\alpha = 0$$

(Yano).

2. Use Prob. 1 to prove that, if $R_{ij}v^iv^j$ is positive-definite everywhere on V^n, there exists no nonzero harmonic vector on V^n (Myers).

3. Prove that

$$\int [(v^i{}_{;}{}^j{}_{;j} - R_j{}^iv^j)v_i + \tfrac{1}{2}(v^i{}_{;}{}^j - v^j{}_{;}{}^i)(v_{i;j} - v_{j;i}) + v^j{}_{;j}v^i{}_{;i}]\alpha = 0$$

(Yano).

4. Use Prob. 3 to prove that w_i is harmonic on a compact orientable Riemann manifold if and only if $w_{i;j;}{}^j - R_i{}^jw_j = 0$.

5. Discuss the cases covered by Yano's theorem 13-33 that are not contained in Nomizu's theorem 13-31 and vice versa.

6. A harmonic vector on a compact orientable Riemann manifold is invariant under the action of any one-parameter group of motions (Yano). Use (9-25) to show that the Lie derivative vanishes.

7. (See exercise 13-1, Probs. 33 and 36.) The *scalar product* of two p forms is $(\sigma^{(p)},\rho^{(p)}) = \int \sigma^{(p)} \wedge *\rho^{(p)}$. From

$$d(\sigma^{(p-1)} \wedge *\rho^{(p)}) = d\sigma^{(p-1)} \wedge *\rho^{(p)} - \sigma^{(p-1)} \wedge *\delta\rho^{(p)}$$

it follows by Stokes's theorem that $(d\sigma,\rho) = (\sigma,\delta\rho)$.

(*a*) Show that the scalar product is a bilinear, positive-definite form on the linear space of p forms on V^n.

(*b*) In the definition of exercise 13-1, Prob. 36,

$$(\Delta\phi,\phi) = (d\phi,d\phi) + (\delta\phi,\delta\phi)$$

Prove this.

(*c*) Defining equation (13-80) may be written $d\omega = \delta\omega = 0$, $\omega = w_i\, dx^i$. Show that a vector on a compact orientable manifold is harmonic if and only if $\Delta\omega = 0$. [Use parts (*b*) and (*a*).]

8. Show that, if w_i and u_i are harmonic vectors and $w_i = f(x)u_i$, then $f(x)$ is a constant.

9. If in every tangent space of a compact orientable Riemann manifold V^n there exist $n - 1$ linearly independent Killing vectors, show that there exist on V^n at most n linearly independent harmonic vector fields (Constantinescu). (Show that $n + 1$ fields are dependent by corollary 13-35 and Prob. 8.)

References

General References: The article by Berwald and the book by Favard cited in the references for Chap. 8 and the books by Bianchi and Willmore cited in the references for Chap. 10.

Berger, M.: Sur les groupes d'holonomie homogène des variétés à connexion affine et des variétés riemanniennes, *Bull. Soc. Math. France,* **83**: 279–330 (1955).

Cartan, E.: "Leçons sur la géométrie des espaces de Riemann," 2d ed., Gauthier-Villars, Paris, 1946.

Cartan, E.: Les groupes d'holonomie des espaces généralisés, "Oeuvres complètes," vol. III/2, pp. 997–1038, Gauthier-Villars, Paris, 1955.

Eisenhart, L. P.: "Riemannian Geometry," Princeton University Press, Princeton, N.J., 1949.

Nijenhuis, A.: On the Holonomy Groups of Linear Connections, *Nederl. Akad. Wetensch., Proc. Ser. A,* **56**: 233–249 (1953).

Nomizu, K.: "Recent Development in the Theory of Connections and Holonomy Groups, Advances in Mathematics," vol. 1, fasc. 1, pp. 1–50, Academic Press Inc., New York, 1961.

Schouten, J. A.: "Ricci Calculus," 2d ed., Springer-Verlag OHG, Berlin, 1954.

SEC. 13-4

Eisenhart, L. P.: "Continuous Groups of Transformations," Princeton University Press, Princeton, N.J., 1933.

Kobayashi, S.: A Theorem on the Affine Transformation Groups of a Riemannian Manifold, *Nagoya Math. J.,* **9**: 39–41 (1955).

Lichnérowicz, A.: "Géométrie des groupes de transformations," Dunod, Paris, 1958.

Montgomery, D., and H. Samelson: Transformation Groups of Spheres, *Ann. of Math.*, **44**: 454–470 (1943).

Nomizu, S.: Studies on Riemannian Homogeneous Spaces, *Nagoya Math. J.*, **9**: 42–50 (1955).

For results on Nomizu's and Kobayashi's problems in a more general setting, see the following:

Hano, J.: On Affine Transformations of a Riemannian Manifold, *Nagoya Math. J.*, **9**: 99–109 (1955).

SEC. 13-5

Bochner, S., and K. Yano: "Curvature and Betti Numbers," Annals of Mathematics Studies 32, Princeton University Press, Princeton, N.J., 1953.

14 CONNECTIONS

In this final chapter we sketch generalizations of riemannian geometry to local geometry based on a Lie group G. The *interesting* problems in this geometry soon lead to questions on global theory of Lie groups and algebraic topology; therefore we shall not go into much detail.

Let G be a subgroup of GL_n. At each point P of an open set V in R^n we have defined a vector space T_P. In each T_P we choose a frame $\mathbf{e}_\alpha = \mathbf{e}_\alpha(P)$ in a differentiable way. The *admissible* frames in the geometry will be all the frames $g\mathbf{e}$, where g is a matrix in G.

The fundamental notion of Riemann geometry is that of a connection. This notion carries over to G geometries.

Let $\mathbf{c}_P{}^Q$ be a continuous curve from P to a point Q in the neighborhood, that is, $\mathbf{c}_P{}^Q(0) = P$, $\mathbf{c}_P{}^Q(1) = Q$ in some parametrization. $\mathbf{c}_P{}^P$ will be a closed curve on which is given a definite orientation. A *connection* is a linear map

$$M(\mathbf{c}_P{}^Q)\colon T_P \to T_Q$$

In riemannian geometry, $M(\mathbf{c}_P{}^Q)$ is the result of a parallel translation of the vectors of T_P along the curve $\mathbf{c}_P{}^Q$. The set of all continuous curves $\mathbf{c}_P{}^Q$ may be turned into a metric space. The distance between two curves $\mathbf{c}_P{}^Q$ and $\mathbf{c}'_{P'}{}^{Q'}$ may be defined, for example, as

$$\inf \int_0^1 |\mathbf{c}_P{}^Q(t) - \mathbf{c}'_{P'}{}^{Q'}(t)|\, dt$$

where the distance is measured in the euclidean geometry of the neighborhood and the greatest lower bound is taken over all admissible parametrizations of the curves. Many other definitions are possible.

The product of two curves has been defined in Eq. (13-28). The *restriction* $\mathbf{c}_P{}^{Q(t_0)}$ is the arc from P to $\mathbf{c}_P{}^{Q(t_0)}$ obtained from $\mathbf{c}_P{}^Q$ for $0 \le t$

$\leq t_0$. The connection will be subject to the following postulates, essentially due to Knebelman:

I. $M(\mathbf{c}_P{}^Q) \varepsilon G$.

II. $\lim_{t \to 0} M(\mathbf{c}_P{}^{Q(t)}) = U$.

III. $M(\hat{\mathbf{c}}_Q{}^R \cdot \mathbf{c}_P{}^Q) = M(\hat{\mathbf{c}}_Q{}^R)M(\mathbf{c}_P{}^Q)$.

IV. $M(\mathbf{c}_P{}^Q)$ is differentiable in the following way. Given any vector $\mathbf{x}^{\cdot} \varepsilon T_P$, let $\mathbf{c}_P{}^Q$ be a curve in the equivalence class of $\mathbf{x}^{\cdot}$. Then

$$dM(P,\mathbf{x}^{\cdot}) = \lim_{t \to 0} \frac{1}{t} [M(\mathbf{c}_P{}^{Q(t)}) - U]$$

will exist for all P and all $\mathbf{x}^{\cdot}$ and be independent of the choice of $\mathbf{c}_P{}^Q$ in the class of $\mathbf{x}^{\cdot}$.

NOTE: Axiom IV may be generalized considerably. If $\mathbf{c}_P{}^Q$ is restricted to vary only in the class of curves with fixed first k terms of the Taylor expansion in P, $\mathbf{c} = (a_1^i\sigma + a_2^i\sigma^2 + \cdot\cdot\cdot + a_k^i\sigma^k + R^i(\mathbf{c}))$, the resulting differential $dM(P,\mathbf{x}^{\cdot},\mathbf{x}^{\cdot\cdot}, \ldots ,\mathbf{x}^{(k)})$ is that of a *higher-order connection.* See also exercise 14-1, Prob. 1.

The connection $M(\mathbf{c}_P{}^Q)$ uniquely defines a Cartan matrix

$$C(M) = dMM^{-1} = L(P,\mathbf{x}^{\cdot})$$

L is the *infinitesimal-connection matrix.* If L is a linear function on the vectors of T_P,

$$L = (L^i_{jk}(P)\, dx^k)$$

it is a *linear-connection matrix.* Our previous existence theorems show that the infinitesimal connection L uniquely defines the connection M by integration of $L = dMM^{-1}$ along $\mathbf{c}_P{}^Q$ under the initial condition II, at least if L is a continuous function of P. The holonomy group $H(P)$ is the group of matrices $M(\mathbf{c}_P{}^P)$.

We now restrict our attention to *linear* connections for which the functions L^i_{jk} are C^2 functions of P. In general, $L^i_{jk} \neq L^i_{kj}$. If the connection is symmetric ($L^i_{jk} = L^i_{kj}$), the curvature matrix

$$K = dL - L \text{ "}\cdot\text{" } L \tag{14-1}$$

has all the properties of the curvature matrix of a riemannian metric. Hence, by theorem 13-13, it may be derived from such a metric. In this case nothing new is obtained.

If the tangent spaces are referred to an $\mathbf{i}$ frame derived from a system of parameter lines, the maps M will no longer be in G, but they will have the form $A(Q)^{-1}M(\mathbf{c}_P{}^Q)A(P)$, where A is the matrix which maps the $\mathbf{i}$ frame into the $\mathbf{e}$ frame in one tangent space. The matrices L will no longer be in the Lie algebra of G; they will, however, be derived from those in $\mathfrak{L}(G)$ by a formula

$$L_{(i)} = A^{-1}L_{(e)}A + C(A^{-1}) \tag{1.5}$$

For the subsequent computations we use the connection referred to an **i** frame.

Assume now that we are given a second linear connection L^*. Both L and L^* transform according to (1-5) in any change of coordinates with jacobian J. Hence $L - L^*$ is transformed into $J(L - L^*)J^{-1}$; it is a tensor field.

Theorem 14-1. *The difference of two linear connections on a space V^n is a tensor field. Every linear connection L^i_{jk} may be written as the sum of a symmetric linear connection Γ^i_{jk} and of a tensor field T^i_{jk}, the torsion tensor of the connection.*

$L^i_{jk} = \frac{1}{2}(L^i_{jk} + L^i_{kj}) + \frac{1}{2}(L^i_{jk} - L^i_{kj})$ shows that

$$\Gamma^i_{jk} = \tfrac{1}{2}(L^i_{jk} + L^i_{kj}) \tag{14-2}$$

and

$$T^i_{jk} = \tfrac{1}{2}(L^i_{jk} - L^i_{kj}) \tag{14-3}$$

since, by (1-5), L^i_{kj} is a connection in the **i** frame if L^i_{jk} is one.

Covariant differentiation may now be introduced along the lines explained in Sec. 13-1, with the same formal rule

$$\nabla \mathbf{v} = d\mathbf{v} - L\mathbf{v} \tag{14-4}$$

From the Frenet equations

$$\begin{aligned} dP &= dx^j\,\mathbf{i}_j \\ d\mathbf{i}_j &= L^k_j\mathbf{i}_k \end{aligned} \tag{14-5}$$

it follows that

$$ddP = -T^k\mathbf{i}_k \tag{14-6}$$

where we have put

$$T^k = T^k_{ij}\,dx^i \wedge dx^j \tag{14-7}$$

This means that $ddP \neq 0$ for all non-riemannian connections. This result is not surprising. The Poincaré relation $ddf = 0$ was derived from the euclidean structure of cartesian space, and it is just these properties that we are seeking to destroy in our generalization in order to obtain always "less euclidean" geometries.

An alternative formulation of (14-6) is

$$DDx^i = T^i \tag{14-8}$$

Exterior differentiation of this equation gives the Ricci identity

$$(dT^i) - (dx^j) \text{ ``}\cdot\text{'' } (K^i_j) + (T^j) \text{ ``}\cdot\text{'' } (L^i_j) = 0$$

or

$$DT^i = dx^j \wedge K^i_j \tag{14-9}$$

In the same way, (14-1) gives the Bianchi identity

$$dK = K \text{“.”} L - L \text{“.”} K$$

or

$$DK = 0 \tag{14-10}$$

Geodesics may again be defined as curves whose tangents remain constant in parallel translations along the curve (or whose development onto a constant euclidean space is a straight line), i.e., by

$$\frac{\nabla}{dt}\frac{dx}{dt} = f(t)\,\frac{dx}{dt}$$

or

$$\frac{d^2x^i}{dt^2} + L^i_{jk}\frac{dx^j}{dt}\frac{dx^k}{dt} = f(t)\,\frac{dx^i}{dt}$$

A parameter s is an affine parameter on the geodesic if $f(s) = 0$. Such an s may replace the arc length of riemannian geometry in some problems.

***Example* 14-1.** If parallel transport is independent of the arc $\mathbf{c}_P{}^Q$, the matrix $M(\mathbf{c}_P{}^Q)$ must be a point function $M(P,Q)$. [By axiom II, $M(P,P) = U$; the holonomy group is reduced to the unit matrix.] The converse trivially is also true. Such a space is called a *space of distant parallelism.*

The infinitesimal-connection matrix of a space of distant parallelism is a Cartan matrix of a matrix function of only one point. Let $\{\mathbf{e}(P)\}$ be a given admissible frame in T_P. Since it is a column vector of row vectors it may be written as a matrix $E(P)$. By hypothesis, its transport along any curve $\mathbf{c}_P{}^Q$ results in a frame $E(Q) = M(P,Q)E(P)$; hence

$$M(P,Q) = E(Q)E(P)^{-1}$$

and

$$L(P,dP) = dE(P)E(P)^{-1}$$

A space of distant parallelism is *flat* if its torsion vanishes,

$$\frac{\partial e_j{}^i}{\partial x^k}\,(e^{-1})_l{}^j = \frac{\partial e_j{}^i}{\partial x^l}\,(e^{-1})_k{}^j$$

or

$$e_j{}^i\,\frac{\partial (e^{-1})_l{}^j}{\partial x^k} = e_j{}^i\,\frac{\partial (e^{-1})_k{}^j}{\partial x^l}$$

Multiplication on the left by E^{-1} shows that the column vectors

$$\mathbf{e}^{-1j} = ((e^{-1})_i{}^j)$$

are gradients of functions y^j,

$$(e^{-1})_k{}^j = \frac{\partial y^j}{\partial x^k}$$

E^{-1} is the (nonzero) jacobian of a coordinate transformation $y^j \to x^k$. In the y coordinate system, $E^{-1}(Q) = J_y\mathbf{y} = U$; hence $L = 0$ and $M(P,Q) = U$.

Theorem 14-2. *A space is flat if and only if there exists a coordinate system for which (in the coordinate systems induced in the tangent spaces) the matrix $M(\mathbf{c}_P{}^Q)$ is the unit matrix for any curve $\mathbf{c}_P{}^Q$.*

***Example* 14-2.** In each tangent space T_P let a nonsingular map $\phi(P) = (\phi(P)_j{}^i)$ be given. The field ϕ will be C^1 in $T(V)$. From a given connection we obtain the connection *transformed by* ϕ:

$$M^\phi(\mathbf{c}_P{}^Q) = \phi(Q)M(\mathbf{c}_P{}^Q)\phi(P)^{-1} \tag{14-11}$$

M and M^ϕ have isomorphic holonomy groups $H(P)$ and $H^\phi(P)$ since $H^\phi(P) = \phi(P)H(P)\phi(P)^{-1}$. For the infinitesimal-connection matrices one obtains as usual

$$L^\phi = \phi L \phi^{-1} + C(\phi) \tag{14-12}$$

The difference of the two matrices L and L^ϕ is a tensor. From

$$L + (d\phi + \phi L - L\phi)\phi^{-1} = L + C(\phi) + \phi L \phi^{-1} - L$$

it follows that

$$L^\phi - L = \nabla\phi \cdot \phi^{-1} \tag{14-13}$$

A connection L is said to be *invariant under* ϕ if $L = L^\phi$.

Theorem 14-3. *A connection is invariant under a tensor field ϕ if and only if ϕ is covariant constant in the connection.*

***Example* 14-3.** A connection can be given by its infinitesimal-connection matrix L. This fact can be used to derive from given connections new ones that enjoy special properties.

For any $L \in \mathfrak{L}GL_n$ the matrix

$$L^* = L - \frac{1}{n}(\text{trace } L)U \tag{14-14}$$

is of vanishing trace, hence is in $\mathfrak{L}SL_n$.

Theorem 14-4. *For any infinitesimal connection L, the infinitesimal connection L^* preserves in parallel translation the volume spanned by a frame in T_P.*

Exercise 14-1

1. We have defined the tangent space T_P on a set of equivalence classes of curves. In the same way one may define the kth *prolongation* $T_P{}^{(k)}$ of the tangent space as a set of equivalence classes of curves. Two curves are in the same class if their Taylor expansions in some local system of coordinates

$$\mathbf{x}(t) = \mathbf{x}_0 + (d\mathbf{x})_0 + (d^2\mathbf{x})_0 + \cdots + (d^k\mathbf{x})_0$$

coincide up to the kth term. This property is invariant in all C^k coordinate transformations with a nonvanishing jacobian determinant. $T_P^{(k)}$ is a vector space under componentwise addition of the differentials. Tensors defined on $T_P^{(k)}$ are sometimes called *extensors*, or *jets*.

(*a*) Find the transformation matrix of $T_P^{(k)}$ induced by a jacobian matrix J acting on $T_P = T_P^{(1)}$.

(*b*) A connection defined on $T_P^{(k)}$ is a *Kawaguchi* connection. Show that the Kawaguchi connection $L = L(\mathbf{x}, d\mathbf{x}/dt, \ldots, d^k\mathbf{x}/dt^k)$ is independent of the parameter t if L is homogeneous of degree one in all $d^j\mathbf{x}/dt^j$.

2. Define the holonomy groups of a G connection (*a*) in terms of the connection matrices M and (*b*) in terms of the infinitesimal connection.

3. Prove that $M(\mathbf{c}_P{}^Q)$ depends only on P and Q and not on the path from P to Q if and only if $K = 0$.

4. Let V be a Lie group germ. For any point $\mathbf{x} \in V$ and any direction $d\mathbf{x} \in T_x$ we define two linear connections $L^+ = d\mathbf{x}\,\mathbf{x}^{-1}$ and $L^- = \mathbf{x}^{-1}\,d\mathbf{x}$.

(*a*) Prove that both the $(+)$ and the $(-)$ connections are of vanishing curvature. (Use Prob. 3.)

(*b*) Compute the curvature of the symmetric connection

$$\Gamma = \tfrac{1}{2}(L^+ + L^-)$$

(*c*) Show that the curvature of Γ is covariant constant in the Riemann geometry of Γ (Cartan-Schouten). (A group space is a symmetric space, exercise 13-4, Prob. 8.)

5. Write the identities (14-9) and (14-10) in coordinates.

◆ **6.** Show that the curvature matrix K of a nonsymmetric connection splits into the Riemann-curvature matrix P of the symmetric connection Γ and a matrix S depending *only* on the torsion tensor. Find S.

7. Define normal coordinates for a nonsymmetric connection.

8. Prove that $w_{k;i;j} - w_{k;j;i} = K^s_{k,ij}w_s - 2T^s_{ij}w_{k;s}$.

9. Find a formula for $w^k{}_{;i;j} - w^k{}_{;j;i}$.

10. A space V^{2n} is *almost complex* if in it there is defined a tensor field F^j_i with $F^s_iF^j_s = -\delta^j_i$. Its *Nijenhuis tensor* is

$$N^h_{ji} = F^s_j\left(\frac{\partial F^h_i}{\partial x^s} - \frac{\partial F^h_s}{\partial x^i}\right) - F^s_i\left(\frac{\partial F^h_j}{\partial x^s} - \frac{\partial F^h_s}{\partial x^j}\right)$$

Let g_{ij} be a metric such that $F^s_iF^t_jg_{st} = g_{ij}$, and consider the connections $L^i_{jk} = \Gamma^i_{jk} + T^i_{jk}$, where Γ belongs to (g_{ij}). We use G to raise and lower indices. Define $F_{jih} = \frac{1}{2}(\partial F_{ih}/\partial x^j + \partial F_{hj}/\partial x^i + \partial F_{ji}/\partial x^h)$. Prove that $g_{ij;k} = F_{ij;k} = 0$ for the three connections defined by

(*a*) $T^i_{jk} = \frac{1}{4}N_{hij} - F^s_hF_{ijs} + F^s_iF_{hjs}$
(*b*) $T^i_{jk} = \frac{1}{2}N_{hji} - \frac{1}{2}F^s_jF_{ihs}$
(*c*) $T^i_{jk} = \frac{1}{2}N_{jih} + \frac{1}{2}F^s_jF^t_iF^u_hF_{stu}$
(Yano).

11. Show that the curvature defined by $K = K^i_{j,kl}\, dx^k \wedge dx^l$ is a tensor field.

12. Let an SL_n connection L be given. A matrix $M \varepsilon GL_n$ may be written as $M = \lambda S$, $S \varepsilon SL_n$. A connection is *projective* if it is invariant under GL_n. Show that $L^* = L + (1/(n-1)UL^h_{jh}dx^j$ is projective. Compute the curvature of L^*, given that of L (Knebelman).

13. Show that theorem 13-16 holds for any linear connection.

14. In general relativity a "pseudo-riemannian" geometry is used, based on a metric tensor g_{ij} with $\|g_{ij}\| < 0$ but using the axioms of Sec. 13-1.

(*a*) Show that the developments of Secs. 13-1 to 13-3 remain valid in this more general case.

(*b*) For definiteness, let us take $n = 4$ and g_{ij} orthogonally equivalent to the diagonal matrix X of signature two, i.e., of diagonal elements 1, 1, 1, -1. The group of the pseudo-riemannian geometry is that of matrices A such that $A^tXA = X$. Write the conditions that are obtained for the elements of A.

(*c*) The "light cone" $(x^{\cdot 1})^2 + (x^{\cdot 2})^2 + (x^{\cdot 3})^2 - (x^{\cdot 4})^2 = 0$ defined in each tangent space is invariant under the group of part (*b*). What does this imply for the holonomy groups of the pseudo-riemannian geometry?

15. Prove that for a general linear connection it is *not* true that $K = 0$ means that the space is locally euclidean.

16. A *Cartan-Finsler connection* is one whose Cartan matrix can be written as

$$(\Gamma^i_{jk}(\mathbf{x},\mathbf{x}^\cdot)dx^k + C^i_{jk}(\mathbf{x},\mathbf{x}^\cdot)\, dx^{\cdot k}) \qquad \Gamma^i_{jk} = \Gamma^i_{kj}, C^i_{jk} = C^i_{kj}$$

(*a*) What are the homogeneity relations in $x^{\cdot k}$? (Use Prob. 1.)

(*b*) Define torsion and curvature by formal analogy with the case of linear connections. Show that the curvature tensor is the sum of three formally distinct tensors.

17. In a certain neighborhood consider all Riemann spaces of the metric $c(\mathbf{x})g_{ij}(\mathbf{x})$. The *Weyl space* is the geometry obtained by identification of all these Riemann geometries; its group is $R \times O_n$, where R acts as a group of homotheties $g_{ij} \to cg_{ij}$ in every tangent space. Find the general form of the Cartan matrix $L^i{}_{jk}$ of a Weyl space (in analogy to Prob. 12).

18. Find all groups that may act as holonomy groups of two-dimensional Weyl spaces. (They are Lie groups.)

♦**19.** The holonomy groups of a Cartan-Finsler space (and more generally, of any space with an infinitesimal nonlinear connection matrix) need not be a Lie group. Why?

20. A *tensor connection* is given by $M(\mathbf{c}_P{}^Q)$ acting on a tangent tensor space $\overset{a}{\otimes} T \otimes \overset{b}{\otimes} T^*$. What would be the formula for a linear tensor connection?

21. Let $f\colon V \to V'$ be a C^2 map of a space V onto a space V'. It induces a map $df\colon T(V) \to T(V')$ of the tangent spaces by mapping $(\mathbf{x},d\mathbf{x})$ into $(f(\mathbf{x}),df(\mathbf{x}))$. By theorem 9-7, the diagram

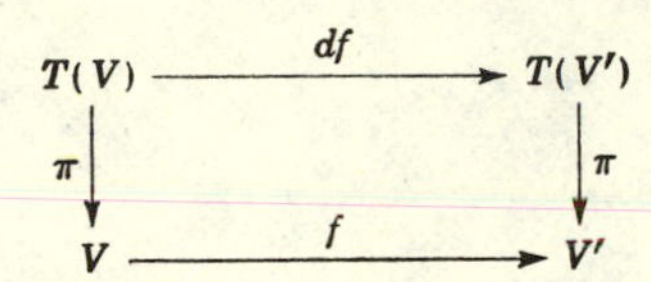

is *commutative*, that is, $\pi\, df = f\pi$ (π is the projection). If a connection M is given in $T(V)$, show that it defines a *unique* connection M' in $T(V')$ by $df\, M = M'\, df$ (Kobayashi).

22. Let ϕ be a tensor field as defined in example 14-2 such that $\phi^p = U$. Prove that, for any infinitesimal connection L, the connection

$$L_\phi = L + L^\phi + L^{\phi^2} + \cdots + L^{\phi^{p-1}}$$

is invariant under ϕ (Tachibana).

23. Prove that an infinitesimal connection L is invariant under ϕ ($\phi^p = U$) if and only if for any connection L^* there exists a tensor $T = T^i_{jk}$ such that $L = (L^* + T)_\phi$ (Tachibana).

References

Schouten's text (see References for Chap. 13) is an exhaustive guide to the computational aspects of linear connections and to the relevant literature. The following references deal with the foundations:

Ehresmann, C.: Les connexions infinitésimales dans un espace fibré différentiable, in "Colloque de Topologie," Centre belge de recherches mathématiques, Thone, Liège, Masson et Cie, Paris, 1950.

Ehresmann, C.: Introduction à la théorie des structures infinitésimales et des pseudo-groupes de Lie, *Colloq. Intern. Centre Natl. Rech. Sci.* (*Paris*), vol. 42, Géométrie différentielle, 1953, pp. 97–110.

Ehresmann, C.: Gattungen von lokalen Strukturen., *Jahresber. Deutsch. Math. Verein.*, **60**: 49–77 (1957).

Knebelman, M. S.: Spaces of Relative Parallelism, *Ann. of Math.*, Ser. 2, **53**: 387–399 (1951).

Kobayashi, S.: Theory of Connexions, *Ann. Mat. Pura Appl.*, Ser. 4, **43**: 119–194 (1957).

Lichnérowicz, A.: "Théorie globale des connexions et des groupes d'holonomie," Edizioni Cremonese, Rome, 1955.

Nomizu, K.: "Lie Groups and Differential Geometry," Mathematical Society of Japan, Tokyo, 1956.

The most important nonlinear connections are those belonging to "Finsler geometry." See the following:

Rund, H.: "The Differential Geometry of Finsler Spaces," Springer-Verlag OHG, Berlin, 1959.

ANSWERS TO SELECTED EXERCISES

EXERCISE 1-2

4. rJ

EXERCISE 1-3

2. Ellipse $x_1 = a \cos t$, $x_2 = b \sin t$: $(a^2 \sin^2 t + b^2 \cos^2 t)^{\frac{1}{2}} \cot t$, $a \cot t$, $(a^2 \sin^2 t + b^2 \cos^2 t)^{\frac{1}{2}} a^{-1}$, $b^2 \sin t/a$; parabola $x_2 = 2px_1{}^2$: $\frac{1}{2}x_1(1 + 16p^2x_1{}^2)^{\frac{1}{2}}$, $\frac{1}{2}px_1$, $2px_1{}^2 (1 + 16p^2x_1{}^2)^{\frac{1}{2}}$, $8px_1{}^2$. **4.** Spirals of Archimedes. **12.** $Cx_2 = e^{x_1/c}$ and x_2 axis. **13.** $x_1 - a = c^{-1}(c^2 - x_1{}^2)^{\frac{1}{2}} + \ln x_2/(c + (c^2 - x_2{}^2)^{\frac{1}{2}})$. **14.** Logarithmic spirals

EXERCISE 2-1

1(b). $12a$. **1(c).** $c^2(m^2 + 1)^{\frac{3}{2}}/m$. **6.** Study, for example, $x_2 = x_1{}^{-2} \sin x_1{}^{-1}$

EXERCISE 2-2

3(a). R^{-1}. **3(b).** $ab(a^2 \sin^2 t + b^2 \cos^2 t)^{-\frac{3}{2}}$. **3(c).** $-2^{-\frac{3}{2}}(1 - \cos u)^{-\frac{1}{2}}$.
3(d). $(2c^2 + r^2)(c^2 + r^2)^{-\frac{3}{2}}$. **3(e).** $\rho = PN$. **3(f).** $a^{-1} \operatorname{Cosh}^{-2} x_1/a$.
3(g). $(x_2{}^2 - x_2(a^2 - x_2{}^2)^{\frac{1}{2}})a^{-3}$. **3(h).** $(3a \cos t \sin t)^{-\frac{1}{2}}$. **4.** Hyperbolic spiral.
16. Write $a^2 = p^2 + q^2$.

$$\begin{pmatrix} (q/a)^2 + (p/a)^2 \cos at & (p/a) \sin at & (pq/a)(1 - \cos at) \\ -(p/a) \sin at & \cos at & (q/a) \sin at \\ (pq/a)(1 - \cos at) & -(q/a) \sin at & (p/a)^2 + (q/a)^2 \cos at \end{pmatrix}$$

21. $s = \int F(k)^{-1}\, dk$, $k = \int F(k)\, ds$; yes

EXERCISE 3-2

2(a). Neill's parabola $27ax_2{}^2 = 8(x_1 - a)^3$. **2(b).** Ellipse.
2(c). Logarithmic spiral (rotated). **2(d).** Cycloid (displaced).
2(e). $x_{1E} = x_1 - x_2{}^2a^{-1}(1 - a^2x_2{}^{-2})^{\frac{1}{2}}$, $x_{2E} = 2x_2$
3. $(h \sin t + h' \cos t, h' \sin t - h \cos t)$. **6.** Astroid. **7.** $x_1x_2 = \text{const.}$
8(b). Projection of the radius of curvature on the tangent to the evolutoid.
8(d). Concentric circles

EXERCISE 3-5

3. $\omega' = -k_f$; $p' = \sin \omega$

EXERCISE 4-1

6. See Sec. 11-1. **8(a).** Straight lines. **8(b).** $a(x - x_0) = (1 + cy)2^{-1}c^{-1}[(1 + cy)^2 - a^2] - a^2(2c)^{-1} \operatorname{Arcosh} (1 + cy)/a$. **8(c).** Catenary

Exercise 5-1

2. $\begin{matrix} e & a \\ a & e \end{matrix}$ $\begin{matrix} e & a & b \\ a & b & e \\ b & e & a \end{matrix}$ **4.** π. **16.** R. **17.** A half plane

Exercise 5-2

21. $-\psi, -\theta, -\phi$ (attention order!)

Exercise 6-1

1. 0. **2.** (α_{ij}), $\alpha_{ij} = -\alpha_{ji} (i \neq j)$, $\alpha_{ii} = c$.
5. $[(\alpha_1,\alpha_2,\alpha_3),(\beta_1,\beta_2,\beta_3)] = (0,\ 0,\ \alpha_1\beta_2 - \alpha_2\beta_1)$.
7. See example 6-6. **17.** Basis *a.* $\alpha_{ii} = 1, \alpha_{m+i\ m+i} = -1$. *b.* $\alpha_{i\ m+j} = \alpha_{m+i\ j} = 1$. *c.* $\alpha_{ij} = 1$. *d.* $\alpha_{m+i\ m+j} = 1, i,j \leq 2$

Exercise 6-3

6. Yes, discrete groups

Exercise 7-1

6. $p_1, p_2, x_1p_1 - x_2p_2,\ x_2p_1,\ x_1p_2$, where $p_i = \partial/\partial x_i$. **9.** Similitudes, composition of translations, and $\mathbf{x}^* = \mathbf{x}_0 \operatorname{Cosh} (t - t_0)$

Exercise 7-2

15. Yes

Exercise 7-3

10. Ratios of areas of two triangles formed from four points

Exercise 8-1

6(a). $tk^2 = 12$. **6(b).** $k^2 = \phi'^2 + \phi^2 + (\phi'\psi + \phi\psi')^2$.

$$\textbf{7.}\qquad A(s) = \begin{pmatrix} \frac{1}{2}\left(\cos\frac{\sqrt{2}}{s} + 1\right) & \frac{1}{\sqrt{2}}\sin\frac{\sqrt{2}}{s} & \frac{1}{2}\left(1 - \cos\frac{\sqrt{2}}{s}\right) \\ -\frac{1}{\sqrt{2}}\sin\frac{\sqrt{2}}{s} & \cos\frac{\sqrt{2}}{s} & \frac{1}{\sqrt{2}}\sin\frac{\sqrt{2}}{s} \\ \frac{1}{2}\left(1 - \cos\frac{\sqrt{2}}{s}\right) & -\frac{1}{\sqrt{2}}\sin\frac{\sqrt{2}}{s} & \frac{1}{2}\left(1 + \cos\frac{\sqrt{2}}{s}\right) \end{pmatrix}$$

12. $\mathbf{x}^{\mathrm{IV}} = (k'' - k^3 - kt^2)\mathbf{n} - 3kk'\mathbf{t} + (2k't + kt')\mathbf{b} = 2kt^{-1}(2k't + kt' - 3kk')\mathbf{x}' + (k''k^{-1} - k^2 - t^2 - 2k'^2k^{-2} + k't'k^{-1}t^{-1})\mathbf{x}'' + (2k't + kt')\mathbf{x}'''$.
16. x_3 independent variable; $k = (x_1x_2)^{-2}(x_1'^2 + x_2'^2 + 1)^{-\frac{3}{2}}2p^2(x_1^2 + x_2^2 + p^2)^{\frac{1}{2}}, t = p(x_1^2 + x_2^2 + p^2)^{-1}$. **18.** *c.* **19.** *c.* **30.** $k_2 = k_1^{-1}|\mathbf{x}''' - (\log k_1)'\mathbf{x}'' + k_1^2\mathbf{x}'|$

Exercise 8-2

5(a). $\cos\phi = 4/78^{\frac{1}{2}}$, $p\sin\phi = -5/78^{\frac{1}{2}}$. **5(b).** $\cos\phi = \frac{1}{2}$, $p\sin\phi = \frac{3}{2}$.
10. $A = (x_1^2 + k^2)^{\frac{1}{2}}[(0,x_1,k) + \tau(0,-kx_1,x_1^2)]$. **14.** Use exercise 2-2, Prob. 16.
21. $d = k_2(\sigma) = k_2(t)/k_1(t)$

Exercise 8-3

4. $\mathbf{x} + (k_2/k_1)\mathbf{x}' - k_1^{-1}\mathbf{x}''' = \mathbf{q}$. **10.** Quartics. **11.** $k_2{}^9k_1{}^{-8}$, $k_2'k_1{}^{-1}$

Exercise 9-1

6. $(0, \ldots, 0,1,0, \ldots, 0)$. **9.** $2x^1 + 5x^2 = 0$. **12.** Tangent plane to level surface

Exercise 9-2

2. $\begin{pmatrix} 1 & 0 & -1 \\ 5 & 0 & -5 \\ 7 & 0 & -7 \end{pmatrix}$ and transpose. **3.** 213

Exercise 9-3

1. $(-1)^{k(n-k)-1}$. **2.** $(-1)^{n(n-1)/2}$. **4(a).** $-(x^1 \sin 3x^1 + 7(x^3)^2)\, dx^1 \wedge dx^2 + x^1\, dx^1 \wedge dx^3 + 7(x^3)^2\, dx^2 \wedge dx^3$. **4(b).** $dx^1 \wedge dx^2$. **4(c).** $-21\, dx^1 \wedge dx^2 \wedge dx^3$. **5(a).** $-(\cos x^1 + \sin x^2)\, dx^1 \wedge dx^2$. **5(b).** 0. **5(c).** $(6 + x^1)\, dx^1 \wedge dx^2 \wedge dx^3$

Exercise 9-4

11. $\mathbf{c}$ skew to x^3 axis. **15.** $\mathbf{c}' \times \mathbf{c}'' = 0$

Exercise 10-1

3. In the notation of Sec. 7-2, $\mathbf{N} = -(1 + p^2 + q^2)^{-\frac{1}{2}}(p,q,-1)$. $K = (rt - s^2)(1 + p^2 + q^2)^{-2}$. $2H = [(1 + q^2)r - 2pqs + (1 + p^2)t](1 + p^2 + q^2)^{-\frac{3}{2}}$.
4. Umbilics $(0,0,0)$ and $x^2 = y^2 = 2a^2$, $x^2 = y^2 = a^2/2$.
7. $2H = u^1\phi''[(u^1)^2 + \phi'^2]^{-\frac{3}{2}}$; $K = (u^1)^2\phi'^2[(u^1)^2 + \phi'^2]^{-3}$

Exercise 10-2

7(c). $K_c = K(1 - 2Hc + c^2K)^{-1}$, $H_c = (H - cK)(1 - 2Hc + c^2K)^{-1}$.
7(d). $c = \pm K^{-\frac{1}{2}}$. **13.** $|x^1| = |x^2| = |x^3| = |a|$. **23.** No condition

Exercise 10-3

1. Notation of exercise 10-2, Prob. 8; $d\phi = (C - \sin^2 u)^{\frac{1}{2}}\, du$. **9.** Yes. **16.** $dx^3 = (1 - (c/a)^2 \operatorname{Sinh}^2 u/a)^{\frac{1}{2}}\, du$, $dx^3 = (1 - (c/a)^2 \operatorname{Cosh}^2 u/a)^{\frac{1}{2}}\, du$, $dx^3 = (1 - (c/a)^2 e^{2u/a})^{\frac{1}{2}}\, du$

Exercise 10-4

15. $z^3{}_{11} + z^3{}_{22} = \Delta z^3 = 0$

Exercise 10-5

17. $\iint(\omega^1 \wedge \omega_2{}^3 + \omega^2 \wedge \omega_3{}^1) = 0$

Exercise 10-6

1. $d\sigma^2 = (du^1)^2$, $\mathrm{II} = -\sin\alpha\, du^1\, du^2$. Rotate the frame by $\pi/4$. Compare Prob. 8.

Exercise 11-1

9. $(x - a)^2 + y^2 = r^2$. **10.** $K = -2$; straight lines $y\, dx - x\, dy = 0$

Exercise 11-2

7. $\Xi^1 \dfrac{\partial\lambda}{\partial x^1} + \Xi^2 \dfrac{\partial\lambda}{\partial x^2} - 2\lambda \dfrac{\partial\Xi^1}{\partial x^1} = 0$; $\dfrac{\partial\Xi^2}{\partial x^1} + \dfrac{\partial\Xi^1}{\partial x^2} = 0$; $\Xi^1 \dfrac{\partial\lambda}{\partial x^1} + \Xi^2 \dfrac{\partial\lambda}{\partial x^2} + 2\lambda \dfrac{\partial\Xi^2}{\partial x^2} = 0$.

19. O_4

Exercise 11-3

1. No, $K = 0$. **2.** 0,1,3

Exercise 12-1

1. $(x^3)^{10}\|L_{\alpha\beta}\| = 4x^1x^2(x^3)^3[2(x^3)^3 - 1]$. **2.** I or J are 0. **6.** $J = 0$.
14. $I,J \to I/c,J/c$

Exercise 12-2

7. $(x_0 + at,\ y_0,\ x_0y_0 - \frac{1}{3}y_0{}^2 + y_0\,at)$. **8a.** $p = e_u(\frac{1}{2}v^2\mathbf{c}_0 + \mathbf{c}_1v + \mathbf{c}_2)$

Exercise 13-1

7. The rotation group may leave no subspace invariant or one of one or two dimensions. **11a.**

$$\Gamma = \frac{1}{\|G\|}\begin{pmatrix} (1 + (x^2)^2)x^1\,dx^1 & -\frac{1}{\sqrt{2}}x^1\,dx^1 & 0 \\ -\frac{1}{\sqrt{2}}x^2\,dx^2 & (1 + (x^1)^2)x^2\,dx^2 & 0 \\ 0 & 0 & 0 \end{pmatrix}$$

$$\mathrm{P} = \frac{1}{2\|G\|^2}\begin{pmatrix} -x^1x^2 - 4(x^1)^3x^2 & \sqrt{2}(x^1)^3x^2 - \sqrt{2}x^1x^2 & 0 \\ \sqrt{2}[x^1x^2 - x^1(x^2)^3] & x^1x^2 + 4x^1(x^2)^3 & 0 \\ 0 & 0 & 0 \end{pmatrix} dx^1 \wedge dx^2$$

Exercise 13-2

5. $d^2x^i/dt^2 - \Gamma^i{}_{jk}\,dx^j/dt\,dx^k/dt = \lambda(t)\,dx^i/dt$. **7.** $[1 + (x^2)^2 + (x^3)^2]d^2x^1/ds^2 - x^1x^3d^2x^2/ds^2 - x^1x^2d^2x^3/ds^2 - x^1(dx^1/ds,dx^2/ds,dx^3/ds)G\{dx^1/ds,dx^2/ds,dx^3/ds\} = 0$ and cyclic equations. **19.** Constant curvature

Exercise 13-4

1(a). Combine action of O_3 in space $x^1 = 0$ and $\Xi = (1,0,0,0)$.
1(b). f must be $\ln\,(c - x)$. **11.** $\mathfrak{L}\Gamma = -\lim t^{-1}(A\Gamma A^{-1} + C(A) - \Gamma)$
$A = U + tJ_x\Xi + \cdots$. **14.** $\mathfrak{L}\Gamma^i{}_{jk} = \delta_j{}^i\phi_k + \delta_k{}^i\phi_j$

Exercise 14-1

6. $T^i_{jl;k} - T^i_{jk;l} + T^i_{sl}T^i_{jk} - T^s_{sk}T^s_{jl} - 2T^i_{js}T^s_{kl}$. **19.** In a linear connection we know that the Lie algebra of the holonomy group is in $\mathfrak{L}(GL_n)$; for infinitesimal connections it may be infinite-dimensional.

INDEX

A CATALOGUE OF SELECTED DOVER BOOKS IN ALL FIELDS OF INTEREST

THE AMERICAN SENATOR, Anthony Trollope. Little known, long unavailable Trollope novel on a grand scale. Here are humorous comment on American vs. English culture, and stunning portrayal of a heroine/villainess. Superb evocation of Victorian village life. 561pp. 5⅜ x 8½.
23801-6 Pa. $6.00

WAS IT MURDER? James Hilton. The author of *Lost Horizon* and *Goodbye, Mr. Chips* wrote one detective novel (under a pen-name) which was quickly forgotten and virtually lost, even at the height of Hilton's fame. This edition brings it back—a finely crafted public school puzzle resplendent with Hilton's stylish atmosphere. A thoroughly English thriller by the creator of Shangri-la. 252pp. 5⅜ x 8. (Available in U.S. only)
23774-5 Pa. $3.00

CENTRAL PARK: A PHOTOGRAPHIC GUIDE, Victor Laredo and Henry Hope Reed. 121 superb photographs show dramatic views of Central Park: Bethesda Fountain, Cleopatra's Needle, Sheep Meadow, the Blockhouse, plus people engaged in many park activities: ice skating, bike riding, etc. Captions by former Curator of Central Park, Henry Hope Reed, provide historical view, changes, etc. Also photos of N.Y. landmarks on park's periphery. 96pp. 8½ x 11. 23750-8 Pa. $4.50

NANTUCKET IN THE NINETEENTH CENTURY, Clay Lancaster. 180 rare photographs, stereographs, maps, drawings and floor plans recreate unique American island society. Authentic scenes of shipwreck, lighthouses, streets, homes are arranged in geographic sequence to provide walking-tour guide to old Nantucket existing today. Introduction, captions. 160pp. 8⅞ x 11¾. 23747-8 Pa. $6.95

STONE AND MAN: A PHOTOGRAPHIC EXPLORATION, Andreas Feininger. 106 photographs by *Life* photographer Feininger portray man's deep passion for stone through the ages. Stonehenge-like megaliths, fortified towns, sculpted marble and crumbling tenements show textures, beauties, fascination. 128pp. 9¼ x 10¾. 23756-7 Pa. $5.95

CIRCLES, A MATHEMATICAL VIEW, D. Pedoe. Fundamental aspects of college geometry, non-Euclidean geometry, and other branches of mathematics: representing circle by point. Poincare model, isoperimetric property, etc. Stimulating recreational reading. 66 figures. 96pp. 5⅝ x 8¼.
63698-4 Pa. $2.75

THE DISCOVERY OF NEPTUNE, Morton Grosser. Dramatic scientific history of the investigations leading up to the actual discovery of the eighth planet of our solar system. Lucid, well-researched book by well-known historian of science. 172pp. 5⅜ x 8½. 23726-5 Pa. $3.00

THE DEVIL'S DICTIONARY. Ambrose Bierce. Barbed, bitter, brilliant witticisms in the form of a dictionary. Best, most ferocious satire America has produced. 145pp. 5⅜ x 8½. 20487-1 Pa. $2.00

AN AUTOBIOGRAPHY, Margaret Sanger. Exciting personal account of hard-fought battle for woman's right to birth control, against prejudice, church, law. Foremost feminist document. 504pp. 5⅜ x 8½.
20470-7 Pa. $5.50

MY BONDAGE AND MY FREEDOM, Frederick Douglass. Born as a slave, Douglass became outspoken force in antislavery movement. The best of Douglass's autobiographies. Graphic description of slave life. Introduction by P. Foner. 464pp. 5⅜ x 8½. 22457-0 Pa. $5.50

LIVING MY LIFE, Emma Goldman. Candid, no holds barred account by foremost American anarchist: her own life, anarchist movement, famous contemporaries, ideas and their impact. Struggles and confrontations in America, plus deportation to U.S.S.R. Shocking inside account of persecution of anarchists under Lenin. 13 plates. Total of 944pp. 5⅜ x 8½.
22543-7, 22544-5 Pa., Two-vol. set $11.00

LETTERS AND NOTES ON THE MANNERS, CUSTOMS AND CONDITIONS OF THE NORTH AMERICAN INDIANS, George Catlin. Classic account of life among Plains Indians: ceremonies, hunt, warfare, etc. Dover edition reproduces for first time all original paintings. 312 plates. 572pp. of text. 6⅛ x 9¼. 22118-0, 22119-9 Pa.. Two-vol. set $11.50

THE MAYA AND THEIR NEIGHBORS, edited by Clarence L. Hay, others. Synoptic view of Maya civilization in broadest sense, together with Northern, Southern neighbors. Integrates much background, valuable detail not elsewhere. Prepared by greatest scholars: Kroeber, Morley, Thompson, Spinden, Vaillant, many others. Sometimes called Tozzer Memorial Volume. 60 illustrations, linguistic map. 634pp. 5⅜ x 8½.
23510-6 Pa. $7.50

HANDBOOK OF THE INDIANS OF CALIFORNIA, A. L. Kroeber. Foremost American anthropologist offers complete ethnographic study of each group. Monumental classic. 459 illustrations, maps. 995pp. 5⅜ x 8½.
23368-5 Pa. $10.00

SHAKTI AND SHAKTA, Arthur Avalon. First book to give clear, cohesive analysis of Shakta doctrine, Shakta ritual and Kundalini Shakti (yoga). Important work by one of world's foremost students of Shaktic and Tantric thought. 732pp. 5⅜ x 8½. (Available in U.S. only)
23645-5 Pa. $7.95

AN INTRODUCTION TO THE STUDY OF THE MAYA HIEROGLYPHS, Syvanus Griswold Morley. Classic study by one of the truly great figures in hieroglyph research. Still the best introduction for the student for reading Maya hieroglyphs. New introduction by J. Eric S. Thompson. 117 illustrations. 284pp. 5⅜ x 8½. 23108-9 Pa. $4.00

A STUDY OF MAYA ART, Herbert J. Spinden. Landmark classic interprets Maya symbolism, estimates styles, covers ceramics, architecture, murals, stone carvings as artforms. Still a basic book in area. New introduction by J. Eric Thompson. Over 750 illustrations. 341pp. 8⅜ x 11¼.
21235-1 Pa. $6.95

TONE POEMS, SERIES II: TILL EULENSPIEGELS LUSTIGE STREICHE, ALSO SPRACH ZARATHUSTRA, AND EIN HELDENLEBEN, Richard Strauss. Three important orchestral works, including very popular *Till Eulenspiegel's Marry Pranks,* reproduced in full score from original editions. Study score. 315pp. 9⅜ x 12¼. (Available in U.S. only)
23755-9 Pa. $7.50

TONE POEMS, SERIES I: DON JUAN, TOD UND VERKLARUNG AND DON QUIXOTE, Richard Strauss. Three of the most often performed and recorded works in entire orchestral repertoire, reproduced in full score from original editions. Study score. 286pp. 9⅜ x 12¼. (Available in U.S. only)
23754-0 Pa. $7.50

11 LATE STRING QUARTETS, Franz Joseph Haydn. The form which Haydn defined and "brought to perfection." *(Grove's).* 11 string quartets in complete score, his last and his best. The first in a projected series of the complete Haydn string quartets. Reliable modern Eulenberg edition, otherwise difficult to obtain. 320pp. 8⅜ x 11¼. (Available in U.S. only)
23753-2 Pa. $6.95

FOURTH, FIFTH AND SIXTH SYMPHONIES IN FULL SCORE, Peter Ilyitch Tchaikovsky. Complete orchestral scores of Symphony No. 4 in F Minor, Op. 36; Symphony No. 5 in E Minor, Op. 64; Symphony No. 6 in B Minor, "Pathetique," Op. 74. Bretikopf & Hartel eds. Study score. 480pp. 9⅜ x 12¼.
23861-X Pa. $10.95

THE MARRIAGE OF FIGARO: COMPLETE SCORE, Wolfgang A. Mozart. Finest comic opera ever written. Full score, not to be confused with piano renderings. Peters edition. Study score. 448pp. 9⅜ x 12¼. (Available in U.S. only)
23751-6 Pa. $11.95

"IMAGE" ON THE ART AND EVOLUTION OF THE FILM, edited by Marshall Deutelbaum. Pioneering book brings together for first time 38 groundbreaking articles on early silent films from *Image* and 263 illustrations newly shot from rare prints in the collection of the International Museum of Photography. A landmark work. Index. 256pp. 8¼ x 11.
23777-X Pa. $8.95

AROUND-THE-WORLD COOKY BOOK, Lois Lintner Sumption and Marguerite Lintner Ashbrook. 373 cooky and frosting recipes from 28 countries (America, Austria, China, Russia, Italy, etc.) include Viennese kisses, rice wafers, London strips, lady fingers, hony, sugar spice, maple cookies, etc. Clear instructions. All tested. 38 drawings. 182pp. 5⅜ x 8.
23802-4 Pa. $2.50

THE ART NOUVEAU STYLE, edited by Roberta Waddell. 579 rare photographs, not available elsew'iere, of works in jewelry, metalwork, glass, ceramics, textiles, architecture and furniture by 175 artists—Mucha, Seguy, Lalique, Tiffany, Gaudin, Hohlwein, Saarinen, and many others. 288pp. 8⅜ x 11¼.
23515-7 Pa. $6.95